Lecture Notes in Business Information Processing 46

Series Editors

Wilhelm Dangelmaier Alexander Blecken
Robin Delius Stefan Klöpfer (Eds.)

Advanced Manufacturing and Sustainable Logistics

8th International Heinz Nixdorf Symposium, IHNS 2010
Paderborn, Germany, April 21-22, 2010
Proceedings

Springer

Volume Editors

Wilhelm Dangelmaier
Alexander Blecken
Robin Delius
Stefan Klöpfer
Heinz Nixdorf Institut
Universität Paderborn
Fürstenallee 11, 33102 Paderborn, Germany
E-mail: {whd,blecken,robin.delius,stefan.kloepfer}@hni.upb.de

Library of Congress Control Number: 2010924281

ACM Computing Classification (1998): J.1, J.3, H.4, I.6, G.1, G.2

ISSN 1865-1348
ISBN-10 3-642-12461-5 Springer Berlin Heidelberg New York
ISBN-13 978-3-642-12461-7 Springer Berlin Heidelberg New York

springer.com

Printed in Germany

Typesetting: Camera-ready by author, data conversion by Scientific Publishing Services, Chennai, India
Printed on acid-free paper 06/3180 5 4 3 2 1 0

Preface

In times of declining economic growth, companies have to control their costs more than ever to save resources needed in the future. Regardless of the economic size of the company, the processes of production and logistics play a decisive role in stabilizing procedures and avoiding waste. Both are important cost drivers in manufacturing companies and therefore they offer large potential savings. Pervasive networking in the last years has contributed to a hitherto unknown transparency of global markets. This harmonization opened up new possibilities of entering foreign markets for procurement and sales to the companies. The emerging global procurement strategy was understood as a chance to rethink the relocation of existing production facilities to profit from existing differences in price and performance as a resource-saving factor. Many companies tended towards a reduction of their vertical integration by outsourcing sections of their value chain. These contracted services of production result in higher transport volumes, increased complexity of supply processes and new requirements on logistic networks. This trend of outsourcing has not stopped, but is slowing down noticeably. Additionally, there is an increasing proportion of companies restoring business units that were outsourced before. Reasons for turning back decisions are often to be found in missed goals. It is not unusual that important cost factors were disregarded in the original basis of decision-making. In the meantime many companies have realized that it is easier to achieve stability of processes and therewith a control of costs by increasing their own contribution to production. Especially in times of under-utilized capacities like in the current crisis, insourcing can be a strategic option. Manufacturing and logistics undergo changes incurred by the development and implementation of advanced information technologies. Network-wide cooperation is required in order to take advantage of these shifting paradigms. Existing potentials such as intra-company knowledge, which are inadequately used so far, can contribute to the development of sustainable master-plans leading to new, learning companies. Corporate networks are strengthened by the application of best practices and by sharing their experience with supply chain partners for achieving integral improvements. Ecological awareness influences existing processes by the application of green principles to logistics and supply chain management. Eventually, everyone will benefit from the fact that the transfer of knowledge by practitioners and researchers is leading to a new quality in cooperation, planning and control.

During April 21–22, the 8th International Heinz Nixdorf Symposium took place in Paderborn. Under the title "Changing Paradigms: Advanced Manufacturing and Sustainable Logistics," the Heinz Nixdorf Institute enabled the discussion between researchers and practitioners about present challenges and possible solutions. On the first conference day, ten speeches were given by the following keynote speakers:

W. Dangelmaier	Heinz Nixdorf Institute
A. Köhler	Knorr-Bremse AG
H. A. Flegel	Daimler AG
E. Gericke	Festo AG & Co. KG
S. Schwinning	Miele & Cie. KG
W. Delfmann	University of Cologne
W. Sihn	Fraunhofer Austria
M. Schenk	Fraunhofer IFF
W. Stölzle	University of St. Gallen
G. Klink	A.T. Kearney GmbH

On the beginning of the second day, six speeches of invited researchers gave insight into the main topics of the symposium:

E. Müller	Chemnitz University of Technology
D. Van Oudheusden	Katholieke Universiteit Leuven
G. Zülch	Karlsruhe Institute of Technology
J. Zak	Poznan University of Technology
S. Wenzel	University of Kassel
A. Blecken	Heinz Nixdorf Institute

Subsequently, international authors presented their contributions in four parallel tracks. Main topics were:

- Production Logistics
- Industrial Engineering
- Operations Research Techniques
- Simulation
- Humanitarian Logistics
- Supply Chain Management

The volume at hand contains the contributions to "Advanced Manufacturing and Sustainable Logistics" presented at the 8th International Heinz Nixdorf Symposium.

April 2010

Wilhelm Dangelmaier
Alexander Blecken
Robin Delius
Stefan Klöpfer

Organization

Conference Chair

Wilhelm Dangelmaier	Heinz Nixdorf Institute

Scientific Organization (Program)

Alexander Blecken	Heinz Nixdorf Institute
Robin Delius	Heinz Nixdorf Institute
Stefan Klöpfer	Heinz Nixdorf Institute

International Program Committee

Wilhelm Dangelmaier	Heinz Nixdorf Institute, Germany
Mauro Dell'Amico	University of Modena e Reggio Emilia, Italy
Ben Immers	Katholieke Universiteit Leuven, Belgium
Herbert Kotzab	Copenhagen Business School, Denmark
Albrecht Köhler	Knorr-Bremse AG, Germany
László Monostori	MTA SZTAKI, Hungary
Jorge Pinho de Sousa	University of Porto, Portugal
Michael Schenk	Fraunhofer IFF, Germany
Wilfried Sihn	Fraunhofer Austria Research GmbH, Austria
Wolfgang Stölzle	University of St. Gallen, Switzerland
József Váncza	MTA SZTAKI, Hungary
Hans-Peter Wiendahl	Leibniz University of Hannover, Germany

Track Chairs

Egon Müller	Chemnitz University of Technology, Germany
Dirk Van Oudheusden	Katholieke Universiteit Leuven, Belgium
Gert Zülch	Karlsruhe Institute of Technology, Germany
Leena Suhl	University of Paderborn, Germany
Sigrid Wenzel	University of Kassel, Germany
Alexander Blecken	Heinz Nixdorf Institute, Germany

Reviewers

M. Aufenanger	N. Boone	A. Cardeneo
C. Bierwirth	D. Brodkorb	C. Danne
C. Böhle	D. Brüggemann	U. Deppe

A. Döring
A. Emmrich
W. Franke
R.W. Grubbström
M. Grunow
H. Hamoudia
H. Holland
D. Huber
N. Ketterer
C. Kilger
B. Klöpper
A. Koberstein
G. Kovcs
C. Kriesel
C. Laroque
A. Lavrov
J. Leukel
H. Meyr
B. Mueck
G. Mummolo
S. Pettit
T. Rust
C. Schindelhauer
A. Schmidtmann
P. Schmitz
K. Spens
L. Suhl
P. Tatham
F. Teuteberg
R. Tomasini
D.L. Van Oudheusden
P. Vansteenwegen
A. Verbraeck
A. Wagenitz
S. Wenzel
H. Winkler
G. Zülch

Sponsor

Stiftung Westfalen, Germany

Table of Contents

Keynotes

Supply Chain Management

Production Logistics/Industrial Engineering

Operations Research Techniques

Humanitarian Logistics

Simulation

A Concept for an Accurate and Closely Coordinated Production

Wilhelm Dangelmaier

Heinz Nixdorf Institute, University of Paderborn, Fuerstenallee 11,
33102 Paderborn, Germany
whd@hni.upb.de

Abstract. Shorter delivery times can be utilized best, if the ordered goods are already in stock. With an increasing product range companies can no longer manage to guarantee the immediate availability of each product variation, which would solve any kind of supply problems. The remaining stocks have to be placed with utmost effectiveness and the available factors of production need to be applied efficiently time and again. This article presents methods of resolution by means of selected practical examples.

Keywords: production planning, variant complexity, decomposition approach, online optimization.

1 Introduction

Shorter delivery times can be utilized best, if the ordered goods are already in stock. With an increasing product range companies can no longer manage to guarantee the immediate availability of each product variation, which would solve any kind of supply problems. The remaining stocks have to be placed with utmost effectiveness and the available factors of production need to be applied efficiently time and again. This article presents methods of resolution by means of selected practical examples.

2 Reducing the Diversity of Variants in Production Networks

Like many other sectors as well, consumer industry is facing the ever increasing requirements in product differentiation. International manufacturers are attempting to meet the customer demands of the target markets by specially adapted products and packaging. The growing complexity of assortment has a great impact on the underlying production and distribution system. With regard to the costs incurred, particularly the effects on production planning and inventory management have to be mentioned. A higher number of materials on all production and distribution stages create additional setup and scrap costs within the production process as well as higher inventory stocks. The need for higher security stocks can be explained due to the fact, that sales forecasts are becoming more difficult with increasing diversity of variants and the greater level of demand uncertainty.

W. Dangelmaier et al. (Eds.): IHNS 2010, LNBIP 46, pp. 1–14, 2010.

Existing approaches for evaluating the cost effects may not reflect the fact, that a modification of the assortment at first affects the optimal configuration of the existing production and distribution processes. By standardizing the products as well as by extending the assortment, parameters like lot-sizes, production cycles, stock points and stock levels have to be readjusted to ensure the optimal configuration of the underlying production and distribution network. Only if this configuration is known, the correct quantification of all costs caused by a certain assortment can be given.

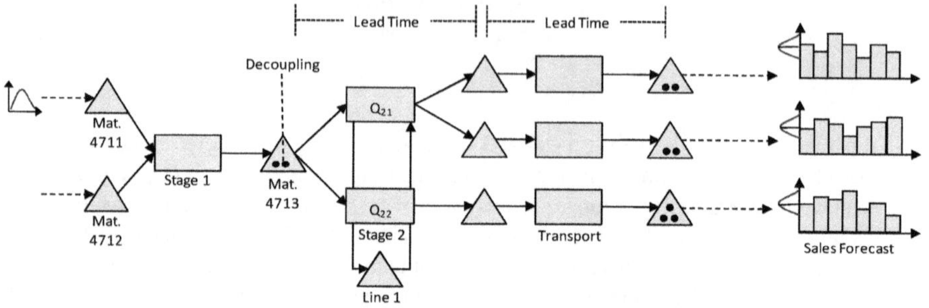

Fig. 1. Effects of Assortment Reduction in MTS Production Networks

Based on this background, in cooperation with Freudenberg Household Products (FHP)[1], the chair in Business Computing, especially CIM developed a new approach for evaluating the assortment complexity. This approach implies precise as well as heuristic optimization techniques with the goal to adapt several production parameters, stock points as well as stock levels to an existing assortment [1]. This enhances the prerequisites for substantiated if-then analyses.

A formal model of the assortment with its product structure and the associated production and distribution processes serves as a basis. With regard to the product portfolio, alternative scenarios can be defined in a second step. After applying the mentioned optimization techniques to the initial model and the several scenarios, a set of optimally configured networks is available which can be used for comparative cost analysis.

The multilevel optimization problem includes the following decision variables:

- Location of stocks for decoupling demands and production
- Level of safety stock, based on the expected sales forecast and the striven ability to supply
- Average lot-sizes or production cycles, resulting cycle stocks and predefined delivery times by the subsequent production stage

The decision regarding these production parameters and stock levels has to be made on each production stage, so that costs for stock management, setup effort and scrap are minimal throughout the whole network. By doing so, the effects of in- or decreasing assortment complexity can be precisely determined. The optimized models provide information on the expected cost effects as well as the necessary changes to the

[1] FHP is a globally active company, distributing its products under the Vileda brand.

network configuration. For example, it is possible to make a statement whether a change from decentralized to centralized warehousing will be rewarding after standardizing the products.

This concept has been prototypically implemented and is currently tested as a pilot project at Freudenberg Household Products. Here, the arrangement of the global supply and production network is significantly defined by the product portfolio and particularly the local product and packaging variants. It is analyzed how changes in the assortment of cleaning cloths are affecting the stocking strategies and the production processes at the manufacturing facility.

3 Planning of Multilevel Flow Production – System Structures in Bordered Workstations

When planning flow production lines, two questions arise:

- Where to separate the line usefully into pre- and final-assembly?
- How to distribute the production quantity to the parallel production lines and how to equip the lines?

These decisions have an impact on the structuring and assignment of staff, resources and space. The chosen production structure has to meet as well the process variety, induced by the production program by avoiding redundancy of the personnel and technical resources.

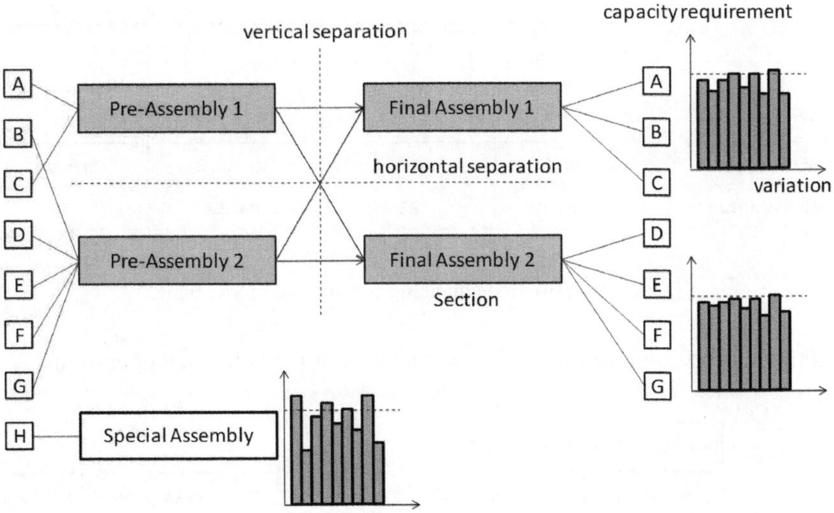

Fig. 2. Dimension of decision in production structuring

Horizontal Separation. The production cycle is determined, when assigning production quantity to a production line:

$$cycle = \frac{working\ time\ [TU]}{production\ quantity\ [pcs.]}$$

The number of stations is obtained by the variant's working time, which is the highest in a potential section. For balancing the line, this describes the lower bound of stations in the corresponding section; exceeding the production cycle in bordered workstations is excluded. The total number of stations and the number of assigned personnel can be determined as follows:

$$stations = \left\lceil \frac{\sum_{sections} MAX(working\ time_{section})}{cycle} \right\rceil$$

The example in Figure 3 is given for clarification: operations 1 to 3 and 4 to 6 can be combined to sections.

	Op. 1		Op. 2		Op. 3		Op. 4		Op. 5		Op. 6	
Variant	V1	V2	V1	V2	V1	V2	V1	V2	V1	V2	V1	V2
Processing Time [TU]	1.2	1	0.5	1	1.9	0	1	1	0.5	0	2	2
Required Resources	R1		R2		R3		R4		R5		R6	

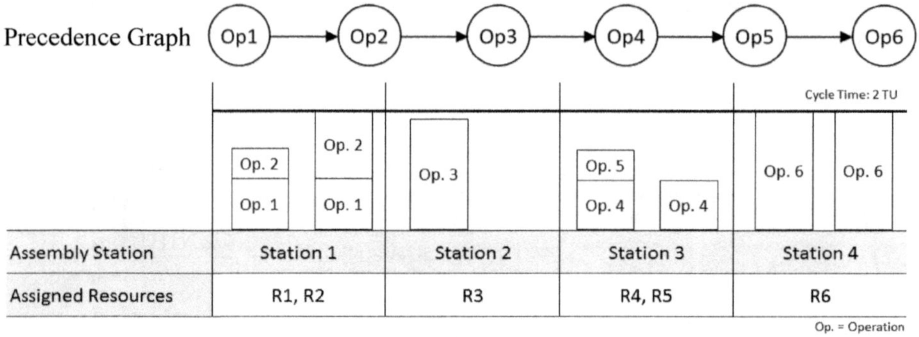

Fig. 3. Assembly line balancing for two product variants

The sections' processing times result from the sum of the contained operations.

	Processing Times in Section [TU]	
	Section 1	Section 2
Product Variant 1	3.6	3.5
Product Variant 2	2	3

	Operations			Resources
Section 1	Op. 1	Op. 2	Op. 3	R1, R2, R3
Section 2	Op. 4	Op. 5	Op. 6	R4, R5, R6

Fig. 4. Assembly times for sections in an example for two product variants

The minimum number of required stations at a tact time of 2 TU results from:

$$Number\ of\ stations = \left\lceil \frac{3.6\ TU + 3.5\ TU}{2\ TU} \right\rceil = \left\lceil \frac{7.1}{2} \right\rceil = 4$$

The maximum number of stations on the production area can be determined together with specifications on the size of the production area and the station size. The maximum number of parallel corridors for assembly lines is calculated depending on the width of the production area, the required width of the aisles, the width of the stating area between assembly lines, as well as the stations' width. Accordingly, the maximum number of work stations can be inferred from the production area's and station's lengths.

Now, a model with a cost based target function (mixed integer programming model; cf. [3]) can be formulated and solved using a branch-and-bound method.

Simultaneous planning of vertical and horizontal division. The assembly system is subdivided into two areas for pre-assembly and final assembly by vertical division. The degree of parallelization is to be determined for both areas independently. The choice of parallelization or horizontal division, respectively, is depicted in Figure 5 and shows possible separation points. For a given number n of assembly sections, 2n-1 optimization runs must be made.

<table>
<tr><td></td><td colspan="4">Assembly Sections</td></tr>
<tr><td></td><td>AS 1</td><td>AS 2</td><td>AS 3</td><td>AS 4</td></tr>
<tr><td>Possible combination 1</td><td>PA</td><td colspan="3">FA</td></tr>
<tr><td>Possible combination 2</td><td colspan="2">PA</td><td colspan="2">FA</td></tr>
<tr><td>Possible combination 3</td><td colspan="3">PA</td><td>FA</td></tr>
<tr><td>Possible combination 4</td><td colspan="4">no division</td></tr>
</table>

PA: Pre-Assembly Area, FA: Final Assembly Area, AS: Assembly Section

Fig. 5. Example of possible combinations for pre-assembly and final-assembly for four assembly sections

In order to identify the best separation point, all combinations have to be examined to determine the solution with the least total costs [3].

4 Reconfiguration of Conveyor Belt Coordination with Open Station Boundaries

Definition of the AIA key indicator. A useful shift of work content reduces the high utilization of the source workplace and results in an increase of low utilization of the target workplace. The AIA key indicator of an operation on a workplace $AIA_{op,wp}$ is the relation of average processing time of all work pieces, which are processed by that operation, and the capacity of the workplace:

$$AIA_{op,wp} = \frac{\frac{1}{|P_{op}|} \cdot \sum_{p \in P_{op}} \tau_{p,wp}}{\zeta_{wp}}$$

with:

P_{wp}	Set of all work pieces $p \in P$, which require operation $op \in OP$
$\tau_{p,wp}$	Processing time of work piece $p \in P$ on workplace $wp \in WP$ in time units
ζ_{wp}	Capacity of workplace $wp \in WP$ in time units

Selection process to calculate the shift by using the AIA key indicator. If the value of AIA is less than 1, the processing times of the work pieces, which require operation op, are in average lesser on workplace wp than the capacity limit of that workplace. If AIA is greater than 1the processing times are greater. The intention is to identify operations with a great AIA (>1) on workplace wp_1 but with a low AIA (<1) on another workplace wp_2. By the shift of these operations, the high processing times on $wp_1$are reduced. At the same time mainly low processing times are build up on wp_2.

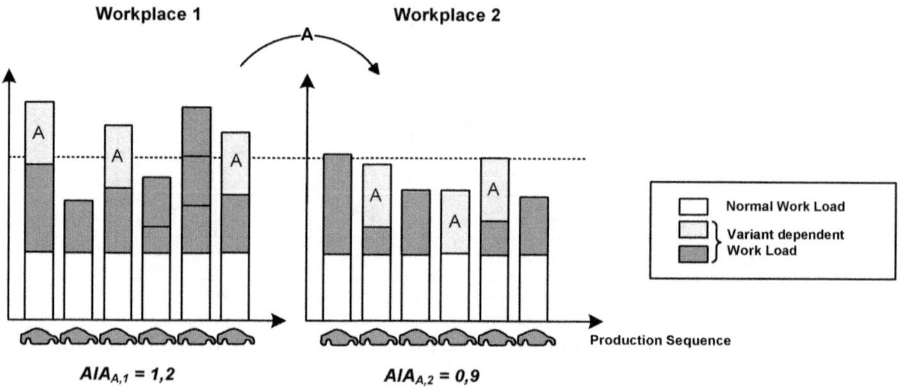

Fig. 6. Identification of a shift by using AIA key indicator

AIA is calculated for all operations of interest on the current workplace wp and all alternative workplaces. The resulting list of possible shifts is sorted by the effect of AIA-reduction.

Optimization approach for creating shift combinations. In the following optimization approach is added to the procedure. It is able to evaluate a sequence of shift combinations. Based on an existing solution a local search method is employed and a set of alternative solutions is generated [2]. This algorithm searches in a limited area defined by the user and presents the best solutions found. The following parameters allow the control of the search process:

- Definition of target value Δ^{LinPos} or $\Delta^{QuadPos}$
- Desired number of solutions generated
- Allowed length of the sequence of shifts
- Number of pursued solution for each shift level. Defines the precision of the algorithm.
- Upper limit of mean utilization for each workplace
- Lower limit of mean utilization for each workplace
- Fixing of operations on specific workplaces
- After evaluating the initial solution, the algorithm evaluates the complete neighborhood reached by the shift of an operation. Some shifts are invalid, defined by the following restrictions:
- Precedence restriction: Each operation can have one or more predecessors which have to be finished before this operation can start.
- Limits of load for workplaces: Upper and lower limits for mean utilization of workplaces can be defined.
- Work piece position: Each process has a list with allowed work piece positions. All other shifts are invalid.
- Fixed Operations: Some operations must remain on specific workplaces because they require special assets or special personnel available at these workplaces. These operations are fixed on these workplaces.
- Grouping of operations: The user is able to define groups of operations, which have to be shifted together. These groups are treated as a cumulated combined operation.

The best solution in a neighborhood is saved. Each of these solutions differs only in one single shift from the initial solution. Every saved solution is the basis of another neighborhood analysis on the next level of the search tree. After this, the best solutions are saved again and are used as initial solutions for the next analyzed shifts. This procedure is repeated until the defined maximal level in the search tree is reached.

Finally the β best solutions are selected. Identical solutions, only differing in the order of shifts, are ignored. Based on these solutions further solutions can be generated by using visual decision support, the AIA key indicator or the approach of optimization.

5 Two Level Multi - criteria Lotsizing

A supplier of the automotive industry which produces on the stages press shop[2], surface, assembly and distribution is at the centre of consideration. Demand data is available on a short term only. The press shop is brought into focus here.

Target hierarchy. Supply availability is of top priority. Allowances for current order procedures and related delivery terms and fluctuations in production have to be made. Capacity utilization of machines and workers are on second and third priority of the target hierarchy. The production of a machine is continued until the end of the setup of the machine which is the antecessor within the setup cycle. The design of the setup

[2] In the examined practical case, compound and transfer dies are used so that produced parts normally only have to be topically treated.

cycle guarantees that only one setup team is required at a time and a minimal number of shifts per week are necessary.

Lots are an integer multiple of the steel coils as the exchange of steel coils imperiled workers and due to technical reasons, detached steel coils cannot be used any more. On the fifth rank of this target hierarchy is the consideration of maintenance intervals of dies[3]. A larger life time and constant part quality is guaranteed. Not till then other technological aspects as well as the minimization of capital commitment are kept in mind.

Decomposition Approach. It does not make any sense to use complex models if data changes permanently. Therefore the planning horizon is subdivided into two sections:

Day	1	2	3	4	5	6	7	8	9	10	11	12	13	14
Method	DLSP			CLSP										

DLSP Discrete Lotsizing Problem
CLSP Capacitated Lotsizing Problem

Every day the planning horizon is shift one day and a new plan is generated with additional or updated demand data. In order to minimize the probability to obtain infeasibilities because of updated demands, an inventory level at the end of the planning horizon is calculated and set for each part. The calculated inventory level should not be mistaken for security inventory levels which have to be kept in the inventory during all the time. The inventory for each part is rather adapted dynamically at each planning process in accordance with last demand dates. Missing availability and the related "nervousness" which are caused exclusively by the planning system can then be reduced.

Gross Planning as CLSP. The planning horizon between day 4 and 14 is modeled with an adapted CLSP. Production plans are calculated on a big bucket basis using optimization software daily. Demands are extracted directly from the ERP and then reallocated within days 4 and 14. Date dependent production capacities, the production of full steel coils as well as maximum die life, which can be seen as lot-size limits, are considered. Setup and adjustment times, coil change times and maintenance times are generalized and subtracted from available capacity.

Scheduling with DLSP. The first three days past T_{now} are modeled by a small bucket model derived from the DSLP. Planning is done in time units of half an hour length. The stock keeping costs and the production costs are to be minimized. Tools which can produce different parts simultaneously are considered. Setup costs are very important, because setup times can last from one hour up to a whole shift, dependant on the sequence of tools. Additionally the personnel required for setup is kept to the minimum by avoiding simultaneous setups on different machines. Lots are defined as

[3] Dies are maintained after each setup in the present situation. In future, dies are only maintained after exceeding half of the rest of the die life in order to reduce maintenance costs at equal quality of parts. An alternative method could be to await the culmination of the maximum die life. But this would limit lot sizes and is contrary to the presented target hierarchy.

n-folds of raw material units. The necessary coil changes are planned. The maximal lot size, defined by tool life, is included in the model.

Plan Update / Feedback. Because of the daily planning update, only one day can be set as fixed – as a production task. The rest of the planning period is fixed for day 2 and 3, not taking backlog and total disaster into account. Days 4 to 14 are the best possible forecast of the future. But the monitoring of production is simplified by the daily planning update: There is a feedback of the number of produced pieces for each part number and day[4]. Taking the outflow to customers additionally into account, the total amount of pieces in the production chain of the considered company can be calculated precisely[5]. This stock level is considered the actual stock for the CLSP or DLSP. Thus, no backlog must be monitored and no precarriage must be considered.

First Results. Especially the comparison of the scheduling with the planning methods used before show significant improvement: Total availability is guaranteed under excessive concentration on machine uptime and personnel employment in normal working times.

6 Rolling Horizon and Online Optimization

The production system. We consider a production system for producing a certain set of products. The products are known in advance and can therefore be produced prior to an order.

Orders are only given for a certain time horizon. By the course of time new orders are revealed. No possible plan can take all orders in consideration, because only a certain, restricted part of the future is known in advance, the planning horizon. Although a scheduling strategy cannot provide an optimal solution, due to a lack of complete knowledge about future orders, it should still avoid circumstances that will handicap future production processes. Hence, our basic rule: Independent of what the future will bring, the performance of the production, respectively the scheduling strategy is not allowed to fall below a certain bound. Of course, this bound should depend on the value of an optimal solution. Unfortunately this bound can only be determined as soon as all orders are known- and that is in retrospect. We are facing the challenge: How should a company produce today, even though only orders for the near future are known, or even worse, only the next order? Problems like this are handled within the framework of online optimizing [5]. Here, in contrast to classical offline optimization the input is revealed piece by piece and hence the optimal solution is unknown in advance, but still certain bounds are guaranteed to be kept independent of future events. The considered production system works under the assumptions of the Discrete Lot sizing and Scheduling Problems; the production of a product covers always complete time segments.

[4] After setup and adjustment of a machine the first produced parts are checked and it is waited until the clearance for serial production is given. In the given example we assume that the clearance is given for all parts. Semi-finished punched parts are counted directly after the punching station. A blocking or charge of faulty single parts is done in a retrograde manner.

[5] If one piece is part of more than one end product this is also considered a customer. The definition of stock delimitation is not considered in this paper.

A schedule covers several periods with several time segments each. It is updated periodically. At any one time the first period is considered to be the schedule implemented. The realization of a schedule is therefore the sustained implementation of respective first periods. Updates to the schedule only occur at proper points of time that are previously fixed for scheduling. Those are beyond the calendar (that is e.g. not during the week if a weekly planning cycle is considered). The scheduled demands (reduction in inventory) are already known "today" and can be postponed or changed in value until their final fixation (latest by the time they are scheduled in the first scheduling period).

Approach for on-line scheduling. Online the following approach is chosen [4]:

- Optimization of the first period + optimization of the remaining periods of the planning horizon, considering the joint inventory at the end of the first period.
- Schedule DLSP1 has to guarantee a (minimum) profit
- Weighting the two (partial) solutions
- Dependent on the chosen weighting, a cost minimal solution for the corresponding planning horizon is chosen.

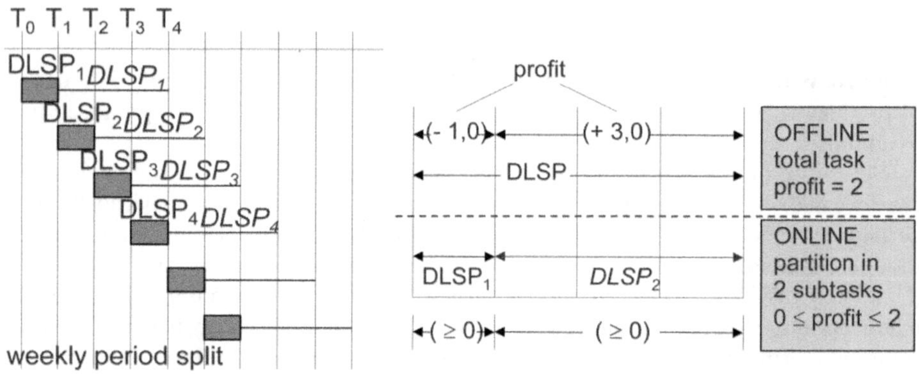

Fig. 7. Partition in two subtasks; an example where DLSP solutions differ

For instance, Figure 7 displays a weekly continued planning, where always one week is fixed respectively. This week is divided for instance in days or shifts (4 week horizon, the schedule is repeated weekly, each week consists of 10 shifts).

At point of time T_0 the schedules DLSP_1 and DLSP_1, are calculated. The schedule that maximizes (profit) respectively minimizes (costs) with $\alpha * \mathrm{DLSP}_1 + (1-\alpha) * \mathrm{DLSP}_1 = \min$ with $0 \leq \alpha \leq 1$ is chosen. This way

- at the end of the first period disprofit is avoided, respectively a (minimum) profit is realized
- a feasible schedule for the planning horizon is created
- due to the DLSP, a mathematical formulation of the problem is enabled

Of course, the solution for the DLSP over the (total-) horizon and the solutions to the two DLSP partial problems that are coupled by the inventory need not to be identically, because each partial solution has to guarantee a positive value.

Competitiveness of the online-approach. The best case for the off-line inventory, concerning a complete balancing of capacity from one period into another one, is the case where there are z empty and cheap time segments at the end of a period followed by a period with z expensive time segments that are covered. Here an off-line algorithm can completely balance its capacity, but at the price of additional inventory costs. For the on-line algorithm holds: In order to shift all expensive time segments to a previous period, the factor of the minimal profit has to be set to "zero". Then the on-line algorithm aligns with the offline algorithm (for this case). If the factor of the minimal profit is set to "1" no shifting to the previous period is possible and the whole costs for the expensive time segments have to be applied.

Therefore, we chose the following approach: We use the "pull ahead" of the respective off-line solution in the current period, unless there is cumulated profit larger than zero. In this case only a certain fraction is used. That means: once there is a cumulated profit, it is not completely spend anymore. If there is no profit so far, we balance and try to keep the costs as low as possible:

- We shift to the front, if the off-line algorithm was unable to make any profit in the past
- We shift to the front, if the off-line algorithm made a "large" profit in the past

For that purpose an off-line solution for the entire past is calculated. The current period is integrated in this off-line solution and is accounted the average value of shifting forward in the past (average use of capacity in periods that were balanced). Then we have for the shifting forward of x periods:

$$\begin{aligned} Savings &= x * expensive\ period \\ &\quad - \sum_{s=1}^{x} (2(s-1)+1) * inventory\ costs\ per\ period \\ &= x * expensive\ period - x^2 * inventory\ costs\ per\ period \end{aligned}$$

From the first derivate to x

$$x/dx\ (savings) = (expensive\ period) - (2 * inventory\ costs\ per\ period)$$

follows: The savings are maximal for

$$x^* = (expensive\ period)/(2 * inventory\ costs\ per\ period)$$

x^* is the best possible horizon for shifting to the front with maximal savings "$savings^*$".

On the other hand, the on-line algorithm cannot avoid the expensive periods:

$$costs_{onl} = x * normal + x * expensive$$
$$(considering\ normal\ time\ slots\ as\ given)$$

$$costs_{off} = costs_{onl} - savings^*$$

This leads to the ratio ($costs_{onl}/costs_{off}$). In general holds:

- Best possible shifting: $x^* = (k_p^t - k_p^n)/(2 * k_B)$
- Maximal savings*=$x^* * \left(k_p^t - k_p^n\right) - (x^*)^2 * k_B$

- Ratio on-line/off-line $\frac{k_{onl}}{k_{off}} = \frac{x^* * k_p^n + x^* * k_p^t}{x^* * k_p^n + x^* * k_p^t - x^* * (k_p^t - k_p^n) + (x^*)^2 * k_B} = \frac{x^* * (k_p^t - k_p^n)}{2 * k_p^n x^* + x^{*2} * k_B}$
- Example
  (cost for producing in an expensive period) : (inventory costs per period) = 10:1
  normal period= 1/2 expensive period
  Savings:$x * expensive\ period - \sum_{s=1}^{x}(2(s-1)+1) * expensive\ period$

  $k_p^t = 1;\ k_p^n = 0{,}5; k_B = 0{,}1; x^* = 2{,}5$

$Savings^* = 1{,}25 - 0{,}625 = 0{,}625$

$$\frac{k_{onl}}{k_{off}} = \frac{2{,}5 * 1{,}5}{2 * 0{,}5 * 2{,}5 + 6{,}25 * 0{,}1} = \frac{3{,}75}{3{,}125} = 1{,}2$$

7 Sequencing of Production Orders / Open Station Boundaries

By rearranging the job steps in a reconfiguration a reduction in the utilization variance can be achieved, but not eliminated completely. However, to ensure a most efficient and smooth production process, the sequence planning has to balance the utilization of the workstations. Vehicles with high production time must be mixed with vehicles with lower production time in the way that the overload for the workers at all stations is minimized.

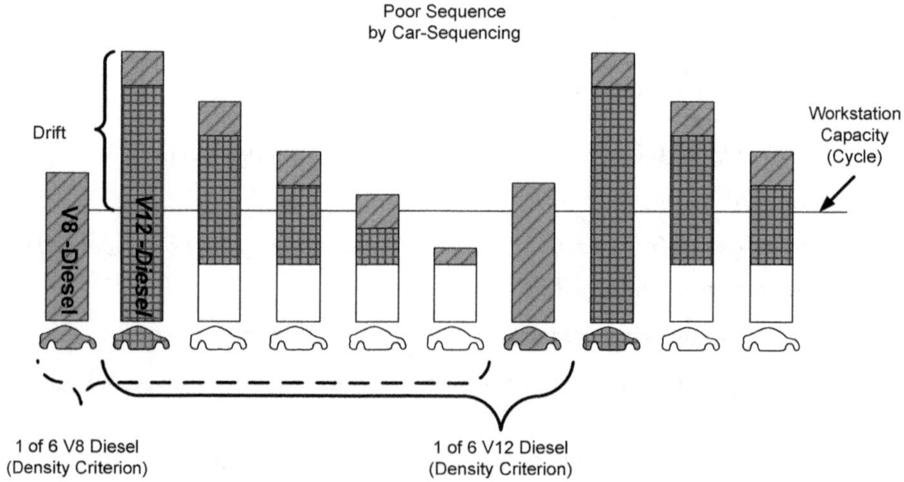

Fig. 8. Example for a poor sequence despite of Car – Sequencing

The common approach which is also used in practice is the so called Car-Sequencing-Approach and is based on density criteria. “One of two” sunroofs would signify, that at maximum every second car could have the option “sunroof”. However, practice shows that this approach is not sufficient for balancing the utilization of workers. The biggest criticism here is the fact that within a rule, only one option can be controlled but the combination of several options is neglected. Figure 8 demonstrates how the overload, which is caused by the first vehicle, remains and cannot be reduced until the sixth vehicle.

The method discussed here [2] is based on the idea of Mixed Model Sequencing [6] and provides as uniform as possible load of workmen without neglecting all relevant practical constraints. In particular, the dependences between workers are considered, since downtime occurs, if a worker is still on the vehicle and the subsequent worker can not begin. (see Figure 9)

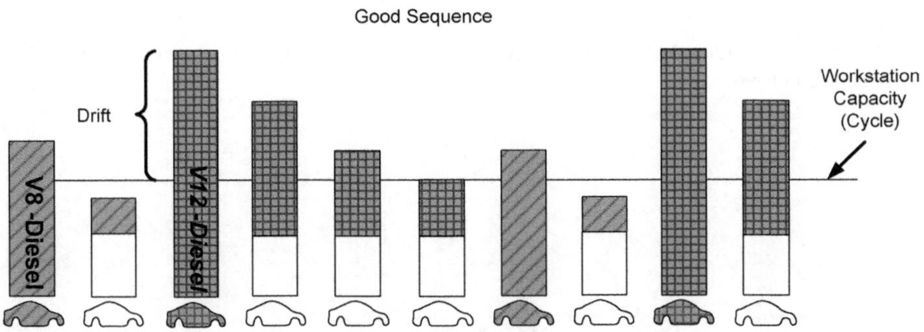

Fig. 9. Example for a production sequence with minimal stress

For this purpose a scoring system was developed in the first step to assess respectively simulate an assembly line setup with a certain production sequence realistically. Key result figures are particularly the average drift and the number and duration of necessary support operations [7]. The average drift defines the position of the worker caused by the overloading vehicles relative to the original starting point position.

In the second step this assessment system is used in a two-step procedure to form a production sequence with minimal stress. Here, the drift curves of the individual jobs are used to form an aggregate curve that allows a statement about which positions within the sequence caused high drift positions and / or supporting operations. The corresponding positions are marked and exchanged with those, which are leading to idle time.

Based on the new drift curves, the simulation is now being re-launched and identifies the vehicles for exchange. This process is repeated until no significant changes in the objective function value can be achieved anymore.

The method is tested in prototype software using real data in cooperation with a practice partner.

References

1. Danne, C.: Assessing the Cost of Assortment Complexity in Consumer Goods Supply Chains by Reconfiguration of Inventory and Production Planning Parameters in Response to Assortment Changes. HNI-Verlagsschriftenreihe, Paderborn (2009)
2. Altemeier, S.: Kostenoptimale Kapazitätsabstimmung in einer getakteten Variantefließlinie unter expliziter Berücksichtigung des Unterstützereinsatzes und unterschiedlicher Planungszeiträume. HNI-Verlagsschriftenreihe, Paderborn (2009)

3. Gans, J.E.: Neu- und Anpassungsplanung der Struktur von getakteten Fließproduktionssystemen für variantenreiche Serienprodukte in der Montage. HNIVerlagsschriftenreihe, Paderborn (2008)
4. Dangelmaier, W., Degener, B.: Production planning and control with discrete lotsizing and a rolling horizon. In: Dagstuhl, Seminar 09261 – Models and Algorithms for Optimization in Logistics (2009)
5. Krumke, S.O., Rambau, J.: Online Optimierung. Vorlesungsskript, Technische Universität, Berlin (2005)
6. Boysen, N., Fliedner, M., Scholl, A.: Sequencing Mixed Model Assembly Lines – Survey, Classification and Model Critique. Technischer Bericht, Friedrich-Schiller-University, Jena (2007)
7. Mollemeier, A.: Integrierte Steuerung getakteter Variantenfließlinien. University of Paderborn, Paderborn (1996)

A Mesoscopic Approach to the Simulation of Logistics Systems

Michael Schenk, Juri Tolujew, and Tobias Reggelin

Fraunhofer Institute for Factory Operation and Automation IFF Magdeburg,
Sandtorstr. 22, 39106 Magdeburg, Germany
{michael.schenk,juri.tolujew,tobias.reggelin}@iff.fraunhofer.de
http://www.iff.fraunhofer.de

Abstract. This paper presents a mesoscopic simulation approach in addition to the prevailing continuous and discrete event simulation approaches used for logistics systems. In terms of level of detail, the mesoscopic approach falls between these two approaches. Mesoscopic models represent logistics flow processes through piecewise constant flow rates. The resulting linearity of the cumulative flows allows for event scheduling and the use of mathematical formulas for recalculating the system‘s state variables at every simulation time step. The simulation time step is variable and the step size depends on the occurrence of scheduled events. This leads to a high computational performance. The mesoscopic approach distinguishes between different parallel product types. The modeling components are multichannel funnels, delays, assemblies and disassemblies.

Keywords: modeling, simulation, mesoscopic, logistics systems.

1 Introduction

Enterprises have to be able to react dynamically to changing market environments, disturbances and unforeseen events. This necessitates suitable tools that immediately show the effects of changing conditions on the time-dependent behavior of the observed system so that qualified measures such as modification of a control strategy can be devised and implemented. Simulation models are the best solution to such problems (cf. [1], [2]).

Two classes of simulation models exist, namely continuous and discrete. Continuous models are based on differential equations and most frequently applied as system dynamics models to reproduce manufacturing and logistics processes [3], [4], [5], [6], [7]. Since these models typically work with aggregated data on a strategic level, they are also referred to as macroscopic models (cf. [2], [3], [8]). Their level of aggregation renders them incapable of accurately representing the numerous logistics objects (products, resources, etc.) and control strategies, which demand consideration when resolving tactical or operational problems (cf. [3], [8]).

The principles and tools of discrete event simulation [9], [10], [11], [12] are utilized to implement discrete models. Since event-oriented models are able to

W. Dangelmaier et al. (Eds.): IHNS 2010, LNBIP 46, pp. 15–25, 2010.

represent workstations, technical resources, carriers and units of goods as individual objects, they are also referred to as microscopic models (cf. [2], [3]). Models in this class can be very complicated and slow and their creation and implementation can be time and labor consuming (cf. [3], [11], [12], [13], [14]).

The mesoscopic simulation approach is situated between continuous and discrete event approaches in terms of level of modeling detail. It supports quick and effective execution of analysis and planning tasks related to manufacturing and logistics networks. The mesoscopic approach is consistent with the principles of the discrete rate simulation paradigm implemented in the simulation software ExtendSim [15], [16] and resembles the principles of hybrid simulation described by Kouikoglou and Phillis [17]. Piecewise constant flow rates and the resulting linear cumulative flows facilitate event scheduling and high computational performance.

Even when the term mesoscopic is not explicitly applied, a mesoscopic view often already exists from the start of flow system modeling and simulation. Many practical analysis and planning problems describe performance requirements, resources and performance results in an aggregated form that corresponds to a mesoscopic view (cf. [15], [18]). Mesoscopic models are particularly suited for the analysis of large-scale logistics networks and processes with a homogenous flow of a large number of objects. In most cases, an item-based discrete event simulation would be overly complex for these applications because of the disproportionate amount of computation required. The principles of mesoscopic simulation of processes in logistics networks described here have been derived from the actual development of several mesoscopic models [18], [19], [20], [21], [22], [23].

2 Forms of Flow Process Representation

A comparison of mesoscopic flow modeling with the two conventional types of modeling elucidates its distinctive features [24] (Fig. 1). A system dynamics model reproduces continuously changing processes and advances the simulation time in constant time steps. The state variables of the observed processes are recalculated at each time step. For the sake of accuracy, the time steps must be relatively small. Since one single variable reproduces the flow between two nodes of a network structure, a flow's individual segments are neither identifiable nor traceable. Discrete event modeling on the other hand represents a discrete flow as a sequence of events. Every event is distinctly identifiable by the time of its occurrence and the objects involved in it. Mesoscopic models represent material flows as:

- piecewise constant flow rates and
- sequences of timeless events (impulse-like flows).

Piecewise constant flow rates have been adopted in order to apply event scheduling (see section 3).

Planners often define piecewise constant processes, e.g. performance requirements (input flows and system load) or resource capacities, without explicitly labeling them as such. Typical statements furnish examples:

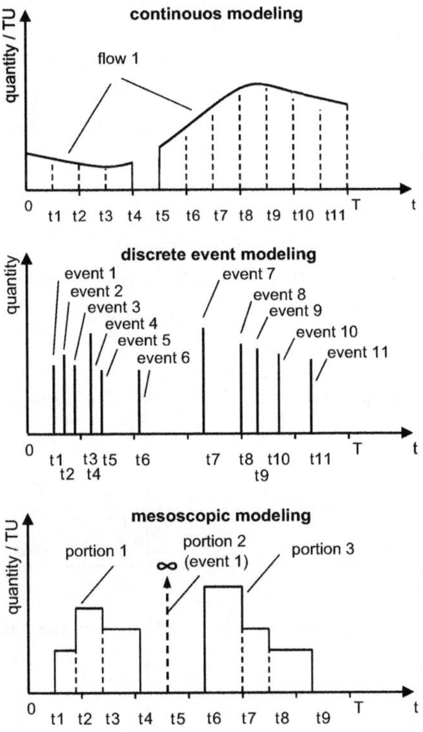

Fig. 1. Different types of flow process modeling

- An average of 12 trucks arrives at the warehouse premises per hour from 8 a.m. to 12 p.m. and approximately 8 trucks per hour from 12 to 4 p.m.
- Forklifts transport pallets from the ramp to the storeroom from 8 a.m. to 12 p.m., every forklift moving an average of 35 pallets per hour; stock is retrieved from 12 to 4 p.m., every forklift being able to transport an average of 25 pallets to the ramp per hour.

A timeless event is defined when, for example, only the total quantity of goods a truck delivers to receiving is represented and unloading time is not factored in.

3 Event Scheduling

Event scheduling in mesoscopic models is based on the calculation of a linear time function's future development [25]. Fig. 2 presents an example of event scheduling for continuous processes in a funnel. Event 1 and 2 are assumed to be independent of the system's current state. Specific conditions are specified for all other events:

- Event 3: The funnel's output flow can only start when its stock has reached the critical level 2.
- Event 4: The funnel's input flow is terminated when the entire quantity of products in the flow has reached the critical level 1.
- Event 5: The funnel's output flow is terminated when its stock has reached the critical level 3.

Times t3, t4 and t5 can be calculated precisely and entered into the simulator's chain of future events. Ultimately, only five events must be processed when a variable time step is used. In terms of performance and accuracy, the advantage of this type of time advance over a constant time step is obvious.

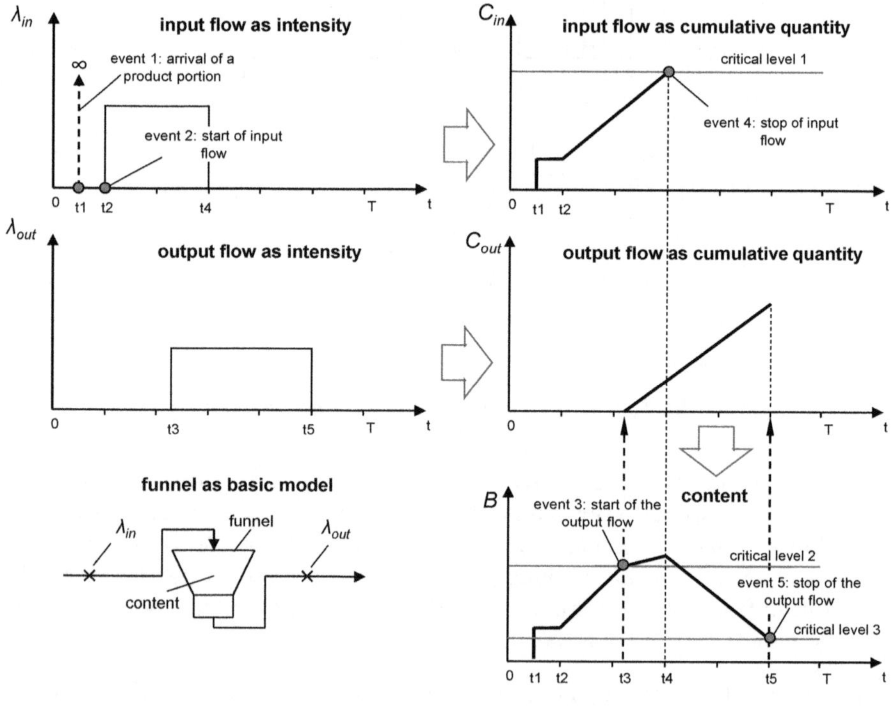

Fig. 2. Event scheduling

4 Mesoscopic Product Model

Mesoscopic models call the objects of material flows, e.g. goods, materials or persons, products. A mesoscopic model may employ different product types [25] in parallel in order to differentiate between flow objects with different characteristics. Physical properties or other relevant features distinguish the individual product types from one another. Every product type is assigned to its own channel at a model component (Fig. 3).

Furthermore, so-called product portions [25] are introduced in order to sequentially differentiate a flow of a product type. Their number is specified during the conceptual modeling phase. Certain quantities of products, e.g. lot size, cargo size, number of goods in a shipment or number of people in a group, may be modeled as product portions. Thus, the path from individual product portions that may be spatially distributed throughout the network can be tracked and relevant events that may occur along this path can be captured.

5 Multichannel Funnels as Main Components of a Mesoscopic Model

A mesoscopic model may employ the basic components of source, sink, funnel and delay to represent a material flow structure (Fig. 4). Flows may be additionally modified with the components of assembly and disassembly.

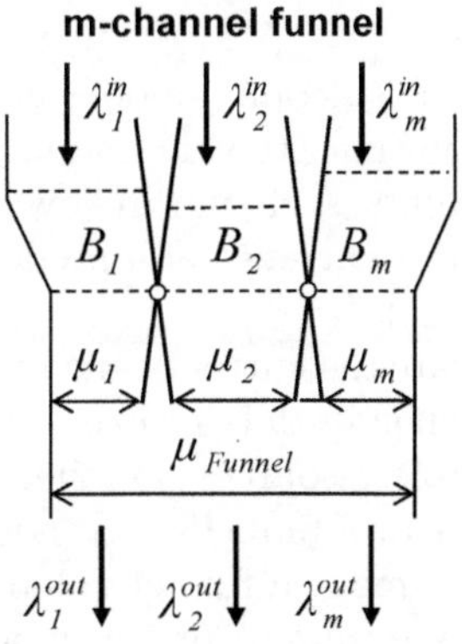

$$\lambda_i^{out} = \begin{cases} 0, \text{ if } \lambda_i^{in} = 0 \text{ and } B_i = 0 \\ \lambda_i^{in}, \text{ if } \lambda_i^{in} > 0 \text{ and } \lambda_i^{in} \leq \mu_i \text{ and } B_i = 0 \\ \mu_i, \text{ if } B_i > 0 \end{cases}$$

Fig. 3. Mathematical model of the multichannel funnel

Multichannel funnels [26], [27] (Fig. 3) are a mesoscopic model's main component because they properly represent the processes of parallel or sequential processing and storage of several product types and product portions in a real area of operations.

The variables λ_i^{in} and λ_i^{out} are the intensities of incoming and outgoing flows of a funnel's channel i ($i = 1, \ldots, m$). B_i denotes the content of a funnel's channel i. The model's control block determines the current value of each channel's maximum throughput μ_i . These values are normally the results of the allocation of different resources available to the entire area of operations represented by the funnel. Standard rules of resource allocation are described below. Fig. 3 presents the mathematical model for calculating λ_i^{out} .

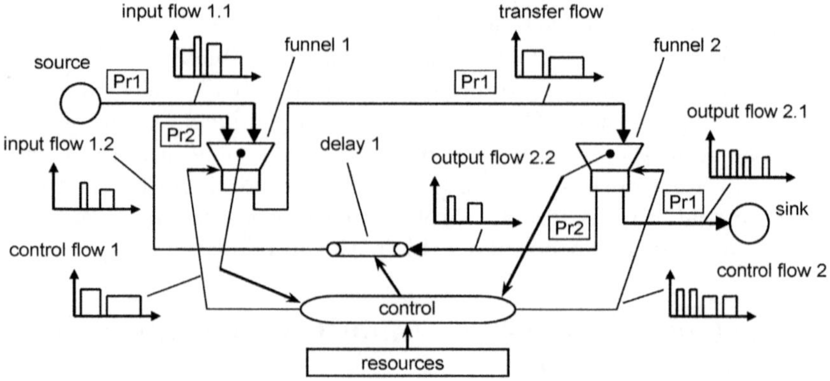

Fig. 4. Mesoscopic model layout

6 Standard Models for Allocation of Resources at a Funnel

Resource allocation algorithms are activated in the mesoscopic model's control block (Fig. 4) at the beginning of every new simulation time step. The simulation time advance in mesoscopic models is controlled by events. A change in flow rates, a change in the quantity of available resources or the occurrence of a predefined condition such as crossing a threshold can cause events.

Rigid resource allocation algorithms only incorporate the current time but not the current status of individual components, e.g. when shift data or other predefined time intervals determine the quantity of available resources. Furthermore, standard solutions that incorporate the current state of a funnel's channel (λ_i^{in} and B_i) may be applied to flexibly allocate resources among a funnel's channels.

A multichannel funnel represents a homogeneous resource[1] of the modeled system and has n resource holders at its disposal. Every resource holder k ($k = 1, \ldots, n$) has a potential performance pl_k. Three standard potential performances pl_k may be distinguished in terms of units of measurement:

- Case 1: pl_k [unit of quantity/unit of time]
- Case 2: pl_k [operating time/unit of time]
- Case 3: pl_k [operations/unit of time].

The total potential performance PL is the sum of all resource holders' potential performances:

$$PL = \sum_{k=1}^{n} pl_k \tag{1}$$

[1] Here, a homogeneous resource refers to identical types of resources that may be arbitrarily allocated among a funnel's different channels.

The coefficient of consumption vk_i converts fluxes and maximum throughputs measured in [unit of quantity/unit of time] into the variables of resource consumption and back. In Case 1, a coefficient of consumption is obviously not necessary: $vk_i = 1$. In Case 2, the coefficient of consumption vk_i [operating time/unit of quantity] specifies the operating time required to generate one unit of quantity for the output flow λ_i^{out} of a funnel's channel. In Case 3, the coefficient of consumption vk_i [operations/unit of quantity] specifies the number of operations required to generate one unit of quantity for the output flow λ_i^{out} of a funnel's channel.

The demand for resources bd_i [unit of resource/unit of time] for all types of products is calculated with the coefficient of consumption vk_i :

$$bd_i = vk_i \cdot \lambda_i^{in} \tag{2}$$

The total demand BD is the sum of the demands for all product types:

$$BD = \sum_{i=1}^{m} bd_i \tag{3}$$

The maximum throughput μ_i [unit of quantity/unit of time] of every channel determines the allocated resources r_i [unit of resource/unit of time] for every channel:

$$r_i = \mu_i \cdot vk_i \tag{4}$$

The total amount of allocated resources R is the sum of the allocated resources for every channel:

$$R = \sum_{i=1}^{m} r_i = \sum_{i=1}^{m} (\mu_i \cdot vk_i) \tag{5}$$

The maximum throughput μ_i of a channel may be calculated when r_i is known:

$$\mu_i = \frac{r_i}{vk_i} \tag{6}$$

The following examples show how μ_i is calculated at every time step for eight different resource allocation rules. The following examples display the complete calculation for rule 1, but only the results for the other rules. The different resource allocation rules generate different development of channels' contents (see Fig. 5).

Resource allocation rule 1

The maximum throughput is identical for every product type: $\mu_i = \mu$.
The total potential performance is distributed among the channels. Therefore $R = PL$ applies.

$$R = \sum_{i=1}^{m} r_i = \sum_{i=1}^{m} (\mu_i \cdot vk_i) = \mu \cdot \sum_{i=1}^{m} vk_i = PL \tag{7}$$

$$\mu_i = \mu = \frac{PL}{\sum_{i=1}^{m} vk_i} \tag{8}$$

Resource allocation rule 2

The maximum throughput is proportional to the input flow: $\mu_i \sim \lambda_i^{in}$.

$$\mu_i = \frac{PL}{BD} \lambda_i^{in} \tag{9}$$

Resource allocation rule 3

The maximum throughput is proportional to the content of the channel: $\mu_i \sim B_i$.

$$\mu_i = \frac{PL}{\sum_{i=1}^{m} (B_i \cdot vk_i)} B_i \tag{10}$$

Resource allocation rule 4

The maximum throughput is proportional to the demand: $\mu_i \sim bd_i$.

$$\mu_i = \frac{PL}{\sum_{i=1}^{m} (bd_i \cdot vk_i)} bd_i \tag{11}$$

Resource allocation rule 5

The rates of change of the contents of all the channels are the same: $\lambda_i^{in} - \mu_i = const$.

$$\mu_i = \lambda_i^{in} - \frac{BD - PL}{\sum_{i=1}^{m} vk_i} \tag{12}$$

Resource allocation rule 6

The content for all product types may not change: $B_i = const$.

$$\mu_i = \lambda_i^{in} \tag{13}$$

This rule is applicable only when the total potential performance is at least as great as the total demand for resources: $PL \geq BD$.

Resource allocation rule 7

The maximum throughput is proportional to the priorities of the product types: $\mu_i \sim pr_i$.

$$\mu_i = \frac{PL}{\sum_{i=1}^{m} (pr_i \cdot vk_i)} pr_i \tag{14}$$

Resource allocation rule 8

Top priority is assigned to one product type p. Demand will be completely satisfied when $bd_p \leq PL$ such that

$$\mu_p = \frac{bd_p}{vk_p} \tag{15}$$

with the remaining resources being allocated following one of the allocation rules described above.

The demand for the product with top priority cannot be completely satisfied when $bd_p \geq PL$ such that

$$\mu_p = \frac{PL}{vk_p} \tag{16}$$

with the other product types not receiving any resources.

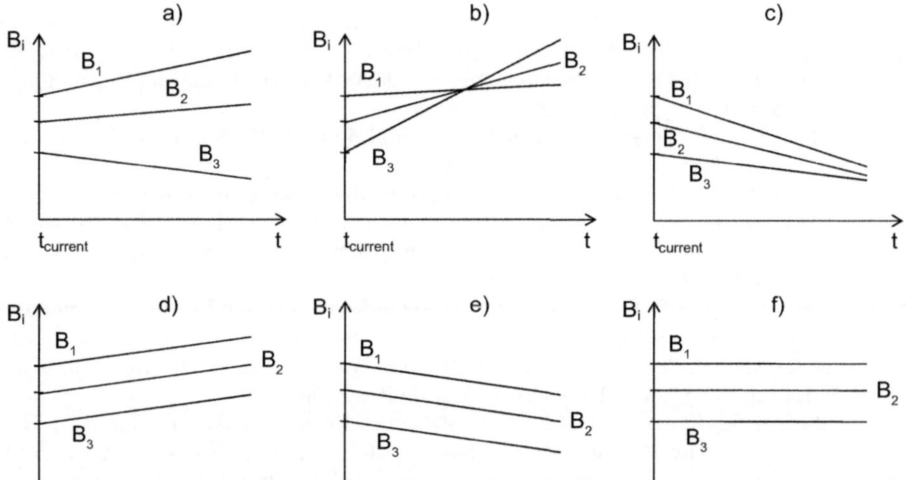

Fig. 5. Developments of content for different resource allocation rules. a) and b) possible behavior for rules 1, 2, 4, 7 and 8. c) possible behavior for rule 3. d) and e) possible behavior for rule 5. f) possible behavior for rule 6.

7 Conclusion

The approach to mesoscopic simulation described here has two main advantages:

- Flow processes are modeled with a simple and universal form of representation (piecewise constant flow rates and impulse-like flows) that is suitable for many real manufacturing and logistics processes.
- Models employing variable time steps perform significantly better and require less computing time than continuous or discrete event models.

This approach expands the present set of tools for logistics systems simulation.

References

1. Aufenanger, M., Varnholt, H., Dangelmaier, W.: Adaptive Flow Control in Flexible Flow Shop Production Systems: A Knowledge-Based Approach. In: Rossetti, M.D., Hill, R.R., Johansson, B., Dunkin, A., Ingalls, R.G. (eds.) Proceedings of the 2009 Winter Simulation Conference, pp. 2164–2175. IEEE, Inc., Piscataway (2009)
2. Borshchev, A., Filippov, A.: From System Dynamics and Discrete Event to Practical Agent Based Modeling: Reasons, Techniques, Tools. In: Proceedings of the 22nd International Conference of the System Dynamics Society, Oxford, pp. 25–29 (2004)
3. Pierreval, H., Bruniaux, R., Caux, C.: A Continuous Simulation Approach for Supply Chains in the Automotive Industry. Simulation Modelling Practice and Theory 15(2), 185–198 (2007)
4. Scholz-Reiter, B., Delhoum, S., Zschintzsch, M., Jagalski, T., Freitag, M.: Inventory Control in Shop Floors, Production Networks and Supply Chains Using System Dynamics. In: Wenzel, S. (ed.) Tagungsband 12. ASIM-Fachtagung Simulation in Produktion und Logistik, pp. 273–282. SCS, Erlangen (2006)
5. Angerhofer, B.J., Angelides, M.C.: System Dynamic Modeling in Supply Chain Management: A Research Review. In: Joines, J.A., Barton, R.R., Kang, K., Fishwick, P.A. (eds.) Proceedings of the 2000 Winter Simulation Conference, pp. 342–351. IEEE, Inc., Piscataway (2000)
6. Sterman, J.D.: Business dynamics: System Thinking and Modeling for a Complex World. Irwin, McGraw-Hill, Boston (2000)
7. Anderson, E.G., Morrice, D.J.: A Simulation Model to Study the Dynamics in a Service-Oriented Supply Chain. In: Farrington, P.A., Nembhard, H.B., Sturrock, D.T., Evans, G.W. (eds.) Proceedings of the 2009 Winter Simulation Conference, pp. 742–748. IEEE, Inc., Piscataway (1999)
8. Größler, A.: System Dynamics zur Strategiesimulation im Produktionsmanagement. In: Specht, D. (ed.) Strategische Bedeutung der Produktion: Tagungsband der Herbsttagung 2006 der Wissenschaftlichen Kommission Produktionswirtschaft im VHB, pp. 83–87. Dt. Univ.-Verl., Wiesbaden (2007)
9. Schriber, T.J., Brunner, D.T.: Inside Discrete-Event Simulation Software: How it Works and Why it Matters. In: Mason, S.J., Hill, R.R., Mönch, L., Rose, O., Jefferson, T., Fowler, J.W. (eds.) Proceedings of the 2008 Winter Simulation Conference, pp. 182–192. IEEE, Inc., Piscataway (2008)
10. Banks, J.: Discret-Event System Simulation. Pearson, Prentice Hall, Upper Saddle River (2005)
11. Law, A.M.: Simulation Modeling and Analysis. McGraw-Hill, New York (2007)
12. Košturiak, J., Gregor, M.: Simulation von Produktionssystemen. Springer, Wien (1995)
13. Huber, D., Dangelmaier, W.: Controlled Simplification of Material Flow Simulation Models. In: Rossetti, M.D., Hill, R.R., Johansson, B., Dunkin, A., Ingalls, R.G. (eds.) Proceedings of the 2009 Winter Simulation Conference, pp. 839–850. IEEE, Inc., Piscataway (2009)
14. Scholz-Reiter, B., de Beer, C., Freitag, M., Hamann, T., Rekersbrink, H., Tervo, J.T.: Dynamik logistischer Systeme. In: Nyhuis, P. (ed.) Beiträge zu einer Theorie der Logistik, pp. 109–138. Springer, Berlin (2008)
15. Krahl, D.: ExtendSim Advanced Techology: Discrete Rate Simulation. In: Rossetti, M.D., Hill, R.R., Johansson, B., Dunkin, A., Ingalls, R.G. (eds.) Proceedings of the 2009 Winter Simulation Conference, pp. 333–338. IEEE, Inc., Piscataway (2009)

16. Damiron, C., Nastasi, A.: Discrete Rate Simulation Using Linear Programming. In: Mason, S.J., Hill, R.R., Mönch, L., Rose, O., Jefferson, T., Fowler, J.W. (eds.) Proceedings of the 2008 Winter Simulation Conference, pp. 740–749. IEEE, Inc., Piscataway (2008)
17. Kouikoglou, V.S., Phillis, Y.A.: Hybrid Simulation Models of Production Networks. Kluwer Academic/Plenum Publishers, New York (2001)
18. Schenk, M., Tolujew, J., Reggelin, T.: Mesoskopische Modellierung für die schnelle und aufwandsarme Planung und Steuerung robuster und sicherer Logistiksysteme. In: Pfohl, H.-C., Wimmer, T. (eds.) Robuste und sichere Logistiksysteme, pp. 263–292. DVV Media Group, Deutscher Verkehrs-Verlag, Hamburg (2008)
19. Schenk, M., Tolujew, J., Reggelin, T.: Comparison of Three Methods of Implementation of Mesoscopic Flow Models. In: Ivanov, D., Meinberg, U. (eds.) Logistics and Supply Chain Management: Modern Trends in Germany and Russia, pp. 36–44. Cuvillier Verlag, Göttingen (2009)
20. Reggelin, T.: A Mesoscopic Approach to Modeling and Simulation and its Application for a Logistics Hub. In: Internationale Maschinenbaukonferenz Comec 2008. Universidad Central "Marta Abreu" de Las Villas, Santa Clara (2008)
21. Savrasov, M., Tolujew, J.: Mesoscopic Approach to Modeling a Traffic System. In: Kopytov, E., Pranevicius, H., Zavadskas, E., Yatskiv, I. (eds.) International Conference Modelling of Business, Industrial and Transport Systems, pp. 147–151. TSI, Riga (2008)
22. Tolujew, J., Alcalá, F.: A Mesoscopic Approach to Modeling and Simulation of Pedestrian Traffic Flows. In: Horton, G. (ed.) Proceedings of the 18th European Simulation Multiconference, pp. 123–128. SCS, Ghent (2004)
23. Hanisch, A., Tolujew, J., Richter, K., Schulze, T.: Online Simulation of Pedestrian Flow in Public Buildings. In: Proceedings of the 2003 Winter Simulation Conference, pp. 1635–1641. IEEE, Inc., Piscataway (2003)
24. Tolujew, J., Reggelin, T.: Mesoskopische Simulation: Zwischen der kontinuierlichen und ereignisdiskreten Simulation. In: Rabe, M. (ed.) Advances in Simulation for Production and Logistics Applications: Tagungsband zur 13. Fachtagung, pp. 585–594. Fraunhofer IRB-Verlag, Stuttgart (2008)
25. Schenk, M., Tolujew, J., Reggelin, T.: Mesoscopic Modeling and Simulation of Logistics Networks. In: Preprints of the 13th IFAC Symposium on Information Control Problems in Manufacturing, Moscow, pp. 586–591 (2009)
26. Schenk, M., Tolujew, J., Reggelin, T.: Mesoskopische Simulation von Flusssystemen: Algorithmisch steuern und analytisch berechnen. In: Nyhuis, P. (ed.) Beiträge zu einer Theorie der Logistik, pp. 463–485. Springer, Berlin (2008)
27. Schenk, M., Tolujew, J., Reggelin, T.: Mesoskopische Modellierung und Simulation von Flusssystemen. In: Ivanov, D., Mueller, E., Lukinsky, V. (eds.) Logistics Collaboration DR-LOG 07, pp. 40–49. Publishing House of the Saint Petersburg State Polytechnical University, Saint Petersburg (2007)

Development of a Simulation Model for Multimodal, Cross-Company Logistics Networks

Wilfried Sihn[1,2], Christian Hillbrand[3], Felix Meizer[2], René Leitner[2], and Margarethe Prochazka[1]

[1] Institute for Management Science, Technical University of Vienna, Theresianumg. 27, 1040 Vienna, Austria
sihn@imw.tuwien.ac.at, prochazka@imw.tuwien.ac.at
[2] Division Production and Logistics Management, Fraunhofer Austria Research GmbH, Theresianumg. 7, 1040 Vienna, Austria
felix.meizer@fraunhofer.at, rene.leitner@fraunhofer.at
[3] V-Research, Center for Tribotronic and Technical Logistics GmbH, Stadtstr. 33, 6850 Dornbirn, Austria
christian.hillbrand@v-research.at

Abstract. Continuous cost pressure causes companies to move production sites to low cost countries. Although production costs decline, logistics costs are negatively affected. Individual companies with limited shipment volumes do not have access to cost efficient and highly productive transport networks. Multimodal, cross-company logistics models are one approach to open up the potentials of transport networks for companies. This paper deals with the conceptual design and evaluation of cross-company logistics models. A new, holistic simulation and evaluation model was designed for development support of multimodal logistics concepts and potentials are validated instancing an example region.

Keywords: Logistics, Simulation, Evaluation.

1 Introduction

Progressive globalization and rising cost pressure in the automotive industry are increasingly forcing car manufacturers and component suppliers to set up new production sites in or move existing locations to low-wage region [1], [2]. It is for this reason that Central and Eastern Europe (known as CEE) became a popular target for relocation. These new production sites often just assume the supplier and customer structures of the parent plants. Approximately two thirds of suppliers as well as customers of Eastern European Tier 1 suppliers are still situated in Western Europe (Figure 1), [3].

However, benefits tend to fall short of predicted cost advantages, due to rising wage costs (in particular in industrial regions). The automotive industry in particular is thus focusing on the costs of transport logistics arising from intense transportation between Eastern and Western Europe.

The trend towards relocation has shown that the exchange of goods leads to new demands and challenges for transportation and logistics [4]. Business networking

W. Dangelmaier et al. (Eds.): IHNS 2010, LNBIP 46, pp. 26–36, 2010.

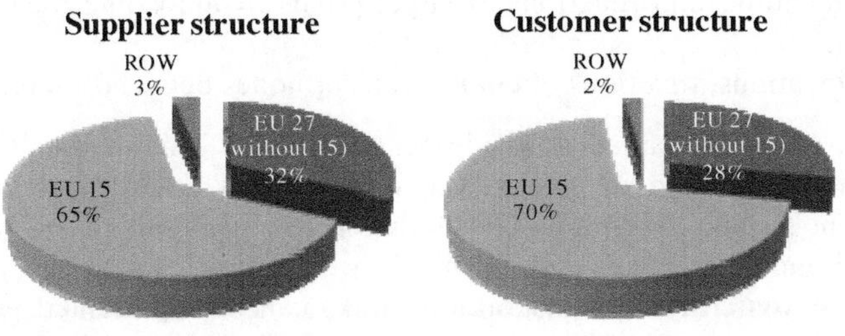

Fig. 1. Demonstration of supplier and customer structure of suppliers in the CEE region[1]

strategies and especially cross-company co-operation is one of the key factors to improve in production issues as well as in logistics and hence to survive in competitive markets [1], [5].

Based on this situation, a new simulation and evaluation model, which supports the development and evaluation of new logistics concepts, was developed. It is used for the validation and evaluation of cross-company logistics models. Due to the new, holistic evaluation approach potentials for optimization in the areas emissions, costs and logistical competitiveness are targeted on developing new sustainable and energy-efficient logistics models.

2 Cross-Company Logistics Models

The currently applied logistics processes, especially for the specific needs of individual enterprises in automotive industry do not appear optimal from a holistic point of view[2]. Deficits might emerge from direct transport running far beyond capacity, use of small transport carriers, less-than container load (LCL) with long running times or multiple handling steps as well as bad transportation tariffs due to small quantities. High stocks and capital tied up are results of this inefficiency. Since many companies have a similar source-target-behavior the potential of cross-company bundling to optimize transport efficiency is high.

2.1 Logistics Networks

There are various approaches for cross-company logistics models that conform to the general network model of logistics. These models represent networks transporting rights, goods, finance and information where spatial, quantitative, informational and temporal differences as well as company boundaries are crossed [6]. Parameters defining the structure of a logistics network are paramount [7]:

- Number, locations and functions of source points (= loading locations, making goods available),

[1] EU 15: member countries of the European Union prior to May 2004, EU27: current member countries of the European Union.

[2] This is part of the result of a questionnaire of 7 automotive enterprises in the near region of Timis in Romania in the course of the research project Trans Austria.

- Number, locations and functions of target points (= unloading locations, points of reception),
- Number, locations, functions of connections or nodes between sources and targets.

The network nodes are called transshipment terminals. This implies that only transshipment but not storage in general (no inventory) is foreseen at these locations. Transshipment terminals serve as consolidation terminals where the flows of goods are collected and/or as break-bulk terminals where the flows are in turn distributed [7]. The **basic structure** of transportation links can be represented either as **direct connection** ("point-to-point" transport) in its simplest form (single-stage, uninterrupted transport chain) or as a **multi-stage system** with preliminary leg, main leg and subsequent leg with transshipment terminals where the network nodes serve as consolidation terminals where the flows of goods are collected and/or as break-bulk terminals where the flows are in turn distributed.

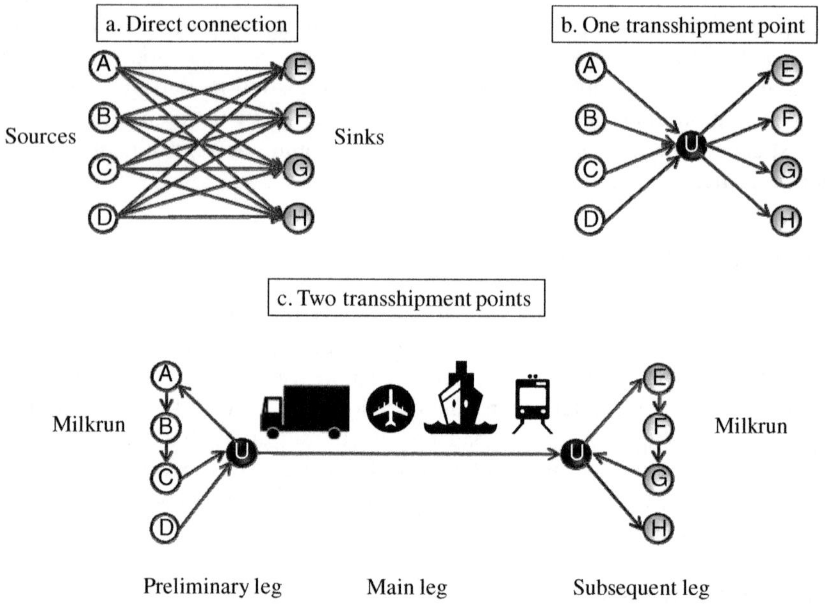

Fig. 2. Illustration of consolidation by means of transshipment points [8]

The mixture of logistics systems made up from the given basic structures is decided in the logistical network structure. The processes are designed when the logistical capacities are superimposed on this. The logistical capacity can be subdivided into transport capacity, warehousing capacity and information capacity. In addition to the basic structure of the systems, the speed of traffic flowing between the individual points in the system must be taken into account [9]. The network strategy is also based on geo-economic considerations such as the long-term development of customer demand or the development of the required delivery time.

Summing up, the criteria logistics costs, supply service, adaptability, susceptibility to interference, transparency and time for planning and establishment of the system are important in the moment of developing and evaluation logistics models [9].

2.2 Consolidation of Shipments

As described in the initial situation, optimization of transports for individual businesses does not appear ideal; therefore companies can align with partners to a logistical cooperation and bundle transport volumes. Bundling, also referred to as consolidation, happens when transport volumes are combined to form larger transport batches in order to allow more efficient and more frequent shipping by concentrating large flows onto relatively few links between terminals, thus lowering transport unit costs and the unit costs of incoming or outgoing goods at their starting or target points. The starting points for the scenarios for transport bundling are the individual parameters of the logistical network structure. The following forms may thus be used:

- Source-point bundling often following the principle of the "milk run" (the shipments intended for a particular destination are collected from several places of shipment, from neighboring places of shipment or from a shipment region and processed together)
- Target-point bundling, where shipments from one place of shipment intended for several destinations or for a delivery region are processed jointly and transported together and
- Transport bundling, where shipments are collected and delivered in one tour

Further forms of bundling can be inventory bundling or temporal bundling, and vehicle bundling and transshipment point or transit terminal bundling as forms of spatial bundling (Figure 2). The number of transports between sources and targets can be reduced by the setup of transshipment points from m x n to m + n, m and n being the number of source and target points [10], [11].

Bundled transport over the long run between 2 transshipment points can raise high potentials due to low transport costs and efficient use of transport capacities [12]. Logistics performance is improved by the raised frequency of transports. Overall every bundling type must meet the requirements of savings through consolidation of synergy effects to cover higher transport costs, operation costs of handling points or longer distances of time frames in comparison with direct relations.

The goal to reduce logistics costs while keeping logistics quality at the same level or raising the quality (delivery times, adherence to delivery schedule) is the main focus when designing the transport network. An iterative method is needed to evaluate the impacts of modifications in logistics models regarding ecology, economy or logistic competitiveness.

Transport bundling or cross-company logistics networks are originally based on the idea of good distribution in urban centers. The different approaches can be summed up with the term city logistics [13]. Other known developments of transport bundling of different suppliers are area contract freight forwarders, bundling and delivering goods for one plant conjointly. Collaborative approaches and the logistics models in this case are mainly based on the following premises:

- Identification of route sections where transport volumes can be handled with efficient transport carriers
- Availability of adequate partner for transport bundling on route sections (legs)
- Possibility of individual businesses to efficient usage of carriers
- Distance from source to target of possible nodes considering impacts of variance from ideal path
- Prioritization from transport volumes given limited capacities of one carrier in the main run as a result of different impacts on target categories
- Possibility to change transport frequency

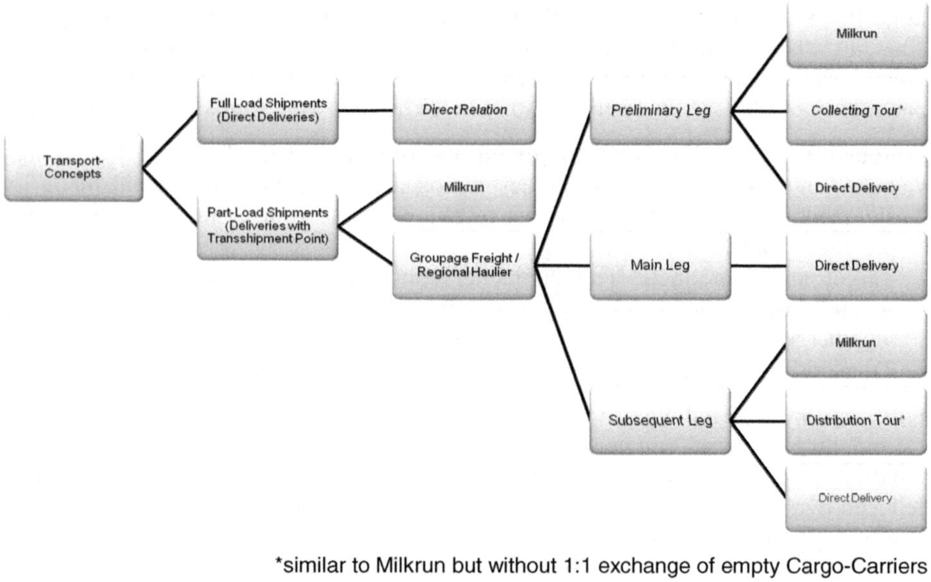

Fig. 3. Overview of logistics models considered

3 The Evaluation Model and the Simulation

The control variables required to achieve the objectives (minimize emissions, reduce costs of logistics and increase logistical competitiveness) are the following:

- Traffic avoidance: organize transport more efficiently (improve vehicle utilization, cut down on transport capacity)
- Bundling of goods flows: consolidate in order to optimize the substitution relationship between transport and inventory costs
- Switching freight transport to other means of transport: inter-modal transport

In the long run, it is also necessary to validate the results of the model's conceptual design. This should be performed in accordance with the main target dimensions - emissions, costs, and competitiveness.

It is in particular the intermodality aimed for in the models that plays an important role in the evaluation of this target dimension. In this point, only a selection of the most harmful emissions - CO_2, NO_X, and amounts of particulates - is analyzed; they are mainly accounted for in the dominant means of transport - road haulage. The emission levels are mainly dependent on the journey, i.e., distance covered by the predefined journey profile and on the allocated transport resource. Diesel or electrical power consumption also plays a decisive role in the output of emissions.

The cost calculation model is somewhat more extensive and can be subdivided into three different categories (Transport costs, transshipment costs, and inventory costs). When it comes to transport costs, it is important that the model is based on the actual costs incurred, i.e., the overhead costs, road charges, customs clearance, and wage costs, and not on the transport tariffs charged by shipping companies. The road charges are particularly difficult to determine due to differing systems in the individual countries, and they play a considerable role in determining the route.

$$C_{\text{Total}} = C_{\text{Transport}} + C_{\text{Handling}} + C_{\text{Inventory}} \tag{1}$$

$$C_{\text{Transport}} = C_{\text{Transport Truck}} + C_{\text{Transport Train}} + C_{\text{Transport Plane}} + C_{\text{Transport Ship}} \tag{2}$$

$$C_{\text{Handling}} = \sum_{i=1}^{n} \text{Quantities}_{\text{Change of Resource}} \cdot C_{\text{Transshipment (Change of Resource)}} \tag{3}$$

$$C_{\text{Inventory}} = C_{\text{Capital}} \cdot C_{\text{Warehousing}} \tag{4}$$

The third criterion in evaluating optimization models is logistic competitiveness, which is made up of the ability to deliver (a measure of the extent to which the company can guarantee the logistical service requested by the customer - short delivery times compared to the competition are especially important for a high ability to deliver) and delivery reliability (delivery reliability rates the service provision of the logistics process - it indicates the proportion of the complete and punctual deliveries compared to all delivery orders) [14].

Against this background, it becomes apparent that the cooperative planning of multimodal transport is affected by a multiplicity of factors. Numerous interdependencies between these parameters and the goal criteria such as costs, emissions, or flexibility exist. The evaluation model thus has to solve the conflict of goals like the trade-off between costs and guaranteeing competitiveness by applying stocks.

Beyond that, the identified parameters are in reality very often afflicted with uncertainty. These affects frequently influence the quality of material planning decisions and transport planning considerably.

Owing to dynamic interactions and taking stochastic phenomena into account, a static estimation of the behavior is difficult or almost impossible. Simulation has satisfactorily demonstrated its ability to illustrate and evaluate systems with dynamic behavior.

In order to provide a simulation model for logistics operators, the system-specific characteristics in a suitable simulation environment have to be illustrated. To achieve flexibility in planning regarding changes of logistics models or changes of the basic conditions, it is useful to provide a generic and easily adaptable model. For this purpose, the models of intermodal transport were illustrated within an application platform for simulation. This simulation environment permits the illustration of various technical models in a modeling environment adaptable to the planning domain. Furthermore, the environment permits an automatic derivation of simulation models and a return of the results to an economic or ecological level. In this section, the fundamental methods of the application platform simulation and the development of a simulation model for the planning of multimodal transport are described.

3.1 Application Platform for Simulation

Depending on which simulation tool is chosen from the set of tools with different universal validities and application references, the simulation expert can access more or less preconfigured building blocks. Special simulators contain solutions matched to the specific area of the domain of application, thus simplifying handling. The higher the degree of universality, the more varied the possibility of creating and depositing own functional building blocks becomes; however, the necessity of a simulation-specified training increases as well. The application environment developed by V-Research deals with simulation models in production and logistics.

To execute a conventional simulation study, a substantial amount of time and money has to be invested, which still prevents small and medium-sized businesses to apply those. To prevent a planner, who is usually well trained in his respective systems domain, from dealing with the simulation expertise itself, it is helpful to develop an instrument that lets him answer upcoming questions without specific simulation know-how in an accessible amount of time. Thus, the planner is able to define planning tasks, generate models, analyze results, and optimize those results by comparing different scenarios.

The core of the platform represent application related and technically oriented components that base on the utilization of open industry standards (.Net Framework 2008) and reliable, well-tested simulation software. The concept embraces the idea that each simulation study needs certain key functions and procedures; contains customer and project specific characteristics and requirements; should be expandable concerning detailing, functionality, and system boundaries; and lets other users (e.g. customers) enhance the model.

The separation between task specific and resource specific components is reflected in the structure of the application environment. Task specific components summarize all aspects concerning process logic of an application that are illustratable by a (production and/or logistic) system. Examples would be the definition of manufacturing technologies used for processing orders, process outcomes, product structure etc.; logistical sequencing strategies (push, pull, KANBAN, …); organizational classifications (employees assigned to department and resources, shift schedules, …); as well as an architecture for administering simulation runs, result data, and shift schedule data.

These components signify the applied level. The technical level comprises the structure of a production and logistics system. The resources (machines, means of transportation, tools, ...) are systematically described in predefined components. In order to create a complete model, these project specific resources are substantiated and incorporated into a basic model. The entire architecture thus differentiates between technical components, application oriented components, and the actual business application. The business application represents the simulation study, which can be made up of a number of simulation experiments. The component model is well suited for customer specific simulation applications within a short period of time.

3.2 Simulation Model for the Planning of Multimodal Transport

By describing the behavior and the possibility of providing process cycle time fluctuations using distribution functions, so-called confidence intervals can be determined through a number of replications (simulation runs with independent random variables) which allow a prediction to be made as to the bandwidths where the target dimensions are likely to be with a given level of probability.

The basis for the simulation model is formed by individual logistical building classes (factory, transshipment centre, etc) that can be combined with one another to represent any desired logistics concept (point-to-point transportation, consolidation terminals, milk run). These building classes are created in the simulation environment Flexsim®. Structural, procedural, and resource-related data are required in order to model with the simulation system. By modeling with building classes, it is possible to describe both the structure and the behavior of individual resources independently, as well as their interaction with other building classes. This inherent knowledge contained in the individual building classes is used and extended by configuring the building classes to form an overall model.

The central technical construct in the domain transport planning is the route, which represents a given start-destination-relation, for example between a loader in Eastern Europe and a Western European production facility. This route is either being served by a direct relation or by intermodal transport, which is usually conducted through certain hubs. In any case, the chosen domain model displays the relations from start to destination between Eastern and Western production locations. The actual routes taken were tracked via GPS and then contributed into the system's data. The information includes not only kilometers driven for the running time calculation, but also other data such as altitude profiles. These profiles further refine the simulation model leading to consumption and emission data of the transport resources to be implemented in the model.

The system load of the simulation model is determined by the output of products on the loader's side, which is being processed in form of a tabulated schedule. A profile of each originating plant is defined therein, stating which products are being put out in which amount at what time. The running time of the simulation is thus supplemented with the corresponding number of product objects that later undergo the defined transport process (stated below).

Transport resources are active parts when it comes to bridging regional distances within the domain model. In the case of the afterwards presented application various types of trucks and trains are displayed in the model. Transport resources are matched with certain routes and characterized by a number of attributes. Besides loading and unloading times for the transport resource, triggers have to be defined when a transport resource leaves a location (e.g. every Monday 8 a.m.). Furthermore, additional data can be added to the model, such as cost, consumption, emissions, or any local- or time-specific data. Therefore, the model is formed in a way that specifies consumption data for different types of trucks and diesel locomotives. In case of electrified railways, the proportionate emissions for the necessary power are included as ecologic target factors in the simulation for each country.

In the domain model, a plant can serve as a starting point or as a destination, or even as a hub to add products to or discharge items from the transport. Product objects are created in the course of simulation, beginning at the starting plant with a time-amount-profile, and they are dissolved after reaching the destination.

The handling point describes a special location along the route, with incoming transports being unloaded and leaving transports prepared for the next trigger with their inherent loading and idle times (see transport resource).

In contrast to the handling point, no change of transport resource and no transshipping take place at a processing point. This simulation block simply represents a location where the transport is halted for a certain handling time, calculated as a stochastic variable. This method enables the display and calculation of customs clearance at national borders. The handling itself is not directly calculated due to model simplification, and it is thus represented by idle time.

Beside of these simulation blocks, the transport process is an integral part of the simulation model. The later presented application includes a process that starts off with the production of products at a starting point and ends at the destination point, where the products are destroyed. On closer inspection, the transport starts, when a specified amount of products is available at a certain location and the starting trigger for a certain means of transportation is activated. Consecutively, the transport process is being planned according to the route; the products are loaded on the transport resource and then transported to the next stop on the route. If the next stop is a plant or handling point, the products are unloaded and, if needed, prepared for the next shipment. In case of the next stop being a handling point, the idle time is applied and the transport continues. Upon arrival and unloading of the transport, the respective products are destroyed in the simulation environment and the simulation run is ended.

Throughout this process, the running time and its contributing factors of each process entity, such as time needed for loading procedures, handling activities, driving times, and emissions depending on the road profile (distance, incline, street conditions), are logged. After the simulation, this logged data supplements the results analysis. A historiography of the granulated simulation data in a database enables detailed and flexible analyses and a deduction of key performance indicators. The (logistic) results drawn from the simulation can consequently be applied to calculations with data sources concerning emission behavior, for cost analyses, and for comparisons whether logistics targets have been met. The model structure simplifies changes made to the parameters (e.g. varied transport cycles, use of various traffic resources and carriers etc.), and a comparison of the effects on the three target areas economy, ecology, and competitiveness is enabled. Based on the opposed target dimensions, a decision can only be made through comparing different scenarios and considering the pros and cons of all compromises.

4 Application of the Modeling Approach to a Romanian Automotive Cluster

In a research project funded by the Federal Ministry for Transport, Innovation and Technology (BMVIT) as well as the Austrian Research Promotion Agency (FFG) the developed logistics models are demonstrated by means of the region Timis in Romania. Focusing on 7 automotive companies volumes of outgoing transports were analyzed. Starting from the current state of individual transports, different scenarios were defined. The scenarios are aimed at cost reduction and sustainability in using modes of transport for high volumes like rail traffic.

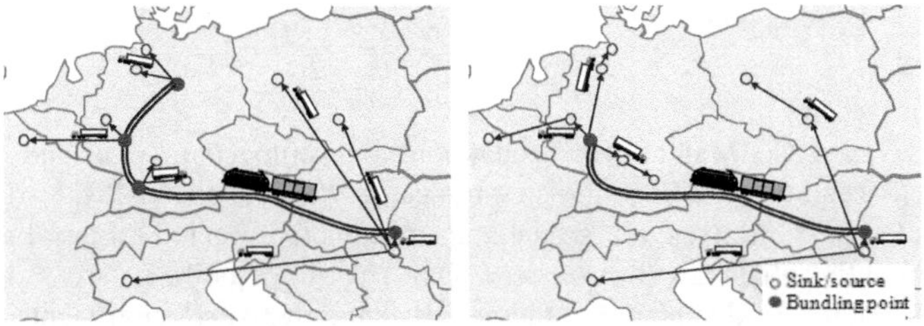

Fig. 4. Scenarios 1 and 2 [15]

Figure 4 shows 2 defined scenarios of transport bundling for the Timis Region. Scenarios 1 using block train with 3 stops and direct relations from the end of the train not considering locations in Poland and Italy that cannot profit from consolidation with the block train. Scenario 2 limits the block train to one stop but bundles transports further leaving the train to their final destinations. Destinations not considered in the main leg bundling where consolidated as well. Shifting the main leg to railway and optimizing the collection and distribution of goods from and to transshipment points logistics costs could be reduced by 15 % in the given case. The ecological impact in reduction of CO2 emissions by 40 %, cutting fuel consumption in half, shows the success in more than one target dimension. The main deficit of the models is overcoming the doubled lead time coming from the ceteris paribus inspection of transports.

In addition to the simulation and evaluation of scenarios a sensitivity analysis was executed to cover the ecological and economical results. Therefore the evaluated scenarios indicated were simulated with lower basic loads keeping all other factors stable. At a level of 70% of the load, block train concepts as well as the transfer of 66% of transports to railroad could be maintained. Negative effects of the change in basic loads were determined in the capacity utilization of transport capacities and the flexibility especially for block trains. Nevertheless the developed transport concepts and cross-company models can stand up to the actual transport handling. Economic considerations show lower costs of scenarios compared to the actual situation. Therefore the model indicated shows full functionality even with fluctuation of volumes and prices.

5 Concluding Remarks

This paper visualized models to simulate and evaluate potentials of transport bundling for cross-company logistics networks to identify those potentials given a specific case. The research project Trans Austria demonstrated the similar source-sink-behavior for companies and thus the potential of cross-company transport bundling.

The empirical analysis showed the great complexity of the problem that was built in a simulation framework and therefore the challenge for possible implementations which constitute further research developments. Optimization of the given set of problems would be a possible expansion keeping in mind the time and cost needed to set up the base and ensuring the transferability of an optimization model.

References

1. Kinkel, S., Lay, G., Maloca, S.: Produktionsverlagerungen ins Ausland und Rückverlagerungen. Report on the BMF research project (8/95), Fraunhofer-ISI, Karlsruhe (2004)
2. Sihn, W., Palm, D., Matyas, K., Kuhlang, P.: Automotive Region Eastern Europe –AREE: Chancen und Potentiale des "Detroit des Ostens" für Automobilzulieferer, Vienna (2006)
3. Sihn, W., Palm, D., Schmitz, K., Leitner, R.: Automotive Region Central and Eastern Europe – Produktionsstrukturen von Automobilherstellern und ihrer Zulieferer, Vienna (2009)
4. Feldmann, K., Rottbauer, H., Roth, N.: Relevance of Assembly in Global Manufacturing. Annals of the CIRP 45(2), 545–552 (1996)
5. Wiendahl, H.-P., Lutz, S.: Production in Networks. Annals of the CIRP 51(2), 573–586 (2002)
6. Vahrenkamp, R.: Logistik: Management und Strategien, 6. Auflage, Oldenbourg Wissenschaftsverlag GmbH, Oldenbourg (2007)
7. Rösler, O.M.: Gestaltung von kooperativen Logistiknetzwerken: Bewertung unter ökonomischen und ökologischen Aspekten. Deutscher Universitäts-Verlag/GWV Fachverlage GmbH, Wiesbaden (2003)
8. Brauer, K.M., Krieger, W.: Betriebswirtschaftliche Logistik, Berlin (1982)
9. Pfohl, H.-C.: Logistikmanagement: Konzeption und Funktionen. Springer, Berlin (2004)
10. Campbell, J.: Freight consolidation and routing with transportation economies of scale. Transportation Research Part B 24(5), 345–361 (1990)
11. Simchi-Levi, D., Kaminsky, P., Simchi-Levy, E.: Designing and Managing the Supply Chain – Concepts, Strategies and Case Studies. Irwin McGraw-Hill, Boston (2000)
12. Trip, J., Bontekoning, Y.: Integration of small freight flows in the intermodal transport system. Journal of Transport Geography 10, 221–229 (2002)
13. Taniguchi, E., Thompson, R.G., Yamada, T., Van Duin, R.: City Logistics: Network Modelling and Intelligent Transport Systems, Oxford (2001)
14. Wiendahl, H.-P.: Erfolgsfaktor Logistikqualität: Vorgehen, Methoden und Werkzeuge zur Verbesserung der Logistikleistung. Springer, Berlin (2002)
15. Research Report: Trans Austria, Vienna (2009)

Building Blocks as an Approach for the Planning of Adaptable Production Systems

Egon Müller

Chemnitz University of Technology, Department of Factory Planning and Factory Management, 09107 Chemnitz, Germany
egon.mueller@mb.tu-chemnitz.de

Abstract. Global trends demand the design of adaptable production systems. The department of Factory Planning and Factory Management at Chemnitz University has proven its extensive competences in this field. On this basis the concept of Building Blocks for Adaptable Production Systems was developed. It is implemented in the existing Experimental and Digital Factory of the department. This includes an integrated design of product and production system. The approach is modular. In result, planning/visualization modules of the Digital Factory as well as modules for machining, material flow, information flow and energy flow are investigated. Fields of application are shown.

Keywords: Adaptability, changeability, Digital Factory, Component-based Planning.

1 Introduction

Such trends as the reduction of product life cycles, individualization of customer demands and globalization lead to increasingly dynamical and incalculable markets. In result, the design of adaptable production structures is required from factory planning.

The department of Factory Planning and Factory Management at the institute of industrial sciences and factory systems of Chemnitz University of Technology has proven its extensive competence in this field (e.g. Component-based Development [1], Component-based Planning [2], PLUG+PRODUCE [3], HIPER [4], Collaborative Research Center (CRC) 457 [5], Project Cluster 196 [6]). The results of this research are now consistently implemented in the project "Building Blocks for Adaptable Factory Systems". Further basic research and a transfer to industrial practice will be realized at selected factory subsystems. The general target is to achieve an adaptability, which meets the requirements, by elaborating the so called change drivers universality, mobility, scalability, modularity and compatibility.

Therefore, the mechanisms of the change drivers have to be understood and made assessable. Concepts and methods for the design of future adaptable production systems need to be developed.

The main component of the building block is an assembly/production plant, which will be connected to the logistics system of an already existing Experimental and Digital Factory (EDF) through appropriate interfaces.

W. Dangelmaier et al. (Eds.): IHNS 2010, LNBIP 46, pp. 37–45, 2010.

The adaptability of the system shall be supported amongst others by tools of the Digital Factory, innovative identification and positioning systems, concepts for the supply of material and autonomous control of logistical objects as well as man-machine interfaces.

2 Basics, Problem, Scientific Questions and Need for Research

2.1 Basics

The process of adaptation results from the permanently changing production environment. From a system theoretical point of view it is mainly driven by the three factory properties dynamics, complexity and cross linking, which catalyze adaptation. Adaptability can be understood as a meta-property which bundles these three system properties [7]. Change drivers are amongst others universality (neutrality of function and usage), mobility, scalability (extensibility and reducibility), modularity and compatibility (ability for networking, ability to (dis-)integrate) [7-9].

Different types and dimensions of adaptability are distinguished; spatial adaptability, temporal adaptability, structural adaptability and technical adaptability.

Adaptability can be also connected in an object oriented way to products and services, technologies and processes as well as to production systems (man, technology, organization);

- Adaptability with regard to products/services, which have to be generated
- Adaptability with regard to functions and processes which can be realized
- Adaptability with regard to elements and structures of the factory

2.2 Problem, Scientific Questions and Need for Research

The research on production plants and factory planning is lead by increasingly urgent requirements like flexibility and reactivity or adaptability, agility, handling of complexity as well as orientation on the demands of customers/markets. Decisive driving factors for the satisfaction of such requirements are amongst others universality, modularity, mobility, compatibility/ability for networking, scalability and availability. (see e.g. [3, 10-22])

A study about adaptable production systems [9] identified the following demand for research and development;

- Optimization of interfaces(robust, inexpensive and standardized interfaces, plug & produce strategies, and methods are needed to reduce complexity.)
- Harmonization and holistic design of the value chain(methods for identifying the need for adaptability in the whole value chain and for the design and continuous optimization of adaptability are needed.)
- Making adaptability assessable(Evaluation models , which enable every enterprise to execute a cost-benefit calculation for planned investments in adaptability, are needed.)

In addition to the expressed need for research, this study led to a publication of the Federal Ministry of Education and Research (BMBF) from March 30, 2009 to the topic "Safeguarding Competitiveness by adaptable production systems" [23].

Building blocks for adaptable production systems are meant to contribute to the answering of the scientific questions.

In order to investigate adaptability, the factory building blocks are so designed that

- The change drivers universality, mobility, scalability, modularity and compatibility as well as
- The types and dimensions of adaptability can be comprehensibly and systematically researched,
- A large scope of adaptable production systems can be provided by freely combining a selected number of components,
- Conclusions about the design of factory systems, which meet the requirements and the current situation, can be drawn.

3 Concept Building Blocks for Adaptable Production Systems

Building blocks for adaptable production systems should make it possible, to investigate almost any factory configuration (e.g. order driven, flexible layouts of machines (figure 1) in connection with production and logistics concepts for optimal lot sizes such as one-piece-flow) prototypically and emulate them close to reality. Conclusions about the behavior of the factory and its components shall be drawn in anticipation of an actual realization. Innovative concepts, technologies and solutions shall be evaluated with regard to their suitability for industrial practice. New and further developments of components of factories, such as energy-saving solutions for materials handling technology, and their interaction with other components, shall be tested. Last but not least the interaction and the behavior of humans in the factory as well as their acceptance of technical innovations will be researched.

In addition to adaptability the following trends and research areas of industrial science will be covered with the factory building blocks;

- Demography, group work and working environment
- Information and communication, man-machine interfaces
- Handling and supply concepts
- Energy efficiency
- Unity of planning and operation of a factory
- Digital factory, further development of methods and tools for factory planning.

According to the approach of the Digital Factory, the research will be carried out on one hand virtually at the computer and on the other hand real in a small model factory. In result, the factory building blocks, which enhance and complete the existing Experimental and Digital Factory (EDF), consist of the components Digital Center (DC) and Experimental Center (EC). The components can be combined via defined interfaces (figure 1).

The following modules are planned to be integrated in the building blocks (figure 1);

- Components of the digital center (DC)
  - Modules for planning/visualization (component configurator, visualization (VR, interaction))
- Components of the experimental center (EC)
  - Machining modules (manufacturing/assembly modules (4+1), test module)
  - Material flow modules (handling module (gantry robot), transport module)
  - Information flow modules (control center, control, interface modules)
  - Energy flow modules (supply, energy management modules)

Figure 1 illustrates the concept of the integration of building blocks for adaptable production systems in the EDF.

The planning and visualization modules are integrated in the Digital Center (room D19). The machining modules as well as the material, information and energy flow modules are integrated in the experimental center (rooms D17 and D18).

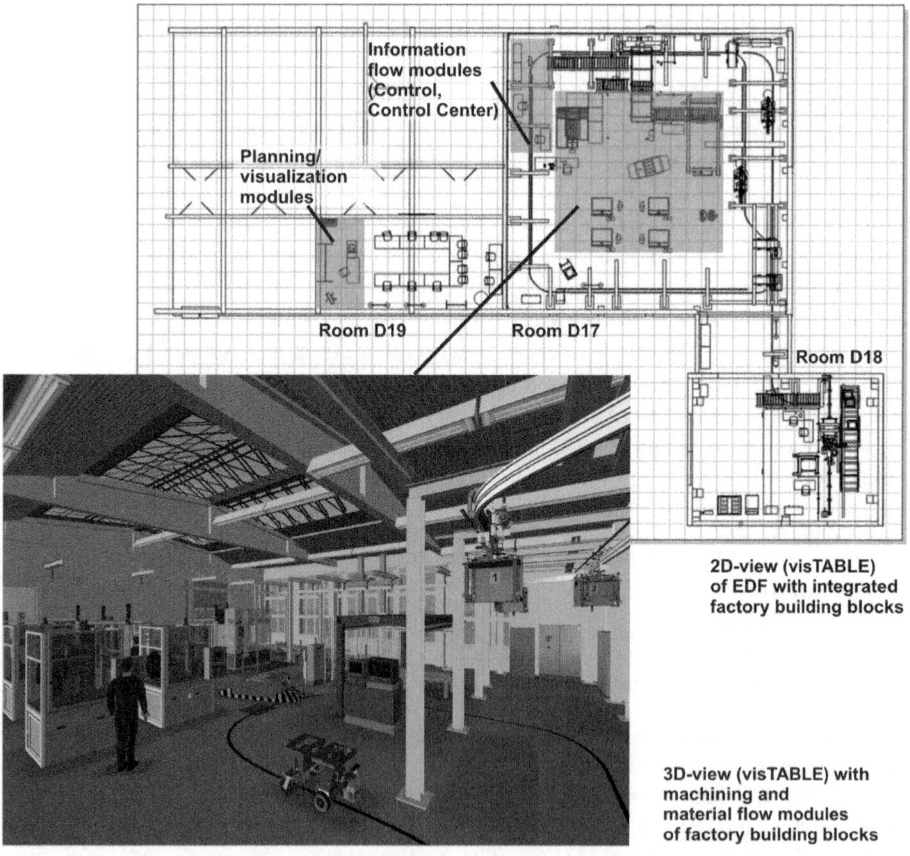

Fig. 1. Concept of the integration of building blocks for adaptable production systems in the EDF

3.1 Planning/Visualization Modules

The planning and visualization modules component configurator and visualization base on such concepts as Component-based Development [1], PLUG+PRODUCE [3] as well as Component-based Planning [2, 5, 24, 25]. The modules ensure that the designed factory can be planned and displayed efficiently, rapidly and in high quality on a computer.

The general idea of the method of Component-based Planning (figure 2) is that catalogues of components are not only provided for the object domain/production process (object components) but also for the method domain/ planning process (planning components). A standardization of the planning process and the production process as well as a reduction of the complexity can be achieved by the stringent application of the building block principle. Hence, through the use of proven and tested components, a rationalization and acceleration of the planning process is targeted, at the same time increasing quality.

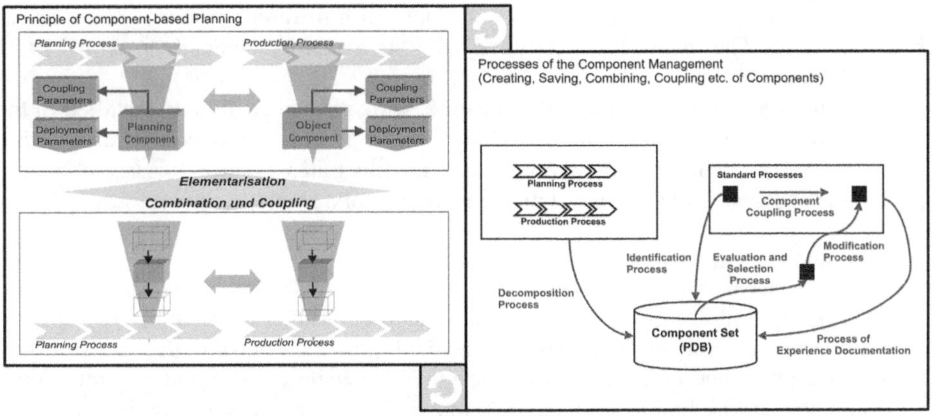

Fig. 2. Method of Component-based Planning (Principle) with component management processes [5]

Components of the planning process (planning components) are standardized units of the process of transformation of information. These components mainly integrate the planning functions, the relevant planning methods and competences as well as input and output information. On the contrary components of the production process (object components) refer rather to physical objects as well as the necessary competences. There are predefined procedures for the identification, selection and coupling etc. of building blocks. On the base of these procedures generated building blocks can be used for the Component-based Planning of production plants. They can be combined to obtain planning processes as well as production processes for production plants. The experiences from the application of the building blocks are used to improve the building blocks. That leads to the optimization of the building blocks, which at the same time store the acquired knowledge.

Therefore, not only the particular parts of factory equipment (object components of the Component-based Planning), but also the interdependent steps of the planning process (planning components of Component-based Planning) itself are cataloged in the component configurator (figure 2). In addition, it must be possible to parameterize and modify both types of components in a comprehensive way. Components of lower order should be possibly joined to components of higher order. The configuration of the planning process should be done graphically. Its implementation should be guided by a planning manual. The visualization should provide a realistic mapping of the factory in an easy way. In addition to technical considerations, also issues of work science (e.g. working environments, interactions of humans) should be addressed.

3.2 Machining Modules

Four functionally universal manufacturing and assembly modules are the core of the building blocks for production systems. They can be placed arbitrarily in the layout and are meant to represent the machining equipment of a factory. Figure 1 illustrates one possible layout variant in the EDF.

That way it becomes possible to produce different products (e.g. variants of a wing mirror as an automotive mechatronical product) with different process chains on laboratory scale. The vision is to copy the EDF as a factory model (e.g. with LEGO building blocks).

In addition an automatic assembly machine, which has been already installed in the EDF (as fifth cell), should be integrated in the overall concept as well as an existing test station. The automatic assembly machine illustrates control, disassembly and packing processes, while the test station illustrates control and test processes.

3.3 Material Flow Modules

Material flow modules have to connect the manufacturing and assembly modules as well as already existing components such as the high-rack storage in the material flow technical aspect. Therefore, an automatic guided vehicle (AGV) and a gantry robot are integrated in the building blocks for adaptable production systems. By means of its controlling unit, the AGV can recognize the cells, which can be arbitrarily arranged in non central configurations, at their position and supply or collect material. The gantry robot can handle the material supply of the four cells in a central configuration. It can as well be used for complex storage and commissioning assignments. Additional handling equipment supports amongst others the handling to manufacturing and assembly modules as well as the implementation of supply concepts.

3.4 Information Flow Modules

The information flow modules ensure the interconnection of the manufacturing/assembly, material and energy flow as well as the planning and visualization modules in the information aspect. This includes the architecture of the control center module, the installation of additional components as part of the control module as well as the incorporation of innovative identification systems (e.g. RFID) in the machining and material flow modules (amongst others definite identification and routing of the load units).

The adaptability of products, processes and equipment, will be supported amongst others by information and data flows for;

- Factory planning (CAD (product) <-> process design <-> CAD (equipment)),
- Automation (CAD (product) <-> process design <-> CAD (equipment) <-> NC/PLC/…),
- Order Processing (CAD (product) <-> ERP <-> NC/PLC/…).

3.5 Energy Flow Modules

A supply module, which was designed as an overhead media line, has to assure the flexible discharging of energy by the mobile manufacturing and assembly cells. Especially the media electrical energy and compressed air should be provided area-wide.

An energy management module provides the elicitation of energy-related information from the machining, material flow and energy flow modules. Inversely a well-directed simulation and regulation of energy-relevant parameters is enabled by the energy management module. The energy management control will be incorporated in the control module. It will help to investigate the interdependencies between adaptability and energy efficiency as part of the building blocks for adaptable production systems.

4 Fields of Application and Deployment

The building blocks for adaptable production systems are primarily used for basic research about adaptability. At Chemnitz University of Technology the factory building blocks serve as research environment.

Secondarily the factory building blocks shall however establish the transfer of results and technologies to industrial practice at an early stage. Hereby the factory building blocks serve as test environment for regional system producers, manufacturers and service providers.

References

1. Wirth, S., Zeidler, H.: Untersuchungen zur Entwicklung und Anwendung von Bausteinen in der Technologischen Betriebsprojektierung unter besonderer Berücksichtigung ihrer Schnittparameter. Dissertation B, Dresden (1974)
2. Näser, P.: Bausteinbasierte Planung von Produktionsstätten in Hierarchielosen regionalen Produktionsnetzwerken. In: Zukunftsweisende Konzepte für Produktions- und Logistiksysteme. Begleitband zu den Kolloquien des SFB 457 mit dem SFB 559 und dem SFB 582 am 20.04.2004 und 15.06.2004. Wissenschaftliche Schriftenreihe des IBF, Sonderheft 10, pp. 86–92. TU Chemnitz, Chemnitz (2005)
3. Hildebrand, T.: Theoretische Grundlagen der bausteinbasierten, technischen Gestaltung wandlungsfähiger Fabrikstrukturen nach dem PLUG + PRODUCE Prinzip. Dissertation, Wissenschaftliche Schriftenreihe des IBF (44). TU Chemnitz, Chemnitz (2005)
4. Müller, E., et al.: High Performance Ramp-up – Gestaltungsstrategien für beschleunigte Serienanläufe (HIPER). Reihe Schneller Produktionsanlauf in der Wertschöpfungskette, Band 1. VDMA Verlag GmbH, Frankfurt/Main (2007)

5. Müller, E., et al.: Sonderforschungsbereich (SFB) 457: Hierarchielose regionale Produktionsnetze - Theorien, Modelle, Methoden und Instrumentarien. Abschlussbericht 2006. TU Chemnitz, Chemnitz (2007)
6. Müller, E., et al.: Paketantrag (PAK) 196: Kompetenzzellenbasierte Produktionsnetze - Entwicklung einer Methodik zum Bilden und Betreiben von kompetenzzellenbasierten Produktionsnetzen. Neuantrag, TU Chemnitz, Chemnitz (2006)
7. Hernández Morales, R.: Systematik der Wandlungsfähigkeit in der Fabrikplanung. Dissertation, Fortschrittberichte VDI Reihe 16, Technik und Wirtschaft 149, VDI, Düsseldorf (2003)
8. ElMaraghy, H.A.: Flexible and reconfigurable manufacturing systems paradigms. International Journal of Flexible Manufacturing Systems 17, 261–276 (2006)
9. Nyhuis, P., Reinhart, G., Abele, E.: Wandlungsfähige Produktionssysteme. Heute die Industrie von morgen gestalten. Studie zum Forschungsprojekt Wandlungsfähige Produktionssysteme des BMBF/Forschungszentrum Karlsruhe (PTKA) innerhalb des Rahmenkonzeptes Forschung für die Produktion von morgen, IFA Hannover (2008)
10. Dove, R.: Agile Production: Design Principles for Highly Adaptable Systems. In: Maynard's Industrial Engineering Handbook, pp. 9.3–9.26. McGraw-Hill, New York (2001)
11. Kobylka, A.: Simulationsbasierte Dimensionierung von Produktionssystemen mit definiertem Potential an Leistungsflexibilität. Dissertation, Wissenschaftliche Schriftenreihe des IBF(24). TU Chemnitz, Chemnitz (2000)
12. Kuhn, A., et al.: Sonderforschungsbereich 559: Modellierung großer Netze in der Logistik. Finanzierungsantrag 2004/2, 2005, 2006, 2007, 2008/1. Universität Dortmund, Dortmund (2004)
13. Müller, E.: Production Planning and Operation in Competence-cell-based Networks. Distributed operations, simulation and decision making. Production Planning & Control. The Management of Operations Special Issue 17(2), 99–112 (2006)
14. Reinhart, G.: Virtuelle Fabrik: Wandlungsfähigkeit durch dynamische Unternehmenskooperationen. Transfer-Centrum GmbH, München (2000)
15. Reinhart, G., Berlak, J., Effert, C., Selke, C.: Wandlungsfähige Fabrikgestaltung. ZWF 97(1-2), 18–23 (2002)
16. Schenk, M., Seelmann-Eggebert, R.: Die Kundenwunschfabrik – Anforderungen und Lösungsansätze. In: Vortragsband IBF-Fachtagung: Vernetzt planen und produzieren, pp. 39–43. TU Chemnitz, Chemnitz (2000)
17. Schenk, M., Wirth, S.: Fabrikplanung und Fabrikbetrieb wandlungsfähiger und vernetzter Fabriken. In: Jahrbuch der Logistik 2005, pp. 40–43. Handelsblatt, Düsseldorf (2005)
18. Schuh, G., Wiendahl, H.-P., et al.: Komplexität und Agilität. Steckt die Produktion in der Sackgasse? Springer, Berlin (1997)
19. Westkämper, E., et al.: Sonderforschungsbereich 467: Wandlungsfähige Unternehmensstrukturen für die variantenreiche Serienproduktion. Ergebnisbericht 1997, 1998, 1999. Universität Stuttgart, Stuttgart (1999)
20. The Digital Factory: Visions & Action. In: ManuFuture Workshop 2004 (2004), `http://portal.lfo.unidortmund.de/content/mf/mf001195/MANUFUTURE_2004_WS_Dortmund_Digital%20Westkaemper.pdf`
21. Wiendahl, H.-P., Hernández, R.: Fabrikplanung im Blickpunkt. Herausforderung Wandlungsfähigkeit. wt Werkstattstechnik-online 92(4), 133–138 (2002)
22. Wiendahl, H.-P., Nofen, D., Klußmann, J.H., Breitenbach, F.: Planung modularer Fabriken. Vorgehen und Beispiele aus der Praxis. Carl Hanser, München (2005)
23. BMBF, `http://www.bmbf.de/foerderungen/13526.php`

24. Näser, P.: Component based planning of manufacturing plants. In: Proceedings of the 3rd Annual Conference on Systems Engineering Research – CSER, pp. 393–402. Stevens Institute of Technology, Hoboken (2005)
25. Müller, E., Rösch, N., Ackermann, J.: Component-based Planning of flexible Assembly Systems: a Methodical Approach and its Implementation. In: Proceedings of the 2nd CIRP Conference on Assembly Technologies & Systems – CATS 2008, Toronto, pp. 41–49 (2008)

Challenges for the Provision of Process Data for the Virtual Factory

Gert Zülch[1], Martin Waldherr[1], and Michael Zülch[2]

[1] Ifab-Institute of Human and Industrial Engineering,
Karlsruhe Institute of Technology (formerly University of Karlsruhe),
Kaiserstrasse 12, D-76131 Karlsruhe, Germany
{gert.zuelch,martin.waldherr}@kit.edu
[2] Gefora-Beratungs-Gesellschaft für Organisation und Arbeitswirtschaft GmbH,
Silvanerweg 10, D-76356 Weingarten, Germany
gefora-beratung@t-online.de

Abstract. During practical simulation projects, the required model data must be imported from existing data bases. The focus is to keep the amount of effort required to a minimum. The main issue is the provision of process data in particular concerning work operations that require timing and resources to be accounted for. This task is comparatively easy in the field of parts manufacturing as the operations are sequential in nature. However, with assembly processes requiring interlinked operations, this task turns out to be considerably more complex. A detailed discussion of differences between the relevant process data and the relevant time data required for simulation is followed by the illustration of two examples showing how the interface problem can be solved for the export of sequential and interlinked process data to a simulation procedure. The difficulties which usually arise during such a project and future approaches to the problem are also discussed.

Keywords: Time management, Process data, Simulation, Modelling, Virtual factory, Interface, Data export.

1 Time Management Views on Production Processes

1.1 The Need for Time Management Data

The volume of time management data required for the tools used in the virtual planning of production systems is dependent on the complexity of the respective production processes. This data is used to generate planned processes and resources which are modelled through simulation procedures as realistically as possible or to illustrate them in the form of animations using the tools of the virtual factory concept [1]. The requirements for providing the respective time management data increase with the complexity of resources to be modelled. If it is only required to simulate material movements or mechanical manufacturing processes, it is usually sufficient to use kinematically determined time data which can be defined through the use of formulae.

W. Dangelmaier et al. (Eds.): IHNS 2010, LNBIP 46, pp. 46–56, 2010.

This method cannot be used if the processes are carried out by people instead of machines, due to the fact that the cooperation between machinery and personnel resources must be reflected within the time data [2]. Time data concerning the work object are often also required, for example when surfaces need drying time or when adhesive needs time to set.

1.2 The REFA View Concept

For decades, the concept of different views on production processes has been a well-known part of the REFA methodology. REFA focuses mainly on the views on the person, the means of production and the specific work task or order ([3], pp. 20). REFA focuses to a lesser extent on the work object and only very little on work information ([4], p. 12).

The process types resulting from the different views and the corresponding types of time data must be taken into consideration when virtual factory tools are used to model the complex interaction of all resources involved as well as mechanized or automated production processes. The REFA view concept states that only some process types or types of time data can be used as input data required to simulate and animate processes ([5], p. 12-98) compared to other types of time data derived from the same resource that constitute output data in the form of simulated production process results (Fig. 1).

One typical example can be seen in the transition of a manufacturing order between two machines (Fig. 2): when the first work process is completed, the order must first wait for transportation to the next step, where it will generally then be placed in the queue for the next machine. Although it is possible to determine in advance the

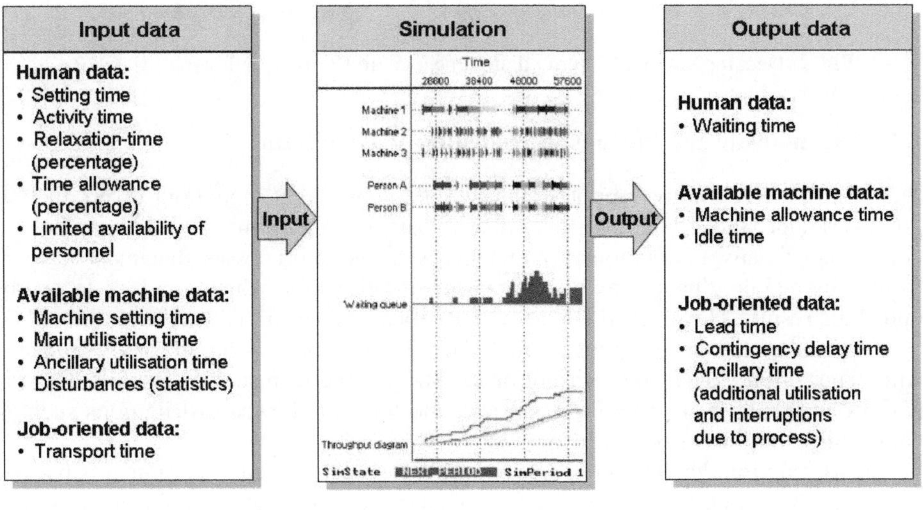

Fig. 1. Input and output data of organisation-centred simulation (following [2], p. 14)

transportation time to a good level of accuracy with deterministic methods and use the data as an input value for the model, the two waiting periods are a result from simulation of the dynamic production process with all its resource limitations.

Consequently, the common approach to this problem, which consists of the simplification of material handling processes through the use of global values for the transition time data (see e.g. [6], p. 319), leads to simulation errors. However, the time decomposition described in this article, what REFA refers to as the 'intermediate time', has not yet become a standard of the REFA methodology ([7], pp. 8).

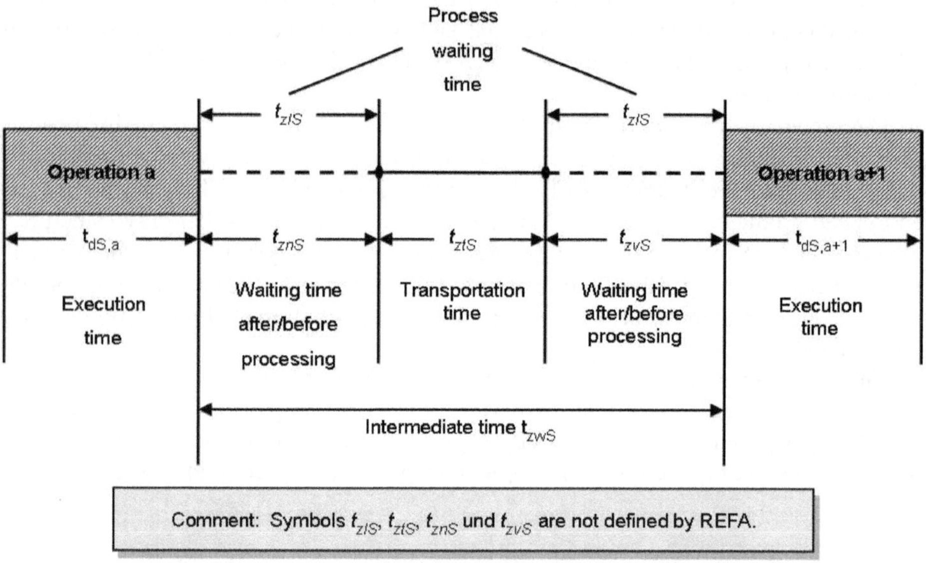

Fig. 2. Decomposition of intermediate time (acc. to [7], p. 9; following [4], p. 17)

1.3 Deterministic and Stochastic Modelling of Time Data

The REFA methodology supports analysis and subsequent modelling of production processes under the assumption that the time data are deterministic, and therefore, are not subject to any fluctuations. With regards to work processes that must be performed by people, this assumption implies a simplification which can lead to overly optimistic results. However, it is known that processes performed by people should be assumed to be stochastically distributed. This distribution is generally not symmetric with expectation values and measure of statistical dispersion such as in normal distributions and uniform distributions but takes the form of skewed distributions such as gamma distributions or beta distributions.

The assumption that human operation times are stochastically distributed is backed by the fact that the time data required for planning are often not available at all, or are of poor quality. The assumption of beta distributed operation times ([8], pp. 620; [9], pp. 814) linked to the *PERT* method (*P*rogram *E*valuation and *R*eview *T*echnique) since the beginning of the network technique ([10], pp. 545) provides a solution that is suitable for real life application. Especially projects within the service sector have shown that corporate partners involved in joint simulation projects find it relatively easy to estimate the parameters required for beta distributions, i.e. the optimal and pessimal values, the mean or the mode [11].

In manufacturing and logistics required time data often may be represented as statistical distributions, which simplifies process modelling as it is then not necessary to allocate a deterministic time value to every single process of an order. Experience shows that industry tends to consider this as a "band-aid solution" and is more inclined to change to a deterministically defined order programme. The latter also represents the only viable solution if the simulation model is to be examined for validity through the use of historical data ([12], pp. 111).

2 Modelling Sequential and Interlinked Production Processes

2.1 Provision of Time Management Data

So far this article has been limited to stating the need for time modelling for individual operations. It is also necessary to model the logical sequential interdependencies between operations within a work order. This does not only raise the question of how these relationships between different operations are to be modelled, but also of how the resulting abundance of data can be imported from existing databases with minimum effort.

This problem is discussed below and is illustrated by the example of incorporating sequential and interlinked processes from existing data bases into a simulation procedure. As the target system, the *FEMOS* simulation procedure developed by the Institute of Human and Industrial Engineering (ifab) of Karlsruhe Institute of Technology (formerly University of Karlsruhe) is used (see [13]; alternatively see *OSim* [14]). This procedure resorts to the traditional graph-based modelling of processes. The import of time management data from existing data bases is illustrated using the example of completed projects. In the case described, the data had to be imported in order to handle the abundance of data within the specified project timelines.

2.2 The Export of Sequential Process Data to the *FEMOS* Simulation Procedure

The example discussed in this chapter describes the export of sequential process data from a company's proprietary enterprise resource management system (ERP) to the *FEMOS* simulation procedure (see also [15]). In practical projects of today, the problem is on how to create an interface between two data bases while keeping the required effort to a minimum (following [16], p. 54). The challenge involves importing existing data from a production planning and control system (PPC) or from a more comprehensive ERP system, to a simulation procedure in very little time.

Applications from the virtual factory concept tend to be based on high-performance, decentralised applications combined with centralised data management. This approach reduces both the effort and the time required while avoiding the risk of information loss. However if there is no centralised data management, then there must be an interface between the different data bases. In order to facilitate this, some manufacturers have included an export data option. Siemens PLM Software simulation tools, for example, offer interfaces in the form of text files, object files, excel sheets and data base interfaces as well as XML format interfaces that allow the import of data ([17], pp. 259).

However, for the project discussed in this article this was not an option. The project was initiated when the production of precision parts had to be relocated to a new building. In order for production logistical, staff-related and economic targets to be met, production had to be restructured. One of the main objectives included a reduction of the machinery by approximately one third and a significant production area reduction of approximately 50 % while retaining the permanent staff. Other objectives involved an increase in flexibility due to the expected batch size reduction and a major reduction in manufacturing order lead times.

Manufacturing was subdivided into sequential machining steps performed by cutting processes such as turning, drilling, and grinding. Additional steps consisted of special machining processes, and the cleaning of the manufactured parts. The complete machining cycle consisted of between 5 and more than 30 individual steps. Fig. 3 shows the production of two sample parts, including the machining steps, in the form of nodes of a sequential graph.

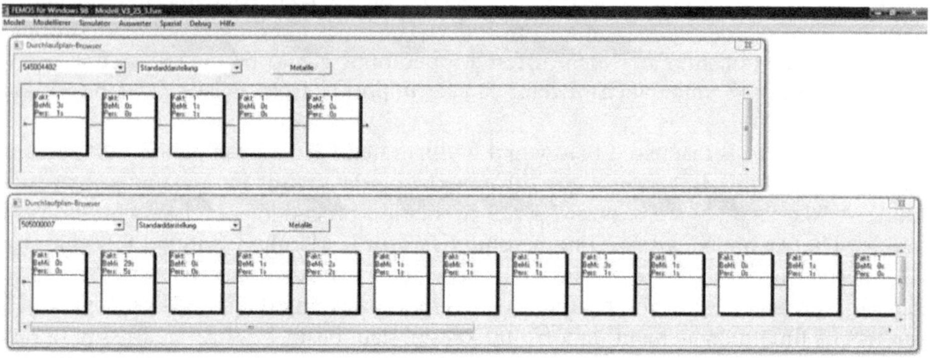

Fig. 3. Operation sequence for manufacturing a precision part in *FEMOS*

The layout of the operating equipment was previously based on technology requirements instead of being targeted towards the process, i.e. the machining steps. Only rarely was it possible to machine a complete order at the same machining centre – an option which was meant to be utilised more following restructuring, i.e. after the reduction in production area and the intended reduction in lead times.

As it was not possible to predict the target achievement rate of possible restructuring measures by using conventional planning methods, a simulation project was initiated to evaluate potential planning solutions. The evaluation should be based on the assumption of a new production programme stretching over several months. It soon became obvious that the manual entry of required model data would take longer than the intended duration of the project, meaning that an IT solution was required for data import.

The *FEMOS* simulation procedure was used for the project. In addition to the evaluation of planning solutions through the use of common production logistics indicators such as lead time and service ability, *FEMOS* offers other evaluation options. For example, staff-related indicators such as work load, and economic indicators such as personnel costs, manufacturing costs and net present value can be determined from the results of a simulation study.

The requirement that entries of the order-specific data needed to be highly detailed added complexity to the project. When the simulation model was validated against a real order programme from the existing production system, it became evident that the lower level of detail of the data caused excessive deviations between the simulation results and the actual situation. The reason for this was that during the specific period that was looked at, orders were often temporarily discontinued for order sequencing reasons because high-priority rush orders had to be completed first.

The existing *FEMOS* text interface was used to export data from the company's ERP system (Fig. 4). This interface was particularly suitable due to its ability to be easily adapted to the ERP system data format. Therefore, it was possible to link the ERP system and the simulation system within a week.

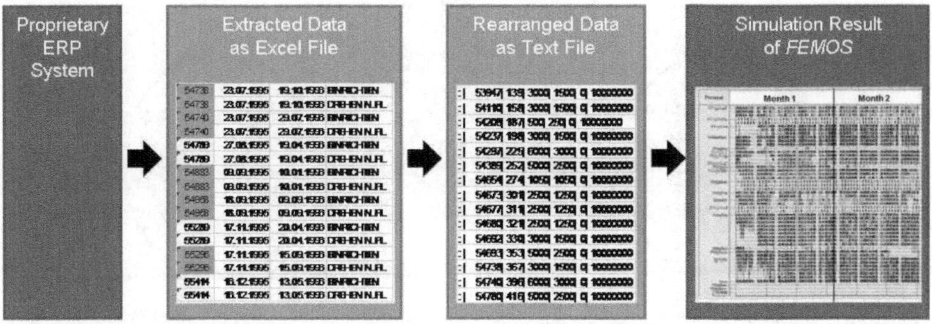

Fig. 4. Concept for exporting sequential orders from the ERP system into the *FEMOS* simulator (fictive data)

With the exception of some inconsistencies within the existing data, the export of the machining sequences was shown to be relatively easy, which resulted, in part, from the fact there were no fork nodes or join nodes to be modelled by the net graph. The following example describes such a case from a different project.

2.3 Export of Interlinked Process Data to the *FEMOS* Simulation Procedure

Problem Statement. The second project involved an order sequence simulation for an extensive production programme of one-of-a-kind products (for the operations research solution for this problem refer to [18]). The process data was available in the form of net graphs from the *MS-Project 2003* procedure. The production programme encompassed a total of 41 products and approximately 4,500 operations. The production system consisted of 30 different types of operating equipment with a total of 41 machines and manual work stations. The initial solution provided a makespan for the whole production programme of approximately 1 year. The simulation-based improvement of the sequence finally led to a makespan of approximately 8 months. For reasons of confidentiality, the details of this production programme cannot be disclosed.

The intention was to achieve an automated export of net graphs from *MS-Project* to the *FEMOS* simulation procedure in order to reduce the modelling effort required. This data then needed to be completed with the project data that was either not modelled in *MS-Project,* or that was not contained in the MS-Project representation. Furthermore, the basic data regarding personnel and operating equipment also needed to be exported. Finally, the graphs had to be verified as the data basis could not be considered error-free according to the *FEMOS* modelling requirements.

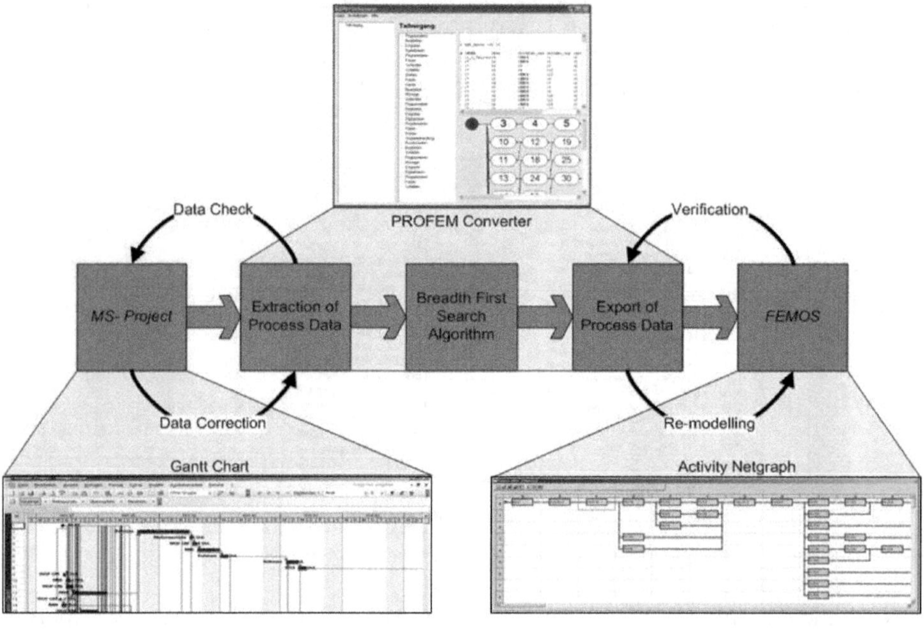

Fig. 5. Interface for exporting process data from *MS-Project* to *FEMOS*

Following this approach, the interface problem was subdivided into the three sub-aspects extraction of data from MS-Project, interface algorithm programming, and data import into *FEMOS* (Fig. 5). Three possible solutions were highlighted for the implementation of the required algorithm:

- Implementation using a stand-alone programme which reads the data in the *MS-Project* MPP file format, converts the data and then exports the data in the *FEMOS* FSM file format.
- Implementation of an import filter for *FEMOS* which creates the database for the simulation procedure from the MPP file format.
- Use of an add-in programme for *MS-Project* which exports the data in the FSM file format.

Given that the *MS-Project* file format is a proprietary binary file format that has yet to be disclosed, the implementation of the two scenarios using this format for the exchange of data would require considerable effort and the use of complex reverse engineering techniques. Consequently, an add-in programme for *MS-Project* was chosen, the so-called *PROFEM* converter. The original data is accessed via the component object model interface (COM) of *MS-Project*, which means that the MPP file format can be evaded.

Extraction and Completion of the Basic Data. The different approaches by *MS-Project* and *FEMOS* for the modelling of a work system had to be considered for the extraction of the process and resource data. *MS-Project* follows a strictly resource-oriented approach which means that every operation is allocated to the resource earmarked for it. Depending on the respective bottleneck, that resource could be either operating equipment or the personnel required to perform the operation. *FEMOS* in turn, breaks down the work system to be simulated into the element categories "operation", "personnel type" and "workplace type" which are interlinked via matrices. The relations between the element categories, i.e. "ability", "responsibility" and "feasibility" are modelled within these matrices (Fig. 6).

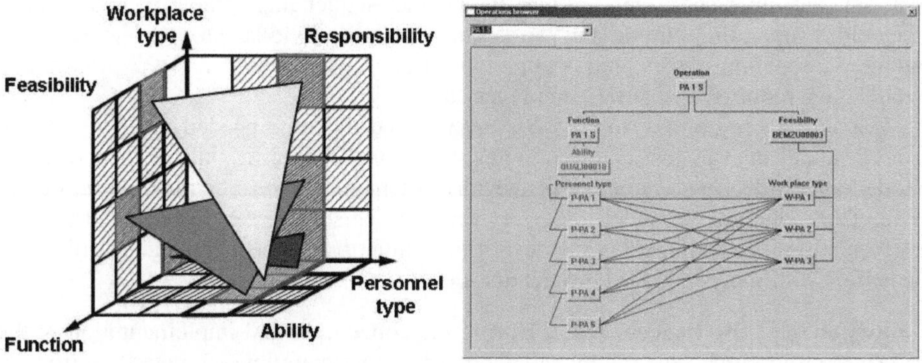

Fig. 6. Concept for modelling qualifications and its representation in *FEMOS*

In order to use the more generic approach of the *FEMOS* simulation procedure, the "resources" from *MS-Project* were integrated into the *FEMOS* personnel list and workplace list. The element categories were assigned to one another via a 0-1 allocation within the matrices. Subsequent examinations of the dynamic behaviour of the work system considering multiple qualifications or backup machines could only be performed through manual remodelling in *FEMOS*. Due to the missing subdivision in *MS-Project*, the time data of each operation had to be allocated to both the relevant operator (as order time) and to the operating equipment (as occupation time) in *FEMOS*, and had to be subsequently modified as required.

Remodelling and Verification of the Net Graphs. Also, the basic data of the orders that had to be simulated was derived from the *MS-Project* database and was modelled in *FEMOS*. This aspect of the interface problem involved the export of the net graphs of the individual production orders modelled as Gantt charts in *MS-Project*. Due to the *FEMOS* modelling approach it was required to arrange and link the individual operations in a planar way, i.e. two-dimensionally and without intersections. The operations were arranged through the use of a breadth-first-search algorithm [19] that runs through the graph breadthways. The nodes found are added to a two-dimensional grid, and nodes that have already been placed are moved around to comply with the requirement of having a non-intersecting graph.

The data base was insufficient to ensure that all net graphs would allow for planar arrangement and would be error-free in accordance with the *FEMOS* modelling approach. Therefore, the model was subjected to a verification process which identified typical modelling errors and eliminated them by inserting additional dummy operations. The verification process also identified and highlighted graphs that were unsuitable for planar arrangement. These graphs were either remodelled manually or were modified in the *MS-Project* database in such a way that it would be possible for them to become planar through the use of the respective algorithm.

Efficiency of the Export Procedure. The above-described software-based solution to the interface problem reduced the modelling effort considerably. The exemplarily performed manual modelling of parts of the production programme showed that without the interface, the complexity of the sequencing problem with all its operations, operating equipment, work stations and individual products would have required modelling times that would have been unacceptable for the project and would have required a qualified employee fulltime.

The add-in programme for *MS-Project* reduced the time needed to create the required model data to just a few minutes. Only minor specific interference with the modelling process was required in the form of manual modifications of the output data. The reduction in modelling effort was the decisive factor that enabled the implementation of a comprehensive design of simulation experiments against different scenarios within the specified project duration.

Consistency of the Process Data. During the conception and implementation of the interface programme, several core challenges relating to automated modelling from an *MS-Project* database were identified. The most important aspect was data inconsistency in *MS-Project*. The lack of restrictions for a user who enters data into the programme often poses an obstacle to the automated export of data.

Simple deficiencies such as typing errors can inadvertently lead to the creation of an additional operating resource requiring manual elimination from the database that would be time-consuming. Also, non-standardised designations in *MS-Project* made by different users can lead to the incorrect modelling of the work system. Therefore, it is indispensable to work according to a standardised and error-free designation system when entering and maintaining the original data and to thoroughly check the data prior to its export.

3 The Need for Further Developments

The interface programmes developed here for specific applications represent individual solutions that are targeted towards specific data sources and to the *FEMOS*

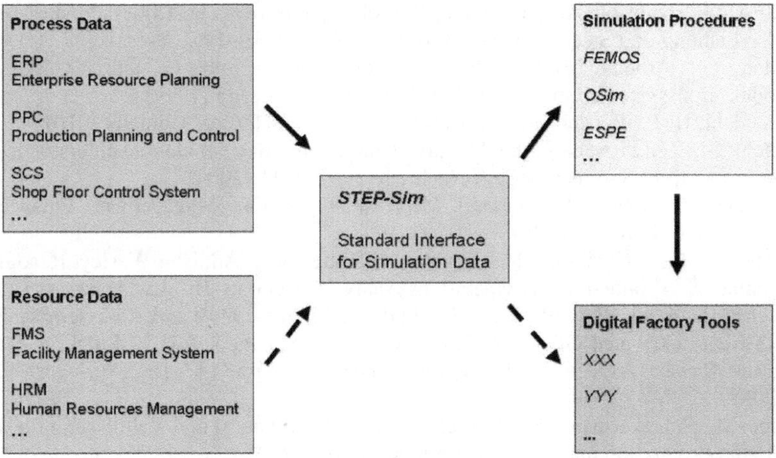

Fig. 7. Concept for a standardized interface for exporting process data into simulation procedures

simulation procedure as the target system. Therefore, no conclusions can be drawn concerning their portability to other applications. However, it can be stated that the modelling work required for complex problems is unacceptable without such an interface, specifically if actual production programmes are to be simulated.

The common approach of resorting to stochastically distributed times and arrival intervals for comprehensive process data, and of generating orders of a production programme stochastically even if they have a network-like structure (see e.g. [20], pp. 67; [21], pp. 185; [22], pp. 531) is not always an appropriate solution, as can be seen from the example of sequencing for one-of-a-kind products. Furthermore, partners from industry often refuse to accept such an approach because they prefer to see the effects that structuring measures have on actual orders.

The existing problem of importing process data from ERP and PPC systems will continue to exist in the future and will require the creation of individual solutions for interface programmes. As a future solution, the definition of standard interfaces is requested to allow for the provision of at least a basic pool of process data for simulation applications through the use of IT (Fig. 7). The interoperability of applications is supported by the fact that the modelling principles of both application areas are often based on net graphs. Data interfaces like *STEP* (*St*andard for *E*xchange of *P*roduct Model Data; [23] [24]) show that such a solution is feasible in principle. There can be no doubt about the fact that this would increase the use of simulation procedures and would include a huge savings potential for project implementation.

References

1. VDI 4499, part 1: Digitale Fabrik - Grundlagen. Beuth, Berlin (2008)
2. VDI 3633, part 6: Simulation von Logistik-, Materialfluss- und Produktionssystemen - Abbildung des Personals in Simulationsmodellen. Beuth, Berlin (2001)
3. REFA - Verband für Arbeitsstudien, Betriebsorganisation und Unternehmensentwicklung (ed.): Datenermittlung. Hanser, Munich (1997)

4. REFA - Verband für Arbeitsstudien, Betriebsorganisation und Unternehmensentwicklung (ed.): Planung und Steuerung, Teil 3. Hanser, Munich (1991)
5. Zülch, G.: Arbeitswirtschaft. In: Eversheim, W., Schuh, G. (eds.) Gestaltung von Produk¬tionssystemen, pp. 12–94, 12–124. Springer, Berlin (1999)
6. Wiendahl, H.-P.: Betriebsorganisation für Ingenieure. Hanser, Munich (2010)
7. Zülch, G.: Zeitwirtschaftliche Voraussetzungen für die simulationsunterstützte Planung von Produktionssystemen. REFA-Nachrichten 57, 4–11 (2004)
8. Slack, N., Chambers, S., Harland, C., et al.: Operations Management. Financial Times Management, London (1998)
9. Krajewski, L.J., Ritzman, L.P.: Operations Management. Addison-Wesley, Reading (1999)
10. Nahmias, S.: Production and Operations Analysis. McGraw-Hill Irwin, Boston (2001)
11. Fischer, J., Stock, P., Zülch, G.: Simulation of Disassembly and Re-assembly Processes with Beta-distributed Operation Times. In: Zülch, G., Jagdev, H.S., Stock, P. (eds.) Integrating Human Aspects in Production Management, pp. 147–156. Springer, New York (2005)
12. Rabe, M., Spieckermann, S., Wenzel, S. (eds.): Verifikation und Validierung für die Simulation in Produktion und Logistik. Springer, Berlin (2008)
13. Zülch, G., Grobel, T.: Simulating alternative organizational structures of production systems. Production Planning & Control 4, 128–138 (1993)
14. Zülch, G., Fischer, J., Jonsson, U.: An integrated object model for activity network based simulation. In: Proceedings of the 2000 Winter Simulation Conference. WSC 2000 The New Simulation Millenium 1, pp. 371–380. The Institute of Electrical and Electronics Engineers, Piscataway (2000)
15. Zülch, G., Fischer, J.: Datenbankunterstützte Modellbildung. Industrie Management 18, 57–60 (2002)
16. Kühn, W.: Digitale Fabrik. Fabriksimulation für Produktionsplaner. Hanser, Munich (2006)
17. Bangsow, S.: Fertigungssimulation mit Plant Simulation und SimTalk. Hanser, Munich (2008)
18. Zülch, G., Steininger, P., Gamber, T., Leupold, M.: Generating, benchmarking and simu¬lating production schedules – From formalisation to real problems. In: Proceedings of the 2009 Winter Simulation Conference: Energy Alternatives, pp. 2238–2249. IEEE Service Center, New York (2009)
19. Heun, V.: Grundlegende Algorithmen. Einführung in den Entwurf und die Analyse effizienter Algorithmen. Vieweg, Wiesbaden (2003)
20. Grobel, T.: Analyse der Einflüsse auf die Aufbauorganisation von Produktionssystemen. Diss., Karlsruhe (1993)
21. Brinkmeier, B.: Prozeßorientiertes Prototyping von Organisationsstrukturen im Produktionsbereich. Shaker, Aachen (1998)
22. Zülch, G., Brinkmeier, B., Rinn, A.: Design of Experiments for Investigating Departmental Structures and Control Strategies. In: Horton, G., Möller, D., Rüde, U. (eds.) Simulation in Industry 1999, pp. 530–534. Society for Computer Simulation International, San Diego (1999)
23. STEP Tools (ed.): What is STEP?, `http://www.steptools.com/library/standard/step_1.html`
24. ISO 10303-1: Industrial automation systems and integration – Product data representation and exchange. Part 1: Overview and fundamental principles (1994)

Application of Operations Research Techniques to the Redesign of the Distribution Systems

Jacek Zak

Poznan University of Technology, 3 Piotrowo street, 60-965 Poznan, Poland

Abstract. The paper presents the application of varies OR techniques to the redesign of the distribution systems. Two real-world case studies are considered. In the first case study the optimization (mathematical programming) approach is discussed. In the second case study heuristic design of different variants of the distribution system supported by object oriented simulation combined with their multiple criteria evaluation is carried out. The following OR techniques and tools are applied: single and bi-criterion mathematical programming, object oriented simulation, multiple criteria ranking methods. The results of computational experiments are presented.

Keywords: multiple criteria decision making/aiding, distribution systems, redesign.

1 Introduction

Operations Research (OR), also called Management Science (MS) is a quantitatively oriented field focused on: analyzing, mathematical modelling and solving various categories of decision problems that arise in organizations, processes and systems [1]. As its name implies OR involves "research on operations". Thus, it is applied in many situations in which the decision maker (DM) faces a complex decision problem that involves planning, control and/ or coordination of various operations (activities) within a certain organization, process and/ or system. OR can be classified as an interdisciplinary field of knowledge located at the boundaries of such areas as: mathematics, probability theory and statistics, computer science, economics and management, engineering and physical sciences, behavioural sciences. The major features of OR are as follows [1]:

- It uses scientific methods and rigid, systematic way of thinking to investigate the problem of concern;
- It looks at the problem at stake from a broad perspective; it applies system approach to problem analysis and solution; thus, OR searches for solutions that attempt to resolve conflicts between different components of the organization / system and satisfy its overall objectives;
- It attempts to find the best/optimal solution for the problem under consideration; instead of improving the status quo OR is focused on identifying the best possible course of action;
- It requires a variety of skills and competences and thus a team approach to carry out a complete study of a certain decision problem.

W. Dangelmaier et al. (Eds.): IHNS 2010, LNBIP 46, pp. 57–72, 2010.

The beginnings of OR are associated with the analysis of military operations during World War II. The first British and US OR teams applied different scientific, analytical tools and methods to optimally allocate scarce resources to various military operations (activities). Thanks to their efforts many tactical and strategic military problems have been solved and complex military operations have been rationally planned. Some of their successes include: developing effective methods of using radar, improvement of convoy management and optimization of antisubmarine operations.

Operations Research is a field that developed a spectrum of quantitative methods that support optimal decision making. The following techniques belong to traditional OR techniques [1]: linear and non-linear programming, integer programming, transportation and assignment methods, network algorithms (shortest path and critical path methods; maximum flow and minimum spanning tree algorithms), dynamic programming, game theory algorithms, decision analysis methods (decision trees and decision tables), Markov chains and Markov decision processes, queuing theory methods, inventory control methods and simulation algorithms.

Operations Research also created background for the development of the neighbouring fields, such as: probability theory and statistics, multiple criteria decision making/ aiding, artificial intelligence, data mining, machine learning and many others. In this study the interface between traditional OR techniques and multiple criteria decision making/ aiding (MCDM/A) methods is presented.

MCDM/A is a field which aims at giving the decision maker (DM) some tools in order to enable him/her to solve a complex decision problem where several points of view must be taken into account. MCDM/A concentrates on suggesting "compromise solutions", taking into consideration the trade-offs between criteria and the DM's preferences [2]. The decision processes based on multiple criteria analysis involve the following parties: the decision maker (DM), stakeholders and analyst. MCDM/A methods are computer based tools that assist DMs in solving multiple criteria decision problems. Those problems are the situations in which having defined a set of actions (decisions, alternatives) A and a consistent family of criteria F the DM tends to: define a subset of actions (decisions, alternatives) being the best on F (choice problematic), divide the set of actions (decisions, alternatives) into subsets according to certain norms (sorting problematic), rank the set of actions (decision, alternatives) from the best to the worst (ranking problematic). The classification of MCDM/A methods corresponds to the above classification of multiple criteria decision problems. Thus, one can distinguish MCDM/A choice (optimization) methods, MCDM/A sorting methods and MCDM/A ranking methods. Many MCDM/A specialists suggest also the division of MCDM/A methods based on their approach to aggregating global preferences of the DM [17]. Based on that division criterion one can distinguish two major streams of methods i.e.: the American school based on multiatribute utility theory and the European school based on the outranking relation. Well-known representatives of those streams are: AHP [3] and UTA [4] methods versus ELECTRE [2][5], Oreste [6] and PROMETHEE [7] methods, respectively.

The paper presents the application of OR techniques to the redesign of a certain distribution system. The distribution system is defined [8] as a set of such components as: infrastructure (roads, warehouses), human resources, fleet, business processes and organizational rules which are responsible for planning, implementing and controlling the physical flows of products from points of origin to points of destination. Those components should match together to assure the efficiency and effectiveness of the whole distribution system. That is why its design and redesign is a very complex task [9][10]. The redesign process is connected with significant changes in the structure of the distribution system [11] i.e. changes in the location of warehouses, reassignment of tasks and redistribution of inventory between warehouses, reassignment of roles and responsibilities among supply chain points, changes in a transportation network etc.

Based on the classical division of the decision problems [2][5] one can perceive the redesign of the distribution system as a choice (mathematical programming) problem or as a ranking problem. In the first case the decision maker (DM) is looking for the optimal structure (design) of the system and the decision problem is formulated as an optimization problem. In the second case different development scenarios (alternative solutions) are designed, evaluated and ranked from the best to the worst. In the design phase, different solutions concerning transportation, organizational rules, material and information flow, human resources, infrastructure etc. are proposed by the experts. In the evaluation phase the selection of the best candidate (distribution system) is carried out. Both of the above approaches are presented in this study.

2 Methodological Background

In general, practically oriented OR studies usually involve the following phases [1]:

- Definition and thorough recognition of the problem of interest combined with its precise, verbal description;
- Formulation of the mathematical model representing the problem and collection of the relevant data required to run the model;
- Selection / customization and/or development of computer-based procedures capable to generate solutions to the problem considered;
- Testing and adjusting the model, often called model validation; running a series of computational experiments to generate different samples of solutions;
- Implementation of the selected, most desired (optimal) solution.

This scheme has been applied in the below characterized case studies. Due to specific character of the decision problems considered in particular case studies the above mentioned generic phases of OR – oriented projects have been customized. In both cases the decision problems have been recognized and verbally described. Different decision models have been formulated to represent decision situations in case studies 1 and 2, respectively. In case study 1 the decision problem has been formulated as a single and bi – criterion mathematical programming problem. It involved the definition of the decision variables and parameters, objective function(s) and constraints. In case study 2 a combination of two OR approaches has been utilized. The first model of the object oriented simulation has been applied to generate different variants of the distribution system, while the second one – of the multiple criteria ranking problem has been used to evaluate and rank the considered distribution systems. In both case studies appropriate computer – based procedures have been utilized to solve the respective decision problems. In case study 1, computer implementations of simplex [1] and gradient methods [1] have been utilized to solve single criterion, linear and non-linear formulations of the decision problem, respectively. To this end a commercial solver MS Excel Solver – Premium Solver Plus by Frontline Systems has been applied. In case study 2, stochastic simulation procedures implemented in the computer package ExtendSim have been applied to observe the operations of different variants of the distribution system. Those variants have been evaluated and ranked with the application of the multiple criteria ranking method – ELECTRE III and AHP (Analytical Hierarchy Process).

3 Case Studies

Case study 1 – Single and bi – criterion optimization of the distribution system. Application of mathematical programming [12]. The first case study focuses on the redesign of the distribution system for an international company based in Warsaw, Poland. The primary concern of the project is to define the number and the location of the warehouses as well as the service area of each warehouse. The considered distribution system is managed by the manufacturer of cosmetics, detergents and washing articles. The annual turnover of the company is roughly 100 million Euro (400 million Polish zloty – PLN). 85% of the turnover is generated at the Polish local market and the remaining 15% is an export to Eastern European countries. At the Polish market the company's products are mainly sold to wholesalers (60% of sales) and the chains of large retailers (hipermarkets, supermarkets) – 20% of sales. The company is concerned about its costs, including distribution costs, which amount to 6.5% of sales. Its major focus, however, is the enhancement of the customer service level. The company would like to increase the market share and make the company's products available at each retail shop. In the opinion of the company's top management this goal can be obtained through the redesign of the distribution system, which should be more reliable and more flexible. The target for the distribution system is to fulfill each customer's order and deliver the required products within 24 hours. Delivery time should be further decreased in the next years. This should satisfy customers' expectations and requirements and make the company's slogan: "Our products in each Polish home", rational.

The existing distribution system is based on two production plants and two warehouses located in cities A and B, next to the production facilities. Due to the fact that the production profile and the product portfolio are different in the production plants in A and B trunking between warehouses is required. It amounts to 45% of the total number of tkm covered by the distribution system. Each warehouse has a certain area to cover, which corresponds to a concrete number of customers served. The total number of customers is about 400. The products are delivered by 24 tractor – semi trailer units with a capacity of 33 Europallets each.

As far as delivery time is concerned, the existing distribution system guarantees a 48 hour deliveries in the majority of cases and 72 hour deliveries to the most distant customers. The customer service level (defined as a product availability percentage) is 95%. Customers are not satisfied with timeliness and frequency of deliveries. Thus, the first objective of the project is to minimize the total distribution costs, including the warehousing, transportation and capital in stock costs. The second objective is to minimize the delivery time, which may be reduced by many actions influencing different stages of the order fulfillment process. The actions considered in this study may influence riding time. Thus, the minimization of the riding time is the second objective of the study.

The decision problem is formulated in terms of multiple objective mathematical programming with binary and continuous variables.

Data. The following data is used in the model:

I - the number of potential locations of warehouses,

J - the number of regions that have to be assigned to the warehouses,

DA_j - annual demand of region j for products of production plant A in [pallets], $j = 1,...,J$,

DB_j - annual demand of region j for products of production plant B in [pallets], $j = 1,...,J$,

TC_{ij} - average transportation cost from warehouse at location i to region j in [PLN/pallet], $i = 1,...,I$, $j = 1,...,J$,

TCA_j - average transportation cost from production plant A to warehouse at location i in [PLN/pallet], $i = 1,...,I$,

TCB_j - average transportation cost from production plant B to warehouse at location i in [PLN/pallet], $i = 1,...,I$,

TT_{ij} - average travel time from warehouse at location i to the first customer in region j in [min], $i = 1,...,I$, $j = 1,...,J$,

PHC_i – cost of pallet handling in warehouse at location i, $i = 1,...,I$,

CRT - current average pallet rotation time in [days], i.e. the average number of days that a pallet spends in the current distribution system,

CCA - average daily cost of capital in stock per pallet produced in production plant A in [PLN/day],

CCB - average daily cost of capital in stock per pallet produced in production plant B in [PLN/day],

MCC_i - minimum annual cost of capital in stock in warehouse at location i related to the safety stock of pallets,

ML - maximum load of vehicles used for transportation from production plants to warehouse,

DY - average number of working days in a year.

Decision variables. Two groups of decision variables are considered:

$y_i \in \{0, 1\}$, $i = 1,...,I$, equals to one if warehouse at location i is included in the plan, and 0 otherwise.

$x_{ij} \in \{0, 1\}$, $i = 1,...,I$, $j = 1,...,J$, equals to one if region j is assigned to warehouse at location i, and 0 otherwise.

Constraints. The first group of constraints assures that regions are assigned only to warehouses included in the plan:

$$x_{ij} \leq y_i, \qquad i = 1,...,I, j = 1,...,J. \qquad (1)$$

The second group of constraints assures that each region is assigned to exactly one warehouse:

$$\sum_{i=1}^{I} x_{ij} = 1, \qquad j = 1,\ldots,J. \tag{2}$$

Some additional constraints added to the model in order to obtain a linear model are described below.

Objective functions. Two objectives are considered in the model:

- *TDC* - total annual cost of distribution,
- *MRT* - maximum riding time to the first customer on a route.

Both objectives are minimized.

The first objective is defined in the following way:

$$TDC = TTC + TPHC + TCC,. \tag{3}$$

where: *TTC* denotes total annual transportation cost, *TPHC* denotes total annual cost of pallets handling in the warehouses, and *TCC* denotes total annual cost of capital in stock.

Total annual transportation cost takes into account transportation from production plants to warehouses and transportation from warehouses to the customers in the regions. It is defined in the following way:

$$TTC = \sum_{i=1}^{I} y_i \left(TCA_i \sum_{j=1}^{J} x_{ij} DA_j + TCB_i \sum_{j=1}^{J} x_{ij} DB_j \right) + \sum_{i=1}^{I} \sum_{j=1}^{J} x_{ij} TC_{ij} (DA_j + DB_j) \tag{4}$$

Total annual cost of pallets handling in the warehouses is defined in the following way:

$$TPHC = \sum_{i=1}^{I} y_i PHC_i \left(\sum_{j=1}^{J} x_{ij} DA_j + \sum_{j=1}^{J} x_{ij} DA_j \right), \tag{5}$$

Total annual cost of capital in stock is defined in the following way:

$$TCC = \sum_{i=1}^{I} y_i \max\left\{ MCC_i, \sum_{j=1}^{J} x_{ij} DA_j CCA(CRT + DHA_i) + \sum_{j=1}^{J} x_{ij} DB_j CCB(CRT + DHB_i) \right\} \tag{6}$$

where: DHA_i and DHB_i is an average headway of deliveries for production plants *A* and *B*, respectively, defined in the following way:

$$DHA_i = ML / \left(\sum_{j=1}^{J} x_{ij} DA_j / DY \right), \tag{7}$$

$$DHB_i = ML / \left(\sum_{j=1}^{J} x_{ij} DB_j / DY \right). \tag{8}$$

Total annual cost of capital in stock is not linear as it uses operator max. In order to obtain linear model we add I continuos variables cc_i interpreted as cost of capital in stock in warehouse at location i. Furthermore, we add two groups of constraints:

$$cc_i \geq MCC_i, \quad i = 1,...,I, \tag{9}$$

$$cc_i \geq \sum_{j=1}^{J} x_{ij} DA_j CCA(CRT + DHA_i) + \sum_{j=1}^{J} x_{ij} DB_j CCB(CRT + DHB_i) \quad , i = 1,...,I. \tag{10}$$

Total annual cost of capital in stock is defined as:

$$TCC = \sum_{i=1}^{I} y_i cc_i . \tag{11}$$

Maximum riding time to the first customer on a route is defined in the following way:

$$MRT = \max\{x_{ij} TT_{ij}\}. \tag{12}$$

Again, this objective is not linear. In order to obtain linear model we add a continuos variable *mrt* and the following group of constraints:

$$mrt \geq x_{ij} TT_{ij}, \quad i = 1,...,I, j = 1,...,J. \tag{13}$$

Maximum riding time to the first customer on a route is then equal to a new variable:

$$MRT = mrt. \tag{14}$$

Finally, we obtain a mixed binary bi-objective mathematical programming problem with $I \times J + J$ binary variables and $I + 1$ continuous variables defined in the following way:

minimize: $TCC = \sum_{i=1}^{I} y_i cc_i$

minimize: $MRT = mrt$

subject to:

$$
\begin{aligned}
& x_{ij} \le y_i, \qquad i = 1,\dots,I, j = 1,\dots,J \\
& \sum_{i=1}^{I} x_{ij} = 1, \qquad j = 1,\dots,J \\
& cc_i \ge MCC_i, \quad i = 1,\dots,I \\
& cc_i \ge \sum_{j=1}^{J} x_{ij} DA_j CCA(CRT + DHA_i) + \sum_{j=1}^{J} x_{ij} DB_j CCB(CRT + DHB_i) \\
& , i = 1,\dots,I \\
& mrt \ge x_{ij} TT_{ij}, \quad i = 1,\dots,I, j = 1,\dots,J \\
& y_i \in \{0, 1\}, i = 1,\dots,I \\
& x_{ij} \in \{0, 1\}, i = 1,\dots,I, j = 1,\dots,J
\end{aligned} \tag{15}
$$

In the analyzed case $J = 39$, $I = 49$, thus the problem has 1900 binary variables and 50 continuous variables.

Case study II – heuristic redesign of the distribution system supported by object oriented simulation and multiple criteria evaluation of the variants of the distribution system [14][15]. The problem considered in case study II focuses on the design and evaluation of alternative logistic solutions for the distribution system, which consists of 24 warehouses/ distribution centers (DCs) uniformly spread all over Poland. The system is owned and operated by a medium sized trade-distribution company generating an annual turnover of 55 mln Euro (210 mln PLN). The company, with a market share of 11% and the overall labor size of 160 employees belongs to major players at the Polish electrotechnical market. The analyzed distribution system can be divided into 5 echelons (Figure 1): a suppliers' level (SL) – 75 manufacturers and distributers, a central level (CL) – 1 central warehouse (CW), a regional level (RL) – 12 DCs, a local level (LL) – 11 DCs and a customers' level (CuL) – 400 customers. It distributes and delivers for sales a full range of electrotechnical products from fuses and bulbs to pylons, and from sockets and switches to electrical wires, with a total number of 38,5 thousand units, divided into 56 groups. The DCs are differentiated by their locations and areas to serve, building structures and equipment, warehousing capacity, inventory portfolio, crew size, etc. All the goods (products, materials) are moved in the distribution system by road transportation, which is partially outsourced and partially carried out as in-company activity by a fleet of 55 vehicles including 38 vans and trucks.

From the OR perspective the analysis of case study II can be split into two major phases. In phase I the variants – development scenarios of the distribution system are designed / constructed and in phase II they are evaluated and ranked from the best to the worst.

In phase I, based on the comprehensive diagnosis (SWOT analysis) of the above described existing distribution system (variant 1-AI) its strengths and weaknesses have been recognized. To reduce disadvantages of the distribution system its redesign has been proposed. The system is restructured heuristically, with the application of common sense and expert knowledge. In the redesign process major objectives of the management team of the trade-distribution company, including: improvement of the customer service (minimizing delivery time) and enhancement of efficiency (maximizing rotation level of inventories and fleet, infrastructure utilization) are taken into account. The restructuring of the distribution system consists in the introduction of its certain improvements and changes, including: redefinition of the location and number of the DCs, reassignment of their customers and service areas, enhancement of the logistic infrastructure and equipment used in the DCs, redesign of the warehousing space and changes in warehousing capacities, changes in the organization of the transportation system, redefinition of the fleet composition, reshaping of the inventory portfolio and balancing the inventory levels, reconstruction and improvement of the information and material flows, reassignment of tasks and responsibilities in the distribution system, redefinition of duties for the employees, rationalization of the crew size. As a result of the proposed changes 6 alternative variants - development scenarios of the distribution system (variant 2 - AII, variant 3 - AIII, variant 4 - AIV, variant 5 - AV, variant 6 - AVI, variant 7 - AVII) have been constructed. The heuristic construction of variants is supported by object-oriented simulation carried in the computer-based tool ExtendSim. The operations of the designed variants are simulated and certain parameters characterizing them are generated. The structure of the simulation model is presented in Figure 1. The model reflects the 5 echelon structure of the distribution system, with five levels: SL, CL, RL, LL and CuL, four of which include the hierarchical objects denominated by $G_i^{SL}, G_i^{CL}, G_i^{RL}, G_i^{LL}$. These objects are responsible for the generation of information characterizing the structure and the course of actions of the ordering process at particular levels. They are linked with another set of hierarchical objects $N_N^{SL}, N_N^{CL}, N_N^{RL}, N_N^{LL}, N_N^{CuL}$, representing suppliers (SL), distribution centres (CL, RL and LL) and final customers N_N^{CuL}, respectively. They are supplied with the information regarding the content and characteristics of the orders, e.g.: number of pallets ordered, type of the products ordered, name of the customer, distance between supplier / distribution centre and the customer. The physical flow of products is represented by the hierarchical objects denominated by $G_k^{SL}, G_k^{CL}, G_k^{RL}, G_k^{LL}$. The arrows present the connections between suppliers, distribution centers and customers. This generic structure of the simulation model is customized to specific features of variants A_{II}, A_{III}, A_{IV}, A_V, A_{VI} and A_{VII}. The designed and simulated variants represent different level of changes from superficial and evolutionary to very comprehensive and radical. They differ for example by the following features:

- number of echelons (SL, CL, RL, LL and CuL) ranging from 3 to 5,
- total number of warehouses in the distribution system ranging from 1 (supported by 49 retail shops) to 32; different locations of the warehouses;
- number of warehouses at the central level ranging from 1 to 4;
- share of the company owned and managed warehouses ranging from 0% (complete outsourcing) to 100% (full ownership);
- share of transportation activities carried out at different echelons of the distribution system by the company, ranging from 0% to 100%;
- number of employees ranging from 120 to 250.

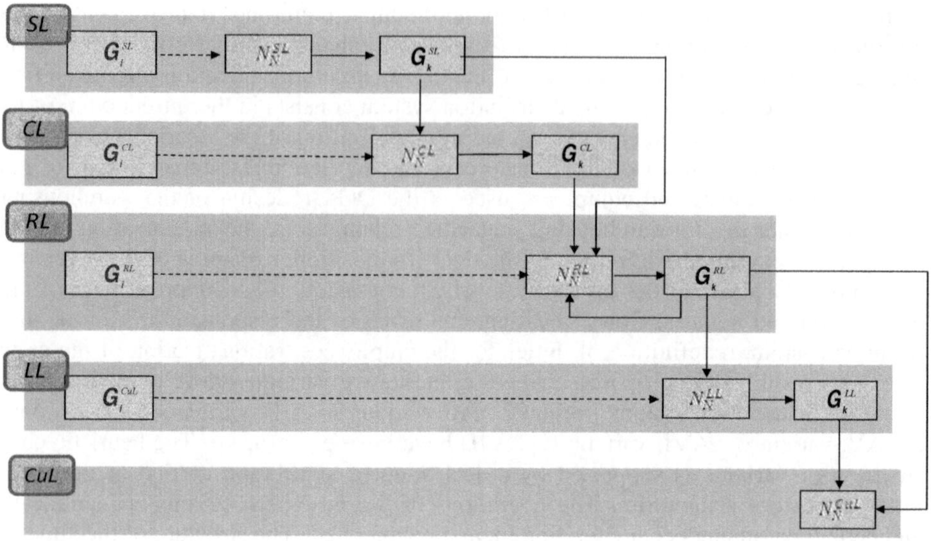

Fig. 1. The generic structure of the simulation model for the distribution system

As a result the variants of the distribution system are characterized by different measures of performance/ evaluation criteria that represent the interests of different stakeholders, including: owners/managers of the distribution system, final customers, haulers, employees involved in the distribution process.

In phase II, the evaluation process of the proposed variants is carried out. It is based on the application of MCDM/A methodology. Two MCDM/A methods: ELECTRE III [2][13] and Analytic Hierarchy Process - AHP [3] are used to rank the variants. Phase II involves the definition of criteria, construction of the evaluation matrix, definition of the DM's preferences and computational experiments resulting in the generation of final rankings.

In the analyzed case the following evaluation criteria, constituting a so called consistent family of criteria [2][13], are taken into account:

- C_1 – average delivery time [days], minimized criterion;
- C_2 – dispersion of warehouses [%], maximized criterion;
- C_3 – share of distribution costs in the total sales [%], minimized criterion;
- C_4 – average rotation level of inventories in a distribution system [days], minimized criterion;
- C_5 – difference between the levels of investments and divestments [PLNx1000], minimized criterion;
- C_6 – the company's ability to accept changes [%], maximized criterion;
- C_7 – turnover per employee [PLNx1000], maximized criterion;
- C_8 – MIRR (Modified Internal Rate of Return) [%], maximized criterion;
- C_9 – NPV (Net Present Value) [PLNx1000], maximized criterion;
- C_{10} – market share [%], maximized criterion.

The evaluations of alternatives on all criteria are presented in the matrix of performances (see Table 1).

Table 1. Matrix of performances for alternative distribution systems

Alternative distribution systems	Criteria									
	C_1 [ds]	C_2 [%]	C_3 [%]	C_4 [ds]	C_5 [PLN x 10^3]	C_6 [%]	C_7 [PLN x 10^3]	C_8 [%]	C_9 [PLN x 10^3]	C_{10} [%]
A_I	4	33	4.7	32	0	52.7	1313	31	6750	11
A_{II}	2	33	5.5	43	4855	60.1	1528	15	1188	14
A_{III}	3	27	4.3	31	-50	61.6	2148	172	4321	11
A_{IV}	2	33	5.2	45	2610	55.3	1838	9	8940	16
A_V	4	33	4.9	46	10475	58.8	1989	36	430	14
A_{VI}	3	50	4.4	28	10320	59.5	3049	11	28684	21
A_{VII}	1	60	4.6	33	-385	83.0	2064	41	21200	24
								65		

The phase of defining the DM's preferences allows for taking into account specific and subjective aspirations and expectations of the DM. In general, the majority of the DM's models of preferences include: the importance of criteria and the DM's sensitivity with respect to the changes of the values of criteria. The models of the DM's preferences differ in both methods. In the ELECTRE III method it is determined by weights w_j for each criterion and the indifference q_j, preference p_j, and veto v_j thresholds. In the AHP method pairwise comparison judgements between criteria and variants are carried out. These pairwise comparisons are quantified by the standard "one – to – nine" measurement scale: 1 – equally preferred; 3 –moderately preferred; 5 – strongly preferred; 7 – very strongly preferred; 9 – extremely strongly preferred. The intermediate judgements like: 2, 4, 6, 8 can be used if necessary The model of the DM's preferences for both methods is presented in Table 2.

Table 2. Model of the DM's preferences in ELECTRE III and AHP methods

Criteria	ELECTRE III w_j	q_j	p_j	v_j	AHP Criteria C_1	C_2	C_3	C_4	C_5	C_6	C_7	C_8	C_9	C_{10}
C_1	9.8	0.6	1.1	2.2	1	2	2	1	5	7	5	5	5	4
C_2	8.1	8.9	16.1	30.4	1/2	1	1	1/3	4	5	4	3	3	2
C_3	8.2	0.2	0.5	1.1	1/2	1	1	1/3	4	5	4	3	3	2
C_4	10.0	2.0	4.7	8.3	1	3	3	1	7	7	7	5	5	3
C_5	6.1	1964.3	4071.5	9785.7	1/5	1/4	1/4	1/7	1	1	1	1/2	1/2	1/4
C_6	6.0	4.7	11.7	23.1	1/7	1/5	1/5	1/7	1	1	1/2	1/2	1/2	1/4
C_7	6.3	142.9	303.6	503.6	1/5	1/4	1/4	1/7	1	2	1	1/2	1/2	1/3
C_8	6.7	25.4	50.7	98.6	1/5	1/3	1/3	1/5	2	2	2	1	1	1/3
C_9	6.6	2458.4	5000	9833.4	1/5	1/3	1/3	1/5	2	2	2	1	1	1/3
C_{10}	7.6	2.1	5.2	10.2	1/4	1/2	1/2	1/3	4	4	3	3	3	1

4 Computational Experiments

Case study I. In case study I the optimization of the distribution system has been performed with the application of the extended version of MS Excel Solver – Premium Solver Plus by Frontline Systems. It solves linear problems composed of up to 2000 variables and 400 constraints. The experiments have been divided into two steps: the first one, based on single objective optimization and the second one focused on bi-objective optimization.

In the first step, only one objective - total distribution costs was used, to demonstrate the comparison between the existing and the optimal distribution systems (see Table 3.). The optimal system is composed of 7 warehouses.

Table 3. The comparison between the existing and the optimal distribution system (single objective optimization)

Distribution system	Number of warehouses	Total annual distribution costs [PLN]	Ridind time [h:mm]
Existing	2	9 924 300	9:22
Optimal	7	9 357 784	6:09

The interesting observation is that single objective optimization resulted in reducing both total distribution costs (by 566 516 PLN) and the riding time (by 3 hours and 13 minutes). Thus from the multiple objective point of view the optimal distribution system dominates the current one. As a result of the optimization process new definition of service areas (new assignment of 49 regions to 7 warehouses) for each warehouse was obtained.

In the bi-objective optimization ε-constraints method [16][17] has been used to generate a representative sample of Pareto-optimal solutions. In the computational procedure riding time was being constrained from 6 to 2 hours. The result of the bi-objective optimization is a set of Pareto-optimal distribution systems presented in Figure 2. The generated distribution systems are composed of 7 – 23 warehouses. The results show the existing cost-time trade-offs. For instance, riding time reduction from 6:09 to 5:23 (by 46 min) results in 2 more warehouses in the distribution system and additional costs of 186 000 PLN, while riding time reduction from 2:44 to 2:41 (by 3 min) also results in 2 additional warehouses in the distribution system and corresponding increase in total distribution costs of 548 000 PLN.

Another interesting observation comes from the comparison of the existing distribution system (see Table 3) with a set of Pareto-optimal distribution systems (see Figure 2). Maintaining the same level of the total distribution costs (around 9 900 000 PLN) one can replace the existing distribution system (with 2 warehouses) by the Pareto – optimal one (with 10 warehouses) and shorten the riding time from 9:22 to less then 4:20 (55% reduction). The distribution system with 10 warehouses has been finally selected as the most desired option and practically implemented.

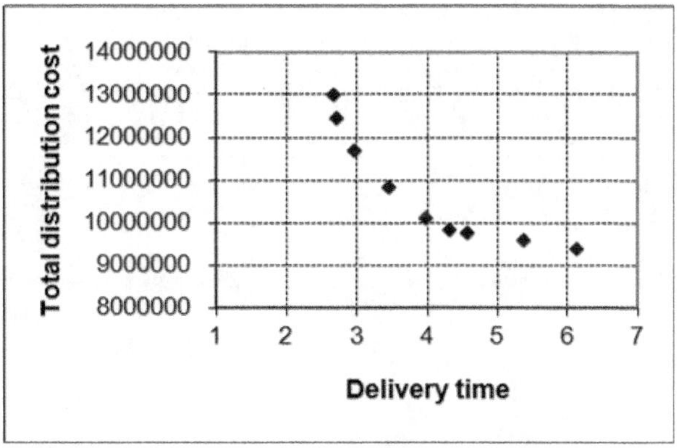

Riding time [h:mm]	Total distribution costs [PLN]	No. of warehouses
2:41	12 972 507	23
2:44	12 424 210	21
2:59	11 653 423	18
3:28	10 813 246	15
4:00	10 090 964	12
4:20	9 802 832	10
4:35	9 746 413	10
5:23	9 543 711	9
6:09	9 357 784	7

Fig. 2. Pareto-optimal set of the distribution systems in the numerical and graphical form

Case study II. In case study II computational experiments refer to two phases, i.e. designing and evaluating the variants of the distribution system. In the design phase an object-oriented simulation tool ExtendSim is applied. This is a user friendly package which can be used to construct operational model of complex systems in a graphical form without advanced programming skills [18]. It is typically used in transportation, logistics, business processes redesign, manufacturing, as well as in healthcare, service and communications industries. This tool is based on the application of continuous and discrete-event simulation. It provides hierarchy structure of a model and a comprehensive library of objects that can be adjusted to specific environments and concrete decision problems.

The example of simulation modeling of the distribution system is presented in Figure 3. One can distinguish two areas A and B, which correspond to the software functionalities of information generation and information memorization, respectively. In the area B parameters of incoming orders, such as: starting distribution center n_{NS} as a place of incoming order; ending distribution center/ final customer n_{NE}, which placed an order for products; number of ordered products $\tilde{q}_{kn}$ within the assortment k; distance $s^{n_{NS}n_{NE}}$ between n_{NS} and n_{NE}, speed value $\tilde{v}^{n_{NS}n_{NE}}$ between n_{NS} and n_{NE}, are modeled. This information is generated by objects presented in the area A and sent to the

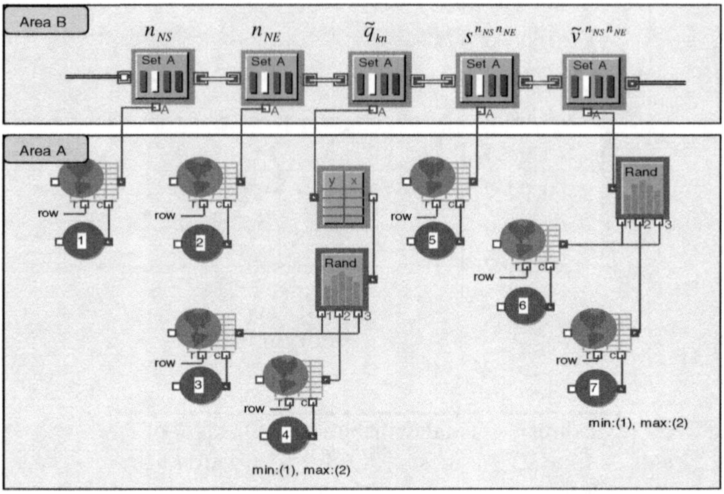

Fig. 3. Sequence of objects representing information flow in the simulation model of the distribution system

objects in area B. Objects in the area A are connected with spreadsheets of MsExcel including data collected during the analysis of the distribution system. Parameters of a random character, such as: $\tilde{q}_{kn}$ and $\tilde{v}^{n_{NS}n_{NE}}$ are modelled by objects Rand, which are responsible for applying different patterns of random variables distributions.

During the simulation process different parameters of the designed variants of the distribution system are generated (estimated), including: daily number of departures of vehicles and EURO pallets dispatched at different levels of the distribution system, unit transportation and warehousing costs, total distance covered by transportation fleet per day, utilization of warehouses, forklifts and operators, average queue length of vehicles waiting for unloading, inventories rotation indexes, etc. Those parameters allow to compute the values of criteria C_1 to C_{10}, presented above.

In the evaluation phase two MCDM/A methods, including ELECTRE III and AHP are applied. The algorithm of the ELECTRE III method computes the following parameters[13]: concordance index, discordance index and the outranking relation. Based on that, two complete preorders (descending and ascending) are generated[13]. Their intersaction constitutes the final ranking. In the AHP method[3] the matrixes of normalized absolute weights are constructed at each level of the hierarchy and an eigenvalue problem[3] is solved for each matrix. In the next step global consistency indexes[3] are computed and checked for feasibility. In the last step utilities[3] for each variant are calculated and based on that their ranking is constructed. Figure 4 presents the final rankings of the distribution system development scenarios generated by ELECTRE (a) and AHP (b) methods.

The comparative analysis of both rankings reveals a high degree of similarity between them ELECTRE III method indicates that the best solution is the alternative A_{III}. This alternative involves the minimum number of changes in the distribution

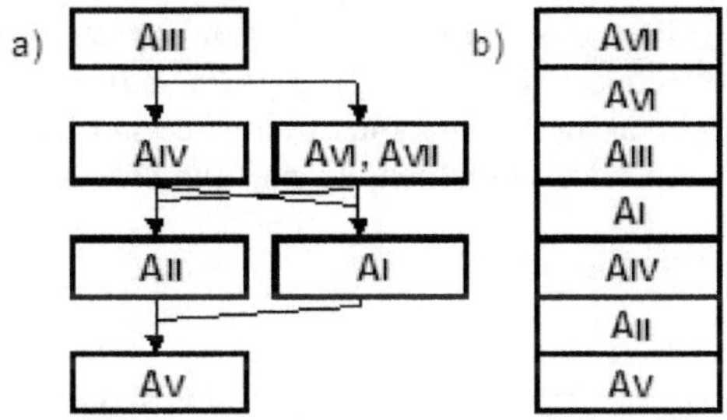

Fig. 4. Final rankings of the development scenarios of the distribution system generated by: a) ELECTRE III method, b) AHP method

system. Its advantage is a high MIRR index and the lowest (in comparison with other alternatives) distribution costs. The main disadvantage of A_{III} is a low market share (almost 11%) and long average delivery time (3 days). The AHP method suggests that alternative A_{VII} is the best solution. A_{VII} is characterized by the most radical changes in the distribution system. In this case the redesign guarantees a high market share (24%) and low distribution costs. The average delivery time is 1 day. Both final rankings reject alternatives A_V and A_{II}. The evaluation of alternatives A_I and A_{IV} is ambiguous.

Based on the results of the computational experiment, the author of the paper suggests the following stepwise path of changes:

- Introduction of the evolutionary changes represented by AIII in the first phase.
- More radical transformation from AIII to AVII in the second phase.

5 Conclusions

The paper presents practical application of OR techniques to the redesign of the distribution system. The classical OR methodology is presented, including: problem recognition and verbal description, formulation of the mathematical model, selection of computer-based methods capable of solving the considered decision problem, computational experiments resulting in the selection of the most desired solution, implementation of the selected solution.

The problem of redesigning the distribution system has been presented in two alternative ways:

- In case study I as a single and bi-objective choice problem, formulated as a mathematical programming problem;
- In case study II as a multiple criteria ranking problem in which the definition of the variants (distribution systems) has been supported by object – oriented simulation.

The article demonstrates the application of the following techniques: single and bi-objective optimization, object oriented simulation, multiple criteria ranking methods.

The paper shows practical applicability of the proposed approach. In case study I the optimal distribution system is 6% more efficient than the existing one (cost-wise) and the finally selected Pareto – optimal distribution system provides an enhanced by 55% customer service (reduction of the delivery time). In case study II both selected solutions (A_{III} and A_{VII}) assure noticeable financial and market-oriented improvements. An interesting step-wise implementation of the changes is proposed.

References

1. Hillier, F.S., Lieberman, G.J.: Introduction to Operations Research. McGraw-Hill, New York (2005)
2. Vincke, P.: Multicriteria Decision-Aid. John Wiley & Sons, Chichester (1992)
3. Saaty, T.L.: The Analytic Hierarchy Process: Planning, Priority Setting, Resource Allocation. McGraw-Hill, New York (1980)
4. Jacquet-Lagreze, E., Siskos, J.: Assessing a Set of Additive Utility Functions for Multicriteria Decision-Making, the UTA Method. European Journal of Operational Research 10, 151–164 (1982)
5. Roy, B.: Decision-Aid and Decision Making. European Journal of Operational Research 45, 324–331 (1990)
6. Roubens, M.: Preference Relations on Actions and Criteria in Multicriteria Decision Making. European Journal of Operational Research 10, 51–55 (1982)
7. Brans, J., Vincke, P., Mareschal, B.: How to Select and How to Rank Projects: The PROMETHEE Method. European Journal of Operational Research 24, 228–238 (1986)
8. Kotler, P.: Marketing Management. Prentice Hall, Englewood Cliffs (1994)
9. McKinnon, A.: Physical Distribution Systems. Routledge, New York (1989)
10. Ross, D.F.: Distribution, Planning and Control. Kluwer Academic Publishers, Boston (1996)
11. Kopczak, L.R.: Logistics Partnerships and Supply Chain Restructuring: Survey Results from the VS Computer Industry. Production and Operations Management 6(3), 226–247 (1997)
12. Hapk, M., Jaszkiewicz, A., Zak, J.: The Design of the Physical Distribution System with the Application of the Multiple Objective Mathematical Programming. Case Study. In: Trzaskalik, T., Michnik, J. (eds.) Multiple Objective and Goal Programming. Recent Developments. Advances in Soft Computing, pp. 297–309. Physica-Verlag, Heidelberg (2002)
13. Roy, B.: The Outranking Approach and the Foundations of ELECTRE Methods. In: Bana e Costa, C.A. (ed.) Readings in Multiple Criteria Decision Aid, pp. 155–183. Springer, Berlin (1990)
14. Zak, J., Wlodarczak, H., Kicinski, M.: The MCDM Based Redesign of the Distribution System. In: 9th Meeting of the Euro Working Group on Transportation: Intermodality, Sustainability and Intelligent Transportation Systems, pp. 665–670. Bari (2002)
15. Sawicka, H., Zak, J.: Object-Oriented Modeling and Simulation of the Transportation System. In: 11th Meeting of the EURO Working Group on Transportation and Extra EURO Conference on Transportation, pp. 312–322. Bari (2006)
16. Steuer, R.: Multiple Criteria Optimization: Theory, Computation and Application. John Wiley & Sons, New York (1986)
17. Figueira, J., Greco, S., Ehrgott, M. (eds.): Multiple Criteria Decision Analysis. State of the Art Surveys. Springer, New York (2005)
18. Krahl, D.: Extend: An Interactive Simulation Tool. In: 2003 Winter Simulation Conference, pp. 166–196. The Fairmont, New Orleans (2003)

Simulation in Production and Logistics: Trends, Solutions and Applications

Sigrid Wenzel[1], Pinar Boyaci[2], and Ulrich Jessen[1]

[1] Universität Kassel, Institut für Produktionstechnik und Logistik, Fachgebiet Produktionsorganisation und Fabrikplanung, Kurt-Wolters-Straße 3, 34125 Kassel, Germany
{s.wenzel,jessen}@uni-kassel.de
[2] Robert Bosch GmbH, Materialwirtschaft und Logistik, 70442 Stuttgart, Germany
Pinar.Boyaci@de.bosch.com

Abstract. By reason of increasing product complexity and diversity of variants and increasing system and network complexity as well as increasing demands on quality, flexibility, service time, and costs companies need efficient methods and procedure models for planning, redesigning and improving their production and logistics systems. The management of logistics networks like supplier, production, distribution, and transportation networks has become an important strategic task due to ongoing globalization of economy, too. Nowadays, simulation is a well-established model-based analysis method for a lot of applications in this domain. This paper discusses the trends and solutions in development and research for simulation technology in production and logistics and highlights some current logistics applications.

Keywords: Simulation, Production, Logistics, Manufacturing, Digital Factory.

1 Introduction

For more than twenty years the Discrete Event Simulation (DES) (cf. e.g. [1]-[3]) has been used as a professional method for analyzing dynamical interdependencies in production and logistics systems and for proving and improving design, control, and operation of material flow, human resources and information flow. The simulation term in production and logistics accords with the definition given by the professional committee "Modelling and Simulation" of the VDI[1] Society Production and Logistics. "Simulation is the representation of a system with its dynamic processes in an experimentable model to reach findings which are transferable to reality. In particular, the processes are developed over time. In the broader sense, simulation refers to the preparation, execution and evaluation of targeted experiments with a simulation model."[4], p. 2. Additionally, we would like to point out that for a wide range of production and logistics applications the restricting term *material flow simulation* is often misleadingly a synonym for DES in German-speaking regions in Europe.

[1] VDI is the association of German engineers.

W. Dangelmaier et al. (Eds.): IHNS 2010, LNBIP 46, pp. 73–84, 2010.

The applications of DES relate to the model-based analysis of all production and logistics processes within one company and between two or more companies and to the phases of the system life cycle like system acquisition, system design and planning, implementation, start of operation, ramp-up as well as the operation itself. The industrial application domains range from automotive, aircraft and shipbuilding industry, semiconductor industry, plant engineering and construction to supply chain management and healthcare logistics or call centres. They are supported by a great variety of saleable high-quality DES tools for production and logistics applications, which differ in the scope of application, the modelling concept, the software functionality and the performance. More details about applicability as well as costs and benefits of DES in production and logistics you can find inter alia in [4]-[8].

Although DES is a well-established model-based method for investigating production and logistics systems there are a lot of current activities to enhance the interoperability, the applicability and the usability of the simulation technology. In the next section, we discuss some current development and research activities with regard to the simulation technology and its applications (Section 2). Section 3 addresses the developments resulting from the trends in the application field of production and logistics. Finally, a summary concludes the paper.

2 Research and Development Trends

The current state of development trends in production and logistics described as follows bases on papers in national and international proceedings of simulation conferences (cf. e.g. [9], [10], [11], [12]) as well as papers which discuss trend statements like [13], [14] and [15]. Additionally, activities of the ASIM[2] Working Group "Simulation in Production and Logistics", current guidelines of the professional committees "Modelling and Simulation" and "Digital Factory" of the VDI Society Production and Logistics, industrial requirements (cf. e.g. [16]) and identified research fields of the Automotive Working Group "Simulation" [17] were taken into account. For example, the ASIM Working Group "Simulation in Production and Logistics" intends to recollect quality aspects in simulation project implementation and discusses quality aspects ([18], [19]) and verification and validation ([20], [21])] as essential for high-quality simulation projects and credible simulation results. The cooperation of simulation users in the automotive industry would like to jointly adapt and standardize the DES methods, tools and interfaces for their applications. The activities include the standardization of simulation element libraries, the implementation of standard delivery specifications for simulation studies, the improvement of verification and validation methods for simulation models and the management of simulation data and models (cf. [17], pp. 425-426). These activities correspond to a lot of development activities using digital planning methods within the Digital Factory (cf. e.g. [22], [23]).

Discussing all current research and development activities will not be possible in detail. Therefore they will be subsumed under the categories

[2] ASIM is the "Arbeitsgemeinschaft Simulation", which is the simulation society of the German-speaking regions in Europe.

- positioning DES within the Digital Factory
- implementing quality in simulation studies
- assisting in performing simulation tasks

These categories will be discussed in the following sections.

2.1 Positioning DES within the Digital Factory

Continuous planning and operation of complex structures and processes – in line with the fulfilment of more challenging requirements for detailed and profound planning – can only be achieved with a suitable digital tool infrastructure used by a collaboration network of decentralized teams participating in the planning and operation processes. Such an infrastructure for an interdisciplinary collaboration over several organizational units and even beyond company borders represents the Digital Factory which is defined as "a generic term for a comprehensive network of digital models, methods and tools – including simulation and 3D visualisation – integrated by a continuous data management system. Its aim is the holistic planning, evaluation and ongoing improvement of all main structures, processes and resources of the real factory in conjunction with the product." [22], p. 3. A continuous data management system is characterized by a transfer between all application programs without media discontinuity (cf. [23], chapter 5).

The current development trends in the context of the Digital Factory influence inter alia the research activities in DES, for example tool integration, (semi-)automatic model generation, collaborative modelling and simulation as well as simulation and visualization.

Tool Integration. In recent years development activities for simulation tools have been focussing on embedding of the simulation tools in the enterprise tool sets. Spieckermann [15] also discusses the tool integration as an important development field, but he rightly points out, that this is not really a research subject of simulation. In order to achieve interoperability of the tools and the models within the Digital Factory, the integration has to comprise a syntactic, a semantic and a pragmatic integration [24]. The syntactic integration deals with questions regarding data integration and data maintenance as well as the used exchange formats, communication protocols and architectural concepts. Methods to solve the semantic integration include modelling conventions, standardized domain specific libraries (e.g. like the Automotive library [17]), the (semi-) automatic model generation (see below), concepts for collaborative modelling and simulation (see below), as well as meta-modelling or ontology concepts. Last but not least the pragmatic integration solves tasks such as the execution of business processes and the detailed specification of roles as well as the definition of security services such as authorization and authentication.

(Semi-)Automatic Model Generation. Relating to the request for an efficient build-up of cost-effective models the (semi-)automatic model generation by utilization of CAD data and also working plan data is more often up for discussion. Automation allows not only the reduction of effort to build up models but also produces standardized models, which eliminate the individual facets of model building. Already in the 90s first results regarding the model generation are published (cf. [25][26]), which have not achieved practical relevance because of necessary rework, e.g. modeling of control rules, as well as the absence of change management between different data sources and of the simulation model. As dissertations within the scope of Digital Factory document, automatic model generation triggered by application seems to be requested again (cf. [27][28][29]).

Collaborative Modelling and Simulation. The development of so-called model management systems, which amongst others control the utilization of models and guarantee their consistency (cf. e.g. [30][31][32]) are the base for new approaches for collaborative modelling and simulation tasks (cf. e.g. [30][33]) as well as for implementation of methods and functions for realization and management of simulation studies [34]. The target is the precise connection of single steps of planning and operation with simulation to shorten planning and development periods, to improve the production process, to enhance quality of results during system planning and operation and – last but not least – to reduce the costs. Within this context, based on sophisticated data and model management, web-based tools allow the synchronous (common model utilization) as well as the asynchronous (reference model and component catalogues) project related collaboration.

Visualization and Virtual Reality. Application fields of visualization within the scope of simulation in production and logistics can be found in all phases of model building and simulation. Beside data preparation the main emphasis lies on the one hand on validation, experiment realization as well as analysis and interpretation of results with the objective of gaining insight and on the other hand on knowledge transfer, in particular communication within interdisciplinary teams, presentation and training. Today, extensive 3-D-visualizations already allow a demonstrative presentation of design, structure and dynamics of a system, virtual reality even of a quasi-walkable plant. Despite high expectance of different target groups, from operation to management, regarding the utilization of visualization, amongst others influenced by developments of consumer electronics, 3-D-visualization is not the suitable method for each case of simulation result presentation. In VDI 3633, part 11 [35] and extended for input data e.g. in [36] and [37] a comprehensive classification of visualization methods as well as the information to be presented are developed.

2.2 Implementing Quality in Simulation Studies

Unfortunately, the matter of course in using simulation methods leads to underestimate the time and manpower requirements for a simulation study. Neither statistical verification of the simulation results needed for a high-quality planning nor the relevance of the simulation results for the planning task are considered sufficiently. To eliminate quality deficiencies and to consider quality criteria in modelling and simulation the necessity for using standardized procedure models is identified.

According to [18] and [19] five basic quality criteria have to be fulfilled: Accurate project preparation, consistent documentation, integrated verification und validation (V&V), continuous participation of the client and systematic project implementation. Based on a procedure model for simulation including V&V published in [21] checklists (depicted as circles in Fig. 1) are defined which support the work of the project team in all phases of a simulation project and are a fundamental part of the quality improvement philosophy discussed in [19]. In contrast to other publications on procedure models for simulation projects, the Project Definition as a pre-project phase and the Re-Use as a post-project phase are explicitly referred. Starting from the Sponsor Needs (like initial situation, scope of the project, and constraints) the phases of the modelling process are defined as Task Definition, System Analysis, Model Formalisation, Model Implementation as well as Experiments and Analysis (depicted by ellipses in Fig. 1). The phases Data Collection and Data Preparation are intentionally pointed out in a second path, as they can be independently processed from model building in content, time and organisation, but iterative matching processes between the results of the phases are necessary.

To each phase a phase result (models and/or documents) is assigned (rectangles in Fig. 1). The document Sponsor Needs is the basis for starting the simulation study.

The V&V of data and models have to be implemented during all phases of model building (bar on the right in Fig. 1). Therefore, the procedure model does not contain a special phase "V&V". But, V&V – both of the data and the models – is an essential part of the whole simulation study.

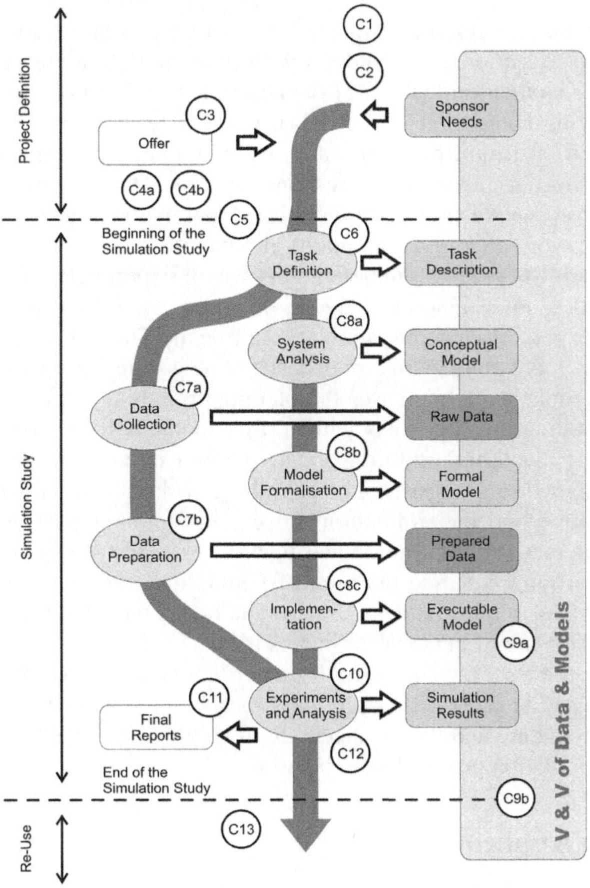

Fig. 1. Extended procedure model [19]

An approach for a consistent documentation and an integrated verification and validation within simulation studies in production and logistics is discussed in particular in [20] and [21]. A detailed procedure model for the information acquisition is developed in [38]. Input data management and validation in simulation studies are presented in [39] and [40].

2.3 Assisting in Performing Simulation Tasks

Although there are procedure models and instructions to support simulation studies, uncertainties about possible imponderables and uncalculated additional costs often deter companies from using simulation technologies. For this reason the applicability of simulation technologies still has to deal with a lot of acceptance problems in many companies, in particular in small and medium-sized enterprises (SMEs). Therefore, a company-wide, process-integrated and web-based simulation services und business processes for SMEs will be developed in a research project [41] funded by the Federal Ministry of Economics and Technology. In that project SMEs will be explicitly supported by special showcases (show me), learning units (coach me), and assisting functions (let me).

Additionally, current research projects intend to assist in collaborative modelling and simulation (see section 2.1) and deals with scenario definitions and the navigation for collaborative simulation tasks [33] or the implementation of organizational functions for managing simulation studies [34]. Current research projects (cf. e.g. [15], [42]) are worked on acquiring, preparing and bundling expert knowledge in checklists. The basic idea is to select corresponding aims of investigation out of a predefined list. These checklists include meta-information about the data required for the aims of investigation, data given an additional benefit or non-required data. Furthermore, these checklists give recommendations which experiments should be made for the selected aims of investigation. They are based on expert knowledge acquired from the simulation experts in different simulation projects. The utilization of these checklists results in a reduction of time and cost for acquiring information and planning simulation experiments and increases the planning reliability. A current research project aims at bundling the checklists with further methods for statistical experimental design and experimentation and consolidating them into a software tool [43]. This tool will be designed to give assistance beginning with the experimental design up to the experimentation and the evaluation process in a simulation study. Additionally, based on this a method for the standardization of effort analysis was developed, which, starting from a concrete problem of simulation, chooses a suitable visualization method for possible investigation targets and calculates the corresponding effort, e.g. hardware, software or personal resources [44].

Substantially the research activities discussed above show that the current trend as well as the current demand is characterized by assisting simulation non-experts and experts in the procedure and the management of simulation projects and not by changing the way a simulation project has to be done.

3 Trends in Applications

In addition to the traditional applications of simulation in production and logistics (cf. e.g. [4], [7], [8]) new paradigms for logistics systems have to be simulated on the one hand and new application fields for DES in production and logistics have to be made accessible on the other hand. The following section will discuss two representative topics in logistics applications.

3.1 New Paradigm: Autonomous Control in Production and Logistics

The motivation for implementing decentralized control strategies in logistics systems is generated by the fact that in complex and dynamic systems the finding of an optimal solution is hardly possible. The decision problem is too complex and needs therefore long computation times. Furthermore the relevant information is often not available at the right time. Before getting the right decision the restrictions change and therefore the decision is not optimal anymore. To obtain a stable systems behaviour, a suboptimal solution is accepted in centralized systems control. In complex systems this results in lower efficiency and in some cases even in unexpected systems dynamics which affects the stability of the system. A possibility to overcome such lacks of stability and robustness and improve therefore the efficiency of a logistics system is the introduction of autonomous control in logistic processes, which is supposed to cause an increase in the flexibility and adaption ability of the processes. Autonomous control in logistics systems means the dislocation of single decision functions from a central planning and controlling entity to single logistical objects. Logistical objects are physical objects like machines or parts as well as information objects like production orders. Decentralization promises an improvement of the reactivity, concerning changing boundary conditions and results therefore in an improvement of the system. This leads to a better logistical target deployment and in order of that to an improved efficiency of the whole logistics system [45]. In cause of a decentralization of the planning and control the whole system can be designed more flexible to be able to react faster on rescheduling.

Environmental impacts of transport processes gain recently a lot of attention which can be considered by using autonomous control strategies [46]. Due to the communication level, transport logistics are modelled by using agent technologies like multi-agent systems [47].

To keep production profitable inventory costs have to be kept low. Therefore inventory oscillations caused by delay of material and information have to be avoided. For making these oscillations predictable, System Dynamics models are realized to implement autonomous control in shop floor logistics. This method enables modelling of dynamic and uncertain systems due to its inclusion of feedbacks, nonlinearities and shifting loop dominance [48].

For modelling autonomous control the abstraction level has to be defined. DES models allow a very good description of the local behaviour of a system whereas continuous simulation models describe the global state of a system. To obtain the best fit of autonomous control strategies analyzing the interdependencies of local and global behaviour is necessary. Therefore the dual use of both methods is recommended [49].

3.2 New Application Fields: Biomass Conversion Networks

Beside new paradigms also new application fields constantly appear on which DES can prove its efficiency. An example is the simulation of biomass conversion paths representing the complete process of gas production from biomass. Biogas is the result of anaerobic digestion of organic raw material (substrate). Primarily, it is utilized to run combined heat and power units. The intensive utilization of renewable primary products as substrate for biogas production results in increasing logistics requirements. Beside tasks within context of cultivation of energy crops the tasks of logistics comprise the prompt transportation of harvested plant parts and their quick and effective storage and preservation as well as leading residual materials to waste treatment or back to agricultural production circle. In such a way arising networks become increasingly more complex. Thus, holistic simulation can make a valuable contribution to their planning and design.

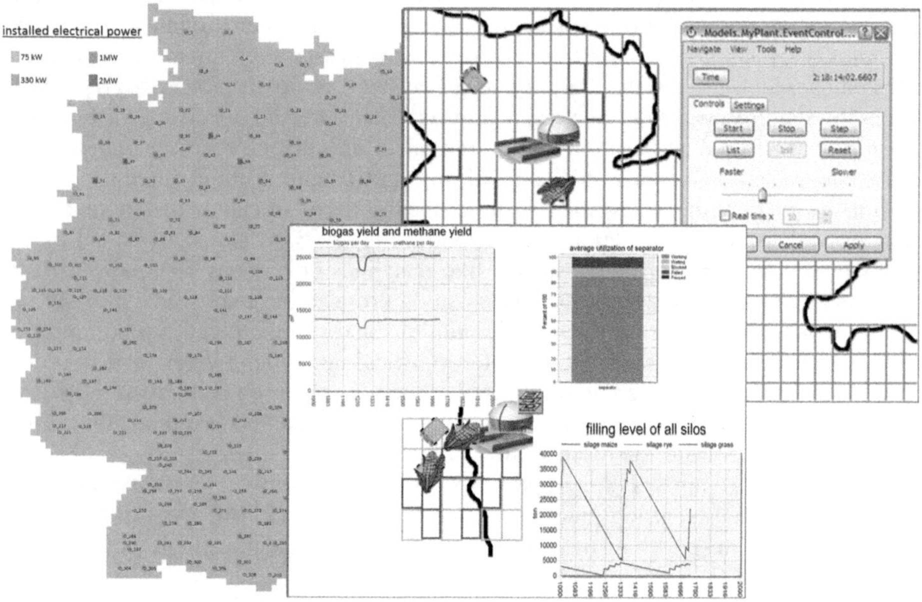

Fig. 2. Simulation model of biomass conversion in Germany

First models are available, which focus on simulative investigation of networks from energy crops cultivation respectively providing substrate including fertilizer supply to biogas conversion to alternative forms of energy. The interesting approach of the University of Kassel using the simulation tool Plant Simulation enables researchers not only to model single biogas plants and their surroundings, but also, based on a raster map, the partly automated modeling of any German region (see Fig. 2). The region model includes areas reserved for cultivation of energy crops, installed biogas plants and, where applicable, waste treatment plants as well as a transportation network [50]. Beside logistics processes also continuous real-world processes like biogas production are considered as a discretized process within the simulation model. External data, in particular data concerning the current land use, are imported by asynchronous interfaces and considered within the model as parameter. Vice versa, the model includes measuring points which record data during simulation to make them available to processes of external tools.

The solution confirms the current trends in research and development as tool integration and (semi-)automatic model generation. The simulation model provides a comprehensive experiment environment. The completed experiments with the simulation model could show among others the effect of crop production variation to supply of biogas plant and, thus, to biogas production. Moreover, the utilization of transport vehicles and the impact of distribution structure variation as well as process dependencies, e.g. between crop production and subsequent transports, could be investigated. Further research tasks exceed the detailed model of biomass conversion and put their focus on planning an overall biomass supply chain.

4 Summary

This paper shows the state of the art in simulation in production and logistics and highlights some uprising trends. Basic definitions along with simulation in production and logistics are introduced, and DES is positioned within the Digital Factory. Models for addressing the quality of simulation studies are presented as well as tools for assisting the user of simulation methods within real-life applications. Trends in the current research and development like implementing autonomous control in production and logistics systems as well as modelling biomass conversion networks stand for ongoing activities concerning simulation in production and logistics.

References

1. Law, A.M.: Simulation Modeling and Analysis. McGraw-Hill, Boston (2007)
2. Robinson, S.: Simulation – The Practice of Model Development and Use. John Wiley & Sons, Chichester (2004)
3. Pidd, M.: Computer Simulation in Management Science. John Wiley & Sons, Chichester (2004)
4. VDI 3633 Part 1: Simulation von Logistik-, Materialfluss- und Produktionssystemen, Grundlagen. VDI-Handbuch Materialfluss und Fördertechnik, Bd. 8, Beuth, Berlin (2010)
5. Liebl, F.: Simulation. Problemorientierte Einführung. Oldenbourg, München (1995)
6. Kosturiak, J., Gregor, M.: Simulation von Produktionssystemen. Springer, Wien (1995)
7. Rabe, M., Hellingrath, B. (eds.): Handlungsanleitung Simulation in Produktion und Logistik. SCS International, San Diego (2001)
8. Kuhn, A., Wenzel, S.: Simulation logistischer Systeme. In: Arnold, D., Isermann, H., Kuhn, A., Tempelmeier, H., Furmans, K. (eds.) Handbuch Logistik, pp. 73–94. VDI Springer, Heidelberg (2008)
9. Wenzel, S. (ed.): Simulation in Produktion und Logistik, Tagungsband zur 12. ASIM-Fachtagung. SCS Publishing House, San Diego (2006)
10. Rabe, M. (ed.): Advances in Simulation for Production and Logistics Applications – Tagungsband zur 13. ASIM-Fachtagung. Fraunhofer IRB Verlag, Stuttgart (2008)
11. Mason, S.J., Hill, R., Mönch, L., Rose, O., Jefferson, T., Fowler, J.W. (eds.): Proceedings of the 2008 Winter Simulation Conference. IEEE Omnipress (2008)

12. Rossetti, M.D., Hill, R.R., Johansson, B., Dunkin, A., Ingalls, R.G. (eds.): Proceedings of the 2009 Winter Simulation Conference. IEEE Omnipress (2009)
13. Barton, R., Fishwick, P., Henriksen, J., Sargent, R., Twomey, J.: Panel: Simulation – Past, Present and Future. In: Chick, S., Sánchez, P.J., Ferrin, D., Morrice, D.J. (eds.) Proceedings of the 2003 Winter Simulation Conference, pp. 2044–2050. SCS Publishing House, San Diego (2003)
14. Spieckermann, S.: Diskrete, ereignisorientierte Simulation in Produktion und Logistik – Herausforderungen und Trends. In: Schulze, T., Horton, G., Preim, B., Schlechtweg, S. (eds.) Simulation und Visualisierung 2005, Proceedings, pp. 3–14. SCS Publishing House e. V. Erlangen (2005)
15. Wenzel, S., Bockel, B., Deist, F.: Die Integration der Produktions- und Logistiksimulation in die Digitale Fabrik – Herausforderungen und Entwicklungstrends. In: Schenk, M. (ed.) Digital Engineering – Herausforderungen für die Arbeits- und Betriebsorganisation. Schriftenreihe der Hochschulgruppe für Arbeits- und Betriebsorganisation e.V. (HAB), pp. 318–339. GITO-Verlag, Berlin (2009)
16. Bayer, J., Collisi, T., Wenzel, S. (eds.): Simulation in der Automobilproduktion. Springer, Heidelberg (2003)
17. Heinrich, S., Mayer, G.: Ablaufsimulation im VDA – Ein Bericht aus der Arbeitsgruppe. In: Wenzel, S. (ed.) Simulation in Produktion und Logistik, Tagungsband zur 12. ASIM-Fachtagung, pp. 423–428. SCS Publishing House, San Diego (2006)
18. Wenzel, S., Weiß, M., Collisi-Böhmer, S., Pitsch, H., Rose, O.: Qualitätskriterien für die Simulation in Produktion und Logistik – Planung und Durchführung von Simulationsstudien. Springer, Berlin (2008)
19. Pitsch, H., Rose, O., Wenzel, S.: Quality Aspects in Simulation Studies for Production and Logistics. Appears in: SNE Simulation News Europe Special Issue Quality Aspects in Modelling and Simulation (2010)
20. Rabe, M., Spieckermann, S., Wenzel, S.: Verifikation und Validierung für die Simulation in Produktion und Logistik – Vorgehensmodelle und Techniken. Springer, Berlin (2008)
21. Rabe, M., Spieckermann, S., Wenzel, S.: Verification and Validation Activities within a New Procedure Model for V&V in Production and Logistics Simulation. In: Rossetti, M.D., Hill, R.R., Johansson, B., Dunkin, A., Ingalls, R.G. (eds.) Proceedings of the 2009 Winter Simulation Conference, December 2009, pp. 2509–2519. IEEE Omnipress (2009)
22. VDI 4499, Part 1: Die Digitale Fabrik, Grundlagen, VDI-Handbuch Materialfluss und Fördertechnik, Bd. 8, Weißdruck, Beuth, Berlin (2008)
23. Bracht, U., Geckler, D., Wenzel, S.: Digitale Fabrik – Methoden und Praxisbeispiele. Springer, Berlin (2010)
24. Wenzel, S., Jessen, U., Bernhard, J.: Classifications and Conventions Structure the Handling of Models within the Digital Factory. Computers in Industry 4 56, 334–346 (2005)
25. Splanemann, R.: Teilautomatische Generierung von Simulationsmodellen aus systemneutral definierten Unternehmensdaten. Bremer Schriften zur Betriebstechnik und Arbeitswissenschaft Band 5, Dissertation, Universität Bremen (1995)
26. Lorenz, P., Schulze, T.: Layout Based Model Generation. In: Alexopoulos, C., Kang, K. (eds.) Proceedings of the 1995 Winter Simulation Conference, pp. 728–735. ACM Press, New York (1990)
27. Jensen, S.: Eine Methodik zur teilautomatisierten Generierung von Simulationsmodellen aus Produktionsdatensystemen am Beispiel einer Job Shop Fertigung: Dissertation, Universität Kassel (2007)

28. Horn, S.: Simulationsgestützte Optimierung von Fertigungsabläufen in der Produktion elektronischer Halbleiterspeicher. Dissertation, TU Dresden (2008)
29. Rooks, T.: Rechnergestützte Simulationsmodellgenerierung zur dynamischen Absicherung der Montagelogistikplanung bei der Fahrzeugneutypplanung im Rahmen der Digitalen Fabrik. Appears as: Dissertation, TU Clausthal-Zellerfeld (2010)
30. Bernhard, J.: Ein Beitrag zur Interoperabilität in der kollaborativen Logistikplanung unter Verwendung der Simulation, Dissertation, TU Dortmund (2009)
31. Bley, H., Franke, C., et al.: Simulationsmanagement in der Produktion. ZWF 10 95, 487–491 (2000)
32. Bernhard, J., Jessen, U., Wenzel, S.: Management domänenspezifischer Modelle in der Digitalen Fabrik. In: Mertins, K., Rabe, M. (eds.) Experiences from the Future, Tagungsband zur 11. ASIM-Fachtagung Simulation in Produktion und Logistik, pp. 289–298. IPK Berlin, Eigenverlag (2004)
33. Bijlsma, R., De Gast, V., Valentin, E.: Deploying Decision Support Systems Based on Simulation Models. In: Rabe, M. (ed.) Advances in Simulation for Production and Logistics Applications – Tagungsband zur 13. ASIM-Fachtagung, pp. 531–544. Fraunhofer IRB Verlag, Stuttgart (2008)
34. März, L., Saler, M.: Ein Analyse-, Planungs- und Entscheidungsinstrument für Lagerlogistikanwendungen. In: Rabe, M. (ed.) Advances in Simulation for Production and Logistics Applications – Tagungsband zur 13. ASIM-Fachtagung, pp. 237–245. Fraunhofer IRB Verlag, Stuttgart (2008)
35. VDI 3633 Part 11: Simulation und Visualisierung. VDI-Handbuch Materialfluss und Fördertechnik. Bd. 8, Weißdruck, Beuth, Berlin (2009)
36. Wenzel, S., Bernhard, J., Jessen, U.: A Taxonomy of Visualization Techniques for Simulation in Production and Logistics. In: Chick, S., Sánchez, P.J., Ferrin, D., Morrice, D.J. (eds.) Proceedings of the 2003 Winter Simulation Conference, pp. 729–736. IEEE Omnipress (2003)
37. Jodin, D., Kuhnt, S., Wenzel, S.: Methodennutzungsmodell zur Informationsgewinnung in großen Netzen der Logistik. In: Buchholz, P., Clausen, U. (eds.) Große Netze der Logistik – Die Ergebnisse des Sonderforschungsbereichs, pp. 1–18. Springer, Berlin (2009)
38. Kuhnt, S., Wenzel, S.: Information Acquisition for Modelling and Simulation of Logistics Networks. Journal of Simulation, Published online 6th November (2009)
39. Skoogh, A., Johansson, B.: A Methodology for input data management in discrete event simulation projects. In: Mason, S.J., Hill, R., Mönch, L., Rose, O., Jefferson, T., Fowler, J.W. (eds.) Proceedings of the 2008 Winter Simulation Conference, pp. 1727–1735. IEEE Omnipress (2008)
40. Skoogh, A., Johansson, B.: Mapping of Time-Consumption During Input Data Management Activities. Appears in: SNE Simulation News Europe Special Issue Quality Aspects in Modelling and Simulation (2010)
41. Research Project simKMU, `http://www.simkmu.de`
42. Csanady, K., Bockel, B., Wenzel, S.: Methodik zur systematischen Informationsgewinnung für Simulationsstudien. In: Rabe, M. (ed.) Advances in Simulation for Production and Logistics Applications – Tagungsband zur 13. ASIM-Fachtagung, pp. 595–604. Fraunhofer IRB Verlag, Stuttgart (2008)
43. Research Project AssistSim (HA-Projekt-Nr. 185/09-15), `http://assistsim.cs.uni-frankfurt.de/`
44. Bockel, B.: Analyse von 3D-Visualisierungsqualität für die Simulation in Produktion und Logistik – Entwicklung und Anwendung von Bewertungskriterien anhand technischer Beispiele. Appears as: Dissertation, Universität Kassel (2010)

45. Scholz-Reiter, B., de Beer, C., Freitag, M., Hamann, T., Rekersbrink, H., Tervo, J.P.: Dynamik logistischer Systeme. In: Nyhuis, P. (ed.) Beiträge zu einer Theorie der Logistik, pp. 109–138. Springer, Berlin (2008)
46. Rekersbrink, H., Ludwig, B., Scholz-Reiter, B.: Entscheidungen selbststeuernder logistischer Objekte. Industrie Management 4 23, 25–30 (2007)
47. Becker, M., Sayyed, G., Wenning, B.-L., Görg, C.: On Mobile Agents for Autonomous Logistics: An Analysis of Mobile Agents considering the Fan Out and sundry Strategies. International Journal of Services Operations and Informatics 1, 114–130 (2007)
48. Scholz-Reiter, B., Delhoum, S., Zschintzsch, M., Jagalski, T., Freitag, M.: Inventory Control in Shop Floors, Production Networks and Supply Chains using System Dynamics. In: Wenzel, S. (ed.) Simulation in Produktion und Logistik, Tagungsband zur 12. ASIM-Fachtagung, pp. 273–282. SCS Publishing House, San Diego (2006)
49. Scholz-Reiter, B., Freitag, M., de Beer, C., Jagalski, T.: Modelling dynamics of autonomous logistic processes: Discrete versus continuous approaches. Annals of the CIRP 55, 413–416 (2005)
50. Jessen, U., Thees, M., Wenzel, S.: Simulation von Logistikstrukturen im Umfeld der Biogasproduktion. In: Gnauck, A., Luther, B. (eds.) Tagungsband zum 20. Symposium Simulationstechnik, Cottbus. Shaker, Aachen (2009) CD-ROM

Logistics in the Context of Humanitarian Operations

Alexander Blecken

Heinz Nixdorf Institute
Fuerstenallee 11, 33102 Paderborn, Germany
blecken@hni.upb.de

Abstract. In the past decade, logistics in the context of humanitarian operations has more and more moved into the focus of logistics practitioners and academics. While humanitarian organisations have realised that a major part of disaster relief spending can be attributed to logistics and supply chain management activities associated with humanitarian operations, researchers have found that the area of logistics in the context of humanitarian operations gives rise to a number of unique and interesting challenges. This paper is meant as an introduction to the track Humanitarian Logistics at the 8th International Heinz Nixdorf Symposium. A recent natural disaster helps to elaborate the domain humanitarian operations. A number of important definitions are presented. Moreover, some of the specific characteristics and challenges of logistics in the context of humanitarian operations are described.

Keywords: Disaster Relief Logistics, Humanitarian Logistics, Supply Chain Management, Humanitarian Operations.

1 Introduction

A powerful earthquake of 7.2 magnitude on the Richter Scale affected Haiti on 12 January, at 16.53hrs local time. The earthquake happened 17km south-west of Port-au-Prince, the capital of Haiti. The initial reports suggested a high number of casualties and widespread damage. The capital of Haiti, Port-au-Prince, had been severely affected including critical city infrastructure components such as, electricity, water and phone services. Electricity was not available and communications were difficult. Immediately, emergency teams from humanitarian organisations, such as United Nations Disaster and Assessment Coordination Team had been mobilized. [1] By January 14, 26 international search-and-rescue teams had arrived on the ground and had been deployed to priority sites. At the peak of the search-and-rescue effort, there were 52 teams on the ground with 1,820 rescue workers and 175 dogs. [2] Within ten days of the incident, these teams had rescued over 130 people. [3]

Although the precise humanitarian needs were not known at first, urgent priorities included search and rescue, medical services and supplies, clean water and sanitation, emergency shelter, food, logistics and telecommunications. In

W. Dangelmaier et al. (Eds.): IHNS 2010, LNBIP 46, pp. 85–93, 2010.

particular, medical assistance was the priority due to a high number of people suffering traumatic injuries. [4] The United Nations estimated that 200,000 families (up to one million people) were in need of immediate shelter and non-food assistance. [5]

By January 15, flights carrying humanitarian aid were arriving in Port-au-Prince with medical supplies, medical teams, food and non-food items. A total of 180 tons of relief supplies had arrived in-country so far. [6] One of the most pressing problems in the humanitarian response of the first days was a shortage of fuel due to the destruction of Haiti's transport infrastructure. By January 16, fuel restrictions had been put in place. [7] Better estimations of the humanitarian needs only became known after some time: According to the Shelter Cluster, the number of people living in temporary shelter sites in Port-au-Prince was estimated to be as high as 800,000, on January 24. Water continued to be distributed daily at 115 sites in Port-au-Prince reaching an estimated 235,000 people. [8] Yet, hundreds of thousands of people remained in need of food and shelter. [9]

More than two weeks after the earthquakes, the numbers illustrate the massive destruction and human suffering caused in Haiti: Of the 3,700,000 people living in the areas affected by the earthquakes, some 112,000 were killed, almost 200,000 had been injured and more than 480,000 had been left displaced. 40,885 patients had been treated by 274 humanitarian organisations on the ground. 2,000,000 people were estimated to be in need of food, 1,100,000 people in need of shelter and 500,000 in need of water. [10]

2 Disasters and Humanitarian Crises

The world today has to face a number of severe problems: Some three billion people live on less than two dollars a day. Moreover, the past decade has seen an increasing number of medium to high impact disasters to which the poor are most vulnerable. The humanitarian crises caused through these disasters are frequently addressed by the international community with concerted efforts and humanitarian operations. Humanitarian operations range from short-term humanitarian relief in response to acute emergencies to medium and sometimes long-term assistance focusing on recovery and reconstruction in post-emergency contexts. Due to the substantial mobilisation and deployment of material and financial resources involved, these kinds of operations rely to a large extent on effective and efficient supply chain management.

Humanitarian operations seek to alleviate suffering and save lives of victims of natural or man-made and human-induced crises. Humanitarian organisations, which carry out these activities, have existed at least since the late 19th century, when the International Red Cross was founded on a battlefield as a reaction to the inhumane suffering and lack of provision given to the wounded. In carrying out their operations, humanitarian organisations adhere to certain principles, such as humanity, neutrality and impartiality, i.e. they deliver their services on a basis of need without any discrimination regarding religion, ethnicity or gender. Since the first humanitarian organisations were founded, the multitude and magnitude of humanitarian operations has risen dramatically. Nowadays, these operations include services and goods to provide food, water, shelter and medical care, to protect victims of natural and man-made disasters as well as bear witness to their plight, to name just a few examples of the broad range of different operational foci.

Between the 1970s and 1990s, the number of disasters has tripled. [11] [12] It has been predicted that the number of both natural and man-made disasters will increase five-fold [13] and that world-wide costs due to these events will sum up to \$64tr in the next 50 years. For Germany alone, the cost will be around \$800bn. [14] In 2008 between 150-220 million people were affected by disasters. These disasters claimed more than 240,000 fatalities and between \$190-270bn financial losses. [12] More than 90% of the victims of natural and man-made disasters live in developing countries. This shows that "poverty, population pressures and environmental degradation exacerbate suffering and destruction." [15] Figure 1 presents some key indicators of the effect of disasters during the past 10 years. [16]

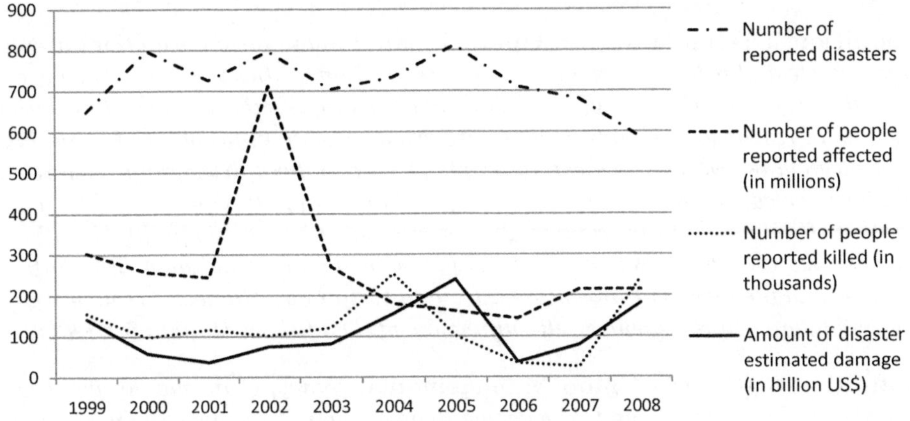

Fig. 1. Disaster indicators

The before-mentioned figures refer to disasters with a natural trigger only, and do not include wars, conflict-related famines, diseases or epidemics. However, humanitarian operations also take place during or in the aftermath of armed conflicts such as civil wars and wars between states. While wars between states have remained at a relatively low level since World War II, intra-state conflicts have risen considerably in the last decades. [17] During the past decade, the number of high intensity conflicts has risen only slightly, whereas medium intensity conflicts have doubled. Since the end of World War II, a total number of 228 armed conflicts have been recorded, of which the majority took place after the Cold War. A total number of 20 million people have been killed and 50 million injured in 160 major armed conflicts. 100 million people were forced to flee. [18]

3 Definitions

The previous two sections have illustrated the application context of humanitarian logistics with the example of a current natural disaster and with a number of figures which show that the triggers for humanitarian operations will further rise in the future. In this and the following section, the tasks and challenges of logistics in the context of humanitarian operations (or humanitarian logistics) will be elaborated. Some definitions are provided below that serve to provide a foundation for and further the understanding of the research area humanitarian logistics.

Definition 1 (Disaster). *A disaster is a serious disruption of the functioning of a community or a society involving widespread human, material, economic or environmental losses and impacts, which exceeds the ability of the affected community or society to cope using its own resources. [19]*

Disasters can be natural or man-made, as well as complex emergencies combining man-made and natural disasters.

Definition 2 (Humanitarian Operations). *Humanitarian operations aim to preserve life and reduce suffering of members of communities in crises. They comprise the provision of material and technical aid as well as the delivery of essential services in response to situations of crises when a communitys ability to cope has been severely impeded. It is given to people in need without distinction as to race, ethnicity, creed, nationality, sex, age, physical or mental disability or political affiliation. Humanitarian operations are not motivated by making profits; however, they are based on basic human rights as formulated in International Humanitarian Law and the Geneva Conventions. Humanitarian operations are temporary in nature and aim to re-establish self-sufficiency of the affected community. [20]*

Definition 3 (Supply Chain Management). *Supply chain management encompasses the planning and management of all activities involved in sourcing and procurement, conversion, and all logistics management activities. Importantly, it also includes coordination and collaboration with partners in the supply chain, which can be suppliers, intermediaries, third party service providers, and customers. [21]*

Definition 4 (Humanitarian Logistics). *Humanitarian logistics is the process of planning, implementing and controlling the efficient, cost-effective flow and storage of goods, materials and equipment as well as related information, from point of origin to point of consumption for the purpose of meeting the end beneficiaries' requirements. [22]*

Interesting to note is the fact that many humanitarian organisations use 'logistics' as an umbrella term encompassing a broad variety of tasks. According to a large humanitarian organisation, the following non-comprehensive list details specific tasks covered by the logisticians in the operations: [23]

- Management and maintenance of hospitals and feeding centres in cooperation with the medical team
- Administration and organisation of the supply chain of required medical goods and equipment and other material (procurement, customs, transport, and stock management)
- Installation and maintenance of technical equipment
- Construction and maintenance of WatSan installations
- Maintenance of computers, communication systems, and IT equipment
- Responsibility for vehicle fleet and fleet management
- Recruitment, training, and supervision of national staff
- Data management
- Security reporting

Definition 5 (Humanitarian Supply Chain Management). *Supply chain management in the context of humanitarian operations encompasses the planning and management of all activities related to material, information and financial flows in disaster relief and humanitarian assistance. Importantly, it also includes coordination and collaboration with partners in the supply chain both horizontally and vertically, third party service providers, and across humanitarian organisations.*

4 Characteristics and Challenges

Anisya Thomas, executive director of the Fritz Institute, makes the case for academic contributions to supply chain management in the context of humanitarian operations, when she states: "Humanitarian logistics has much in common with corporate logistics, yet the best practices from the corporate world [...] have not crossed over. It is paradoxical that a sector, which has such extreme requirements in terms of timeliness, affordability and oversight, is so underdeveloped." [24]

Humanitarian operations with their ultimate stake of saving lives frequently take place in highly unstable and volatile environments, under great time pressure, working on poor infrastructure and under exhausting working conditions. Humanitarian supply chains, which deliver goods and services in response to sudden and slow-onset disasters as well as chronic emergencies and which strive to provide humanitarian assistance both rapidly and efficiently, are subject to specific challenges. These supply chains suffer from frequent breakdowns and interruptions in the material and information flow, which is in contradiction to the extremely short lead times required. In the following, some of the challenges of humanitarian supply chains, which set them apart from commercial supply chains, are presented. [20]

- **Uncertainty:** These supply chains need to operate in highly volatile environments as well as provide assistance in both the short and medium-term time horizons, i.e. in both emergency and post-emergency or more stable contexts. Uncertainty in humanitarian supply chains refers to uncertainty of demand, supply, personnel, and equipment as well as lead time, process instabilities, other actors in the supply chain, and financial resources. Demand patterns are highly irregular and the environment puts unique constraints on the operations. Especially in disasters relief operations, uncertainty also includes the number of people affected, the infrastructure that is still intact, and information uncertainty. Hence, uncertainty is thus inherent to any type of humanitarian operation whether conducted during the emergency or post-emergency phase. However, the areas in which uncertainty is present shift while the humanitarian operation develops from the emergency to the post-emergency phase.

- **Earmarking of funds:** Due to the nature of how humanitarian organisations are funded, investments in research, information systems, infrastructure and other long-term investments which cannot be directly related to any concrete operation, are severely restricted. Donors frequently earmark their funds, i.e. limit the flexibility how humanitarian organisations can use these funds. Hence, donors have influence as to where and how humanitarian operations take place. However, funding for process improvement, installation of state-of-the-art information and communication infrastructure, operations' support with a long-term focus, and organisational learning are often not in the focus of donors. There is consistent under-investment in these areas which have the potential to improve the efficiency and responsiveness of humanitarian supply chains.
- **Infrastructure:** With respect to transportation infrastructure as well as communication infrastructure, it can be observed that disasters tend to happen in areas where the local infrastructure is already in a poor state. This situation can be dramatically deteriorated during and after a disaster when possibly few remaining roads cannot take the number of refugees and disaster management vehicles that pour into and out of these areas. Poor communication infrastructure prevents efficient information management and increases uncertainty. Transportation remains vulnerable to disruptions and needs to take into account the possibility of using multiple modes.
- **Human Resources:** Humanitarian logistics is affected by the lack of professionally trained logistics personnel. Logistics personnel are in the midst of numerous requirements posed by local governments and officials, donors, the media, beneficiaries, and their own headquarters. Despite recent advancements and the introduction of specialised training and graduate university courses in humanitarian logistics and humanitarian supply chain management, professional logisticians remain rare. Retention of personnel is extremely difficult considering the aforementioned challenges and requirements to logisticians. Due to a variety of factors such as a constantly high work load, high stakes, limited privacy, often low remuneration, and security concerns, staff turnover rates are high.
- **Supply Chain life-cycle:** While traditional supply chains focus on the optimisation of the different flows within the network, the challenges in the humanitarian supply chain lie more in the area of establishing and maintaining a (possibly highly vulnerable) supply chain. Some humanitarian organisations may be issue-related and therefore exist only temporarily. Thus, for each operation, a new supply chain may need to be established.

Humanitarian operations frequently take place in emergency environments, thus, supply chain activities are highly non-routine in the sense that actors in the supply chain need to be redefined anew in each context. These supply chains can frequently be positioned in an early stage in the supply chain life-cycle.

- **Coordination:** There are other factors which challenge the coordination and cooperation of organisations involved in humanitarian operations. Some of these factors include:
  - Mandate: Differing mandates and interests of humanitarian organisations, operational objectives and therefore different agendas, target communities and goals. This also includes humanitarian values such as neutrality and impartiality.
  - Structure: The organisational structures of humanitarian organisations can be incompatible and even incomparable.
  - Information: Missing or wrong information concerning programmes and resources of other organisations such as stock levels and items, transport routes and schedules, already carried out assessments as well as missing reports on similar events, lack of institutional learning, missing and inadequate documentation, especially quantitative documentation.
  - Information and Communication Technology: This includes differing frequencies and licences, missing communication technology, outdated and unreliable IT equipment, incompatible software systems etc.
  - Competition: Governmental as well as non-governmental organisations perceive themselves to be competitors of scarce financial resources. This attitude makes cooperation difficult despite being in the best interest of the beneficiaries. On a supply chain level, this problem might not be as relevant since logistics usually performs support functions in humanitarian operations and is therefore not as donor-visible as direct operations.
- **Others:** Some other characteristics and challenges of supply chain management are briefly addressed here. While supply chain management in other industries addresses the areas of procurement, production, and distribution equally, humanitarian organisations focus mainly on either the procurement or the distribution side. Production can be involved in humanitarian operations but is certainly not of major concern. Demand in the humanitarian supply chain is not only the demand for goods and services, but also the demand for personnel. The supply chains are thus involved with mass movement of goods and people. Further challenges are posed to the humanitarian supply chain in terms of complex documentation requirements, corruption, theft, special packaging requirements, and extreme time pressure to carry out the activities.

5 Outlook

During the past decade humanitarian organisations have come to realise more and more that logistics in the context of humanitarian operations is not just a necessary expense but a key lever to achieve operational excellence. However, there is still a long way to go and large potentials for improvement remain untapped. In recent years, first executive and post-graduate courses in humanitarian logistics and supply chain management in the context of humanitarian operations have become available. The cluster approach by the United Nations has been a notable initiative to focus efforts on collaboration and coordination of humanitarian organisations in humanitarian crises. The amount of researchers dedicated to this subject has risen significantly, as has the number of scientific contributions in international conferences and respected journals. A dedicated scientific journal on humanitarian logistics and supply chain management is on the horizon. This paper intends to supplement the contributions to the track Humanitarian Logistics of the 8th International Heinz Nixdorf Symposium and spark new research in this area by delimiting the application domain and illustrate the challenges and tasks that humanitarian logisticians need to address.

References

1. United Nations Office for the Coordination of Humanitarian Affairs (OCHA), Haiti earthquake situation report no. 1, Internet Resource (January 2010), http://www.reliefweb.int/rw/rwb.nsf/db900SID/MUMA-7ZN76X?OpenDocument
2. United Nations Office for the Coordination of Humanitarian Affairs (OCHA), Haiti earthquake situation report no. 8, Internet Resource (January 2010), http://www.reliefweb.int/rw/rwb.nsf/db900SID/SNAA-7ZV5EL?OpenDocument
3. United Nations Office for the Coordination of Humanitarian Affairs (OCHA), Haiti earthquake situation report no. 11, Internet Resource (January 2010), http://www.reliefweb.int/rw/rwb.nsf/db900SID/MYAI-7ZY7VE?OpenDocument
4. United Nations Office for the Coordination of Humanitarian Affairs (OCHA), Haiti earthquake situation report no. 2, Internet Resource (January 2010), http://www.reliefweb.int/rw/rwb.nsf/db900SID/MUMA-7ZP4NF?OpenDocument
5. United Nations Office for the Coordination of Humanitarian Affairs (OCHA), Haiti earthquake situation report no. 5, Internet Resource (January 2010), http://www.reliefweb.int/rw/rwb.nsf/db900SID/MYAI-7ZS2CZ?OpenDocument
6. United Nations Office for the Coordination of Humanitarian Affairs (OCHA), Haiti earthquake situation report no. 4, Internet Resource (January 2010), http://www.reliefweb.int/rw/rwb.nsf/db900SID/MUMA-7ZR22N?OpenDocument
7. United Nations Office for the Coordination of Humanitarian Affairs (OCHA), Haiti earthquake situation report no. 6, Internet Resource (January 2010), http://www.reliefweb.int/rw/rwb.nsf/db900SID/MYAI-7ZSW5R?OpenDocument
8. United Nations Office for the Coordination of Humanitarian Affairs (OCHA), Haiti earthquake situation report no. 12, Internet Resource (January 2010), http://www.reliefweb.int/rw/rwb.nsf/db900SID/MYAI-8225AV?OpenDocument
9. United Nations Office for the Coordination of Humanitarian Affairs (OCHA), Haiti earthquake situation report no. 14, Internet Resource (January 2010), http://www.reliefweb.int/rw/rwb.nsf/db900SID/MUMA-8257LR?OpenDocument
10. United Nations Office for the Coordination of Humanitarian Affairs (OCHA), Haiti situation update 31 january 2010, Internet Resource (January 2010), http://oneresponse.info/disasters/haiti/

11. Burton, I.: Vulnerability and adaptive response in the context of climate and climate change. Climatic Change 36(1-2), 185–196 (1997)
12. Swiss Reinsurance Company: Sigma no. 2 / 2009: Natural catastrophes and man-made disasters in 2008. north america and asia suffer heavy losses, Swiss Reinsurance Company, Zurich, Tech. Rep. (2009), http://www.swissre.com/
13. Thomas, A.S., Kopczak, L.R.: From logistics to supply chain management: The path forward in the humanitarian sector. Fritz Institute, Tech. Rep. (2005)
14. Kemfert, C.: Fünfzig jahre nach morgen - wo werden wir sein? – die oekonomischen kosten des klimawandels. In: Mlynek, J. (ed.) Oeffentliche Vorlesungen, vol. 145, Humboldt-Universitaet zu Berlin, Wirtschaftswissenschaftliche Fakultaet (2005), http://edoc.hu-berlin.de/humboldt-vl/145/kemfert-claudia-3/PDF/kemfert.pdf
15. United Nations, Basic facts, Internet Resource (2008), http://www.un.org/aboutun/basicfacts/haction.htm
16. IFRC, World disasters report 2009. focus on early warning, early action, International Federation of Red Cross and Red Crescent Societies, Geneva, Switzerland, Tech. Rep. (2009)
17. HIIK, Conflict barometer 2007. crises - wars - coups d'etat. negotiations - mediations - peace settlements. 16th annual conflict analysis. Heidelberg Institute for International Conflict Research, Tech. Rep. (2008)
18. McGuire, G.A.: Development of a supply chain management framework for health care goods provided as humanitarian assistance in complex political emergencies, Dissertation, Wirtschaftsuniversitaet Wien (September 2006)
19. International Strategy for Disaster Reduction (ISDR), Terminology: Basic terms of disaster risk reduction, Internet Resource (February 2009), http://www.unisdr.org/eng/library/lib-terminology-eng
20. Blecken, A.: A reference task model for supply chain processes of humanitarian organisations. PhD Thesis, University of Paderborn (November 2009), http://ubdok.uni-paderborn.de/servlets/DocumentServlet?id=11585
21. Council on Supply Chain Management Professionals (CSCMP), Supply chain management and logistics management definitions,Internet Resource (August 2008), http://cscmp.org/aboutcscmp/definitions/definitions.asp
22. Thomas, A., Mizushima, M.: Logistics training: necessity or luxury? Forced Migration Review 22, 60–61 (2005), http://www.fmreview.org/mags1.htm
23. Medecins Sans Frontieres - Aerzte ohne Grenzen e.V., Logistiker august 2009.ppt (internal presentation) (August 2009)
24. Thomas, A.: Humanitarian logistics: Enabling disaster response, Fritz Institute, White Paper (2003)

Sustainable Process Management - Status Quo and Perspectives

Dennis Kundisch, Philipp Herrmann, and Christian Meier

University of Paderborn,
Chair of Business Information Systems - Information Management & E-Finance
{Dennis.Kundisch,Philipp.Herrmann,Christian.Meier}@wiwi.upb.de
http://www.upb.de/winfo2

Abstract. Sustainability in the context of information and communication systems is often discussed using the technical term "Green IT". However, Green IT-concepts often just address the infrastructure level of an information and communication system architecture. Acknowledging the potentially huge contribution to sustainability improvements that lies in the adequate design of processes, we discuss concepts to evaluate and compare different process designs with respect to sustainability considerations.

Keywords: Process management, process design, sustainability, economic dimension, ecological dimension, social dimension, Green IT.

1 Sustainable Processes?

Sustainability has become "en vogue" both in research as well as in business and is used inflationary – sometimes in a correct way, most of the times not. Hardly any new product can be found that is not marketed with the adjective "sustainable" and there is hardly any company that does not praise itself as being sustainable and producing sustainably.

Sustainability is often associated with the level of consumption of natural resources causing emissions of so-called "greenhouse gases" like carbon dioxide and equivalent gases (in the following: CO_2e gases). CO_2e gases are believed to cause global warming, which in turn – if too high – may have devastating consequences on the biospheres of human beings, animals and plants [1]. Although the UN Climate Conference in Copenhagen in December 2009 was not concluded with a globally binding treaty concerning the reduction of emissions, sustainability will still remain on the political agenda for the coming years.

In this context Information Technology (IT) is discussed from two perspectives: On the one hand, the usage of IT currently generates almost 2% of the total emissions of CO_2e gases – and this usage is believed to grow further in the coming years [2]. Thus, IT-induced emissions of CO_2e gases have become a relevant factor in the overall emissions production and sustainability considerations have to be applied to the employment and usage of IT. On the other hand, IT is often cited as one of the most promising factors contributing to savings

W. Dangelmaier et al. (Eds.): IHNS 2010, LNBIP 46, pp. 94–106, 2010.

in emissions of CO_2e gases. Still, research with respect to IT and sustainability is often just concerned with efficiency considerations on an infrastructure level so far. These activities are subsumed under the key word "Green IT". The potential for savings in emissions of CO_2e gases, however, seems much more promising on higher levels of an information and communication system architecture. Especially business processes as the glue between the business model of a company and its implementation using IT- and also non-IT resources appear as an attractive starting point for such considerations.

Research has done comparably little so far to define characteristics of a sustainable process. Based on a literature review and conceptual considerations, we discuss and analyze opportunities for a process design that improves sustainability. We find the following: (1) There are already some valuable instruments available that may help a company to improve its processes in terms of sustainability, however, the business case for the employment of these instruments is all but clear. (2) The usage of such instruments would be dispensable if prices reflected all costs associated with the usage of a specific resource. (3) Strong governments and global agreements seem to be the dominant, yet tough to implement strategy to internalize currently external costs and, thus, to provide economic incentives for companies to produce and act in a more sustainable way.

The remainder of this paper is organized as follows. In the next section sustainability is briefly defined. Afterwards, an introduction to "Green IT" as the umbrella term for sustainable activities in the area of IT is provided as a starting point for the discussion. Subsequently, sustainability is discussed on the process level, which is the main focus of this paper. We propose an idealized model to support the decision on alternative process designs and present selected instruments to facilitate this task. We conclude with a discussion and summary of our findings.

2 Sustainability - Definition and Description

Originally, sustainability relates to forestry. In a very broad und fuzzy definition, forestry is called sustainable, if just as much timber is cut down as can replenish to maintain the basis of life for future generations [3]. A definition that is not just related to forestry can be taken from the Brundtland report authored by the World Commission on Environment and Development published in 1987 [4]:

> *Sustainable development is development that meets the needs of the present without compromising the ability of future generations to meet their own needs.*

Based on this view, the adopted "Agenda 21" at the Rio Summit in 1992 broadly distinguishes three dimensions of sustainability and types of capital that shall be preserved [5],[3]:

- Ecological sustainability - natural capital
- Economic sustainability - economic capital
- Sociopolitical sustainability - social capital

These types of capital often exhibit characteristics of complements rather than substitutes. Even if substitution appears to be possible for a specific function, due to the multi-functionality of each type of capital, substitutability is often limited. Therefore, the central idea of sustainability evolves from the integrated view on these dimensions.

Corporate sustainability – doing business in a sustainable way – is concerned with balancing these dimensions in managerial decision making. Hence, it is called the *three-pillar-model* or the *triple-bottom-line-theory* in literature [5]. The requirement to have an integrated view on the three dimensions necessitates the availability of instruments and key figures to evaluate and, ideally, compare the performance within and between these dimensions. While the literature with respect to economic capital is rich and comprehensive, natural and social capital have been neglected so far.

The economic dimension is primarily concerned with economic efficiency, i.e. the ratio of monetarily evaluated inputs and outputs and the creation of shareholder value [6]. Due to this normalization on monetary terms, the comparison between alternatives is relatively easy. This is not true for the ecological and social dimensions. E.g. the social dimension is concerned – among others – with equal rights and opportunities, prohibition of child labor, job security, "good" management principles, salary structure and employee benefits, reduction of health and safety risks at the workplace, continuing education, fighting corruption, and cultural engagement [5].

Obviously, to achieve "social efficiency" – or at least an improvement in "social efficiency" – is much more fuzzy and complicated compared to the economic dimension, given all the different aspects listed above and taking into account the partially conflicting goals of different stakeholders. The same holds for the ecological dimension.

For our further argumentation it is important to be precise about the semantics, when we use the terms *economic*, *ecological*, and *social*. For each resource somehow consumed in the production process (input) and each output – which may be products, services or waste – we theoretically distinguish between an economic, ecological, and social part. The economic part comprises all effects that are mediated by market or market-like prices, whereas all effects that are not mediated by market prices are either ecological or social. Here, the distinction is made based on whether the input or output directly (social) or indirectly (ecological) affects a human being.

Two examples shall illustrate this special view: (1) Take the example of labor as input factor. If we talk about the economic dimension of this factor, we refer to all aspects of this input factor that are reflected by its market price. This includes the manpower, the social benefits that may be legally prescribed or provided in order to motivate the employee and measures that are taken to increase employee satisfaction in order to increase labor productivity. However, all measures that aim at employee satisfaction as an end in itself constitute the social part of labor. (2) The consumption of electricity as an input factor is another example. All effects of the pollution and potential consumption of natural resources while producing electricity that are reflected in the market price – e.g. due to emissions trading – are denoted economic, whereas any remaining pollution in this production process is denoted ecological. Obviously, over time, social and ecological parts of an input or output may become economic and vice versa – dependent e.g. on legislation, consumer buying behavior, strategy, and other influence factors.

Based on the definitions above we turn now to the concept of "Green IT" as the starting point for our considerations and discussion.

3 Green IT

In 2007, the European Union (EU) announced their plan to reduce the emissions of greenhouse gases by 2020 to approximately 70-80% of the emission level in 1990. Almost 2% of the total emissions of 45 billion tons of CO_2e gases are generated by IT. The use of IT has increased drastically over the last years and it will continue to grow in the future. In 2020, the world-wide use of IT is expected to be three times higher than today and the share of IT in the total CO_2e emissions will rise to approximately 3%, which equals the amount of CO_2e emissions of today's world-wide air traffic. Not even the expected improvements in energy efficiency of IT are able to compensate for this effect. Approximately 40,000 terawatt-hours are required to operate today's IT-infrastructure in the EU, which produces energy costs of about €6 billion. In addition to increasing energy costs, companies have to expect penalty payments raised by the EU, if the reduction of CO_2e emissions cannot be met.[2]

Because of these reasons, the demand for a more energy-efficient IT has arisen under the key word "Green IT" and attracts more and more practitioners and researchers. A survey [7] showed that the definition of "Green IT" in most companies is blurry, but it seems that an agreement can be achieved by understanding "Green IT" as a collective term for IT-products and -services which reduce energy consumption. In addition, "Green IT" contains all activities and solutions that support a more environmentally friendly production of IT-hardware and reduce the energy consumption of the IT-infrastructure employed, as for example, the use of more energy-efficient Thin-Clients. "Green IT" of today is focused on office environments and electronic data processing centers (EDPC). Server- and memory virtualization as well as the optimization of network structures are typical "Green IT"-activities in the field of EDPCs. The company IBM, for example, reorganized an EDPC radically and thereby reduced the number of servers required from 3,900 to 33. This decreased the energy consumption of this EDPC by 85% [2].

In office environments, "Green IT" focuses, for instance, on shared-desk concepts and the use of more energy-efficient laptops rather than on conventional desktop computers. Most of today's "Green IT"-activities and -solutions are related to just the infrastructure level of IT. A survey of McKinsey [8] indicates, that infrastructure-related "Green IT"-activities mentioned above can significantly reduce the resource consumption of an individual, but that the global CO_2e emissions can just be reduced by 0.5 billion tons maximum. According to

McKinsey, the highest potential for reducing CO_2e gases lies in the comprehensive use of IT in order to enhance the efficiency of business processes. McKinsey estimates that the use of IT to support sustainable process management can reduce the CO_2e gas emissions by 7.3 billion tons (in just four selected industries) which equals 15% of the expected emissions in 2020 [2].

A few activities – by some authors still subsumed under "Green-IT"– already try to bridge the gap between the infrastructure level and the level of business processes by e.g. substitution of business trips with videoconferencing or the use of electronic mail instead of paper-based mail to reduce the consumption of raw materials. By reducing the energy costs (economic capital) and thereby the amount of resources consumed, these activities implicitly make a contribution to the ecological aspect of sustainability. Usually the possible social impacts – positive or negative – are not considered at all.

In the following, we will discuss concepts to evaluate and compare different process designs with respect to sustainability considerations.

4 Sustainability on the Process Level

As pointed out in the last section, IT has a great potential to enhance the efficiency of business processes. We describe an idealized model of how the three dimensions of sustainability can be integrated into management decisions on this level. Due to restrictions which hinder the real-world application of our model on a global level, we present different instruments to – at least partly – measure and integrate these dimensions on the company level.

4.1 Idealized Model

Based on our definition of sustainability we present an idealized model of how sustainability can be implemented on a process level. Following [9] and [10] we define a process as the self containing sequence of temporal and factual connected activities that are necessary to handle an economically relevant object. A process can be seen as the transformation of predefined inputs through a production function into predefined outputs. In contrast to the position of the classical production theory and in analogy to our definition of sustainability we do not only consider the economic, but also social and ecological parts of process in- and outputs for our model (see figure 1 left hand side). Therefore, we have an economic, ecological, and social in- and output dimension for every process, where the measuring units of the ecological and social dimension do not have to be the same as the measuring unit of the economic dimension. For that reason, the improvement of a process in at least one dimension of sustainability could lead to two different situations:

1. It is possible to improve a process in at least one dimension without worsening one of the other dimensions (Pareto improvement).
2. It is possible to improve a process in at least one dimension while at least one other dimension deteriorates.

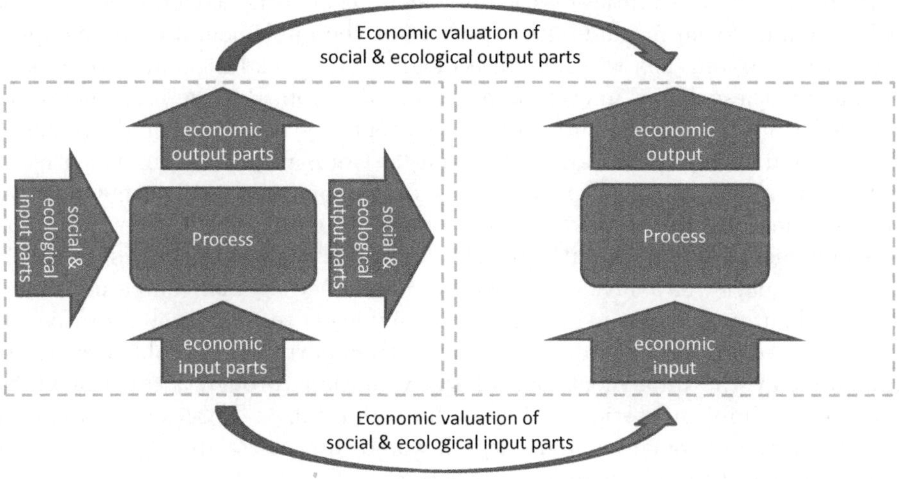

Fig. 1. Idealized sustainable process management model

If the possibility of a Pareto improvement exists, this improvement leads to an increased sustainability of the process and should always be implemented. The second situation leads to a multidimensional decision problem with a tradeoff between at least two of our sustainability dimensions. In this case, a decision maker has to decide in which proportion a gain in one dimension could compensate for a loss in at least one other dimension; let alone the question of compensation within one dimension itself. If the accumulated gains compensate for the accumulated losses, this modification improves the process in terms of sustainability and should be implemented. For accumulated gains that do not compensate for accumulated losses a modification would impair the process in terms of sustainability and should not be implemented. This compensation decision includes an implicit normalization of the ecological and social in- and output parts to the economic dimension. This normalization is a key factor of our idealized model and we present some instruments to support this exercise in the next section.

Assuming that every ecological and social in- and output part is economically valued results in an explicit normalization to the economic dimension for all ecological and social in- and output parts. According to our definition that every economically valued part of the in- and output factors belongs to the economic part, the decision problem is reduced to a single dimension as presented in figure 1 right hand side. Therefore, every process modification with a positive improvement of the economic dimension should be implemented and leads to an improved sustainability of this process. This approach is identical to the internalization of external effects where every ecological and social part of the in- and outputs is interpreted as external effect. [11] The complete internalization of all ecological and social in- and output parts leads to increased process sustainability if the companies maximize shareholder value [6] in the long run.

There are two possibilities for the implementation of the economic valuation of ecological and social in- and output parts. For the implementation on company level each company has to value all ecological and social in- and output parts and include these values in each management decision. But the implementation at this level contains the problem of a potentially deliberate undervaluation of social and ecological in- and output parts. Companies have the incentive to undervalue social and ecological in favor of economic in- and output parts to increase their profits. If the valuation of these factors is not identical across all companies, companies with a lower valuation of ecological and social in- and output parts may have a competitive advantage against companies with a higher valuation of these parts.Over time this leads to additional incentives to undervalue ecological and social in- and output parts. This problem could be eliminated by the implementation of the valuation through a superior global instance. An implementation with globally standardized charges and premiums for ecological and social in- and outputs would internalize all externalities and as a result lead to more sustainable processes.

As the results of the Climate Conference in Copenhagen show, a global internalization of ecological and social in- and output parts is very hard to realize due to economic and political constraints. In the following we present some instruments, which – at least partly – contribute to solve the normalization problem on the company level.

4.2 Available Instruments

In the literature, more than 40 approaches are discussed, which consider corporate sustainability (e.g. [5],[12],[3]). Many of the existing approaches are isolated applications with respect to only one of the presented dimensions of sustainability where the majority of these approaches aims at the ecological dimension of sustainability [12]. In the field of environmental management, for example, some promising process-related approaches like the Life Cycle Assessment [13] approach can be found. However, an integrated view on sustainability necessitates the consideration of all three sustainability dimensions. [12] presents six approaches which comply with this integrated view. Three of these six approaches are especially applicable to all – and not just some – of a company's processes. In the following, we present the basic ideas of these three concepts.

EFQM-Sustainable Excellence model. The classical EFQM-Excellence model is a quality management system developed by the European Foundation for Quality Management (EFQM) to support organizations to be more competitive. The model represents a non-prescriptive concept for the self-assessment of organizations. The classical EFQM-Excellence model already contains dimensions of sustainability, but it has been complemented by social and ecological sustainability dimensions by the Sustainable Excellence Group (SEG) in 2003. Besides, the complemented model includes "know-how" and special procedures to implement sustainability into small and medium-sized organizations. The adjustments have been summarized in the EFQM-Sustainable Excellence model

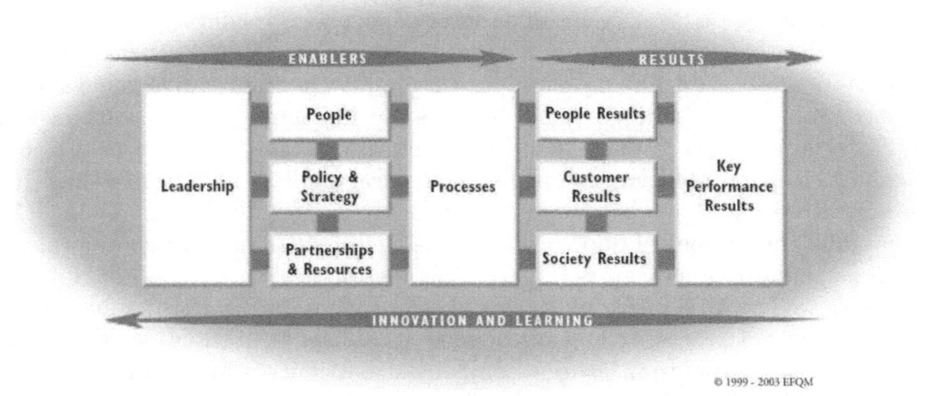

Fig. 2. Criteria of the EFQM-Excellence model [14]

(EFQM-SE model) [5]. Based on eight basic principles, nine criteria have been derived and for each of these criteria (see figure 2), a list of questions has been developed. Five of these criteria are "Enablers" and the remaining four are "Results". The enabler-criteria cover what an organization does, while the result-criteria cover what an organization achieves. "Results" are caused by "Enablers" and "Enablers" are improved using feedback from "Results" [15]. An organization's status quo on its way to sustainable excellence is determined by applying the so-called RADAR-Logic. RADAR means "Results", "Approach", "Deployment" and "Assessment & Review". By continuously iterating these four steps, an organization's strengths and potentials for improvement become visible [15].

For sustainable process management, the German Foundation for Environment promoted a project for the Benchmarking for sustainability (Be.st). In this project, based on the EFQM-SE model a tool has been developed to benchmark business processes. This tool considers business processes as an independent unit and thereby evaluates the processes on the basis of about 40 Questions (based on the nine criteria of the EFQM-SE model). With this method, the convenience and the focus on sustainability of processes can be assessed [16]. This method implies a certain process maturity and process orientation of the organization assessed. By applying the RADAR-Logic, points are assigned to the processes and concrete actions for process optimization are formulated [16]. The Be.st process assessment is a powerful tool to evaluate processes with respect to sustainability. By using the criteria of the EFQM-SE model ecological, economic and social dimensions of sustainability can be considered. Furthermore, the relevant stakeholders can be involved. It is possible to measure and compare a process' contribution to sustainability applying a point based system. By analyzing strengths and potentials, concrete actions to improve sustainability can be formulated. Yet the EFQM-SE model cannot consider interdependencies between the three dimensions of sustainability. Within the model, the ecological and social dimensions are not directly and explicitly related to an organization's success.

Indicator concept. Indicators are measures of performance containing condensed information. It can be differentiated between qualitative and quantitative indicators. Qualitative indicators are descriptive, whereas quantitative indicators are numerically measurable. To amplify the informative value and to reduce ambiguity, qualitative and quantitative indicators can be combined in indicator systems [12]. While quantitative indicators have been used to measure the financial profitability of companies ever since, qualitative indicators like customer satisfaction have become more and more important for companies [17]. Especially the ecological and social dimensions of sustainability contain many of these qualitative indicators [18]. For this reason, the consideration of quantitative as well as qualitative indicators is of great importance for measuring the sustainability of processes. One of the most established guidelines for the development of indicators to measure a company's contribution to sustainability was created by the Global Reporting Initiative (GRI) [19]. In their guidelines the GRI suggests 30 ecological (e.g. CO_2e gas emissions), 40 social (e.g. employee satisfaction) and nine economic indicators for measuring the contribution of the company's processes to sustainability.

Indicators are an adequate tool to measure the ecological, social and economic contribution to sustainability of (mainly less complicated) processes. Yet, indicators cannot consider the interdependencies between the three dimensions of sustainability. Concepts like the Sustainability Balanced Scorecard can integrate indicators and indicator systems to measure sustainability on the process level ([17],[12]).

Sustainability Balanced Scorecard. The Balanced Scorecard (BSC) formulated by Kaplan and Norton is a management tool to translate an organization's strategy into actions [20]. The BSC contains an indicator system that considers financial as well as non-financial quantitative and qualitative indicators. The qualitative and non-financial indicators are related to the financial indicators of the organization. Thereby they can be aligned to the long-term business performance (economic capital) of the organization [3]. The BSC differentiates four perspectives as shown in figure 3.

The four perspectives stand in a cause-effect-relationship to each other. For each of the perspective's goals, performance indicators, and actions to achieve the goals can be derived from the organization's strategy. The indicators used in the BSC concept can be split into lagging and leading indicators. The lagging indicators describe when the goals are achieved, whereas the leading indicators give information about how the goals can be achieved [18]. Because of the relationship between the perspectives, the lagging indicators of a subordinate perspective become leading indicators for the superordinate perspective. Due to the ability to consider quantitative as well as qualitative indicators, the BSC can be used in form of the Sustainability Balanced Scorecard (SBSC) to embed sustainability into the management processes of organizations [22]. If the reduction of CO_2e gases, for example, is part of the strategic goals of an organization, the

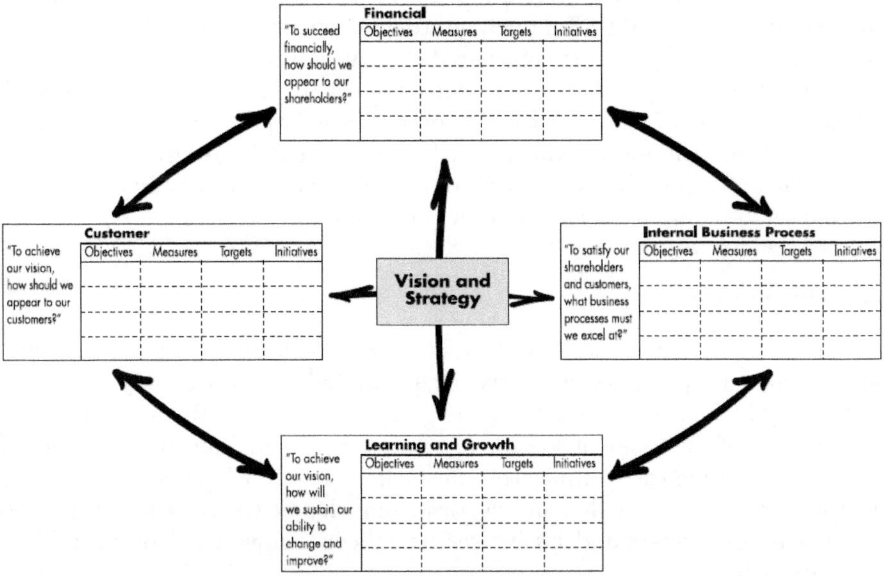

Fig. 3. Perspectives of the Balanced Scorecard [21]

maximum amount of CO_2e gases emitted can be a lagging indicator and energy efficiency can be used as a leading indicator. The more ecological and social dimensions are considered during the formulation of the corporate strategy, the more sustainability aspects gain relevance in management [22].

The traditional BSC concept aims primarily at the economic market environment whereas most of the ecological and social dimensions of sustainability are placed outside the markets, for example in society. The literature provides different approaches to integrate sustainability into the classical BSC concept [18]. To formulate a SBSC, basically, three steps are necessary. At first, the business unit has to be chosen for which the SBSC is to be formulated. Then the relevant ecological and social aspects and at last the strategic relevance of these aspects have to be determined [22]. Catalogues with potentially relevant ecological aspects are provided by [23], for example. The literature also provides catalogues of potential direct and indirect stakeholders, from which relevant social aspects for an organization can be derived [18]. The strategic relevance of the identified aspects is analyzed by classifying the identified aspects into lagging and leading indicators as well as hygiene-factors [22]. For larger organizations, a hierarchical concept of SBSCs is appropriate. From the BSC of the top level of the organization, subjacent scorecards for business units and business processes can be derived. This approach allows to break down the corporate strategy to the level of business processes [10]. For each Process Balanced Scorecard a so-called Sustainability Process Balanced Scorecard (SPBS) can be developed in analogy to the previously described procedure.

The SPBS provides a powerful tool for sustainable process management. In comparison to the EFQM-SE model, interdependencies between the three dimensions of sustainability can be considered. Yet, it can be used beneficially only if the corporate strategy is formulated with respect to sustainability aspects. Furthermore, because of the multitude of processes in organizations, only business processes with strategic relevance can be considered. In addition to that, this concept lacks the possibility to measure the degree of sustainability of processes and thereby to compare them to other processes.

4.3 Discussion

The idealized model illustrates that as long as a normalization to a unit of comparison is not possible, decisions on sustainability improvements may just be accidentally rationale and often may be even counterproductive when important consequences of the usage of specific input factors are not considered holistically. Certainly, revealing the complicated cause-effect-relationships between measures and their outcomes is a demanding task, but this is where research has been contributing over years and businesses have been innovative – often in a trial-and-error fashion.

The presented instruments may help in different ways to improve the sustainability of an organization through a better design of business processes. If the implementation and maintenance of these instruments came at no costs, the usage of all three seems advisable. Rooted in total quality considerations the EFQM-SE model primarily helps at identifying processes that should be improved, while the indicator list by the Global Reporting Initiative supports in selecting a few relevant indicators to measure a process' status quo and respective improvements. Still both of these instruments do not support a decision maker in balancing improvements and worsenings in and between different sustainability dimensions that are rooted in theory. It remains more the "gut feeling" or "strategic considerations" that may guide these decisions. This problem may be partially solved when implementing a BSC concept and using specific SPBSs for the management and controlling of each relevant process in the organization. However, even though the BSC concept explicitly considers the relations and interdependencies between different perspectives, dimensions and indicators, it does not provide for clear guidance of how to select the "right" indicators. Thus, the challenge of identifying the relevant cause-effect-relationships remains prevalent. But at least, over time an implemented BSC concept will show whether the assumed relationships are tendentially correct.

To make things worse, the implementation and maintenance of the described instruments require considerable resources and the business case for these investments has still to be made. However, most companies already have some management concept – quite often the BSC concept – in place. Hence, the investment to integrate sustainability governance into a corporate governance concept may be less expensive compared to a stand-alone implementation.

5 Summary and Outlook

Sustainability in our everyday life as well as in business gained in importance in recent years. As discussed in this paper, IT has a great potential to contribute to corporate sustainability. While the widely discussed "Green-IT"-concepts can make a modest contribution to sustainability mostly on an infrastructure level, the highest potential for sustainability improvements lies in the comprehensive use of IT to enhance the efficiency of business processes. We find that the implementation through globally standardized charges and premiums for all externalities seems to be the best way to incorporate all dimensions of sustainability into management decisions. As the results of the Climate Conference in Copenhagen show, such a global solution cannot be expected in the near future. Yet, there exist some instruments to incorporate sustainability into the management process on the company level. Companies still have to balance costs and benefits of the implementation and usage of these instruments. Moreover, the available instruments focus primarily on the measurement *or* management of sustainability on the process level but hardly provide for any guidance with respect to the actual necessary process (re-)design in detail.

Numerous perspectives for further research arise from our analysis. The cause-effect-relationships between the three dimensions of sustainability need to be further investigated to understand how they can be normalized to the economic dimension. This is also a necessary requirement for an implementation of sustainable process management through charges and premiums on a global level. Furthermore, the presented instruments may serve as a starting point for the development of an integrated management concept that covers all three dimensions of sustainability on the process level. Such a management concept should also comprise an adequate decision support for design choices when redesigning the processes in order to improve sustainability. This could also result in enhancing existing reference models for process design with respect to sustainability considerations.

References

1. Gore, A.: Our Choice. St. Martin's Press, New York (2009)
2. Buhl, H.U., Laartz, D.J.: Warum Green IT nicht ausreicht - oder: Wo müssen wir heute anpacken, damit es uns übermorgen immer noch gut geht? Wirtschaftsinformatik 50(04), 261–264 (2008)
3. Burschel, C., Losen, D., Wiendl, A.: Betriebswirtschaftslehre der nachhaltigen Unternehmung. Oldenbourg Wissenschaftsverlag, München (2004)
4. World Commission on Environment and Development: Our Common Future. Oxford University Press, Oxford (1987)
5. Fischler, U.: Corporate Sustainability. VDM Verlag Dr. Müller, Saarbrücken (2007)
6. Rappaport, A.: Creating shareholder value: A guide for managers and investors. Rev. and updated. Free Press, New York (1998)
7. Kolbe, L., Zarnekow, R., Schmidt, N.H., Erek, K.: Nachhaltigkeit und Green IT in IT-Organisationen. Technical report, Georg-August-Universität Göttingen; Technische Universität Berlin (2009)

8. McKinsey: SMART 2020: Enabling the low carbon economy in the information age. Technical report, The Climate Group and McKinsey (2008)
9. Becker, J., Kugeler, M., Rosemann, M.: Prozessmanagement: Ein Leitfaden zur prozessorientierten Organisationsgestaltung. Springer, Berlin (2005)
10. Schmelzer, H., Sesselmann, W.: Geschäftsprozessmanagement in der Praxis. Carl Hanser Verlag, München/Wien (2006)
11. Pindyck, R.S., Rubinfeld, D.L.: Microeconomics. Pearson Education, New Jersey (2009)
12. Schaltegger, S., Herzig, C., Kleiber, O., Klinke, T., Müller, J.: Nachhaltigkeitsmanagement im Unternehmen. Technical report, Universität Lüneburg (2007)
13. Klöpffer, W., Grahl, B.: Ökobilanz (LCA) – Ein Leitfaden für Ausbildung und Beruf. Springer, Heidelberg (2009)
14. EFQM: Excellence einführen (2003)
15. EFQM: EFQM Excellence Model. Technical report, European Foundation for Qualitymanagement (2003)
16. Merten, T., Baur, A.: Nachhaltigkeitsorientierte Selbstbewertung von Geschäftsprozessen. Technical report, future e.V.
17. Jankulik, E., Pfiff, R.: Praxisbuch Prozessoptimierung. Publicis Publishing, Erlangen (2009)
18. Figge, F., Hahn, T., Schaltegger, S., Wagner, M.: Sustainability Balanced Scorecard - Wertorientiertes Nachhaltigkeitsmanagement mit der Balanced Scorecard. Universität Lüneburg (2001)
19. GRI: G3 Leitfaden zur Nachhaltigkeitsberichterstattung. Web ressource (2006)
20. Kaplan, R.S., Norton, D.P.: The Balanced Scorecard: Translating Strategy Into Action. Harvard Business School Press, Boston (1996)
21. Kaplan, R.S., Norton, D.P.: Using the Balanced Scorecard as a Strategic Management System. Harvard Business Review (07-08), 75–85 (1996)
22. Hahn, T., Wagner, M.: Sustainability Balanced Scorecard - Von der Theorie zur Umsetzung. Universität Lüneburg (2001)
23. Heijungs, R., Guinée, J., Huppes, G., Lankreijer, R., de Haes, U.A., Wegener Sleeswijk, A.: Environmental life cycle assessment of products: Guide. Technical report, Centrum voor Millieukunde, Leiden (1992)

A Model for Quantifying Impacts of Supply Chain Cost and Working Capital on the Company Value

Marcus Brandenburg and Stefan Seuring

Department of International Management, Faculty of Organic Agricultural Sciences, University of Kassel, 37213 Witzenhausen, Germany
brandenburg@uni-kassel.de
www.uni-kassel.de/agrar/ima

Abstract. Supply Chain Management (SCM) is identified and accepted as a competitive advantage. Nevertheless holistic approaches for value-based SCM to leverage this advantage in a value adding way are missing so far. Efficient approaches to quantify and compare value contributions from Supply Chain (SC) value drivers are needed. This paper contributes to this need by proposing a model to efficiently quantify and compare value contributions from SC cost and working capital, that affect the profitability and asset performance. Properties and characteristics of the model, which is based on the Discounted Cash Flow concept, are illustrated by an industrial example of a single company. In this example, the relevance of timing and continuity of developments of SC cost and working capital for value creation is pointed out.

Keywords: supply chain management, supply chain cost, working capital, value-based management, discounted cash flow.

1 Introduction

Supply Chain Management (SCM) has become a competitive advantage for companies from various industries (Mentzer et al. [20], Christopher [2]), and competitive advantage is even considered as a defining characteristic of SCM (Handfield and Nichols [10]). Supply Chain (SC) performance improvements (Shepherd and Günter [28]) are initiated to increase this advantage. Improvements focusing on financial SC value drivers most often show conflicting effects on cost and capital respectively: methods to improve working capital components (inventory, trade payables, trade receivables) for instance often result in cost increases (e.g. negative effects on cash discount or production cost) and hence reduce the profitability, measured by e.g. earnings before interest and taxes (EBIT). One key question is how value contributions from SC value drivers can be quantified efficiently and how financial comparability of EBIT relevant changes of SC cost and asset relevant changes of working capital can be ensured.

To answer this question, a model based on a Discounted Cash Flow (DCF) approach is suggested to quantify value contributions generated by reductions

W. Dangelmaier et al. (Eds.): IHNS 2010, LNBIP 46, pp. 107–117, 2010.

of those two SC value drivers by one aggregated figure. The conceptual elements are described in section 2. Section 3 introduces the quantification model, which is illustrated and discussed based on an industrial example given in section 4. The paper concludes with summarized findings and future prospects for research in section 5.

2 Conceptual Elements

Value-Based Supply Chain Management. The idea of value-based management (VBM) is strongly linked to the term shareholder value, which stipulates all parts of a company to be managed in such a way that the equity value is sustainably increased (Rappaport [23]). Different valuation approaches, such as Economic Value Added (EVA) (Stewart [30]), Cash Flow Return on Investment (CFROI) (Madden [18]), Cash Value Added (CVA) (Ottosson and Weissenrieder [21]), Earned Economic Income (Grinyer [9]), have been developed and increased the popularity of VBM in academic science and managerial practice (Damodaran [7], Copeland et al. [6], Malmi and Ikäheimo [19], de Wet [31]). One well known valuation approach is the Discounted Cash Flow (DCF) model, which is broadly accepted in academic research (Hawawini and Viallet [11], Ross et al. [25], Weber et al. [33]) and often applied industrial practice (Geginat et al. [8]).

Existing literature emphasizes that the question of value-based SCM is of considerable interest for academic research for the last ten years. A thorough literature review would exceed the extent of this paper. Hence, a survey on selective literature indicating the heterogeneity of approaches and concepts for value-based SCM is depicted in Table 1. Value-based SCM concepts vary from functional or company specific perspective to network considerations. Core concepts and identified drivers for value-based SCM, which is most often linked to shareholder value and EVA, range from single functions, operational SC activities and processes to SC strategy. Empirical aspects are reflected differently - some rather theoretical papers give few references to industrial examples, some papers comprise case studies from one or few companies, some papers extensively evaluate several hundreds of observations.

Although it is taken for granted that SCM influences the cash flow and hence the shareholder value of a company (Christopher and Ryals [3], p. 5-7; Hendricks and Singhal [12], p. 503), value-based SCM approaches based on DCF models are found rather seldom. Furthermore little evidence exists that effective SCM is linked to shareholder value creation (Hendricks and Singhal [12], p. 502). Besides, future prospects in research address the need of cost effective ways to measure shareholder value impacts (Lambert and Burduroglu [15]) and the improvements regarding data used for evaluation (Hofmann and Locker [14]).

This paper contributes to these gaps and future prospects in academic research on value-based SCM. An approach is proposed which links SC cost and working capital to the company value. A simple quantification model based on DCF allows for measuring shareholder value impacts of those two SC value drivers in a simple and efficient way by using published data. The approach is illustrated by a company example from Fast Moving Consumer Goods (FMCG) industry.

Table 1. Value based SCM in selected research papers

Paper	SCM approach	VBM approach	Concept, value driver	Empirical aspects
Christopher and Ryals [3] (1999)	network of linked companies with internal processes and interfaces to suppliers and customers	shareholder value, mainly EVA	SCM affects four drivers of shareholder value (revenue, cost, fixed capital, working capital)	12 observations from industrial or consulting practice
Lambert and Burduroglu [15] (2000)	functional (logistics)	different approaches, shareholder value as most comprehensive	logistics influence on revenue, cost, fixed capital, working capital	none
Lambert and Pohlen [16] (2001)	integration of 8 key processes across companies to add value for customers and stakeholders	shareholder value, EVA	map 8 key processes to shareholder value	none
Hendricks and Singhal [12] (2003)	company-internal supply chain	market value of a firm (stock price)	examine impacts of production or shipment delays on stock price; SC strategy is linked to cash flow, earnings and assets	519 observations from industrial practice
Sridharan et al. [29] (2005)	functional (IT for SCM)	market value of a firm (stock price)	examine effects from implementation of IT for SCM on the market value of a firm	three company case studies
Hofmann and Locker [14] (2009)	inter-organizational management of flows of goods and information to ensure performance and customer achievements	shareholder value, EVA	SCM processes linked to EVA categories via KPI	case study from packaging industry

SC Cost. Research in SC cost management ranges from concepts, instruments and models (see e.g. Seuring and Goldbach [27]) to the link to other conceptual approaches, e.g. value based pricing (Christopher and Gattorna [4]) or logistics cost management (Suang and Wang [32]). SC cost comprise cost of goods sold (COGS), which reflect the direct cost and overhead associated with the physical production of products for sale (Poston and Grabinski [22]), and logistics cost for transportation, distribution logistics, inventory carrying and administration (Cohen and Roussel [5]):

$$SC\ cost = COGS + Logistics\ cost\ . \tag{1}$$

Furthermore, SC risks have a considerable influence on SC cost and SC performance (Winkler and Kaluza [34], Ritchie and Brindley [24], Lee [17]).

Working Capital. SC strategy and logistics management are linked to requirements of working capital (Christopher [2]), which has a strong influence on the liquidity position and the economic value of a company (Schilling [26]). In some definitions working capital comprises other components such as cash, prepaid expenses, accrued expenses (see e.g. Ross et al. [25] p. 176 or Hawawini and Viallet [11] p. 73). For simplicity reasons, these components are neglected in the working capital definition applied this paper. Working capital is defined as the sum of inventories and trade receivables reduced by trade payables (Brealey et al. [1] p. 145):

$$Working\ capital = Inventory + Trade\ receivables - Trade\ payables\ . \tag{2}$$

3 The Quantification Model

The DCF method, which is introduced thoroughly by e.g. Hawawini and Viallet [11] or Ross et al. [25], is the basis for the proposed quantification model. DCF determines the company value V, which is generated during time periods $p = 1, \ldots, P$, by the sum of the discounted free cash flows FCF_p:

$$V = \sum_{p=0}^{p} \frac{FCF_p}{(1 + WACC)^p}\ . \tag{3}$$

The free cash flow FCF is defined as difference between EBIT and expenses for tax, depreciation and net capital adjusted for working capital changes (Hawawini and Viallet [11] p. 399):

$$\begin{aligned} FCF = {} & EBIT - Tax\ expenses + Depreciation\ expenses \\ & - Net\ capital\ expenditures - \Delta\ Working\ capital\ . \end{aligned} \tag{4}$$

The free cash flow of each period is discounted by the weighted average cost of capital ($WACC$), which represents the minimum rate of return that must be generated in order to meet the return expectations of shareholders (Hawawini and Viallet [11] p. 329).

The key question is how to quantify value contributions that arise from changes of working capital or SC cost over a defined period of time periods $p = 1, \ldots, P$. As we see from formula 4, changes of working capital as well as EBIT relevant changes of supply chain cost affect the FCF and thus contribute to the company value. Therefore the value contribution VA^{WC} or VA^{SCC} generated by the respective value drivers over a time horizon of periods $p = 1, \ldots, P$ is the sum of the value contributions, or precisely value contributing FCF effects, VA_p^{WC} or VA_p^{SCC} of each period discounted by $WACC$:

$$VA^{WC} = \sum_{p=0}^{P} \frac{VA_p^{WC}}{(1 + WACC)^p} \ . \tag{5}$$

$$VA^{SCC} = \sum_{p=0}^{P} \frac{VA_p^{SCC}}{(1 + WACC)^p} \ . \tag{6}$$

To quantify those value contributions of each period, the respective FCF effects are calculated under consideration of three propositions:

1. In case that the development of a SC value driver (working capital or SC cost) is proportionate to sales development, the resulting value contribution is credited against sales development. Only in case that a SC value driver develops disproportionate to sales, the resulting value contribution is credited against the respective SC value driver.
2. Changes in working capital are one-time effects and thus result in a value contribution in only one period.
3. Changes of SC cost are recurring effects and thus result in value contributions in all subsequent period.

The first proposition is illustrated in Fig. 1 based on the example of working capital development.

To realize the second proposition, the value contribution VA_p^{WC} arising from working capital development in a time period p is defined by the difference of working capital WC_p at the end of period p and working capital WC_{p-1} at the beginning of period p adjusted for sales development $\frac{S_{p-1}}{S_p}$ in period p:

$$VA_p^{WC} = WC_{p-1} - \frac{S_{p-1}}{S_p} \cdot WC_p \ . \tag{7}$$

Similar approaches apply to quantify effects from supply chain cost under consideration of the third proposition. The value contribution arising from changes of SC cost SCC_p in a period p is defined by the difference of SC cost SCC_0 of period 0 and SC cost SCC_p of period p adjusted for sales development $\frac{S_0}{S_p}$ and tax rate T:

$$VA_p^{SCC} = (SCC_0 - \frac{S_0}{S_p} \cdot SCC_p) \cdot (1 - T) \ . \tag{8}$$

Illustrative example[1]:

[1] Note: In this paper, all figures are rounded to maximum one decimal digit.

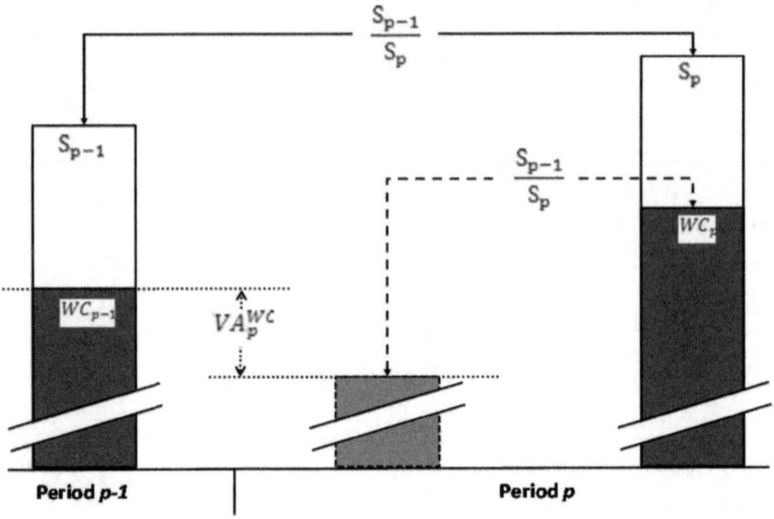

Fig. 1. Value contribution from a disproportionate working capital development

Let a fictitious company of 10,000 m EUR sales have a working capital of 1,600 m EUR and SC cost of 7,000 m EUR at a WACC of 10.0% and tax rate of 35.0%. Two example cases A and B are considered for developments of working capital and SC cost, each case split in two scenarios for sales development (with constant sales and with growing sales).

Case A: The developments of the two SC value drivers, working capital and SC cost, are proportionate to the sales development. Therefore they do not result in value contribution which is credited against those value drivers, although in scenario 2 the absolute figures for both SC value drivers grow (as required by proposition 1).

Case B: A value adding disproportionate working capital development in period p=1 is saved in period p=2 without additional value contribution (as required by proposition 2). The disproportionate development of SC cost in period p=1 results in value contributions in both periods p=1 and p=2 (as required by proposition 3).

Table 2 depicts the results of the model for this example.

4 Illustration Example from Industry

To illustrate and discuss the proposed quantification model, its properties and functionality are shown based on an example of a single company. Henkel KGaA from FMCG industry is selected as example case, because all required data is published in the annual reports [13] of this company: Working capital figures were

Table 2. Example for quantification of value contribution

			Constant sales			Growing sales		
	p		0	1	2	0	1	2
	S_p	m EUR	10,000	10,000	10,000	10,000	20,000	30,000
Case A	WC_p	m EUR	1,600	1,600	1,600	1,600	3,200	4,800
	VA_p^{WC}	m EUR	-	0	0	-	0	0
	VA^{WC}	m EUR	-	0	0	-	0	0
	SCC_p	m EUR	7,000	7,000	7,000	7,000	14,000	21,000
	VA_p^{SCC}	m EUR	-	0	0	-	0	0
	VA^{SCC}	m EUR	-	0	0	-	0	0
Case B	WC_p	m EUR	1,600	1,500	1,500	1,600	3,000	4,500
	VA_p^{WC}	m EUR	-	+100.0	0	-	+100.0	0
	VA^{WC}	m EUR	-	+90.9	+90.9	-	+90.9	+90.9
	SCC_p	m EUR	7,000	6,900	6,900	7,000	13,800	20,700
	VA_p^{SCC}	m EUR	-	+65.0	+65.0	-	+65.0	+65.0
	VA^{SCC}	m EUR	-	+59.1	+112.8	-	+59.1	+112.8

Table 3. Industrial example from Henkel KGaA

p	UOM	2003	2004	2005	2006	2007	2008
S_p*	m EUR	9,436	10,592	11,974	12,740	13,074	14,131
SCC_p*	m EUR	4,965	5,615	6,533	6,963	7,013	8,190
WC_p*	m EUR	1,845	1,840	1,693	1,699	1,500	1,651
Inventory*	m EUR	1,053	1,196	1,232	1,325	1,283	1,482
Receivables*	m EUR	1,581	1,743	1,794	1,868	1,694	1,847
Payables*	m EUR	789	1,099	1,333	1,494	1,477	1,678
$WACC$*	%	8.0	7.0	7.0	7.0	7.0	7.5
T*	%	35.0	35.0	35.0	30.0	30.0	30.0
VA_p^{SCC}**	m EUR	-	-24	-119	-135	-68	-353
VA^{SCC}**	m EUR	-	-23	-127	-236	-288	-534
VA_p^{WC}**	m EUR	-	206	342	96	237	-28
VA^{WC}**	m EUR	-	192	491	570	751	732
VA^{Total}	m EUR	-	170	365	333	463	198

* source: annual report of respective period
** calculated with described formulas

calculated from balance sheet data, for simplicity reasons SC cost are limited to cost of goods sold (COGS) figures taken from P&L statements, figures for tax rate and $WACC$ are published in the respective annual reports, too. The input data and results of the quantification model are depicted in Table 3.

The total value contribution VA^{Total}, i.e. the sum of value contributions VA^{WC} and VA^{SCC} from each SC value driver, is positive. Hence, SCM at Henkel KGaA added value to the company. Two other observations of this case can easily be made and examined further by a root cause analysis and a simulation scenario:

Table 4. SC cost and working capital ratios at Henkel KGaA

p	UOM	2003	2004	2005	2006	2007	2008
SC cost ratio	%	52.6	53.0	54.6	54.7	53.6	58.0
Working capital ratio	%	19.6	17.4	14.1	13.3	11.5	11.7

1. In no period of time, any value was generated by SC cost.
2. Value was generated by working capital in each period of time except in 2008.

Analyzing the developments of SC cost ratio, i.e. SC cost in % of sales, and working capital ratio, i.e. working capital in % of sales, helps explaining why value was added or lost by the developments of those value drivers (Table 4).

The SC cost ratio shows a continuous and deteriorating trend, the level of 52.6% in 2003 was never achieved in the subsequent periods 2004-2008. Hence no value was added from SC cost. The working capital ratio on the contrary develops continuously and significantly improving between 2003 and 2007, this achieved decline significantly adds value. A comparably small increase of working capital ratio in 2008 does not substantially influence the value contribution achieved so far.

A simulation scenario derived from this example case helps illustrating and discussing value adding developments of SC value drivers. This simulation scenario answers the question how SC cost should have developed in 2007 and 2008 to avoid the value loss of -534 m EUR. Similar approaches are possible for working capital. Based on defined target value contributions va_p^{SCC} for 2007 and 2008, the figures of target SC cost scc_p which are needed to achieve those value contributions can be calculated by rearranging formula 8:

$$scc_p = \frac{S_p}{S_0} \cdot (SCC_0 - va_p^{SCC} \cdot (1-T)^{-1}) \ . \tag{9}$$

By 2006 a value contribution of -236 m EUR from SC cost was achieved, hence this value loss would be compensated by achieving +118 m EUR value contribution from SC cost in each of the last two periods. Applying formula 9, the required target SC cost scc_p can be calculated to obtain the simulation scenario depicted in Table 5 (figures obtained from simulation depicted in *italics*).

The example case and the derived simulation scenario illustrate important characteristics of value-adding SC value driver developments:

1. Continuous improvements and sustainable developments of SC value drivers foster value creation: Alternating improvements and deteriorations in SC value driver developments cannot guarantee overall value achievements.
   Proof from example case: Although in 2007 the SC cost improved slightly compared to 2006, this improvement was not sufficient to compensate losses from earlier periods. Furthermore the improvement was not strong enough to achieve the SC cost ratio from 2003 and hence did not add any value.
2. Timing aspects have to be considered: Improvements of SC value drivers achieved in later phases cannot guarantee to fully compensate deteriorations from earlier phases.

Table 5. Results of the simulation scenario for SC cost

p	UOM	2003	2004	2005	2006	2007	2008
S_p	m EUR	9,436	10,592	11,974	12,740	13,074	14,131
SCC_p	m EUR	4,965	5,615	6,533	6,963	*6,572*	*7,072*
VA_p^{SCC}	m EUR	-	-24	-119	-135	*155*	*170*
VA^{SCC}	m EUR	-	-23	-127	-236	*-118*	-
SC cost ratio	%	52.6	53.0	54.6	54.7	*50.3*	*50.0*

Proof from simulation scenario: The deterioration of SC cost in early periods 2003-2006 (SC cost ratio in 2006 deteriorated vs. 2003 by 2.1pp) can only be compensated by an disproportionately high reduction in the late periods 2007 and 2008 (SC cost ratio improvement vs. 2003 by 2.6pp in 2008 needed to compensate value losses from earlier periods).

5 Conclusion

This paper contributes to the development of frameworks and approaches to value-based SCM. A model is proposed to quantify impacts of SC cost and working capital on shareholder value. This model is mathematically derived from the DCF model and hence proofs the link between SC value drivers and shareholder value. The financial comparability of EBIT relevant changes of SC cost and asset relevant changes of working capital by one aggregated figure is ensured. An example of a company from FMCG industry illustrates how to use published data to efficiently measure shareholder value impacts of SC value drivers.

Further research opportunities are offered by the possibility to extend the empirical evaluations to other companies or industries. Furthermore the model can be applied or extended to evaluate specific SC value driver components. The performance of the working capital components inventory, trade receivables and trade payables as well as the value contributions of function-specific cost, e.g. for material acquisition or production, can be analyzed. Besides, the model can be extended to other financial SC value drivers, such as fixed asset performance. The context of SC risk management and value creation allows further research, especially regarding influences of SC risks on SC cost or WACC. Beyond this, a holistic value-based SCM concept must consider non-financial value drivers, such as SC risk, intangibles or intellectual capital, and inter-company aspects to reflect the network approach of SCM.

References

1. Brealey, R.A., Myers, S.C., Allen, F.: Principles of Corporate Finance. McGraw-Hill, New York (2008)
2. Christopher, M.: Logistics and Supply Chain Management - Creating Value-Adding Networks, 3rd edn. Prentice Hall, Harlow (2005)

3. Christopher, M., Ryals, L.: Supply Chain Strategy: Its Impact on Shareholder Value. The International Journal of Logistics Management 10(1), 1–10 (1999)
4. Christopher, M., Gattorna, J.: Supply chain cost management and value-based pricing. Industrial Marketing Management 34, 115–121 (2005)
5. Cohen, S., Roussel, J.: Strategic Supply Chain Management - The five Disciplines for Top Performance. McGraw-Hill, New York (2005)
6. Copeland, T., Koller, T., Murrin, J.: Valuation: Measuring and Managing the Value of Companies. Wiley, New York (1996)
7. Damodaran, A.: Valuation approaches and metrics: a survey of the theory and evidence, `http://www.stern.nyu.edu/~adamodar/pdfiles/papers/valuesurvey.pdf`
8. Geginat, J., Morath, B., Wittmann, R., Knüsel, P.: Kapitalkosten als strategisches Entscheidungskriterium. Roland Berger Strategy Consultants, München (2006)
9. Grinyer, J.: Earned economic income: a theory of matching. Abacus, 43–54 (March 1985)
10. Handfield, R.B., Nichols, E.L.: Introduction to Supply Chain Management. Prentice-Hall, Upper Saddle River (1999)
11. Hawawini, G., Viallet, C.: Finance for Executives - Managing for Value Creation, 2nd edn. South-Western, Mason (2002)
12. Hendricks, K.B., Singhal, V.R.: The effect of supply chain glitches on shareholder wealth. Journal of Operations Management 21, 501–522 (2003)
13. Henkel KGaA, Annual Reports 2003-2008, `http://www.henkel.de/investor-relations/downloads-archiv-10481.htm`
14. Hofmann, E., Locker, A.: Value-based performance measurement in supply chain: a case study from the packaging industry. Production Planning & Control 20(1), 68–81 (2009)
15. Lambert, D.M., Burduroglu, R.: Measuring and Selling the Value of Logistics. The International Journal of Logistics Management 11(1), 1–17 (2000)
16. Lambert, D.M., Pohlen, T.L.: Supply Chain Metrics. The International Journal of Logistics Management 12(1), 1–19 (2001)
17. Lee, T.Y.S.: Supply Chain Risk Management. International Journal of Information and Decision Science 1(1), 98–114 (2008)
18. Madden, B.J.: The CFROI Valuation Model. The Journal of Investing 7(1), 31–44 (1998)
19. Malmi, T., Ikäheimo, S.: Value Based Management practices-some evidence from the field. Management Accounting Research 14(3), 235–254 (2003)
20. Mentzer, J.T., DeWitt, W., Keebler, J.S., Min, S., Nix, N.W., Smith, C.D., Zacharia, Z.G.: Defining Supply Chain Management. Journal of Business Logistics 22(2), 1–25 (2001)
21. Ottosson, E., Weissenrieder, F.: Cash Value Added - a new method for measuring financial performance. Gothenburg Studies in Financial Economics 1. Gothenburg University, Gothenburg (1996)
22. Poston, R., Grabinski, S.: Financial impacts of enterprise resource planning implementations. International Journal of Accounting Information Systems 2, 271–294 (2001)
23. Rappaport, A.: Creating Shareholder Value. The Free Press, New York (1998)
24. Ritchie, B., Brindley, C.: Supply chain risk management and performance - A guiding framework for future development. International Journal of Operations and Production Management 27(3), 303–322 (2007)
25. Ross, S.A., Westerfield, R.W., Jaffe, J.F.: Corporate Finance, 6th edn. McGraw-Hill, New York (2002)

26. Schilling, G.: Working Capital's Role in Maintaining Corporate Liquiditiy. TMA Journal, 4–7 (September/October 1996)
27. Seuring, S., Goldbach, M. (eds.): Cost Management in Supply Chains. Physica-Verlag, Heidelberg (2001)
28. Shepherd, C., Günter, H.: Measuring supply chain performance: current research and future directions. International Journal of Productivity and Performance Management 55(3/4), 242–258 (2006)
29. Sridharan, U.V., Caines, W.R., Patterson, C.C.: Implementation of supply chain management and its impact on the value of firms. Supply Chain Management: An International Journal 10(4), 313–318 (2005)
30. Stewart, G.B.: The Quest for Value: the EVA Management Guide. HarperBusiness, New York (1991)
31. de Wet, J.H.v.H.: EVA versus traditional accounting measures of performance as drivers of shareholder value - A comparative analysis. Meditari Accountancy Research 13(2), 1–16 (2005)
32. Suang, H., Wang, L.: The status and development of logistics cost management: evidence from Mainland China. Benchmarking: An International Journal 16(5), 657–670 (2009)
33. Weber, J., Bramsemann, U., Heineke, C., Hirsch, B.: Wertorientierte Unternehmenssteuerung: Konzepte - Implementierung - Praxisstatements. Gabler, Wiesbaden (2004) (in German)
34. Winkler, H., Kaluza, B.: Integrated Performance- and Risk Management in Supply Chains - Basics and Methods. In: Kersten, W., Blecker, T. (eds.) Managing Risks in Supply Chains - How to Build Reliable Collaboration in Logistics. Erich Schmidt Verlag, Berlin (2006)

Assessing the Effects of Assortment Complexity in Consumer Goods Supply Chains

Christoph Danne[1] and Petra Häusler[2]

[1] Heinz Nixdorf Institute, Paderborn, Germany
cdanne@hni.upb.de
http://wwwhni.upb.de
[2] Freudenberg Household Products, Weinheim, Germany
petra.haeusler@fhp-ww.com
http://www.vileda.com

Abstract. Complexity management and management of assortment variety in particular is the subject matter of many strategic decisions in consumer goods industries. As the assortment of a company evolves continuously by introducing new or discontinuing existing products, the most important question is *what effects on the configuration of the production and distribution network and related costs can be expected if the assortment is changed in a particular way?*. This work presents the concept for a decision support system (DSS) that evaluates the most relevant assortment dependent cost positions for a production and distribution network, focusing on the areas of inventory management and production execution. An overview of the prototype DSS implementation *Complana* and its sample application to the cloths' supply chain of an international household product manufacturer is given.

Keywords: Assortment Complexity, Supply Chain Modelling, Inventory Management, Production Planning.

1 Introduction

1.1 Assortment Complexity in Consumer Goods Supply Chains

Companies today find themselves exposed to an increasing pressure for product differentiation in order to cover a wide range of customer preferences to remain competitive. This often results in a high number of product variants and an increasing assortment complexity. This trend is supported by the fact that most manufacturing companies supply international markets and therefore are required to create product variants tailored to their destination market. This results in additional variety both in product characteristics and packaging in order to meet customer preferences or comply with local standards and legal regulations.

Assortment complexity has a major impact on the complexity of the entire supply chain, as the number of products and product variants affects the complexity of the production and distribution system in several ways. Managerial

W. Dangelmaier et al. (Eds.): IHNS 2010, LNBIP 46, pp. 118–129, 2010.

decisions about changes of the assortment are driven by the trade off between additional benefits in terms of sales and additional costs incurred by the additional complexity. Assortment complexity incurs costs in almost all areas of a company's operations and therefore has been analysed extensively for different functional areas and with different methods. However, assessing the effects of assortment related decisions on the underlying production and distribution network is still an open question ([1, p. 49]). As the assortment of a company evolves continuously by introducing new or discontinuing existing products, the most important question is *what effects on the configuration of the production and distribution network and related costs can be expected if the assortment is changed in a particular way?*

Current approaches cannot answer this question satisfactorily, mainly because they neglect the non-linear interrelations between complexity and related costs. Methods from costing, especially activity-based costing, seek to assign costs to single product variants fairly according to the input involved. This approach is not suitable for the aspired what-if analysis, even if the cost assignment was perfectly fair, because of interdependencies between the single product variants that cannot be mapped into a single cost value per product. For example, if a certain degree of standardisation of packaging options across several countries is reached, it becomes favourable to keep inventories at a central warehouse instead of local stocks. Therefore, statements about potential savings derived from assortment reduction can only be made for complete assortment change scenarios rather than on a per product basis. Furthermore, statements about cost effects of these changes can only be made on the basis of a production and distribution network adapted to the new assortment. This leads to the requirement for methods to adapt the planning and control parameters to the new situation in order to assess the optimisation potential offered by assortment changes.

1.2 Problem Statement

This work presents a decision support system to evaluate the most relevant assortment dependent cost positions, focusing on the areas of inventory management and production execution. This is due to the fact that cost effects are particularly expected as a result of changing inventory requirements and related inventory holding costs as well as setup and scrap cost in the production area.

As indicated in Figure 1, the main cost formula developed sums up

- Inventory cost, calculated via the inventory levels at all stockpoints and all time periods
- Cycle stock cost, calculated via the planned production quantities and demand rates over all time periods
- Setup and scrap cost, calculated via the average setup cost with the given production planning parameters and the planned production quantities.

For each cost area, the figure shows the parameters that have to be known to assess the corresponding costs correctly. Considering these parameters, we note

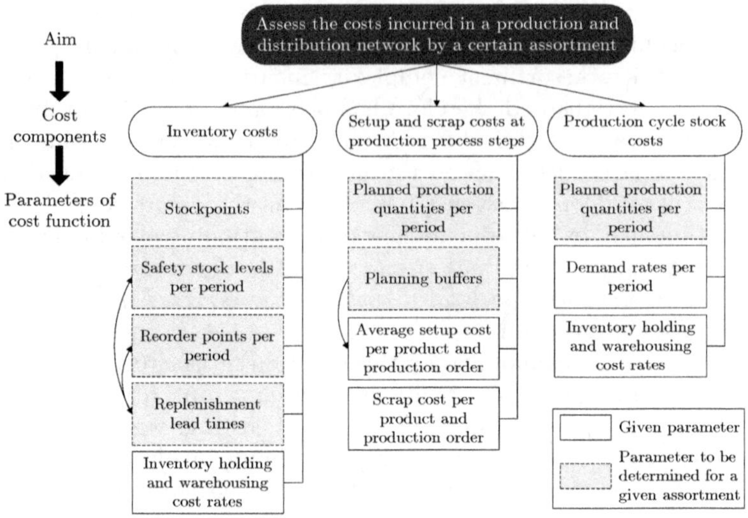

Fig. 1. Cost components and required parameters for their assessment

that some of them are assortment-dependent and should be adapted when the assortment is modified in some way. In order to evaluate concrete assortment scenarios, these parameters of the production and distribution network have to be determined optimally for each considered scenario. As a result, the parameter optimisation for the alternative assortment scenarios adapts the inventory allocation and certain material requirements planning (MRP) parameters for the production process steps of the network, which are:

- Inventory allocation, i.e. determination of products at physical locations where safety stock is held and corresponding safety stock levels for each period such that a given service level is fulfilled despite the uncertainties that result from deviations from forecasted demand volumes. The positioning of inventories also determines the replenishment lead times of the materials.
- Production planning buffers, a buffer time planned at each production stage between the provision of components and the requirement date of orders. The longer this time buffer, the more possibilities to create sequence optimised production plans exist and the lower the average setup costs. However, increasing throughput times also increase safety stocks, a trade off that has to be solved optimally.
- Planned production quantities in each time period such that the expected demand is covered and cycle stocks are kept low.

Figure 2 outlines the steps to be performed in the proposed analysis process. First, the current assortment is represented in a formal model along with its production and distribution structure. On the basis of such a baseline model, theoretical assortment change scenarios with a modified set of materials in the assortment and modified demand and forecast values are derived. Second, each

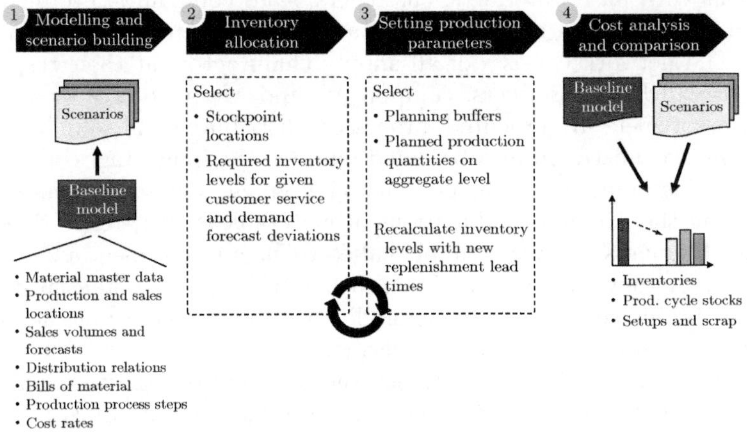

Fig. 2. Analysis process

baseline model and theoretical scenario is optimised with the developed method. In a first step, suitable stockpoints and inventory levels are determined to buffer against the given demand uncertainties. In a second step, planning buffers and planned production quantities are determined. The planning buffers also define the throughput times of the single production stages. Once the stockpoints and the throughput times at all nodes are set, the required levels of inventory are re-calculated to consider the updated throughput times. The results in terms of the considered configuration parameters for the production and distribution network provide the necessary basis for a comparative cost analysis between the baseline model and all alternative scenarios.

2 Solution Approach

This paper addresses the solution approaches for the subproblems 1) modelling aspects, 2) inventory allocation and 3) setting production planning parameters. For (1), a model of the product, production and distribution system is defined. The inventory allocation problem (2) is computationally very complex and not solvable in reasonable time. Therefore, we develop a local search heuristic that makes use of information about the network structure, demand uncertainty information and demand patterns to guide the search. For the production planning parameter subproblem (3), a mixed integer problem is formulated and solved with standard software.

2.1 Modelling Product Assortments as Production and Distribution Networks

An assortment defines a certain product and distribution structure which we represent in a network model. Such a network model thus represents a certain assortment with all its products, customers, sales companies, production facilities and suppliers. We call this model a production and distribution network (PDN). The network consists of all supply chain actors at their respective locations in which the products, components and raw materials of a particular product assortment are procured, processed, distributed and sold. The network may be limited upstream to a bounded number of supply stages. Downstream, the network is bounded by the end customer stage. The set N contains tuples representing the existing combinations of materials at their respective physical locations. Elements $i \in N$ are simply called *items* and represent a certain material at a certain physical location. Each such item $i \in N$ is a potential stockpoint, which can be interpreted as the possibility to keep inventory of the respective product at the respective physical location.

In order to represent the assortment and distribution structure, the set $V \subset N \times N$ represents the links between pairs of items $i, j \in N$. Such a link $v \in V$ may either represent a where-used relation as defined in an entry of j's bill of material (BOM), or a distribution relation between two physical locations. We call the resulting network $PDN(N, V)$ a production and distribution network that represents the full assortment and distribution structure for N.

In order to achieve the change of either product characteristics or physical locations between adjacent items, each item $i \in N$ may have some type of process related to it. These processes may either be a production step, transforming a set of input products into a resulting product, or a transportation process, transferring one product from one location to another. With this definitions, the set of items N can be further classified into disjunctive subsets. First, set N^{PROC} contains all items that are procured externally and thereby mark the system boundary at one side of the network. Consequently, nodes $i \in N^{PROC}$ have no predecessors in the network representation. Second, the set N^{PROD} contains all intermediate stage items that have a production process related to them. Such a node can have an arbitrary number ≥ 1 of predecessors and successors. Third, the set N^{DIST} contains all items with distribution processes. With respect to their position in the network it can only be assured that all last stage items with no successors are elements of N^{DIST}, while not all elements of N^{DIST} are necessarily last stage nodes. There may be distributions that span more than one location, as not all products are necessarily shipped directly from production facilities to the sales companies. These items represent the finished products at the sales companies that are requested by and shipped to end customers from there. It is obvious that these three sets form the set of all items $N = N^{PROC} \cup N^{PROD} \cup N^{DIST}$.

Both of the above mentioned optimisation problems require information about the demand volumes and distributions at each item. We assume that information about the demands is available at some timely aggregated level, e.g. months or weeks. This demand information may be derived from historical data and / or assumptions about future demand developments. As the expected demand $d_{i,t}$ for an item i is based on forecasts, a probabilistic model of the uncertainty related to the forecast is defined based on average absolute forecast errors, i.e. the deviations of forecasted from actual demand.

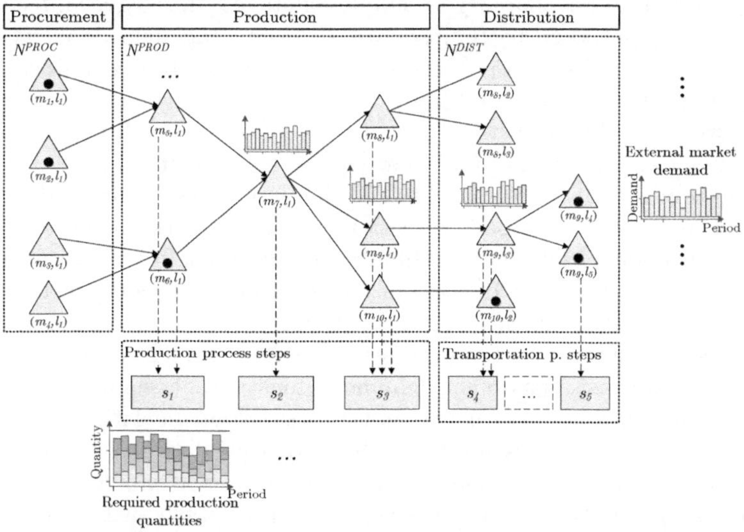

Fig. 3. Simplified example of a production and distribution network

Figure 3 shows a small example of such a PDN model. At all distribution nodes, demand volumes and forecast deviations can originate from external customer demand and from subsequent distribution nodes where appropriate (e.g. node i_9 in LOC_3 in the example). Production and procurement nodes derive their demand information from their successors. The production nodes are each assigned to a production process step (PPS), which may represent a single or a group of production facilities where setup and scrap cost are assessed. At each such PPS, required production volumes can be calculated from the demands at the connected items.

2.2 A Heuristic for Inventory Allocation

The question what savings can be achieved in inventory management by altering the product assortment also leads to the question of optimal inventory allocations in terms of positioning and dimensioning of inventories. While there is a large body of literature on inventory allocation, only very few approaches have been presented for systems with multiple products and locations and arbitrary network structures. The main reason for this is that for general networks, i.e. where each item can have an arbitrary number if predecessors and successors, the possibility to exploit the special characteristics of pure assembly or pure distribution networks vanishes which results in a hard combinatorial optimization problem. Here, two main research streams can be identified which differ in terms of the modelling of replenishment mechanisms and the resulting service time characteristics. In the stochastic service approach (SSA) based on Clark and Scarf [2], the service (=delivery) times at one stage are stochastic and vary based on the material availability at supplying stages. In the guaranteed service approach (GSA) proposed by Simpson Jr. [3], each stage quotes a service time it can always satisfy. Both approaches have been successively extended to work under different surrounding conditions.

In this work, we build on the guaranteed service approach by Simpson and its extensions as summarised by Minner [4], as their assumptions about the underlying system roughly correspond to the characteristics of our production and distribution networks described in section 1.2. They work on network models in which each node i represents an item in a supply chain that performs some processing function like a production or transportation process. Each such item has a known and deterministic throughput time and is a potential stockpoint that can hold inventory after the processing has finished. The only real source of uncertainty is stochastic customer demand, which is represented as some probability distribution with known demand and standard deviation. To our best knowledge, Minner also presents the only heuristic safety stock allocation approach so far. The fact that the safety stock allocation problem can be seen as a combinatorial optimisation problem that can be solved with corresponding heuristics is due to one central result of Simpson's work called the extreme point property, which shows that in an optimal solution

$$ST_i \in \{0, ST_{i-1} + \lambda_i\}, \tag{1}$$

with λ_i being the throughput time of item i. Each item either covers its entire replenishment lead time with safety stock (has a service time ST_i of 0) or does not hold any safety stock and passes its entire replenishment lead time plus its throughput time TT_i to its successors (has a service time of $ST_{i-1} + TT_i$). The remaining decision thus is a binary *stockpoint* or *no stockpoint* decision and the inventory allocation can be mapped to a combinatorial optimisation problem. In order to assess a given solution, the required safety stock level for each stockpoint has to be determined. With a given stockpoint selection, the service times between adjacent items are fixed and a replenishment lead time RLT_i can be calculated for each $i \in N$ as the maximum service time of its predecessors plus its own throughput time. With a given replenishment lead time, expected demand quantity, required service level and the distribution of the forecast error, the required safety stock levels SS_{it} are calculated as a multiple of the demand's forecast deviation over the coverage time:

$$SS_{it} = z_{it} \cdot \sigma_{it} \cdot \sqrt{RLT_i - ST_i} \tag{2}$$

The only unknown parameter here is the multiplier z_{it}, which has to be set such that the safety stock is sufficient to fulfil a certain service criterion. In contrast to many existing works, we do not use the α-service level here but the so called fill rate (β-service level) which is defined as the proportion of demand that can be filled from stock. For a given fill rate β_i, the safety factor is set such that

$$L(z_{it}) = \frac{(1 - \beta_i) \cdot d_{it} \cdot RLT_i}{\sigma_{it} \cdot \sqrt{RLT_i - ST_i}} \tag{3}$$

For the sake of brevity we omit the complete derivation here and refer the reader to [5]. $L(\cdot)$ denotes the standard loss function, for which tabular values exist and thus the corresponding z_{it} values can easily be looked up. Together with the previous formula, the safety stock level is calculated and inventory costs for this stockpoint are incurred linear to this amount. The sum of this inventory costs over all stockpoints and all periods forms the objective value of the safety stock optimisation.

Combinatorial optimisation problems are particularly suited to be addressed with heuristic solution techniques. We have selected tabu search (TS) as a well known and proven meta-heuristic. Instead of random stockpoint changes to the items, we use domain knowledge to guide the search process, i.e. to select those items as stockpoints that are expected to be suitable and to remove safety stock from those items that are probably less suited. These decisions are based on the following rationale:

- Move inventories downstream if an item has only single usage: If an item forms the input to only one product on the successive stage in production or it is distributed to only one location in a distribution stage, stock is probably better kept at the next downstream location to reduce service times to customers or stages further downstream.
- Prefer items with many predecessors as stockpoints: Items that require a large amount of components as input for their corresponding production operation (note that items with multiple predecessors only occur in production stages) are suited to hold safety stock to buffer against their high supply risk.
- Prefer items with many successors as stockpoints: Items that supply many successive items are probably well suited as stockpoints, as they can exploit risk pooling effects by aggregating uncertain demand from their successors.
- Prefer items with low forecast deviations as stockpoints: Cumulative throughput times should be covered with stock at those items that have a low forecast deviation to keep the resulting safety stock levels low.

All these strategies are used to generate neighbourhood solutions in each iteration of the tabu search process. The aim is to include only promising changes in the neighbourhood, rather than testing all possible stockpoint / no stockpoint switches. As some of these strategies describe soft criteria, we calculate a single stockpoint eligibility metric for each item. This metric is normalised on the interval $[0, 1]$. The neighbourhoods in the tabu search process are then constructed to include solutions that add the non-stockpoint items with the highest eligibility ratings to the set of stockpoints, and analogously remove those stockpoints with the lowest eligibility ratings. For a complete description of the tabu-search algorithm, the reader is referred to [6].

2.3 Setting Planning Buffers and Planned Production Quantities

In order to assess the effects of assortment changes in the production stages, an optimisation model is formulated to define appropriate planning buffers and production quantities for each planning period. In the considered environment, all production stages face long and often sequence dependent setup times. Lot sizes can have technical and economic restrictions in terms of rounding values, such that each lot size must be a multiple of this rounding value. Production stages produce scrap as a fixed amount of material waste during the start and end of production runs. Sequence dependent setup and scrap cost affect the planning parameters, as it becomes important that the production planner has the flexibility to create sequence optimised production plans to save setup cost. This implies that each production process step needs some planning buffer, defined as the time buffer between the provision of components for a production order and the requirement date of that order. The shorter these time buffers, the fewer possibilities to create sequence optimised production plans exist and the higher the average setup costs for each production order.

Increasing planning buffers reduces setup costs to some extent, but also increases the throughput times for that production stage, which has negative impact on subsequent stages. A longer planning buffer proportionally increases the throughput time for the item considered and thereby also the replenishment lead time of this item and successor items further downstream. As some nodes in the network keep inventories to buffer against demand variations over their replenishment lead time, these inventories have to be increased to cover this increased time interval (see section 2.2 for the the calculation of inventory levels). Therefore, the additional stock cost incurred on successor stages by increasing planning buffers have to be considered in the determination of planning buffers to yield minimal total cost.

Apart from the planning buffers, planned production quantities are the second class of planning parameters considered. However, the focus of this work is not on lot sizing models that determine production lots and sequences on the detailed planning level. Moreover, we consider planned production quantities for each production stage and the respective products for each of the aggregate time periods mentioned in section 2.1. This approach allows a reasonable assessment of the expected setup costs for different configurations of planning buffers, without the need to know exact order dates and quantities to render a detailed production plan. This view on an aggregate level exposes parallels to existing lot sizing models known as big bucket models that allow production of different products within one time period without necessarily making any statements about their production sequence.

The trade-offs to be solved in this optimisation are as follows: First, reducing setup costs by enlarging planning buffers conflicts with the goal to keep throughput times and thereby safety stocks on successor stages low. Second, increasing lot sizes to reduce the number of changeovers and save setup and scrap cost conflicts with the requirement of keeping cycle stocks low. We address this problem for each production stage separately and therefore formulate a model for a single production stage but with consideration of the safety stocks implications on subsequent stages. The objective function sums up the total cost incurred for production of a certain set of products with known expected demand volumes. The single constituents are

- Setup costs, calculated via the average setup cost for each item for the selected planning buffer and the number of planned production lots
- Scrap costs, calculated via the scrap cost rates and the number of production lots
- Cycle stock holding costs, calculated via the inventory holding cost rates and the cycle inventory that results from the decisions about production lots and the demand quantities
- Cost for inventories incurred by increased planning buffers, either on $i \in N$ if i is a stockpoint itself or on a set of downstream stockpoints if i is not a stockpoint.

Further details on the production parameter optimisation subproblem are beyond the scope of this paper. For a more detailed description of the optimisation model and solution to some the above mentioned subproblems regarding the estimation of model parameters, the reader is referred to [7].

3 Applications and Results

The entire analysis method was implemented as a decision support tool named Complana (COMplexity ANAlyser) that enables the assessment of large production and distribution networks. The implementation is done almost entirely in Java, making use of existing frameworks like Hibernate for data persistence as well as OpenTS as the framework for the tabu search procedure. The main application is developed as a Rich Client Application on the basis of the Eclipse Rich Client Platform (RCP). The optimisation model for production planning parameters is defined using the MathProg modelling language and thus can be solved with any standard solver able to process this language. Currently, the model is being tested with the freely available GNU Linear Programming Kit (GLPK) as well as the commercial CPLEX solver.

The current prototype version implements the full production and distribution network model and provides interfaces to the standard ERP system SAP to automatically generate the network models from existing data to alleviate the effort for model building. This is of special importance as the research project is carried out in cooperation with an international household product manufacturer, whose production and distribution network for the product category of cloths serves as an example application area. In order to show the practical applicability for real-world business problems, the method is validated using two baseline models, one including only the window cloths, and another including the entire cloths assortment. The software allows generating and storing these models from the imported data which then serve as baseline models for the comparisons. Assortment scenarios are then defined and applied to these models to yield new model instances with the assortment changes incorporated.

For illustration purposes, we consider one such scenario where all final products in the window cloth model were replaced with one new standard window cloth variant. For such a scenario, the Complana tool enables the quantitative assessment of the related effects in inventory management and production executing, particularly

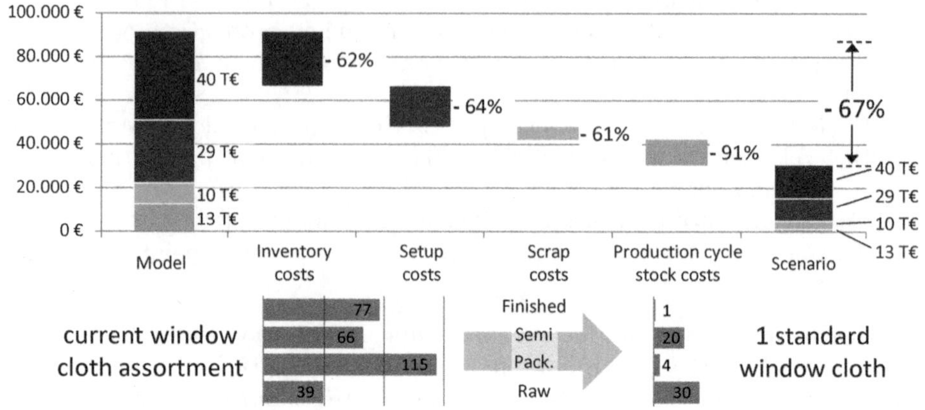

Fig. 4. Example analysis for scenario *standardisation of window cloths*

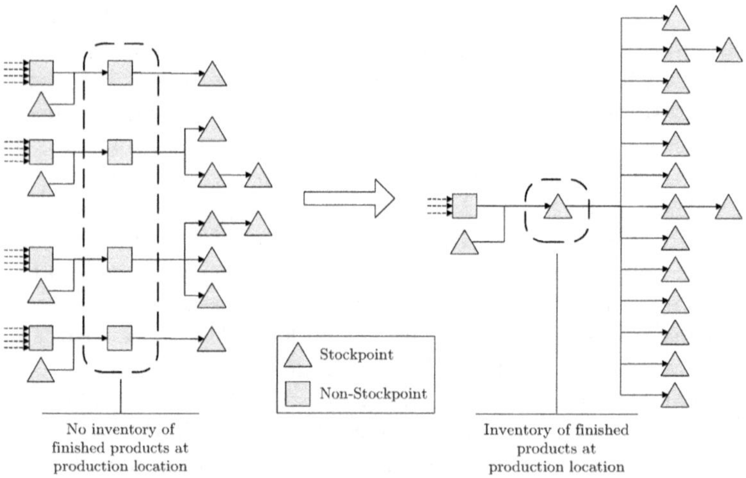

Fig. 5. Centralisation of inventories

- Inventory re-allocation, especially centralisation of inventories
- Changes in setup and scrap costs due to changing lot sizes
- Changes in production cycle stocks due to new production cycles.

Figure 4 shows what such a cost analysis may look like. The stacked bar chart on the left shows the cost composition in the baseline scenario, where total cost comprise inventory, setup, scrap and production cycle stock costs. The middle part shows what changes can be observed in the respective cost areas comparing the optimised versions of the baseline model and the alternative assortment scenario. Deducing these changes from the costs of the baseline model, the resulting cost composition of the alternative scenario is shown on the right. The bottom part of the figure additionally shows what changes in the number of finished, semi-finished, raw and packaging materials could be observed.

It may be noted that the concrete observations are heavily application-dependent. However, it was observed that irrespective of the baseline model and concrete scenario, centralisation of inventories always becomes favourable at a certain degree of standardisation. This effect leads to additional inventories of finished products at the central production location, which serve to reduce replenishment lead times of the subsequent sales companies. Figure 5 shows how these effects can be observed in a visualisation of the network structures and stockpoint status of the individual items.

4 Conclusions

This work has outlined the components and optimisation methods used in the Complana system to support assortment related decisions in the consumer goods industry. This methods and the corresponding tool enable managers to assess the effects of potential assortment changes, referring to both standardisation and amplifications of assortment, in form of *what-if* analyses. The validation with real-world scenarios has proven that the model shows the expected effects of assortment reductions in the areas of inventory management and production execution. For managers, it provides an easy means to quantify these effects automatically, using optimisation methods to adapt the production and distribution system to the new assortment. It thereby provides decision-relevant information both about the expected cost effects as well as about the required changes in the configuration of the production and distribution network.

References

1. Ramdas, K.: Managing product variety: An integrative review and research directions. Production and Operations Management 12(1), 79–101 (2003)
2. Clark, A.J., Scarf, H.: Optimal policies for a multi-echelon inventory problem. Management Science 6(4), 1782–1790 (1960)
3. Simpson Jr., K.F.: In-process inventories. Operations Research 6(6), 863–873 (1958)
4. Minner, S.: Strategic Safety Stocks in Supply Chains. Lecture Notes in Economics and Mathematical Systems, vol. 490. Springer, Heidelberg (2000)
5. Zipkin, P.H.: Foundations of Inventory Management. McGraw-Hill/Irwin, New York (2000)
6. Danne, C.: Assessing the Cost of Assortment Complexity in Consumer Goods Supply Chains by Reconfiguration of Inventory and Production Planning Parameters in Response to Assortment Changes. Dissertation, Universität Paderborn, Heinz Nixdorf Institut, Wirtschaftsinformatik, insbesondere CIM, Paderborn (November 2009)
7. Danne, C., Dangelmaier, W.: Setting planning buffers in MRP driven manufacturing networks considering their implications on safety stock cost. In: Proceedings of the of the 13 IFAC Symposium on Information Control Problems in Manufacturing, Moscow, Russia, IFAC, June 2009. Elsevier Science, Amsterdam (2009)

Dynamic Supply Loops – A Concept for Flexible and Faster Automotive Supply Network Management

Wolfgang Menzel[1], Joachim Lentes[2], Andre Döring[3], Roland Ericsson[4], and Levi Siljemyr[4]

[1] Continental AG, Guerickestrasse 7, 60488 Frankfurt
[2] Fraunhofer IAO, Fraunhofer Institute for Industrial Engineering IAO, Nobelstrasse 12, 70569 Stuttgart
[3] University of Paderborn, Heinz Nixdorf Institute, Fürstenallee 11, 33102 Paderborn
[4] Fujitsu, Kruthusgatan 17, S-411 04 Göteborg
wolfgang.menzel@continental-corporation.com, Joachim.Lentes@iao.fraunhofer.de, adoering@hni.upb.de, {Roland.Ericsson,Levi.Siljemyr}@se.fujitsu.com

Abstract. The situation of automotive industries can be characterized by low and fluctuating demands for final products. Components supply and as well production systems and networks are too often optimized for the operation at high degrees of utilization without explicitly supporting flexibility. Increasingly, production strategies like Build-to-Order (BtO) are applied to enhance the flexibility of the production networks in automotive industries. Customize-to-Order (CtO) is a promising approach to avoid the efforts for coordination and control that result out of BtO. Research and development to bring CtO to life by means of its process and technology enablers is done in the EU-funded project ACDC. ACDC develops a highly dynamic and robust supply loop concept, which is superior to the conventional hierarchic system in reactivity, reliability and costs while maintaining the 100 % guarantee of delivery. For the proof-of-concept a characteristic next generation automotive modular system is being developed, which merges different technologies, mechanics and electronics, into high quality modules to reduce part-count and cost, i.e. first cost, cost for stocks and stocks itself, and to achieve a customer-neutral component/supply concept.

Keywords: Automotive Supply Chain Management, Event Handling, Collaborative Forecasting, Product Development- Customize-to-Order.

1 Introduction

One essential effect of the financial downturn in late 2008 is the general lack of cash and financing sources with serious impacts on the automotive industries, concerning especially sales caused by dramatically decreased market demands and the pressure on suppliers.

The arising new challenges for the automotive industries and especially the suppliers are therefore the substantially reduced predictability, even higher capacity and

W. Dangelmaier et al. (Eds.): IHNS 2010, LNBIP 46, pp. 130–140, 2010.

flexibility demands of the OEM (Orginal Equipment Manufacturer). The ability to operate a supply chain network becomes a decisive role for suppliers.

Goal-oriented strategies to strengthen the European Car Industry are firstly branding & differentiation to deal with the increasing individuality, the needed flexibility and the call for a high-class image. Secondly, the reduction of delivery times resulting in fast responses of a highly flexible overall production system and low working capital. Lastly the questioning of traditional production strategies by introducing approaches like build-to-order.

In the past decades several schemes were proposed to support collaborative planning in supply chains (see [1] for an overview). Actually the supply chain in the automotive industry is organized as a hierarchical upstream planning system, proceeding top-down from the OEM to its suppliers [2]. In [3] we figured out, that the the planning process encodes restrictive planning conditions where the OEM forces the $tier_1$ suppliers to fulfil its specific demands without compromises and delivering the needed information to generate a robust and reliable plan for a longer time period at $tier_1$. The same pattern is repeated at $tier_n$ and $tier_{n+1}$. Hence, $tier_{n+1}$ supplier loses flexibility in its planning pro capacity overloads because of uncertainty according to future demand developments and the needs of an enhanced event handling system to react fast to uncertain demand changes and occurring material shortages.cesses while more often reacting $tier_n$ needs whether then controlling material flow and planning cycles.

Forced through the supply chain, this problem leads to

1. a loss of optimization potentials in local planning decisions, caused by restricted information policies,
2. capacity overloads because of uncertainty according to future demand developments and
3. the needs of an enhanced event handling system to react fast to uncertain demand changes and occurring material shortages.

These circumstances lead to an unstable and nervous system wasting time and money for keeping it running, therefore increase the product price while reduce the accounts.

Summarizing these problems the actual automotive supply chain is operated in local optima according to the specific situation of each partner in the supply chain. E.g., the OEM can optimize his costs unilaterally by forcing the tier1 suppliers to deliver only just in time. [3]

Therfore the vision behind the introduced concept of ACDC (Advanced Chassis Development for 5-Days Cars) is the development of a vehicle production & supply system to deliver a customer ordered vehicle in 5 days. This vision not only targets short order-to-delivery times and low stocks, but the overall flexibility of the automotive production grid. The approach to reach this vision is a dynamic supply network system for the automotive supplier industry that fully supports the "3 H's", i.e. to be "Highly reactive", "Highly reliable" and "Highly flexible".

1.1 Efficient Supply Chains through Customize-to-Order

The enabler for this new supply network system is the Customized-To-Order (CtO) principle. To implement the CtO-principle we propose an integrated change in three major areas of today's automotive supply chain management, illustrated in figure 1. The future automotive supply chain will:

1. deliver new types of products based on mechatronic components that are widely configurable by software, called late-customization, and which can be developed in a distributed development system not limited to a specific location of a development center. From a technical point-of-view, the developed highly mechatronic automotive modules support a late customisation of order-neutral modules towards costumer-specific requirements. Derived from the novel automotive chassis technology developed in ACDC there is plenty of potential for even new drive trains, electrical propulsion, and new wheel systems as well as for radically reducing logistical planning complexity.
2. consist of rather flexible production systems that assure a certain amount of flexibility (plus and minus) according to an average utilization of a plant. The effect is that on the one hand the plants can react very fast to increasing or decreasing demands or supplies without higher stock levels, the delivery of parts can be easily guaranteed and the event handling process during the supply chain operation can be based on more reliable short time production capability data.
3. be based on a more collaborative but strictly organized management system throughout the whole supply chain called Dynamic Supply Loops. The level of collaboration could be determined according to the level of available information. The objective is to enable a fast and reactive demand and supply handling system while regarding certain interest of all partners in the supply chain to shift the efficiency of the supply network to a more global the local maximum.

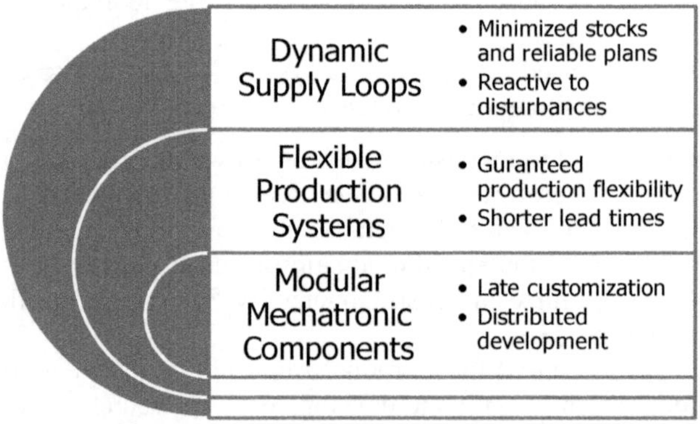

Fig. 1. Integrated view on the new Automotive Supply Network

The project ACDC is split into two incorporating work packages (WP). Work package 1000 implements the technical product oriented principles needed for the late customization approach. Work package 2000 takes care about the supply network planning and control tasks for the ACDC approach. Importantly, ACDC maintains the 100 % guarantee of delivery as an uncompromised constraint. It leaves hierarchic production concepts behind by building on multiple planning loops. This dynamic supply network management is an ideal test case for the integration of both the high-tech modular technology and the appropriate process configuration features.

2 Step-Change in Component and System Technologies

The ACDC objectives in component and system technologies are aiming at the development of component and system technologies, enabling a significant increase in variability and flexibility from a technical viewpoint and at the same time allow a reduction in stocks.

From a technical point-of-view the development of highly mechatronic and individualized automotive chassis modules – in this case a Methods, processes and tools developed in work package n advanced rear-axle module - is an enormous challenge. This enables further developments as of customer-neutral module design methodology to ample applications. Derived from this novel automotive chassis technology there is plenty of potential for even new drive-trains, electrical propulsion and new wheel systems. Existing safety functions will be enhanced and driving comfort options increased. Technical progress in intelligent software and sensor-actuator technology combined in customer-neutral mechatronic chassis modules pave the way to the next generation of automotive chassis, which needs to be taken into account by new automotive production processes.

The potentials and feasibility are exercised on the automotive chassis as a master component system, which comprises the necessary technology convergence of mechanics and electronics to mechatronics that can be featured very late in the production sequence via advanced software technology concepts.

One of the main results is the implementation of the rear axle setup to the ACDC project is to demonstrate the feasibility of the developed modularization concept. The concept is shown at a SUV (Volkswagen Touareg) rear axle. The rear axle is equipped with mechatronic actuators, an active stabilizer, a torque vectoring rear axle differential and electronic dampers. Also the rear axle has two wheel speed sensors. To meet the increase in variability and flexibility all the connections of the rear axle will be consolidated in two central connectors, a power and a signal connector.

Also an impact sound sensor was developed. Essential benefits of this new sensor is that it can be customized late-in-time, that it has a scalable bandwidth and a two band evaluation (LF/HF). By that it is fitting for different tasks at different positions within the vehicle. To meet the desired increase in variability and flexibility the impact sound sensor can be equipped with add-on functionalities like road or manoeuvre detection. The benefit of the sensors flexibility and late-in-time customization ability for the future is a one sensor hardware solution with one single part number which can handle different tasks in different positions in the vehicle.

Algorithms for those add-on functionalities are developed by ACDC as well. The development already includes promising simulation results for the classification of different road conditions with different classification methods, based on the measurements of the impact sound sensor.

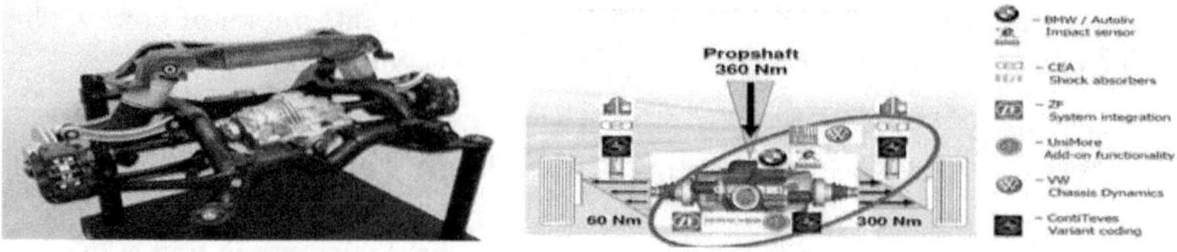

Fig. 2. ACDC Rear Axle Demonstrator (left) – ACDC Rear Axle Partners (right)

Another main result is a security concept for mechatronic automotive ECUs (Electronic Control Unit), that can be individually configured by uploading different software variants, for example for the already mentioned late-in-time customization. The security concept addresses the following threats and other non-desirable actions which include, but are not limited to:

- illegal copying of software from and to the ECU,
- illegal reverse engineering of control program source code,
- use of incompatible variants of software and ECU, and
- secure transmission of software between originator and recipient.

To test the interoperability of various ECUs in individualized automotive chassis modules a hardware-in-the-loop test bench was enhanced to integrate that various ECUs of the different sensors and mechatronic actuators. The enhanced hardware-in-the-loop test bench is able to validate the interoperability of the different ECUs with simulated driving manoeuvres. A focus of that validation is also to ensure that the electrical system within the chassis modules and between the chassis modules and the rest of the vehicle meets the necessary specifications. Part of that was also the development of simulation models for magneto–rheological dampers and an electric power steering system.

To test the interoperability of various automotive chassis modules within a vehicle a remote test framework prototype was developed. The Remote Test Framework creates a virtual car CAN (Controller Area Network) network over standard Internet TCP/IP connections which enables a connected hardware-in-the-loop testing scenario between automotive OEMs and its suppliers.

This new component & system technologies together with the highly reactive supply chain enable a highly flexible and late-in-time customizable vehicle production as well as a more responsive automotive industry.

4 Individually and Highly Reactive Supply Chain

Methods, processes and tools developed in work package 2000 "Dynamic Network Management" have the potential of a wide impact on the automotive supply chain. Their implementation will prove to be of huge benefit for the whole supply grid given the reality of the current unparalleled economic downturn. There is one important prerequisite for the success of the Dynamic Supply Loops (DSL) .Ideally the products have to fulfill the modularity and late customizable aspects, which are researched and realized on a rear axle in the WP1000.

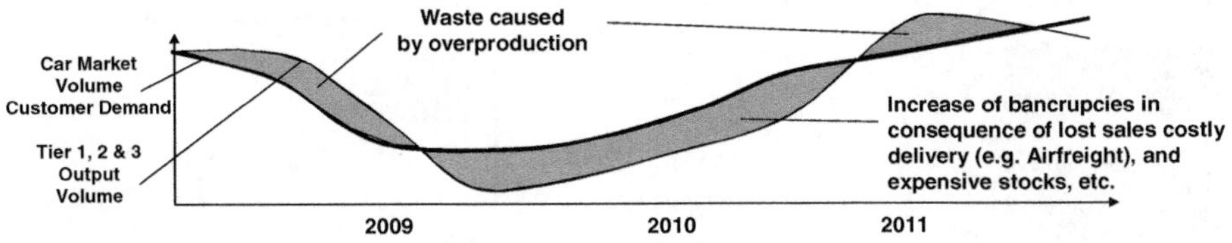

Fig. 3. Economic consequences of the recession on the automotive sector

The figure below mirrors the economic consequences of the recession on the automotive sector and its impact on the supply network. While car makers noticed a massive drop in customer demand in 2008, already, they still maintained their component orders in their systems without giving an early warning. This caused enormous overproduction throughout all supply tiers once the OEMs were forced to suspend or cut production short term.

As emergency measure, state aide in form of scrappage incentive schemes helps to prevent the worst and to provide the necessary kick-start to demand for the automotive industry. Caused by overproduction and lost sales, many suppliers are forced into bankruptcy or radical shrinkage of their companies let alone the cuts in research and development. Once the markets recover, there won't be enough supply capacities left to guarantee the badly needed rapid improvement of the motor industry. This is a typical scenario of the current existing (hierarchical) Supply Chains.

4.1 Main Benefits and Targets of Dynamic Supply Loops

ACDC provides an integrated approach for mid and long term forecasting as well as a highly reactive method to tackle all upcoming events that might hurt the supply of the OEM. The result is a fast, flexible and high reactive process supported via high level of automation. Dynamic Supply Loops create high transparency by combining a central forecasting approach with decentral information and supply loops that make sure events are being considered in real time [4]. Both aspects require a high level of collaboration in SC network. The result is minimization of leadtime to reduce inventory and cost and avoid loss of sales.

The resulting process is a one step negotiation Methods, processes and tools developed in work package process which could be implemented very easy and allows very fast reaction times. The calculation of planning scenarios can be done by using standard planning methods, e. g. offered by SAP or other ERP-Systems. Because of using implicit information of the $tier_{n+1}$ at the scenario generation of $tier_n$ the accepted plans will be more focused to the actual situation of $tier_{n+1}$ in opposite to the classical top-down driven planning procedures in today´s automotive supply chain. This information includes e.g., the knowledge of capacity capabilities, flexibility agreements, quality issues or specializations as well as the frame plan at $tier_{n+1}$. This implicit information will be gained by collecting information about $tier_{n+1}$ at $tier_n$ and by using new protocols based on EDI standards (Electronic Data Exchange).

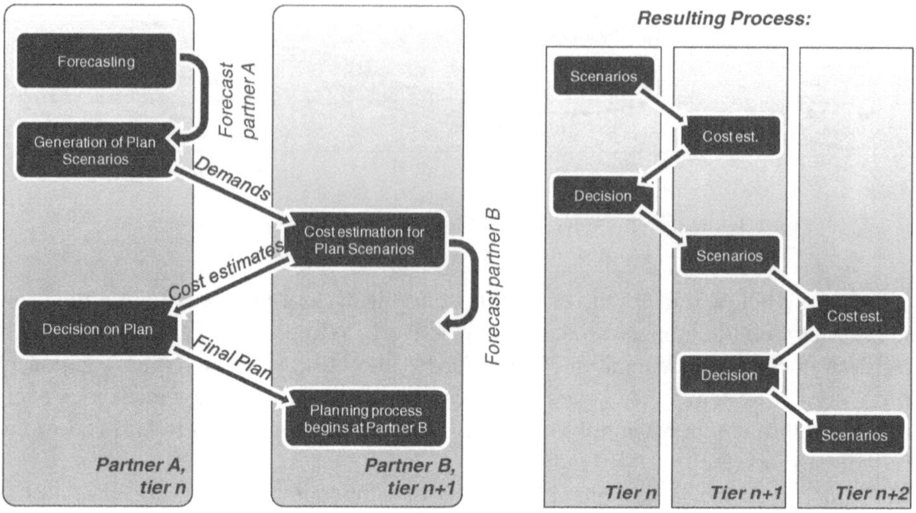

Fig. 4. Principle of the Dynamic Supply Loops

The suppliers are getting more flexibility, now able to reduce costs and stocks because of focused and reliable plans from their customers. Despite costs e. g. the optimization of lead times could also be factor in choosing scenarios.

To offer incentives to take part in this planning concept the $tier_n$ and $tiern_{n+1}$ could define some regulation for balancing the achieved benefits [5]. E. g. the supplier $tier_{n+1}$ offers supplier $tier_n$ a optimal managed vendor managed inventory without including the costs into the products and the $tier_n$ supplier regards the specific $tier_{n+1}$ suppliers situation while generating the planning scenarios. Both have profit in this situation: the inventory level is on the needed level at sufficient costs and the supplier tier-n+1 is able to operate a robust production system relying on expected customer-call-offs. Events will be avoided and costs as well as stocks can be reduced.

4.2 Selected Achievements

Basis (operational, tactical and strategical) Dynamic Supply Loop processes combined with "benefit balancing models" are defined and are implemented in the ACDC use case at Continental supply network. In the actual project phase optimization and adaption processes of the developed methods are started.

The implementation of the Dynamic Supply Loops is based on a toolkit shown in figure 5. There are tools precise processes established for each mentioned planning level and for re-designing and operation flexible production systems.

Major achievements have been actually established in the area of collaborative demand prediction and event management.

The collaborative demand prediction takes its starting point from a model presented by Småros [6]. This model considers levels of complexity in collaboration and in what order to take these steps. In ACDC we have taken this from a "what-to-do"

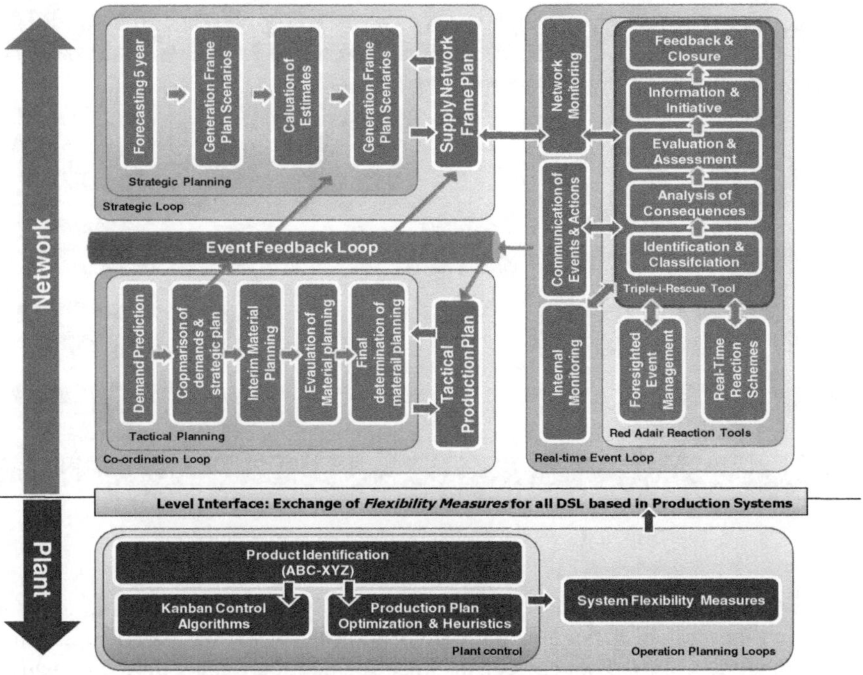

Fig. 5. Basic Process Approach in ACDC

model to an "how-to-do" model by specifying rigid frameworks for each level of collaboration, and the result are quite astonishing (e.g. reduced forecast error by 75%).

Collaboration about demand forecast can be made to the extent that fits the supply chain actors. The collaboration is based on an agreement between a customer and a supplier about the information that may be shared. In traditional methods for collaboration, there is typically a choice for a supply chain actor to share or *not* share information. ACDC suggest methods to limit the visibility of sensitive information without loosing the information that is essential for other actors.

In ACDC, four levels of collaborative demand prediction have been defined as shown in figure 7. Unsynchronized supply chain capacity is one of the major cost drivers in the automotive industry. For each level of collaboration, the rate of synchronized supply chain capacity will increase.

At the first level of collaboration the format for communication of demand data is defined. This includes not only to make the data visible to other actors, but also to clarify the understanding of the data. Already by increasing the understanding about the definitions of data and the rate of uncertainty in the data, the quality of the decisions based on the data will improve. Methods, processes and tools developed in work package.

At the second level demand forecasts are generated by statistical methods. In ACDC a range of time series models and explanatory models have been tested on a use case at Continental. In comparison, an advanced explanatory model can give 2-4 times less forecast error compared to a simpler time series model.

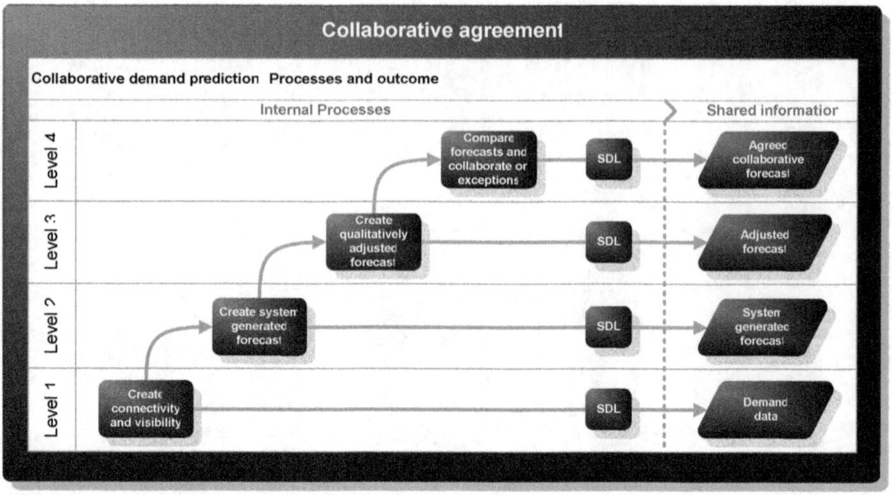

Fig. 6. Levels of Collaborative Demand Prediction

The statistical model cannot possibly cover all aspects that drive the demand. At level three the system generated forecast is adjusted based on expert judgments and management decisions. This can for instance include the impact on the demand from sales actions from a car manufacturer. To start the understanding of causal factors that has not been included in the demand model, ACDC suggest to regularly measuring the forecast precision. This is a factual basis for discussions between the actors, and can also be used to identify reasons to single or systematic deviations in the forecasts.

Finally, at level four, the actors come to closer collaboration by discussing a joint view on the demand, and synchronized capacity decisions. At this level the actors are approaching something that is close a common business plan.

The first two levels are currently being tested and the last two are already realized at Continental.

The topic of event handling is driven by several methods combined under the "Red Adair Reaction Toolkit". Most of them are actually implemented at Continental. The Red Adair Toolkit is a consequence out of the requirement of larger companies to handle disturbance in the supply chain properly.

With the toolkit events can rapidly be identified and analyzed. Former similar cases can easily be used to find a proper solution for the problem. By that the dependence on single individuals becomes less. A "lessons learned" process can be run. Here analogies to the Quality management process established in automotive industry were used.

Solutions on unforeseen events mostly require extra resources. To provide them a proper and fast escalation process has to be implemented. The Red Adair tool systematically and automatically supports the escalation process in a company while by using the tool it is safeguarded that the whole organisation is aware of severe problems.

One speciality is implemented as well. During events that cause a shortage, the available material very often is not sufficient to fulfil the standard plans of internal and external customers. The material therefore has to be allocated to avoid loss of sales for the OEM.

To provide a tool for that process, the Red Adair toolkit contains an Allocation tool that enables allocation managers to handles larger shortages. They can always have an overview about the most critical locations and projects.

Value Stream Mapping and Value Stream Design as the essential steps towards production process optimization are executed in the used case plants at Continental.

The first tests for quasi online testings are done (VW-ATB).

5 Conclusion

The European automotive industry is facing enormous challenges as a result of the financial crisis, the high prices for oil and steel and emission controls. To tackle these challenges, traditional approaches have to be questioned and novel ones have to be applied. The European funded project ACDC supports this transition towards a sustainable automotive economy by means of knowledge-based products and processes by innovative approaches. For the product side, the shift from traditional mechanics towards the combination of mechanics, electronics and software, i.e. mechatronics, will be researched and developed further using the chassis as a master case. Mechatronics supports the customization of neutral components by means of software and parameterization at low costs at a very late stage of product realisation, even at the car dealer. This transformation enables the application of Customize-to-Order as a lean complement to complex BtO-approaches. Basic instrument to exploit the accompanying advantages are Dynamic Supply Loops as an iterative but fast network-based planning and operations approach.

Acknowledgement. The work presented in this paper is funded by the European Commission in the thematic priority "Sustainable development, global change and ecosystems" of the Sixth Framework Programme for Research and Technological Development (contract number FP6-SST-031520). Besides the coordinator Continental AG, the consortium consists in alphabetical order of ATB Bremen GmbH, BMW AG, Centre for Research and Technology Hellas, Commissariat à l'Energie Atomique, ERPC GmbH, Fraunhofer IAO, Fujitsu Services AB, Fundacion CARTIF, INESC Porto, MTA SZTAKI, University of Modena and Reggio Emilia, University of Paderborn, VDI/VDE-IT GmbH, Volkswagen AG and ZF Friedrichshafen AG.

References

1. Stadtler, H.: A Framework for Collaborative Planning and State-of-the-Art. OR Spectrum 31, 5–30 (2009)
2. Meyr, H.: Supply Chain Planning in the German Automotive Industry. OR Spectrum 26, 447–470 (2004)
3. Egri, P., Döring, A., Timm, T., Váncza, J.: Collaborative Planning with Dynamic Supply Loops. Submitted to 43rd CIRP International Conference on Manufacturing Systems, Wien (2010)

4. Dangelmaier, W., Döring, A., Hellingrath, B., Lentes, J., Timm, T.: Integration of Customize-to-Order and Build-to-Order Concepts for the 5-Day-Car. In: Proc. of IFAC Workshop on Manufacturing Modelling, Management and Control, pp. 43–48 (2007)
5. Egri, P., Váncza, J., Gallego, M.A.: Benefit Balancing - Concepts for a Better Collaboration. In: Dangelmaier, W., Blecken, A., Rüngener, N. (eds.) Nachhaltigkeit in Flexiblen Produktions- und Liefernetzwerken, Proc. of the 11. Paderborner Frühjahrstagung, Paderborn, pp. 151–162 (2009)
6. Småros, J.: Information sharing and collaborative forecasting in retail supply chains. Helsinki University of Technology, Doctorial dissertation series (2005)

Development of a Lean Quality Management System: An Integrated Management System

Alexander Blecken[1,3], Alexander Zobel[2], and Evangelos Maurantzas[2,3]

[1] Heinz Nixdorf Institute, University of Paderborn, Fürstenallee 11,
33102 Paderborn, Germany
blecken@hni.upb.de
[2] MBtech Consulting GmbH, Posenerstr. 1,
71065 Sindelfingen, Germany
alexander.zobel@mbtech-group.com
[3] University of Paderborn, Warburger Straße 100,
33098 Paderborn, Germany
evangelos.maurantzas@gmx.de

Abstract. The automotive industry and in particular the automotive supplier industry need to master significant challenges in the future. The Original Equipment Manufacturers (OEMs) require products of highest quality at lowest cost from their suppliers. In order to assure quality, all large automotive manufacturers require an independently certified quality management system, in particular ISO/TS 16949, from their suppliers. These quality management systems often involve a substantial efforts both in time, work and documentation and are thus frequently not consequently followed or face little employee acceptance. On the other hand, companies try to reduce production cost by establishing stable and efficient processes with methods and tools of lean management. This paper deals with an approach which integrates these two management systems, i.e. quality management and lean management, to an integrated and comprehensive Lean Quality Management System. Methods and tools of lean management are used to fulfil the requirements of the quality management system. Two pilot projects served to validate the developed integrated management system.

Keywords: Lean management, ISO/TS 16949, ISO 9001, Integrated management system, Quality management system, Automotive supplier industry.

1 Introduction

In order to increase their efficiency, automotive manufacturers demand from their suppliers products of highest quality at low prices. The effect is that an increasing number of suppliers implement new production strategies such as lean management to yield higher productivity and process stability and hence, lower production cost. Moreover, the quality standard which is required by the large majority of automotive manufacturers is ISO/TS 16949. If suppliers want to continue their relationships with automotive manufacturers, most large OEMs require them to be certified according to

W. Dangelmaier et al. (Eds.): IHNS 2010, LNBIP 46, pp. 141–151, 2010.

this norm [1]. Hence, more and more suppliers need to establish a certified quality management system according to ISO/TS 16949 in the short to medium term.

This paper deals with an approach to combine the quality management system according to ISO/TS 16949 with lean management to an integrated management system which satisfies all requirements of the individual management systems. Based on this integrated management system, a systematic application procedure is developed which can be used when implementing the integrated management system.

The remainder of this paper is structured as follows: Section 2 is dedicated to the problem of lacking integrated management systems as encountered in industry today. In Section 3, the methodology of developing the Lean Quality Management System (LQMS) is presented. Thereupon, Section 4 serves to illustrate the application procedure of the LQMS. In Section 5, two pilot projects are elaborated which proved the validity of the LQMS and the application procedure. Lastly, Section 6 concludes this paper by providing an outlook on further development and application possibilities of the LQMS.

2 Integrated Management Systems in Industry

A management system comprises all measures for design, control and development needed to achieve a company's goals [2]. Management models (such as ISO 9001, for instance) are aggregated illustrations of company-specific management systems. In order to point out cause and effect relationships, the degree of complexity of management models is deliberately chosen to be lower as in reality [3][4].

The introduction of a quality management system frequently involved a substantial amount of efforts. Efforts are incurred by the implementation of the quality management system, but also by the maintenance, upkeep and running of the quality management system which can lead to unsatisfactory employee acceptance [5].

Additional to the quality standard, pressure of competition [6] leads to the requirement of a high degree of flexibility with respect to the fulfilment of the market demand of production at lower and lower costs. In order to achieve this, more and more automotive supplier companies have tried in recent years to introduce the lean management approach in their companies [7]. The concentration on the value-adding activities and the elimination of waste from all company processes is perceived by many companies to be contradicting the implementation of a certified quality management system. This is due to the fact that the implementation of the quality management system itself involves a certain administrative overhead.

Based on an empirical survey, which has been conducted in conjunction with the development of the LQMS presented here, it was found that this is also the reason why integrated management systems for quality management and lean management are not widely spread in industry today. Rather, both systems are often found in parallel and independent from each other.

The parallel introduction of independent management system requires a high degree of professional expertise and stringent project management. If the required professional expertise is missing, redundant work is often carried out, rigid structures may be created, or contradicting requirements at the same processes are formulated.

Hence, the failure of these systems can be anticipated [10]. Thus, the implementation of both management systems without external know-how and respective support is often infeasible. Such support can be given by consulting companies. The benefit of involving consulting companies is that less stress is put on the internal organisation. Also, expert know-how can be brought into the company by the consultants. Through parallel coaching and training of employees the necessary professional expertise is conveyed.

Under consideration of the presumed contradiction as elaborated above, this paper serves to present the development and application of an integrated management system which combines lean management and quality management. This integrated management system, or Lean Quality Management System, can support automotive supplier companies when implementing an inter-divisional integrated management system. Lean methods are used to fulfil the requirements of the norm ISO/TS 16949. A holistic management system is created through the integration of the management and organisation system lean management into the norm ISO/TS 16949. Hereby, the objective is to create a management system which is actually lived and continuously improved by focussing on value-adding activities. The evolving benefit is the unification of functions and data, the unification of the individual management systems and the execution of more efficient processes.

3 Development of a Lean Quality Management System

A number of approaches to combine individual management systems can be found in the literature. According to Seghezzi three separate methods can be applied when combining individual management systems [6]: These methods are the method of addition, the method of merging and the method of integration. The method of addition foresees to add or append the requirements of the newly joined management system to the existing requirements. The two management systems are thus operated in parallel to each other [9]. The method of merging relies on a well structured base system. Contrasting to the method of addition, the contents of all other management systems are built into the base system wherever this is deemed feasible. All other contents, which cannot be feasibly built into the base system, are appended to the base system. Lastly, the method of integration is the most consequent but also the one which involves most efforts. This method foresees that all management systems are transferred into a generic management system. This needs to provide the framework for the implementation of all requirements since the original boundaries of the partial management system will not be perceivable any longer [9].

The challenge of the latter two methods is the selection of a suitable base/generic management system. This management system needs to exhibit a clear structure and offer the possibility to integrated further requirements. Dyllick and Seghezzi examine various models for their suitability as such base/generic management systems. The outcome of their study is that process-oriented models such as ISO 9001:2000, ISO 9004:2000 and the EFQM Excellence-Model are in particular appropriate to be used as base/generic management systems [10]. Consequently, ISO/TS 16949 can also be used as a base system as it is based on ISO 9001.

The integration of the quality management system according to the norm ISO/TS 16949 with lean management to a Lean Quality Management System is achieved here by the method of merging. This is due to two reasons: Firstly, a suitable base system is available in the form of ISO/TS 16949. Secondly, lean management is a management system which is relatively simple to merge as the requirements of ISO/TS 16949 are on a more abstract level as the requirements of lean management. Lean management exhibits possibilities of implementation through concrete requirements. Thus, requirements of lean management can be subordinated to the requirements of ISO/TS 16949.

The development of the Lean Quality Management System is partitioned in two phases, i.e. the definition of requirements and the merging of requirements. During the definition of requirements phase, all requirements of ISO/TS 16949 as well as all requirements of lean management are gathered und documented in the form of requirements catalogues. Subsequently, all requirements of both management systems are contrasted to each other, possible contradictions are analysed and eventually merged on the basis of the management system ISO/TS 16949.

3.1 Definition of Requirements

ISO 9001 and ISO/TS 16949 are grounded on eight principles: Customer orientation, leadership, integration of people, process-oriented approach, system-oriented management approach, content-oriented approach of decision making and mutually beneficial supplier relationships [15]. Every formulated requirement is taken from the IATF guideline to the norm ISO/TS 16949 [11] and further literature [12] and attributed to one of these principles.

The requirements from lean management also originate from literature sources and are partitioned into the four areas as defined by Liker: Long-term philosophy, processes, people & partner and problem solving [13] [14]. An illustrative excerpt of the coded individual requirements is presented in Tables 1 and 2.

Table 1. Requirements of lean management (excerpt)

Code	Requirements
LP 5.4	Visual systems need to be available that point at processes or machines which need support.
LP4.1/M5	Demand spikes and capacity bottlenecks of personnel, machines/equipment and in the production schedule need to be levelled.
LP6.1/M4	Stable and repeatable methods and processes need to be introduced (standards).
E3	The process of product development needs to be levelled after the creation of a base concept and be transferred into a flow.

Table 2. Requirements of ISO/TS 16949 (excerpt)

Code	Requirements
5.6.1	Management assessment needs to contain opportunities for improvement and change requests of the quality management system.
6.2.1	The organisation needs to ensure that the employees are able to execute the tasks which influence the quality.
7.5.1	Production needs to be planned and executed under controlled conditions.
8.5.1.2	The improvement of the production process always needs to be geared towards the control and reduction of variance of product characteristics and production process parameters.

3.2 Merging of Requirements

In the next step, the requirements of lean management and ISO/TS 16949 are merged (cf. Fig. 1). As the ISO/TS 16949 is used as base system here, the requirements of lean management are attributed to the respective elements of the norm ISO/TS 16949. The result of this merging is the LQMS requirements catalogue.

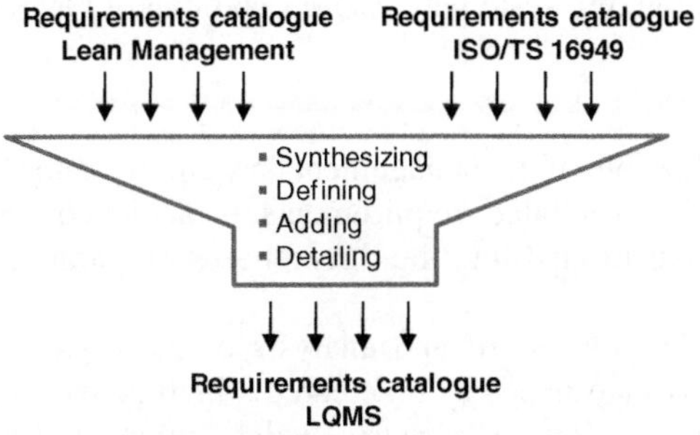

Fig. 1. Merging of requirements

The merging of the two management systems is eventually carried out by means of four different fashions: Synthesizing, defining of new requirements, adding of a new element and detailing. While synthesizing, requirements of ISO/TS 16949 and lean management with complete content congruence have been condensed. Moreover, requirements with respect to the organisation's philosophy existed which could not be attributed to any other chapter of the norm ISO/TS 16949. Here, the new element "5.3 Organisation Philosophy" has been created for the LQMS. Five novel requirements have been identified in order to complement the merging of the two management system. These newly defined requirements are described in Table 4. Lastly, the most frequent way of merging has been through detailing. Here, requirements of lean management, which are often more detailed, have been subordinated to the corresponding requirements of ISO/TS 16949.

Table 3. Novel requirements of the LQMS

Code	Requirements
V1	The organisation needs to appoint a person responsible for lean management, reports on its efficacy and initiates improvements projects.
V2	Internal audits need to be carried out in order to ensure the effectiveness of lean principles.
V3	The focus of internal audits needs to be put on improvement projects which have only recently been launched.
V4	During the beginning of the implementation process, internal audits need to be carried out more frequently.
V5	Senior management needs to consider the efficacy of lean management in the management review.

4 Application Procedure of the Lean Quality Management System

In the previous section, the requirements of the LQMS have been developed. It is the objective of this section to make the LQMS deployable in industry practice. Hence, a suitable application procedure is derived. The components of the application procedure are mainly the definition of a systematic LQMS audit and LQMS implementation.

4.1 LQMS-Audit

The systematic evaluation of a management system with the objective to monitor whether it fulfils all formulated requirements is achieved through an audit. The LQMS audit can be partitioned into the three phases preparation, execution and post-execution.

Preparation includes the record and analysis of the organisation's internal processes. Thus, a process map needs to be created which provides an overview over all processes of the organisation. All control, value-adding and support processes are covered here. Subsequently, all processes are analysis with the help of a turtle model [15]. The turtle model describes every process with all necessary parameters, measure and interfaces. On this basis, audit emphases, i.e. LQMS elements, are defined which are to be covered in particular during the audit.

The execution of the LQMS audit is carried out based on the LQMS question catalogue and the derived turtle models. Foundation for the LQMS question catalogue are the LQMS requirements as developed in Section 3.2. The LQMS question catalogue contains lead questions for the respective elements of the LQMS which are attributed to the analysed processes. The lead questions allow the auditor to assess to which degree each requirement of the LQMS is fulfilled.

If sufficient information can be gathered, the construction of organisation-specific turtle models is recommended. If, however, an organisation does not currently have defined processes, standard processes have been elaborated together with the respective turtle models, which can be used. These standard processes and turtle models serve to enable structured audit procedures and ought to be used when no further information on the organisation is available.

According to Schwendt and Funck, the audit of an integrated management system is subject to specific demands, which have been respected when the LQMS question catalogue has been derived [10]:

- The question catalogue needs to be extendable and flexibly adaptable to changes in the norm.
- The question catalogue's reference to the norm needs to be extendable.
- The organisation's objectives as well as the process objectives need to be monitored in terms of the audit.
- Only one audit protocol, report and review may be created.
- The question catalogue instrument needs to ascertain the audit of the processes.
- The question catalogue instrument needs to support the systematic improvement of the management system.

The assessment of the individual requirements consists of five steps. These five steps include the two assessment steps provided by the IATF "Major Deviation" and "Minor Deviation" [16]. This assessment scale has been extended by the steps "Remark" [17], "Good" and "Best Practice". Through the assessment as "Best Practice" it becomes possible to stress the fulfilment of a requirement which is above average.

4.2 LQMS Implementation

Starting from the results of the audit, solution suggestions which match the requirements are derived for the identified weaknesses and integrated in the to-be processes. [18] The LQMS toolbox functions as an aid for the processing of the identified deviations and improvement potentials. For each element of the LQMS respective tools are available as required. These tools can be used for the analysis and development of to-be concepts.

For the development of the LQMS toolbox, all tools and methods which are used for the two management systems in industry practice today have been collected. Similarly to the formulation of requirements, certain overlaps have also been identified here. Therefore, matching tools and methods have been identified which originate from both areas lean management and quality management. These have been merged in order to avoid duplication and redundancies in the toolbox. Eventually, the tools and methods have been transferred into the LQMS toolbox and have been connected with the elements and standard processes of the LQMS. Hence, it becomes possible that suitable tools and methods to tap the improvement potentials can be selected while viewing the audit results in either an element-oriented or process-oriented perspective.

Fig. 2 displays the three dimensions of the LQMS toolbox. On the axis "QM Elements" the five elements of the LQMS requirements catalogue or respectively the LQMS question catalogue are shown. On the axis "Tools & Methods", an excerpt of the methods and tools which can be used to implement the LQMS is presented. On the axis "Processes" the individual processes of the organisation of the LQMS standard processes are presented depending on the original situation (only indicated here).

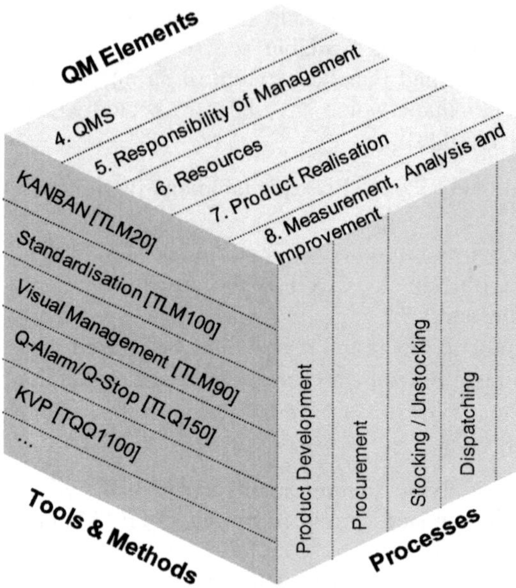

Fig. 2. LQMS toolbox

5 The LQMS in Industry Practice

The LQMS developed here has been deployed in two pilot projects which served to prove the validity of the integrated management system. In order to achieve this, three criteria have been defined: Applicability, adaptability and benchmarking capability. The objective has been to acquire constructive feedback from industry practice already during the development phase of the LQMS. Thus, this feedback could be integrated in the further development of the LQMS itself and its application procedure.

The criteria applicability refers to the degree to which the systematic application procedure, the question catalogue and its assessment procedure are fit for industry practice. The criteria adaptability refers to the degree to which the LQMS can be applied to various organisation structures and models. Here, organisations are distinguished which currently exhibit a different degree of implemented quality and lean principles. Lastly, the two pilot projects have served to set a benchmarking level for the degree to which the requirements have been fulfilled by the two companies (benchmarking capability).

The application procedure described above has been used in two pilot projects to assess the conformance of the company-specific processes with the requirements of the LQMS requirements catalogue. It was also the goal to derive company-specific recommendations for improvements.

Company A is a manufacturer of precision machined parts. With its 120 employees, the company produces small parts for the automotive industry but also for other industries such as medical technology. Company B is market and technology leader in the utility vehicle industry. At a number of production facilities, 3500 employees manufacture trailers, semitrailers and superstructures for utility vehicles.

Both companies together are very well suited to perform as application instance of the LQMS as they are quite different in terms of their structures and processes. Company A is certified according to ISO/TS 16949 and hence operates a well-functioning quality management system. However, lean principles are only scarcely employed. Company B is certified according to ISO 9001 and commands over a strong lean management system, both in the production and in the administrative function.

In the following, the example of company A shall be used to illustrate how the graphical assessment of the LQMS audit can yield recommended actions. The audit results of company A are displayed in Fig. 3 in the form of a spider web diagram. The three major potentials for improvement which have been identified are material logistics, process management and production planning. Thereupon, the following recommendations can be derived:

- The introduction of a KANBAN system leads to transparent disposition processes and the comprehensive application of the FIFO principle.
- The introduction of Shop-Floor Management contributes to stabilise and continuously improve production and work processes. The visualization of current production information over so-called Shop-Floor Boards allows management to take rational and well grounded decisions on a short time horizon, for instance based on shift-related numbers for rejected parts.

The procedure for deriving recommended actions for company B has been carried out analogously. Hence, the recommendations for company B are not further elaborated here.

In conclusion, the application instance has shown that the developed LQMS and its application procedure are well suited for different kinds of organisations. Likewise, the question catalogue is suitable for its intended purpose of conducting and supporting audits. The systematic assessment in the form of spider web diagrams displays the outcome of the as-is analysis in an easily comprehensible manner and highlights areas with specific potential for improvement.

Considering that the LQMS has been employed in pilot projects at two companies with differing prerequisites in the areas of lean management and quality management, it has also been possible to judge the adaptability of the LQMS. During the assessment, the strengths and weaknesses of the respective companies in the areas lean management and quality management have been rediscovered. Company A was able to fulfil all requirements of the norm but not all requirements of lean management. A contrasting result was obtained with company B. Despite this difference, a seamless execution of the LQMS audit has been possible. Therefore, the adaptability of the LQMS could be confirmed.

Through the two application instances, some values could be gathering for setting a benchmark. In order to consider these values as representative, however, an objective assessment according to the same standard needs to be guaranteed. One step into this direction has been taken with the definition of an assessment scale and a detailed description of the assessment steps. Yet, a certain degree of subjectivity remains when the fulfilment of the requirements is evaluated.

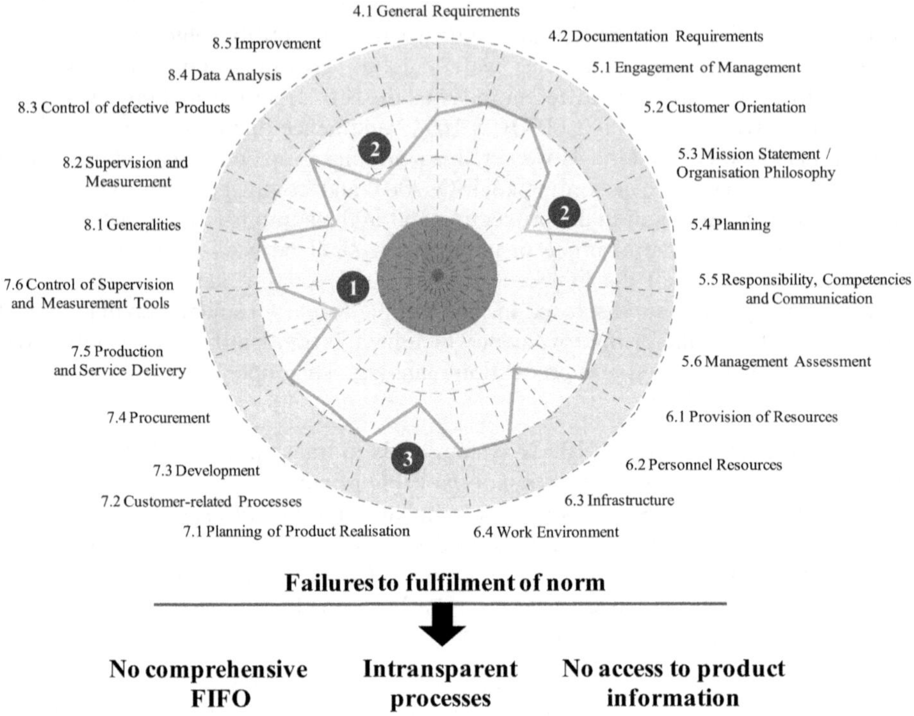

Fig. 3. LQMS audit results company A

6 Outlook

In this paper, it was shown that the requirements of a quality management system are not necessarily conflicting with the lean principles. Through the consequent use of the right lean methods, processes and procedures can be consequently and sustainably optimised at the value-adding locations while requirements from the quality management system are fulfilled. Two pilot projects concerned with implementing the LQMS have served to identify the respective weaknesses of the quality management systems and recommendations for actions have been formulated under consideration of lean principles. The objective is to integrate the requirements of the quality management system into the production and work processes in such a way that no overhead efforts are required by the quality management system. Hence, the introduction of a KANBAN system in the first pilot project helped to significantly improve the material disposition process while simultaneously fulfilling the demand for the comprehensive application of the FIFO principle. Since the KANBAN system represents a consumption-oriented production control method, the FIFO principle has to be adhered to by all employees in order to avoid a production halt.

The LQMS presented here has been specifically developed for the needs of the automotive and automotive supplier industry. Therefore, the requirements of the ISO/TS 16949 have been used as a base system. Yet, the second pilot project has shown that the approach followed here can also successfully be applied to other industries. Currently, efforts are undertaken to adapt the LQMS to the requirements of medical technology. In the future, a general approach unspecific to a certain industry could emerge which harmonizes the requirements of quality management with the requirements of lean management.

References

1. International Organization for Standardization: ISO/TS 16949:2002 - Qualitätsmanagementsysteme - Besondere Anforderungen bei Anwendung von ISO 9001:2000 für die Serien- und Ersatzteilproduktion in der Automobilindustrie (2002)
2. Pischon, A., Liesegang, D.G.: Integrierte Managementsysteme für Qualität, Umweltschutz und Arbeitssicherheit. Springer, Heidelberg (1999)
3. International Organization for Standardization: ISO 8402:1994 - Quality management and quality assurance – Vocabulary (1994)
4. Pischon, A., Liesegang, D.G.: Integrierte Managementsysteme für Qualität, Umweltschutz und Arbeitssicherheit. Springer, Heidelberg (1999)
5. Allweyer, T.: Geschäftsprozessmanagement – Strategie, Entwurf, Implementierung, Controlling. W3I (2005)
6. Becker, H., Stahl, M.J., Dutka, J., Straub, N.: Auf Crashkurs - Automobilindustrie im globalen Verdrängungswettbewerb. Springer, Heidelberg (2005)
7. Pfeiffer, W., Weiss, E.: Lean Management - Grundlagen der Führung und Organisation lernender Unternehmen. Erich Schmidt Verlag (1994)
8. Schwendt, S., Funck, D.: Integrierte Managementsysteme – Konzepte, Werkzeuge, Erfahrungen. Physica Verlag (2002)
9. Seghezzi, H. D., Fahrni, F., Herrmann, F.: Integriertes Qualitätsmanagement - Der St. Galler Ansatz. Hanser Verlag (2007)
10. Dyllick, T., Seghezzi, H.D.: Integrated Standardized Quality and Environmental Management Systems – A Look at the Current Situation. CEM Business Review 2 (1998)
11. International Automotive Task Force: IATF Leitfaden zur ISO/TS 16949:2002. Erste Auflage (2002)
12. Bruckner, C.: ISO/TS 16949:2002 im Klartext – Was Sie bei der Umstellung/Einführung der Technischen Spezifikation beachten müssen. WEKA MEDIA (2003)
13. Liker, J.K.: The Toyota Way - 14 management principles from the world's greatest manufacturer. McGraw-Hill, New York (2004)
14. Morgan, J.M., Liker, J.K.: The Toyota Product Development System – Integrating People, Process, and Technology. Productivity Press (2006)
15. Cassel, M.: ISO 9001 - Qualitätsmanagement prozessorientiert umsetzen. Hanser Verlag (2007)
16. International Automotive Task Force: Zertifizierungsvorgaben der Automobilindustrie zur ISO/TS 16949:2002, Erste Auflage (2002)
17. Gietl, G., Lobinger, W.: Leitfaden für Qualitätsauditoren – Planung und Durchführung von Audits nach ISO 9001:2000. Hanser Verlag (2004)
18. Cassel, M.: ISO/TS 16949 - Qualitätsmanagement in der Autoindustrie umsetzen. Hanser Verlag (2006)

Integrated Adaptive Design and Planning of Supply Networks

Dmitry Ivanov[1], Boris Sokolov[2], and Joachim Kaeschel[1]

[1] Chemnitz University of Technology, 09126 Chemnitz, Germany
[2] Saint Petersburg Institute of Informatics and Automation of the RAS (SPIIRAS), Russia
{dmitry.ivanov,joachim.kaeschel}@wirtschaft.tu-chemnitz.de,
sokol@iias.spb.su

Abstract. In recent years, the works on supply chain management (SCM) have been broadened from isolated static models of supply chain design (SCD) and planning (SCP) to integrated SCD and SCP models. This paper develops a framework for integrated SCD and SCP on the basis of adaptation principles and regarding SC execution dynamics. To achieve this integration, static models of SCD are brought in correspondence to the dynamic models of SCP and control. The adaptive feedback loops between the planning and execution models are established. It becomes possible if traditional operations research (OR) techniques are extended by optimal control theory. We illustrate the general framework with the help of a modelling complex. We show explicitly how to distribute static and dynamics variables and constraints of SCD and SCP by interconnecting static elements in SCD optimization linear programming model with corresponding SCP dynamic elements in optimal control model. This makes it possible to consider conventionally isolated SCD and SCP problems taking into account non-stationarity of supply chain execution along with the adaptive control within a conceptually and mathematically integrated framework. In doing so, the developed framework contributes to the advancing decision-making support for SCM on the basis of interrelating planning and execution levels instead of generating optimal solutions that fail in a real perturbed execution environment.

Keywords: Supply chain, design, planning, operations research, optimal control, adaptation.

1 Introduction

The term "*Supply Chain Management*" (SCM) was coined in the 1980-90s. Presently, SCM is considered as the most popular strategy for improving organizational competitiveness along the entire value chain in the twenty-first century [3, 9, 15, 33].

A *supply chain* (SC) is a network of organizations, flows, and processes wherein a number of various enterprises (suppliers, manufacturers, distributors, and retailers) collaborate (cooperate and coordinate) along the entire value chain to acquire raw materials, to convert these raw materials into specified final products, and to deliver these final products to customers [15]. After long-lasting research on SC optimality

W. Dangelmaier et al. (Eds.): IHNS 2010, LNBIP 46, pp. 152–163, 2010.

from the service level's and costs' points of view, the research community has begun to investigate flexible, agile, adaptable, stable and crisis-resistant processes in SCs to compete in a real perturbed execution environment [8, 28, 38].

With the development of SCM, new integrated problems such as distribution-production or production-inventory problems have arisen. In some cases, it is possible to integrate different problems within one of the modelling techniques, e.g., distribution-production planning with mathematical programming [3, 6, 26, 29, 39]. If this is impossible, integrating elements of different techniques should be considered [15]. In recent years, the works on SCM have been broadened to cover the SC dynamics. In these setting, design of SCs (SCD) and planning of orders in SCs (SCP) is to consider as an *integrated model on the basis of adaptation principles* [15]. To achieve this integration, static models of SCD should be brought in correspondence to the dynamic models of SCP with establishing feedback loops between the SCD and SCP.

It becomes to be possible if traditional operations research (OR) techniques can be extended by means of optimal control theory [15]. This paper joins this research stream and contributes to SCD and SCP from the perspectives of adaptable plans instead of "ideal" optimal solutions which fail in a real execution.

This paper is organized as follows. We start the paper with the state-of-the-art analysis to the SCD and SCP from conceptual and mathematical perspectives. Section 3 discusses the methodical basic of the proposed approach to adaptive SC modelling and optimization. In section 4, a mathematical framework of integrated SCD and SCP is presented. We conclude the paper by summarizing the main findings.

2 State of the Art

SCD is the strategic configuration of supply network structures [18]. SCD has been a very visible and influential topic in the field of OR. The formulation of strategic production–distribution models for SCM has been widely investigated. Most of these formulations are introduced in the form of mixed-integer linear programming (MILP) models. Beginning with the seminal work of [10] on multi-commodity distribution system design, a large number of models have been proposed for the SCD [3, 6, 29, 39]. The drawback of using optimization is the difficulty of developing a model that is sufficiently detailed and accurate in representing complexity and uncertainty of SCM, while keeping the model simple enough to be solved [11].

SCP is composed of tactical problems of SC planning regarding demand forecasts or customers' orders and the most rational use of supply chain resources to meet the market requirements in the cost-efficient manner [22]. SCP depends on the SCD. Tactical goals are subject to such SCD goals as SC costs and service level.

In OR, improvements in SCD and SCP are usually algorithmic and refer to the methods of linear programming, integer programming and dynamic programming [9, 33]. Besides, heuristics are frequently applied [12]. For the last decade, considerable advancements have been achieved in this area [22]. However, a number of limitations can be observed in OR techniques with regard to SCP from the dynamics point of view. First, problems of high dimensionality are whether reduced to a simple dimensionality or heuristics are applied. Secondly, complex dynamics of real SC execution cannot be reflected in single-class models. Thirdly, models of planning and execution are not explicitly interconnected in terms of uncertainty.

Recently, a number of concepts were developed to meet the above-named challenges. In particular, complex adaptive systems (CAS) and control theory (CT) are becoming of even greater interest to researchers and practitioners. The work [5] claimed that emergent patterns in a supply network can be managed much better through positive feedback than through negative feedback from control loops. The work [36] investigated how various concepts, tools and techniques used in the study of CAS can be exploited to characterize and model SC networks. Many other papers have also dealt with agent-based modelling and SC adaptivity [1].

Another research stream has been dealing with control policies and algorithms to adapt SCs by means of different techniques [7, 8, 15, 16, 17, 18, 32, 37]. The study [30] employed a set of neural networks to a combined physical inventory and distribution system in a non-stationary demand environment. The study [31] presented an adaptive control (AC) concept for production networks. In the work [21], centralized and decentralized adaptive inventory-control models have been investigated. The study [19] developed a control architecture originating from modern political systems. The study [4] addressed the problem of short-term SCD using the idle capacities of qualified partners in order to seize a new market opportunity.

The first strong contribution of CT to SCM domain is the interpretation of SCD and SCP not as discrete operations but as a continuous adaptive process. Secondly, the possibility of covering the SC dynamics at the process level and the permanent changes in SC processes and environment is also a strong contribution of the CT in SCM domain. Finally, the CT allows the consideration of a goal-oriented formation of SC structures and the solution of problems in this system as a whole.

However, the application of CT also has its challenges and limitations. First, the decision-making in business systems is of a discrete nature. In technical control systems, it is assumed that the control u can be selected continuously in time. Second, dynamic flow models of optimal program control allow us to model a system behavior in dynamics in order to achieve a certain goal (or a number of goals) performance values of which are accumulated over time. However, the sequencing and resource allocation are not considered in these models [27]. Hence, the mathematical models of classical optimal CT need domain-specific modifications.

In order to conduct research on such complex problems, a combination of different methods becomes necessary. While describing a complex system mathematically, it is almost impossible to consider only one system model [2]. As emphasized in [13, 14, 15, 33], SCD and SCP problems are tightly interlinked with each other and require the application of different integrated frameworks of decision-making support. Isolated application of only one solution method leads to a narrow problem formulation, overdue constraints and sometimes impracticable goals. A combined application of OR and CT may potentially provide new insights into the SCM domain.

The proposed approach is based on fundamental scientific results of the modern optimal CT [15, 27, 32] in combination with the optimization methods of OR. Two basic elements of the SCD and SCP model integration are an adaptation framework and a framework for mathematical model combination. These frameworks will be considered in the Sections 3 and 4 correspondingly.

3 Methodical Basis

The main purpose of the adaptation framework is to ensure a dynamic integrated SCD and SCP model with a parameter tuning with regard to changes in the execution environment and planned values of performance indicators. In the proposed framework, the process adaptation is connected to the model adaptation. In Fig. 1, the adaptive framework of the SCD and SCP is presented.

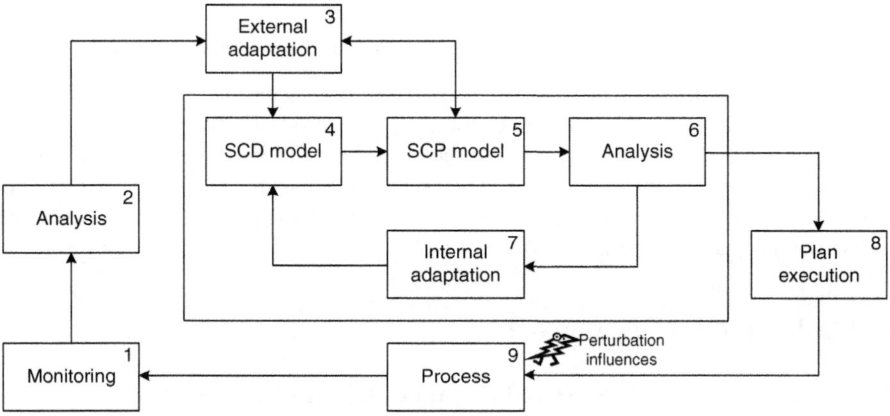

Fig. 1. Integrated adaptive framework of SCD and SCP (based on [34])

The proposed framework includes the following main phases:

- parametric and structural adaptation of SC processes and models (blocks 1, 2, 3);
- integrated SCD and SCP (blocks 4, 5);
- simulation and analysis of SC execution according to different configuration plans, and variants of control decisions in real situations (blocks 6);
- structural and parametric adaptation of the configurations and plans, control inputs, models, algorithms, based on results of simulations experiments (blocks 7); and
- plan realization in real time, correction of the control action (blocks 8, 9).

To implement the proposed concept of adaptive planning and control, let us consider two groups of parameters for SC models and algorithms:

- parameters that can be evaluated on the basis of real data available in the SC; and
- parameters that can be evaluated via simulation models for different scenarios.

The *adaptation procedures* can be organized in two control loops:

- external adapter; and
- internal adapter.

The following parameters belong to the *first group* and can be evaluated through the external adapter:

- the values of SC performance indicators (e.g., minimal admissible service level of 95% can be amended to 93%);
- execution indicators of SC processes (e.g., cycle time); and
- probabilistic characteristics and values of real and observed random processes.

The *second group* of parameters being evaluated through the internal adapter includes such characteristics as:

- a redundancy rate for functional, informational, and temporal reserving;
- priority of SC performance indicators; and
- parameters that define the variants of adaptation in the simulation models.

A hierarchy of adjustment actions is brought into correspondence with different deviations in the SC execution. A state correction takes place with a time horizon of hours, days, weeks or months. In the studies by [13, 14, 18], tools and models for SC structure dynamics control have been presented. It has been proved that structure dynamics considerations may allow the establishment of adaptive feedback between the SC design, tactics, and operations.

4 Mathematical Framework

4.1 Integration of Static SCD and Dynamic SCP Models: Concept

In SCD and SCP models, different goals, variables and constraints can be expressed whether in static or in dynamic form. In applying only one solution method, e.g., mathematical programming or optimal control, significant difficulties in representing both static and dynamic aspects are frequently encountered. Therefore, it can be sensible to to distribute static and dynamics variables and constraints of SCD and SCP between different models where the corresponding static or dynamic elements can be expressed in the best way.

In integrating elements of different modelling techniques, two basic ways exist. First, a basic solution method for all the parts of the integrated problem can be chosen and reinforced by attracting elements from other methods at certain places (e.g., describing scheduling and execution models in terms of optimal control theory [15]). This is the preferable way. If this is impossible, we should use different methods for different parts of the problems and integrate input-output of different models mathematically.

The basics of the SC multi-disciplinary treatment were developed in the DIMA (Decentralized Integrated Modelling Approach) methodology [13]. The mathematical integration and combined application of various models can be implemented by means of multiple model complexes [15, 18, 27, 35], which are based on the application of functors [24, 27, 35]. Functors establish relationships between different models. The multiple-model complexes allow problem examination and solution in different classes of models, and result representation in the desired class of models. The study [35] demonstrated the capabilities of the functorial approach to qualimetry of models. In the work [15] an example of a multiple model complex for the SCM domain has been provided.

Let us discuss an example of how to apply the multiple model complexes [27, 35]. The problem of the SC analysis and synthesis is mostly formalized using either graph (network) models or models of *linear and integer programming*. As a rule, the problem of SC execution is formalized with the help of *dynamic models*. However, the problems of coordination and consistency of the results remain open. To obtain a constructive solution to these problems, we propose to use a functorial transition from the category of digraphs to the category of dynamic models. The mathematical model of the above-described transition is presented in [15]. In the next paragraph, we will consider how explicitly to interconnect a static optimization linear programming model with an optimal control dynamic model within SCD and SCP.

4.2 Integration of Static SCD and Dynamic SCP Models: Formal Statement

The problem consists of the dynamic structural-functional distribution of resources and flows in an SC according to the input data of SCP and regarding the dynamically changing SCD structures. Let A be the set of nodes in SCD. It includes the set A_o of suppliers, the set A_{h} of transit nodes, and the set A_{v} .of customers. The structure of SCD and the parameters of SCP undergo changes at discrete time points $(t_0, t_1, ..., t_k)$. These points divide the planning interval $(t_0, t_k]$ into subintervals. The SC structure does not vary at each subinterval $(t_{l-1}, t_l]$.

We assume that each network element is characterized by the following characteristics: the inventory volume V_j at a node A_j, the intensity of operations execution at *j*-node φ_{jl} abd the capacity ψ_{ijl} of the inter-node channel for the nodes A_i and A_j, where $A_i, A_j \in A$. The following variables are needed for formalization: the amount x_{ijl} of products transmitted from A_i to A_j at time interval number l, product amounts y_{jl}, g_{jl}, z_{jl} relating to the node A_j and to be stored (according to the plan of SC operation), to be processed, and to be delayed (as caused by limited capacity of SC) respectively. Let q_{jl}^{+} be the set of node numbers for the nodes transmitting product to A_j at time interval l and q_{jl}^{-} be the set of node numbers for the nodes receiving products from A_j at time interval *l*.

Now the model of SC planning can be stated as follows (1-3):

$$\alpha_1 \sum_{j=1}^{m} \sum_{l=1}^{k} g_{jl} - \alpha_2 \sum_{j=1}^{m} \sum_{l=1}^{k} z_{jl} = I \to \max, \tag{1}$$

$$x_{jl} = \sum_{i \in q_{jl}^{(-)}} x_{ijl} - \sum_{i \in q_{jl}^{(+)}} x_{ijl} + (y_{j(l+1)} - y_{jl}) + z_{jl} + g_{jl}, \tag{2}$$

$$0 \le x_{ijl} \le \psi_{ijl};\ 0 \le y_{jl} \le V_{jl};\ 0 \le g_{jl} \le \varphi_{jl};\ z_{jl} \ge 0. \tag{3}$$

The SC execution plan should provide the maximal value for the generalized quality index I (equation (1). Here the first component $I_1 = \alpha_1 \sum_{j=1}^{m} \sum_{l=1}^{k} g_{jl}$ of equation (1) characterizes the total amount of customer orders processed in the SC and the second component $I_2 = \alpha_2 \sum_{j=1}^{m} \sum_{l=1}^{k} z_{jl}$ characterizes the amount of not fulfilled customer orders. Both I_1 and I_2 refer to SC service level.

Constraint (2) reflects that the amount of products x_{jl} delivered to the node A_j at the time interval l can be transmitted to other nodes, or processed and stored at this node, or delayed.

Constraint (3) defines the constants of transportation, distribution centers and production intensities regarding the time interval l. As these intensities change with regard to different time intervals, the execution processes become non-stationary.

The planning problem (1)–(3) is a multi-criteria linear-programming (LP) problem of high dimensionality. It includes $m \times k$ equations (2) and $\left(\sum_{l=1}^{k} f_l + 3 \times m \times k \right)$ unknown variables, where m is the number of nodes in SC, k is the number of time intervals and f_l is the number of variables characterizing amounts of received (transmitted) products at the l-interval. An additional peculiarity of the problem is related to two-sided constraints for the variables x_{ijk}, y_{jl}, g_{jl}. A linear convolution of two indices is used to overcome the problem of goal uncertainty. Thus, we assume that $\alpha_1 + \alpha_2 = 1$, $\alpha_1, \alpha_2 \geq 0$ in equation (1).

Analysis of problem (1)–(3) showed that in spite of its high dimensionality it could be solved via special decomposition procedures and the method of successive improvement of plans for LP problems with two-sided constraints [25].

With regard to the SC settings, a special consideration must be given to the fact that real SC operate heterogeneous products of different importance and priority of customer orders. We propose to concretize the obtained plan of SC operation (in terms of model (1)–(3)) via the following differential dynamic model (4):

$$\dot{x}_l^{(o,1)} = u_l^{(o,1)}; \; \dot{x}_{ij\rho l}^{(v,1)} = u_{ij\rho l}^{(v,1)}; \; \dot{x}_{\tilde{j}\rho l}^{(v,2)} = u_{\tilde{j}\rho l}^{(v,2)}; \; \dot{x}_{ij\rho l}^{(v,3)} = u_{ij\rho l}^{(v,1)};$$
$$\dot{x}_{\tilde{j}\rho l}^{(v,4)} = u_{\tilde{j}\rho l}^{(v,4)}; \; i, j, \tilde{j} \in \{1,...,m\}, \tag{4}$$

where $x_l^{(o,1)}$ is a duration of the time interval l, $x_{ij\rho l}^{(v,1)}$ is an amount of product transmitted from A_i to A_j at the time interval l and attributed to the type ρ, $x_{\tilde{j}\rho l}^{(v,2)}$ is an amount of product processed at $A_{\tilde{j}}$ at the time interval l, $u_{ij\rho l}^{(v,1)}(t)$ is a transmission rate in the channel $A_i \to A_j$, $u_{\tilde{j}\rho l}^{(v,2)}(t)$ is a rate of product processing at the node $A_{\tilde{j}}$ on the time interval l for the product, $x_{ij\rho l}^{(v,3)}$ and $x_{\tilde{j}\rho l}^{(v,4)}$ are auxiliary variables denoting correspondingly the time passed from the beginning of product transmission and the time passed from the beginning of product processing or product storage, $u_{ij\rho l}^{(v,3)}(t)$ and $u_{\tilde{j}\rho l}^{(v,4)}(t)$ are auxiliary control inputs assuming values from the set $\{0,1\}$. The control and state variables should meet the following constraints:

$$u_l^{(o,1)}[a_{(l-1)}^{(o,1)} - x_{(l-1)}^{(o,1)}] = 0; \tag{5}$$

$$0 \le u_{ij\rho l}^{(v,1)} \le c_{ij\rho}^{(v,1)} u_{ij\rho l}^{(v,3)};\ 0 \le u_{\tilde{j}\rho l}^{(v,2)} \le c_{\tilde{j}\rho l}^{(v,2)} u_{\tilde{j}\rho l}^{(v,4)}; \tag{6}$$

$$0 \le u_{ij\rho l}^{(v,3)} \le u_l^{(o,1)};\ 0 \le u_{\tilde{j}\rho l}^{(v,4)} \le u_l^{(o,1)}; \tag{7}$$

$$\sum_{i\in q_{jl}^{(+)}} u_{ij\rho l}^{(v,3)} + u_{j\rho l}^{(v,4)} \le 1;\ \sum_{i\in q_{jl}^{(-)}} \sum_{\rho=1}^{P} x_{ij\rho l}^{(o,1)} u_{j\rho l}^{(v,4)} \le \sum_{l=1}^{k} y_{jl} u_l^{(o,1)}; \tag{8}$$

$$\sum_{\rho=1}^{P} u_{ij\rho l}^{(v,1)}(t) \le d_{ijl}^{(1)}(t);\ \sum_{\rho=1}^{P} u_{\tilde{j}\rho l}^{(v,2)}(t) \le d_{\tilde{j}l}^{(2)}(t); \tag{9}$$

$$\psi_{ijl} = \int_{t_{l-1}}^{t_l} d_{ijl}^{(1)}(\tau)d\tau;\ \varphi_{jl} = \int_{t_{l-1}}^{t_l} d_{\tilde{j}l}(\tau)d\tau; \tag{10}$$

$$x_l^{(o,1)}(t_0) = x_{ij\rho l}^{(v,1)}(t_0) = x_{\tilde{j}\rho l}^{(v,2)}(t_0) = x_{ij\rho l}^{(v,3)}(t_0) = x_{\tilde{j}\rho l}^{(v,4)}(t_0) = 0; \tag{11}$$

$$x_l^{(o,1)}(t_f) = a_l^{(o,1)},\ x_{ij\rho l}^{(v,1)}(t_f),\ x_{\tilde{j}\rho l}^{(v,2)}(t_f),\ x_{ij\rho l}^{(v,3)}(t_f),\ x_{\tilde{j}\rho l}^{(v,4)}(t_f) \in \mathbf{R}^1. \tag{12}$$

Constraints (5) specify an ordered sequence of time intervals of SC structure constancy, $a_{(l-1)}^{(o,1)}$ is a given duration of the $(l-1)$ interval of structure constancy. Equations (6) and (7) specify the limits for the intensity of product transmission and processing ($c_{ij\rho l}^{(v,1)}$, $c_{\tilde{j}\rho l}^{(v,2)}$ are given constants). The first inequality in (8) specifies the rules for the operation of the node A_j, that is the ρ-type product either being transmitted to other nodes ($u_{ij\rho l}^{(v,3)}(t) = 1$), or being processed or stored in the same node $u_{\tilde{j}\rho l}^{(\text{п},4)}(t) = 1$). The second inequality in (8) states the limit y_{jl} for the amount of product to be stored (processed) at the node A_j on the interval l. The constraints (9) and (10) specify capacities of channels. The constraints (11) and (12) specify end conditions. The quality indices for the plans of SC operation are expressed as follows:

$$I_3 = \frac{1}{2}\sum_{l=1}^{k}\left[\left(a_l^{(v,1)} - \sum_{i,j,\rho} \gamma_\rho^{(1)} x_{ij\rho l}^{(v,1)}(t_f)\right)^2 + \left(a_l^{(v,2)} - \sum_{\tilde{j},\rho} \gamma_\rho^{(2)} x_{\tilde{j}\rho l}^{(v,2)}(t_f)\right)^2\right], \tag{13}$$

$$I_4 = \sum_{l=1}^{k}\sum_{\rho=1}^{P}\sum_{j=1}^{m} \gamma_\rho^{(2)} x_{j\rho l}^{(v,4)}(t_f), \tag{14}$$

$$I_5 = \frac{1}{2}\sum_{l=1}^{k}\sum_{\tilde{j}=1}^{m}\sum_{p=\tilde{j}+1}^{m-1}\left\{\int_{t_{l-1}}^{t_l}(x_{\tilde{j}}^{(\mathrm{v},4)}(\tau)-x_{p}^{(\mathrm{v},4)}(\tau))^2 d\tau+[x_{\tilde{j}\rho l}^{(\mathrm{v},4)}(t_f)-x_{\tilde{j}\rho l}^{(\mathrm{v},4)}(t_f)]^2\right\}, \quad (15)$$

where $a_l^{(\mathrm{v},1)} = \sum_{i=1}^{m}\sum_{j=1 j=1}^{m} x_{ijl}$; $a_l^{(\mathrm{v},2)} = \sum_{j=1}^{m} g_{jl}$. Performance indicator (13) characterizes the completeness of heterogeneous product processing taking into account priorities of different customer orders. This performance indicator corresponds to the index I_1 of the static model. Indicator (14) expresses total supply delay caused by product processing, storage, and transmission in transit nodes. The given coefficients $\gamma_\rho^{(1)}$ and $\gamma_\rho^{(2)}$ establish priorities for transmission and processing of product of the type ρ . This performance indicator corresponds to the index I_2 of the static model. Indicators (13) and (14) represent the dynamic interpretation of SC service level. Indicator (15) lets evaluate uniformity (irregularity) of resources allocation in SC at each time (integral component of functional (15)) and at the time $t = t_f$ (terminal part of (15)).

5 Discussion of Findings

Continuous optimization is a challenging calculation task. Any sensible judgments on the models and algorithms can be made only by application of special tools. For the experiments, we elaborated a software package [15]. The software has three modes of operation and an additional mode to analyse stability of the plans. This mode is beyond of the scope of this paper.

The first mode includes the interactive generation/preparation of the input data. The second mode lies in the evaluation of SCD and SCP. The following operations can be executed in an interactive regime:

- multi-criteria rating, analysis, and the selection of SC designs and plans;
- the evaluation of the influence that is exerted by time, economic, technical, and technological constraints upon SC execution control; and
- the evaluation of a general quality indicator for SC designs and plans, and the evaluation of particular performance indicators.

The third mode provides interactive selection and visualization of SCD and SCP and report generation. After setting up SC structures, planning goals and environment parameters (customer orders and possible uncertainty impacts), the optimal control planning algorithm is now running. The algorithm of dynamic control addresses the OR model at each point of time during the run of the dynamic model. In combining CT and OR, it becomes possible to approach SCD and SCP problems regardless of the problem dimension and by taking into account non-linearity and non-stationarity of processes along with discreteness of decision-making. The OR models are applied to dimensionally reduced resource allocation problems at each point of time (regarding only those variables and constraints that are relevant at this particular point of time) while the whole dimensionality and dynamics is kept in the CT model. Besides, as both OR and CT are based on optimal principles, the solutions can also be hold for optimal. However, in our opinion, the most important result is even not the optimal solution but the evidence that complex problems of SC adaptive (re)design and (re)planning can be solved in an integrated manner.

Analysis of (5)-(12) shows that the coordination of the planning results of the static and the dynamic model can be carried out through the variables x_{ijk}, y_{jl}, and g_{jl}.

The realization of the system modeling for the considered problem showed the following advantages of the joint use of the static and dynamic planning models:

- Static models of SC operation let take into account the factors (losses, limitation of capacities) which define the state constraints in dynamic models;
- Static models provide input data for the dynamic models while the direct enumeration of variants is impracticable;
- Static models take into account, to a first approximation, the factors of distributed computation and of structure dynamics in SC and provide quantitative estimations for the amounts of supplied, processed, and delayed products.

However, the detailed time-referenced description of product distribution and processing is difficult within the static model. That is why we proposed to use the dynamic model of SC the operation that gives us the following advantages.

- Dynamic models let establish and optimize SC performance indicators that are difficult to express within a static model (for example the indices of uniformity (irregularity) of resources allocation for the product processing in a SC as a requirement on SC collaboration strategy; see equation (15);
- Dynamic models of control theory reflect the non-stationarity of SC execution;
- Dynamic models let use the advanced methods of the optimal control theory for a synthesis of control programs as applied to SC and its subsystems (enterprises), e.g., methods of dynamic stability analysis.

6 Conclusions

In this paper, SCD and SCP problems have been considered in their interlinking and regarding execution control. We showed explicitly how distribute static and dynamics variables and constraints of SCD and SCP by interconnecting static elements in SCD optimization linear programming model with corresponding SCP dynamic elements in optimal control model. Such a problem formulation is near to the real-world settings.

This approach allows the interlinking of planning and execution models within a mathematically integrated framework. Hence, the proposed modelling complex does not exist as a "thing in itself" but works in the integrated decision-support system and guides the SCD and SCP decisions in dynamics on the principles of dynamic optimization and adaptation.

Finally, let us discuss the limitations of the proposed approach. For concrete application cases, it is nevertheless almost impossible to take into account the whole variety of complex interrelated constraints and parameters. For each concrete problem, we should actually build a new complex of constraints. This is a very time-consuming process as we should achieve the strong monosemantic problem formulation to ensure only one solution is gained. Constructive ways to achieve calculation precision with regard to the interrelations of measured and calculated (control) parameters are to be developed further. Finally, the mathematical formalization of uncertainty factors is very complicated. Hence, further developments on the modelling with interval data [15] are needed.

Acknowledgements. The research described in this paper is partially supported by grants from the DFG PAK 196 "Competence Cell-based Production Networks", Russian Foundation for Basic Research (grants 07-07-00169, 08-08-00403, 09-07-00066, 10-07-00311), the Saint Petersburg Scientific Center RAS, and the Department of nanotechnologies and information technologies of the Institute of Systems Analysis RAS (2.3).

References

1. Ahn, H.J., Lee, H., Park, S.J.: A flexible agent system for change adaptation in supply chains. Expert Syst. Appl. 25, 603–618 (2003)
2. Casti, J.L.: Connectivity, complexity and catastrophe in large-scale systems. Wiley Interscience, New York (1979)
3. Chandra, C., Grabis, J.: Supply chain configuration. Springer, New York (2007)
4. Chauhan, S.S., Gordon, V., Proth, J.M.: Scheduling in supply chain environment. Eur. J Oper. Res. 183(3), 961–970 (2007)
5. Choi, T.Y., Dooley, K.J., Rungtusanatham, M.: Supply networks and complex adaptive systems: control versus emergence. J Oper. Manag. 19(3), 351–366 (2001)
6. Chopra, S., Meindl, P.: Supply chain management. Strategy, planning and operation. Pearson Prentice Hall, New Jersey (2007)
7. Disney, S.M., Towill, D.R., Warburton, R.D.H.: On the equivalence of control theoretic, differential, and difference equation approaches to modeling supply chains. Int. J. Prod. Econ. 101, 194–208 (2006)
8. Disney, S.M., Towill, D.R.: A discrete linear control theory model to determine the dynamic stability of Vendor Managed Inventory supply chains. Int. J. Prod. Res. 40(1), 179–204 (2002)
9. Dolgui, A., Proth, J.M.: Supply Chains Engineering: useful methods and techniques. Springer, Berlin (2009)
10. Geoffrion, A., Graves, G.: Multicommodity distribution system design by Benders decomposition. Manag. Sci. 29(5), 822–844 (1974)
11. Harrison, T.P.: Principles for the strategic design of supply chains. In: The practice of supply chain management. Kluwer Academic Publishers, Boston (2005)
12. Hentsch, K., Koechel, P.: Scheduling jobs with forbidden setups with metaheuristics and penalties – a case study at a continuous casting plant. In: Proceeding of the 10th International Symposium on Operational Research, Slovenia, pp. 259–268 (2009)
13. Ivanov, D.: DIMA – A Research Methodology for Comprehensive Multi Disciplinary Modelling of Production and Logistics Networks. Int. J. Prod. Res. 47(5), 1133–1155 (2009a)
14. Ivanov, D.: Adaptive aligning of planning decisions on supply chain strategy, design, tactics, and operations. Int. J. Prod. Res. (2009b), doi:10.1080/002075409028935417 (in press)
15. Ivanov, D., Sokolov, B.: Adaptive Supply Chain Management. Springer, London (2010)
16. Ivanov, D., Arkhipov, A., Sokolov, B.: Intelligent planning and control of manufacturing supply chains in virtual enterprises. Int. J. Manuf. Tech. Manag. 11(2), 209–227 (2007)
17. Ivanov, D., Kaeschel, J., Sokolov, B.: Structure dynamics control-based framework for adaptive reconfiguration of collaborative enterprise networks. Int J Manuf. Techn. and Manag. 17(1-2), 3–41 (2009)

18. Ivanov, D., Sokolov, B., Kaeschel, J.: A multi-structural framework for adaptive supply chain planning and operations with structure dynamics considerations. Eur. J. Oper. Res. 200, 409–420 (2010)
19. Jang, P.Y.: A flexible and adaptive control architecture for the efficient supply chain management (SCM). WSEAS Transactions on Communications 5(6), 1015–1025 (2006)
20. Gunasekaran, A., Kee-hung, L., Cheng, T.C.E.: Responsive supply chain: a competitive strategy in a networked economy. Omega 36(4), 549–564 (2008)
21. Kim, C.O., Jun, J., Baek, J.K., Smith, R.L., Kim, Y.D.: Adaptive inventory control models for supply chain management. Int. J. Adv. Manuf. Tech. 26(9-10), 1184–1192 (2005)
22. Kreipl, S., Pinedo, M.: Planning and Scheduling in Supply Chains: An Overview of Issues in Practice. Prod. Oper. Manag. 13(1), 77–92 (2004)
23. Lalwani, C.S., Disney, S., Towill, D.R.: Controllable, observable and stable state space representations of a generalized order-up-to policy. Int. J. Prod. Econ. 101, 172–184 (2006)
24. Mesarovic, M.D., Takahara, Y.: General systems theory: mathematical foundations. Academic Press, London (1975)
25. Moskvin, B.V.: Optimization of information transmission in packet network. In: Trans. of All-Union conf. "Compac 87", Riga, pp. 168–171 (1987)
26. Müller, E., Horbach, S., Ackermann, J.: Integrative planning and design of logistics structures and production plants in Competence-cell-based networks. Int. J. Services Oper. Informatics 3(1), 40–52 (2008)
27. Okhtilev, M., Sokolov, B., Yusupov, R.: Intelligent technologies of complex systems monitoring and structure dynamics control, Nauka, Moskau (2006)
28. Peck, H.: Supply Chain Vulnerability, Risk, Robustness, and Resilience. In: Global Logistics and Supply Chain management, pp. 229–248. John Wiley and Sons, Chichester (2007)
29. Shen, Z.M.: Integrated supply chain design models: a survey and feature research directions. J Ind. Manag. Optim. 3(1), 1–27 (2007)
30. Shervais, S., Shannon, T.T., Lendaris, G.G.: Intelligent supply chain management using adaptive critic learning. IEEE Trans. Syst. Man. Cybern. Syst. Hum. 33(2), 235–244 (2003)
31. Scholz-Reiter, B., Hoehns, H., Hamann, T.: Adaptive control of supply chains: building blocks and tools of an agent-based simulation framework. CIRP Annals Manufacturing Technology 53(1), 353–356 (2004)
32. Sethi, S.P., Thompson, G.L.: Optimal Control Theory: Applications to Management Science and Economics. Springer, Berlin (2006)
33. Simchi-Levi, D., Wu, S.D., Zuo-Yun, S.: Handbook of quantitative supply chain analysis. Springer, New York (2004)
34. Skurihin, V.I., Zabrodsky, V.A., Kopeychenko, Y.V.: Adaptive control systems in machinery industry. Mashinostroenie, Moscow (1989)
35. Sokolov, B.V., Yusupov, R.M.: Conceptual foundations of quality estimation and analysis for models and multiple-model systems. Int. J. Comput. Syst. Sci. 6, 5–16 (2004)
36. Surana, A., Kumara, S., Greaves, M., Raghavan, U.N.: Supply-chain networks: a complex adaptive systems perspective. Int. J. Prod. Res. 43(20), 4235–4265 (2005)
37. Van Houtum, G.Y., Scheller-Wolf, A., Yi, J.: Optimal Control of Serial Inventory Systems with Fixed Replenishment Intervals. Oper. Res. 55(4) (2007)
38. Van de Vonder, S., Demeulemeester, E., Herroelen, W.: A classification of predictive-reactive project scheduling procedures. Journal of Scheduling 10(3), 195–207 (2007)
39. Vidal, C., Goetschalckx, M.: Strategic production-distribution models: a critical review with emphasis on global supply chain models. Eur. J. Oper. Res. 98, 1–18 (1997)

Lean Intra-corporate Supply Chain Management for Complex Organizations

Andreas Fellhauer[1] and Adam Strożek[2]

[1] Unity AG, Sales Division Supply Chain Management,
Lindberghring 1, 33142 Büren, Germany
andreas.fellhauer@unity.de
[2] Unity AG, Center of Competence Lean Development,
Unter den Linden 21, 10117 Berlin, Germany
adam.strozek@unity.de

Abstract. Existing Supply Chain Management approaches focus predominantly on harmonization of supply processes with numerous external partners. In order to manage the complexity of this objective they assume standardized processes for different companies. This paper proposes a novel approach, which meets the market demands for a proceeding, which is both, more pragmatic and more flexible than common Supply Chain Management reference models. It explains why this demand exists and shows how an adequately differentiated supply chain profile of a company can be derived by analyzing constitutive strategic aspects with impact on supply processes. Finally, it shows how action recommendations for Supply Chain Management can be derived in a lean manner and exemplifies how those recommendations can look like.

Keywords: Supply Chain Management, Supply Chain, SCOR, Supply Modes, Business Structure Analysis, Lean.

1 Introduction

Supply Chain Management (SCM) attracted a lot of attention during the last decade. Numerous research projects have been conducted to define scope and object-matters of SCM and to offer detailed reference models and planning procedures as best practices in SCM. One of the reference models used most frequently is the Supply Chain Operation Reference Model (SCOR), shown in Fig. 1, which has been defined by the Supply Chain Council.

The SCOR is based on the five management processes plan, source, make, deliver and return. It offers relationship frameworks, standard process descriptions and metrics for these processes along the entire supply chain, i.e. from the suppliers' supplier to the customer's customer [1, 2]. Clearly, it covers the most important processes in the supply chain and consequently empowers a company that implements it, to synchronize value creation processes in its supply chain and to balance them with the customers` demands [3]. The SCOR is a very holistic model with strong emphasis on the integration of upstream and downstream partners and the harmonization of communication processes between them.

W. Dangelmaier et al. (Eds.): IHNS 2010, LNBIP 46, pp. 164–174, 2010.

Fig. 1. The SCOR model by the Supply Chain Council [2]

Furthermore it is a general model, which does not account for the specific situation of a company and for its existing structure. It rather implies identical or at least very similar supply chains between the same branches or even between completely different companies [4]. However, this assumption is unrealistic not only within one branch, but also for identical products, if these are distributed within different business areas. For instance, requirements on the delivery time and thus on the supply chain of one product are different if the product is used by different customers with different needs. Those specific conditions can imply a higher importance of several processes described in the SCOR, a particular importance of processes not comprised in it or impede the implementation of the reference processes.

Due to these facts, many companies cannot apply the holistic SCOR approach. Frequent problems are the complexity of ideal SCOR-projects, the long pay back period of investments and the focus on external activities without having sufficient efficiency of internal supply chain processes [5, 6]. Another frequent problem is the change of paradigms, required by the implementation of the SCOR [4]. Confident information has to be exchanged within the supply chain and former functional orientation has to be changed towards a process orientation, what involves a change of roles and responsibilities. Finally, such a strong cooperation between the different partners of a supply chain not only implies undoubtedly high optimization potentials, but also a strong dependency on these partners.

Therefore, many companies confine themselves to a partial implementation of the SCOR model. They focus on strategic problems or solely on distinct operational ones. Those strategic problems might be the relocation of production facilities, the initiation of strategic purchasing activities or the development of strategic supplier partnerships in general. Focusing on operational problems only, many companies start directly with process analysis and reengineering or informational supplier integration in order to improve the efficiency of particular processes [4, 5]. In addition, many companies rely on the theoretic background provided by the vast majority of SCM related publications and regard implementation of suppliers and harmonization of commodity flows outside their company as the main objectives of SCM [1, 2, 7, 8]. These companies, in particular those with complex structure and processes, which have grown over many years, align the external segments of their supply chains to its own inefficient organization and achieve suboptimal results. Finally, these efforts are often done without the necessary differentiation of business areas what leads again to generalized, suboptimal results.

The approach presented in this paper allows avoiding those mistakes, by focusing on the internal supply processes of a company first, and by accounting for the differences resulting from different business areas. Furthermore, it provides a quick analysis, which helps the companies to identify the right levers of SCM activities with focus also on the external elements. Focusing on the important parts, it helps also to reduce the complexity of SCM projects, although a top-down approach is followed and thus supply chain relevant aspects at all management levels are considered.

2 Background – The Idea of Lean Intra-corporate Supply Chain Management

The problems applying SCM methods based on the SCOR have been mentioned in the introduction and can be summarized as following:

- Focus on external activities without efficient internal processes
- Generalization of supply processes
- The projects' complexity
- High investments with long pay back periods

As a result, many SCM efforts remain unsuccessful, focusing on the wrong parts of the supply chain, not considering the existing specific structure and business areas of a company, ending up in a chase for benchmarks for distinct processes without the knowledge if these really are the relevant criteria for the considered company, or overburdening their financial or organizational possibilities. In order to prevent those failures, it is crucial to know own strengths and weaknesses and align all SCM efforts to this awareness.

With this in mind the Lean Intra-Corporate Supply Chain Management has been developed. It allows each company to quickly examine their needs with respect to supply chain management and achieve quick returns on their investments by focusing on their main weaknesses and applying the right supply chain solutions for each part of its supply chain.

The general idea of this method is the creation of a holistic, differentiated overview over a company's supply chain by analysing all constitutive strategic areas of a company, which are depicted in Fig.2. The entire functionality of each company is defined by the way it generates profit, its architecture and the activities it starts in order to fulfil its profit objectives. According to Fig. 2, the first point is defined as the business strategy, the second point is reflected in international strategy, organizational structure and business rules and the last one in objectives, metrics, roles and responsibilities.

2.1 Business and International Strategy

In the first step, business and international strategy are analyzed. The business strategy is reflected in the way the company interacts with the market, i.e. by the

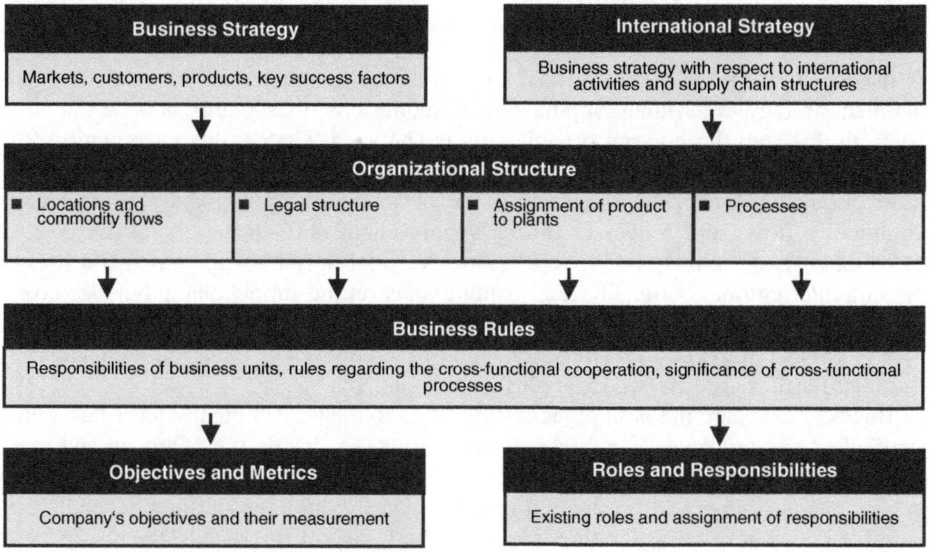

Fig. 2. Constitutive strategic areas with impact on SCM

company's products and clients. These combinations of markets a company serves and the products it delivers to these markets define the company's business areas, i.e. the clusters of its output performance respectively the different ways it generates profit. As it will be shown later each business area may have specific requirements on the supply process. Since those different requirements can also occur for identical products, if these are distributed in different markets, each market-client combinations, i.e. each business area has to be considered. If such an overview has been created the company's key success factors have to be identified. These factors, which determine strengths and weaknesses of the company, are the main points of interest in all corporate processes and thus should define also all supply chain objectives.

The international strategy describes the way in which the company organizes its international activities in order to satisfy the business areas. It represents organizational, distributional and legal supply chain structures, which mostly have grown in the past in the course of the company's, predominantly sales-driven, internationalization. Usually companies start their activities in their home countries and then successively expand abroad by establishing sales dependencies, first in close or well known markets and later in other countries all over the world. In order to supply these dependencies with the initially small demands and for adopting the products to these countries' needs, warehouses are often built in each of these countries. Over the years sales abroad, flexibility of production and distribution increase, and the historically grown structure turns out to not correspond to the customers' needs. Due to this fact, it is an integral part of the Lean Intra Corporate Supply Chain Management to examine the international strategy and verify, if it corresponds to the other strategic areas.

2.2 Organizational Structure and Business Rules

In the second step, the organizational structure is being examined. Whereas the international strategy determines organizational supply chain structures outside the company, in this step the internal organization is analyzed. During this examination four aspects deserve closer attention – locations and commodity flows, legal structure, production sites and processes within the organization. The impact of locations, commodity flows and processes on the supply chain performance has been investigated extensively in the past [1, 9, 10]. However, not only leaving company's sites or passing international borders have a strong impact on the supply chain, but also crossing the borders of legal entities. The additional information required by buying and selling goods compared to a simple change of warehouses or a markup demanded by fiscal authorities are just two examples for this impact.

Business rules are the next object of investigation. Every company has rules, either implicit or explicit ones, which characterize processes inside the company and manage conflicts of interest. Examples are simple logistical criteria, like the date of delivery, or the assignment of production orders to sites, which are defined depending on variable decision criteria like their capacity, the transport distance to the customer or the lot size. Obviously business rules define the path, which commodities and information take across several organizational units and thus are an integral part of any supply processes.

2.3 Objectives, Metrics, Roles and Responsibilities

Based on these results, supply chain related objectives can be derived as well as metrics utilized in order to measure the degree of objective accomplishment by the existing supply chain processes. Those objectives can be a certain order-fill rate for warehouses or a certain on-time delivery rate [11]. However these objectives are, it is crucial to consider, that they have to be derived directly from the key success factors, have be measurable and can be set for business areas or specific processes only.

In the last step, roles and responsibilities are examined as they are existent in the current state. They define responsibility for the achievement of the objectives on an individual basis and thus are an integral part of any supply activities by assigning them one or several executing units [11, 12].

Besides a holistic overview over a company's functionality, the presented strategic analysis offers two important insights. First, it can be immediately reviewed if objectives, metrics and roles derived from the previous analysis steps reflect those set by the company. Second, it can be checked, if the defined objectives are the right ones with respect to the company's supply chain profile.

3 Deriving the Corporate Supply Chain Profile

After having laid out which strategic aspects have to be analyzed and why this has to occur, it will be exemplary shown in the following, how the analysis results form the supply chain profile of the company, the so called Supply Mode Matrix (SMM). As shown in Fig. 6, it shows all different manners of supplying one supply chain element by another which exists in a company and will be explained in the following.

As described in chapter 2, in the first step the company's business strategy, i.e. its business areas as well as its key success factors are identified. Whereas the business areas are constituted by an assignment of all products to the markets in which they are distributed, key success factors are estimated by surveys among management and customers. In each of these surveys management and customers are asked to rank success factors in the two categories importance and company's performance with respect to this factor. Summarized in a portfolio strategic factors, i.e. those with both, high importance and performance, as well as critical factors, i.e. those with a high importance and a low performance can be identified easily. Both compose the company's key success factors, which can be product quality, time of delivery, delivery reliability, price, employee turnover, service reachability and others. Those key success factors provide the only reasonable objectives of all supply processes.

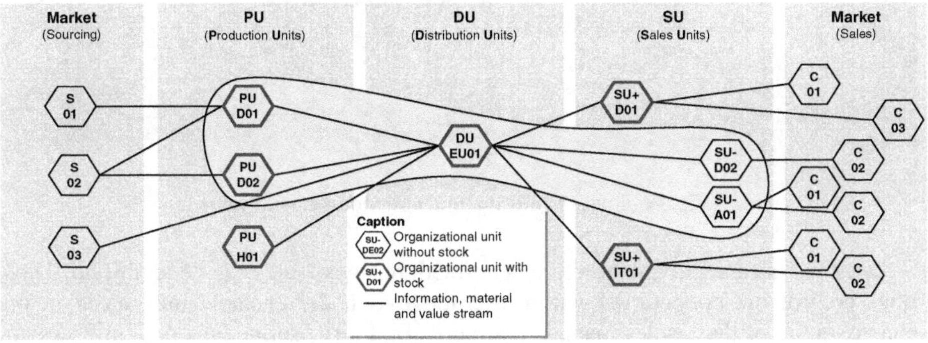

Fig. 3. International Strategy

In order to describe the international strategy it is helpful to create functional clusters and to assign them the company's organizational units. Fig.3 shows an exemplary overview of a company's international supply network, with the functional clusters sourcing market, production, distribution and sales market as well as several organizational units which execute the respective functions. In combination with information like respective countries, relationships between the units and legal differences, as shown in Fig. 3, this supply chain structure forms the international strategy.

In Fig. 3 the considered parent company is encircled, all units outside the encircled area are own legal entities and all units which dispose of own warehouses are framed bold. The arcs within this network represent relationships between the several units. In this manner an assessable overview is gained over the company's organizational units' relationships with each other and with external elements of the supply chain. In the exemplary picture of a international strategy in Fig.3, the distribution unit DU-EU01 supply its demand of commodity goods by the supplier S-03 and products, which are produced in the own company from the production units PU-D01 and PU-D02, respectively from the production site abroad PU-H01.

However, for supply chain optimization not only generalized relationships are interesting, but also the information of the relationships' characteristics. Therefore, not only commodity flows are mapped, however, also information and value streams are being examined for each supply mode and then displayed as shown in Fig.4. As can be seen at this example, commodity, information and value streams can be different for one supply mode as well as their entity can differ depending on different situations. In this manner a company not only has different supply modes, but also can have different peculiarities of each supply mode, as can be seen in Fig. 6 for the supply mode "Supply customer from sales units stock". A process analysis, which can be conducted in the next step, often helps to reveal differences and inconsistencies in the processes occurring in the same organization units for different situations. Then, the company's formal organization is examined and matched with the different supply modes, as it is shown in a simplified manner in Fig.4.

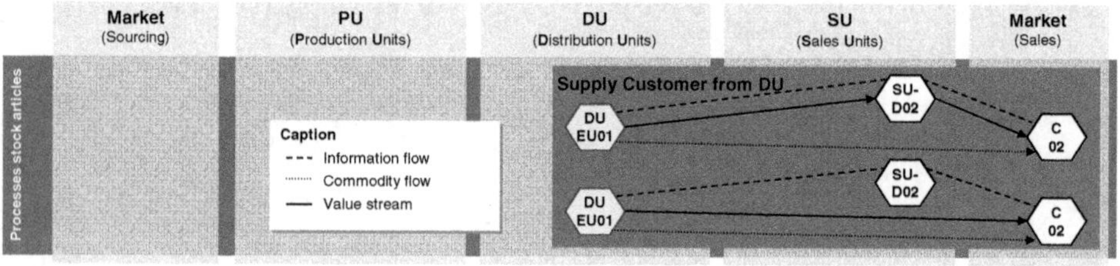

Fig. 4. Flows of information, commodities and value

Once the organizational structure is defined, business rules can be identified. Those rules control the cooperation within a company. In the example introduced in this chapter some of these rules manage conflicts of interest, define information flows and structures or value streams. Whereas distribution units try to minimize their warehouse costs, production units try to operate at full capacity and minimize their costs of production. Correspondingly, definition and maintenance of article master data can either be centralized or be part of the different supply chain units' autonomy. The same question of regulation occurs in many other cases like the definition of internal transfer prices or process descriptions. In each case business rules define the framework for supply processes.

In the next step, rules and responsibilities have to be identified and, if necessary redefined. As mentioned in chapter 2 they assign each process chain and thus each supply mode an executive unit, which is responsible for the achievement of defined supply chain objectives. At this point management attention and an effective change management are indispensable, since all responsibles not only have to be informed about and commit themselves to the scope of their responsibility and their objectives, but also criteria that does not constitute supply chain objectives have to be agreed. In particular in the beginning of changes this may cause considerable resistance among the employees.

In the last step, objectives and metrics have to be defined and visualized for each supply mode. It is crucial that both, objectives and metrics comply with the earlier defined key success factors. Fig.5 shows those metrics for the supply mode "Supply customer from sales unit" of the considered sample company. Delivery time, delivery reliability and payment target have been identified as some of the key success factors and thus serve as possible performance criteria for the supply chain. In the course of a

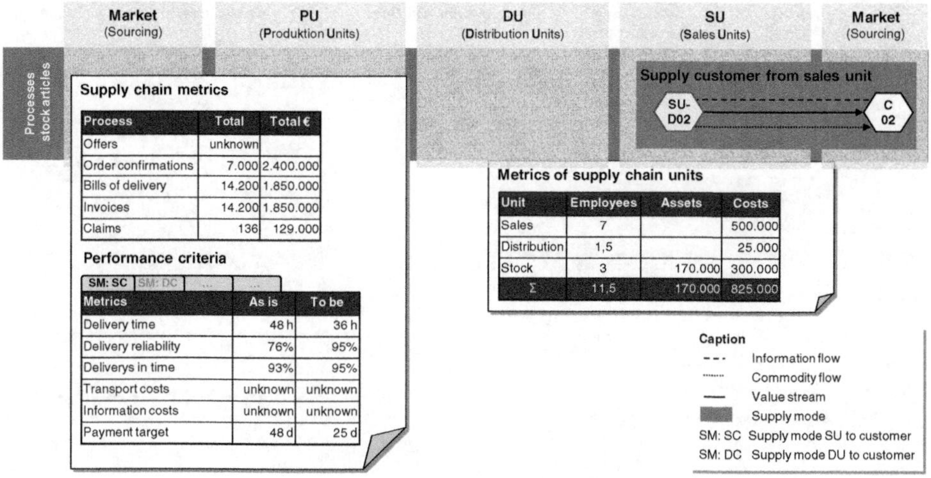

Fig. 5. Metrics and objectives for supply modes

more detailed analysis a delivery time of 36 hours, a delivery reliability of 95% and a payment target of 25 days could been identified as measurable objectives. Other possible criteria, which haven't been identified as key success factors, are not applicable as metrics for the measurement of the supply chain's performance. Some other metrics, like the number of employees, or process costs, are not suitable as control variables, however, are helpful to compare the distinct supply chain elements as well as costs of organizational units with process costs.

With these results, i.e. defined key success factors, visualized international strategy and organizational structure, business rules, roles and responsibilities as well as metrics and objectives for each supply mode, a holistic overview of the company' s supply chain with all its peculiarities, i.e. the Supply Mode Matrix, can be derived as it is shown in Fig. 6.

In this matrix all supply modes of the considered company are visualized. The company depicted in Fig.6 produces its goods according to the two production principles make-to-order and make-to-stock, since this is demanded by its business areas. The supply modes, symbolized by the rectangles in the center are divided in two major groups by these production principles.

As shown in Fig. 6 for the supply mode "Supply customer from sales unit stock", each of these supply modes comprises a different process chain, since for each manner of transferring goods or information between two supply chain elements different actions have to be taken. Furthermore, each supply mode, i.e. also each process chain, can exist in multiple variants. Since each of these variants is different from the other, any objectives, metrics and design activities can be applied to one supply mode variant only. In this manner, the SMM provides a differentiated overview of all existing variants of supply activities with impact on the company's internal processes and offers a tool to realize SCM activities in a lean and tailored manner that prevents typical problems of the SCOR model.

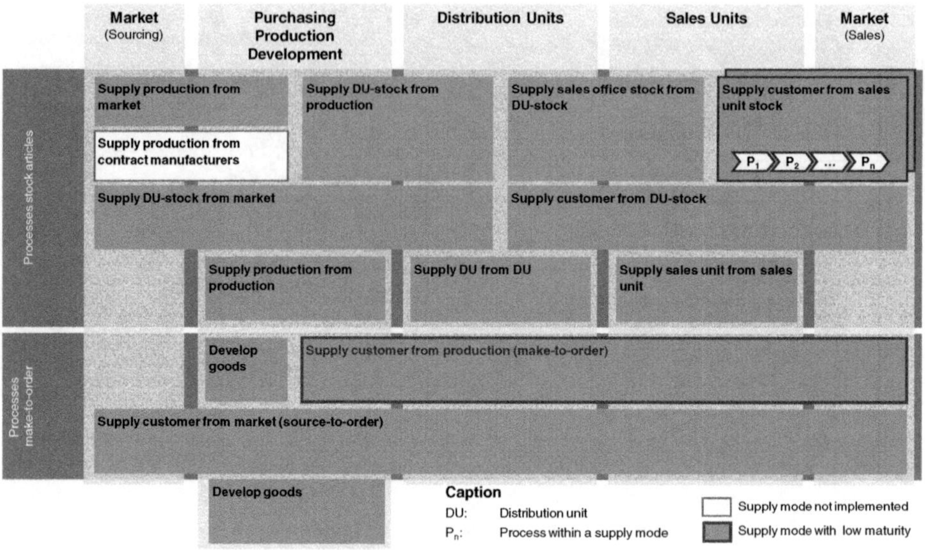

Fig. 6. The Supply Mode Matrix

4 Evaluating the Supply Mode Matrix and Deriving Action Recommendations

Once the detailed description of the company's supply chain profile is available and the characteristics of the company's strategic areas are known, action recommendations with respect to Supply Chain Management can be defined.

In the course of the foregoing analysis it has been found out that the examined company aims to realize the different supply modes depicted in Fig. 6 Comparing this internal supply chain structure with the company's strategic orientation, some inconsistencies could be identified.

First, various supply modes have been identified, which basically comply with the business areas' needs. In some business areas customers expect some of the products to be available at any time, whereas in other a certain waiting time is accepted. For the first markets the company has built up warehouses at the sales units and delivers its products from these warehouses, thus assuring a high availability and short delivery time. In the latter case, where a certain waiting time is accepted, customers are supplied directly from the distribution unit. Both supply modes have been already implemented in the company, including well defined processes. However, both process chains have been controlled by metrics which do not comply with the key success factors and thus both supply modes show gaps between the current and the aspired state of the relevant objectives, as can be seen in Fig. 5.

Second, supply modes have been identified, which are generally defined, however, are realized in an ineffective way with processes, which have a low maturity. For instance, it has been discovered that the company generates 30% of its revenue by highly customized products, which are delivered directly from the production to the customer. However, a process analysis has revealed that the supply mode is realized by ineffective processes due to discontinuities of IT systems and inconsistent article master data. Instead of a real make-to-order process the company assigns article numbers to all make-to-order products and treats them as stock products, i.e. although the production order is activated directly by a customer order, the product makes a detour through the company's stock.

Third, supply modes have been discovered, which are not defined at all. The company aimed to outsource parts of the production, which supply downstream production parts, to contract manufacturers while providing them semi finished parts out of the upstream production. Although this supply mode is realized in some cases, there are neither responsibilities nor processes defined, no rules implemented which define when this has to occur and no support by the current IT systems available.

In this manner different action recommendations could be derived for the examined company. In case that a general lack of a certain supply mode has been identified, some of the measures have been the derivation of objectives and metrics from the company's key success factors, the design of appropriate process chains and definition of roles and responsibilities that support the achievement of objectives with the new processes. In case of insufficiently supported supply modes the harmonization of IT systems, centralization of article master data or implementation of a variant management have been some of the recommendations. Finally, where the supply modes have been controlled by inadequate objectives, these new objectives have been introduced, which comply with the company's key success factor. In the course of the entire analysis many other areas for improvement have been discovered like an inappropriate sales planning, inconsistent internal prices, business rules that do not consider cultural aspects or the lack of a supplier management which accounts for different supply modes.

In applying the Lean Intra Corporate Supply Chain Management approach those action recommendations are obtained in a very pragmatic way. The object of examination can comprise the company's entire supply chain, or only a part of it, still following a top-down-approach and thus assuring that all operational measures comply with the company's strategy. However, not only its pragmatism makes the approach lean. Another important aspect is the focus on the customer, which can be the end-customer or the next supplied organizational unit. In this manner the pull principle can be realized, where it is advantageous, without producing single orders, where it is inappropriate. Finally, the result of the approach presented in this paper is always an adaptation of supply modes and thus of the underlying processes and organization to the respective business areas' requirements. Such a customization of supply modes helps to reduce overproduction, unnecessary transport, inventory, over-processing and waiting and thus to avoid five of the seven types of waste defined in Lean theory [13, 14].

5 Summary

The Lean Intra-Corporate Supply Chain Management approach helps to solve frequent problems with respect to SCM. As has been explained in chapter 2, these problems are an inadequate standardization of supply chain processes, not sufficiently considered internal processes, and thus an inadequately balanced complexity of SCM projects. The key idea of the approach is the focus on the important aspects of a company's supply chain. Therefore an analysis of all relevant strategic aspects is being conducted. Its results are threefold. First, it enables the company to obtain a detailed overview of all peculiarities of its supply chain elements with impact on internal process. Second, it helps to asses if the existing supply structure complies with the needs of its business areas and its strategic objectives. Third, the differentiation of different supply modes allows controlling them by setting objectives, metrics, roles and responsibilities and thus improving their effectiveness and efficiency. As proven in practice for several companies, this proceeding empowers a company to realize quick, measurable, and sustainable improvements of its supply chain and to reduce inefficiency in the supply chain. However, this approach is not only a path to quick success, but also the basis for any further SCM activities, like supplier integration and harmonization of systems, assuring that they follow the right directions and consider an adequate differentiation of supply chain processes.

References

1. Bolstorff, P.A., Rosenbaum, R.G., Poluha, R.G.: Spitzenleistungen im Supply Chain Management. Springer, Berlin (2007)
2. Supply Chain Council: Supply Chain Operations Reference Model – SCOR Version 7 Overview. Brussels (2005)
3. Poluha, R.G.: Anwendung des SCOR-Modells zur Analyse der Supply Chain: Explorative empirische Untersuchung von Europa, Nordamerika und Asien. Josef Eul Verlag, Köln (2006)
4. Becker, T.: Prozesse in Produktion und Supply Chain optimieren. Springer Verlag, Berlin (2008)
5. Poirier, C., Swink, M., Quinn, F.: The Fifth Annual Global Survey of Supply Chain Progress. Computer Science Corporation (2007)
6. Ho, D.C.K., Au, K.F., Newton, E.: Empirical research on supply chain management: a critical review and recommendations. International Journal of Production Research 40, 4415–4430 (2002)
7. Stonebraker, P.W., Liao, J.: Supply chain integration: exploring product and environmental contingencies. Supply Chain Management: An International Journal 11, 34–43 (2006)
8. Rudolph, T., Drenth, R., Meise, J.N.: Kompetenzen für Supply Chain Manager. Springer, Berlin (2006)
9. van Veen-Dirks, P., Verdaasdonk, P.: The dynamic relation between management control and governance structure in a supply chain context. Supply Chain Management 14, 466–478 (2009)
10. Williamson, O.: The theory of the firm as governance structure: from choice to contract. Journal of Economic Perspectives 16, 171–195 (2002)
11. Chae, K.: Developing key performance indicators for supply chain: an industry perspective. Supply Chain Management: An International Journal 14, 422–428 (2009)
12. Shub, A.N., Stonebraker, P.W.: The human impact on supply chains: evaluating the importance of "soft" areas on integration and performance. Supply Chain Management 14, 31–40 (2009)
13. Hines, P.: Value Stream Management: Strategy and Excellence in the Supply Chain. Financial Times, pp. 14–18. Prentice-Hall, Englewood Cliffs (2000)
14. Womack, J.P., Jones, D.T.: Lean Thinking: Banish Waste and Create Wealth in Your Corporation, New York, pp. 15–20 (2003)

Up-to-Date Supply Chain Management: The Coordinated (S, R) Order-Up-to

Salvatore Cannella and Elena Ciancimino

University of Palermo, Faculty of Engineering
viale delle Scienze, Parco d'Orleans,
90128 Palermo, Italy
{cannella,eciancimino}@unipa.it

Abstract. This paper presents the mathematical derivation of a new generation of the most largely used periodic review policy in supply chain: the coordinated (S, R) replenishment rule. We first derive the classical order-up-to model and then we modify it to generate the coordinated decision policy equations. We run a numerical simulation on a serial supply chain model to show differences in the two policies. We conclude on the managerial implications related to coordinated replenishment.

Keywords: multi-echelon; production and inventory management; information sharing; collaboration; synchronization of operations; decision policy; OUT; bullwhip effect.

1 Prologue

In the last decades large-scale collaboration projects have gradually modified the *modus operandi* of global supply chains partners. Efficient consumer response [1], Vendor Managed Inventory (VMI) [2], [3], continuous replenishment [4], collaborative planning, forecasting and replenishment [5], [6], and centralised inventory management [7], have reshaped material and information flows, intensified alliances and transformed strategies, organisations and corporate culture [8]. These structural changes have modified the nature of production planning and control and, more specifically, the dynamics of one of its main objectives: the decision on the order quantity. Real-time sharing of market demand data for the generation of conjoint forecasts and sharing of information on inventory levels and in-transit items for centralised replenishment pushed towards the generation of novel order policies that enable fully coordinated decision-making approach in supply chain. The mathematical formalization of VMI collaboration practices [9], [10], the object-oriented and spreadsheet simulation models of the Electronic Point Of Sales supply chain [11]-[13], the classification framework on the degree of synchronisation of members' operations [14], [15], novel designs of order policies for flexible capacity [16], [17], form a non-exhaustive list of cornerstone publications that prove: *(1)* how classical inventory management policies have definitively evolved to adapt to the increasing proliferation of ICT in logistics

W. Dangelmaier et al. (Eds.): IHNS 2010, LNBIP 46, pp. 175–185, 2010.

operations, *(2)* that formalisation, modelling and clarification of the new coordinated decision-making mechanisms is one of the current paradigms in supply chain management[1].

In this paper we present the evolution of the most largely used periodic review policy [19]-[21] for a make-to-stock multi-echelon supply chain: the (S, R) order-up-to. We first derive the order policy for a traditional serially linked multi-echelon system and then, on the basis of the classical (S, R) archetype, we generate an order-up-to policy for a coordinated multi-echelon supply chain. To undertake our mathematical derivation, we adopt the classical analytical modelling approach for production inventory control [22]-[24]. We run a numerical simulation on a serial supply chain model to show differences in performance of the two policies. We conclude discussing on the benefits deriving from shifting from a classical (S, R) to a coordinated (S, R).

2 Classical (S, R) Policy

Table 1. Notation

W	work in progress	d	customer demand
I	inventory of finished materials	$\hat{d}$	customer demand forecast
O	replenishment order quantity for the classical (S, R)	α	demand smoothing forecasting factor
O'	replenishment order quantity for the coordinated (S, R)	TI	target inventory
C	units finally delivered	TW	target work in progress
λ	production-distribution lead time	I'	multi-echelon inventory
ε	safety stock factor	W'	multi-echelon work in progress
λ'	multi-echelon production-distribution lead time	TI'	multi-echelon target inventory
ε'	multi-echelon safety stock factor	TW'	multi-echelon target work in progress

The classical (S, R) order quantity for a generic echelon is first derived for a traditional multi-echelon model under the following assumptions:

- K-stage production-distribution serial system. Each echelon in the system has a single successor and a single predecessor (Figure1).
- The generic echelon's position is represented by index *i*. Echelon *i*=1 stands for the manufacturer and *i*=K+1 for the final customer.
- Single product. Aggregate production plans are assumed.

[1] Until recently, many firms regarded their knowledge base as a proprietary asset and they hesitated to share it for confidentiality reasons. Established process and models illustrating the positive benefits to information sharing and coordination will help companies to change their attitude towards information integration, meaningfully sharing data and utilising supply chain partners' knowledge [18].

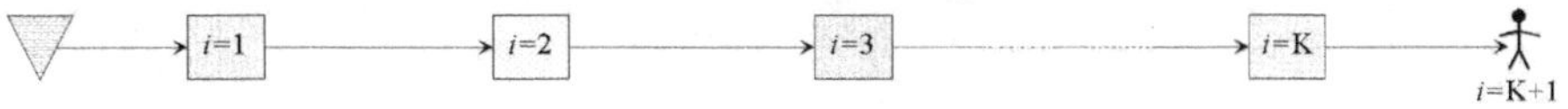

Fig. 1. Serial Multi-echelon supply chain

In the classical (S, R) rule, in a multi-echelon system a generic tier bases its replenishment decision on local information and the order coming from the adjacent downstream echelon. At each review time R, a quantity O is ordered to bring the level of the available inventory up to a level S. The order-up-to order quantity for a generic echelon i is given by eq. 1.

$$O(t) = S(t) - inventory\ \ position\,. \tag{1}$$

The S level for a generic echelon i (eq. 2) is equal to the forecast on the order O_i coming from the subsequent echelon i+1 during the review period R, plus the forecast on the order from echelon i+1 during the production-delivery lead time λ_i, plus a safety stock to prevent shortages. The safety stock depends on a factor ε_i and it is expected to provide sufficient stock to prevent a possible stock-out during the lead time λ_i plus the review period R [20].

$$S_i = R_i\hat{O}_{i+1} + \lambda_i\hat{O}_{i+1} + \varepsilon_i\hat{O}_{i+1}\,. \tag{2}$$

The inventory position (eq. 3) is given by inventory on hand I_i plus pipeline inventory or work in progress W_i (wip).

$$inventory\ \ position_i = I_i + W_i\,. \tag{3}$$

The order quantity for echelon i is herein derived for R=1 (eq. 4, eq. 5 and eq. 6).

$$O_i = \hat{O}_{i+1} + \lambda_i\hat{O}_{i+1} + \varepsilon_i\hat{O}_{i+1} - I_i - W_i\,. \tag{4}$$

$$O_i = \hat{O}_{i+1} + (\varepsilon_i\hat{O}_{i+1} - I_i) + (\lambda_i\hat{O}_{i+1} - W_i)\,. \tag{5}$$

$$O_i = \hat{O}_{i+1} + (TI_i - I_i) + (TW_i - W_i)\,. \tag{6}$$

According to Disney and Lambrecht [20], the safety stock $\varepsilon_i\hat{O}_{i+1}$ can be viewed as a target net stock TI_i and the quantity $\lambda_i\hat{O}_{i+1}$ as a target pipeline stock or target work in progress TW_i

From eq. 4-6 it is possible to identify that, in order to generate an order-up-to level in a traditional production-distribution system, the data to be gathered are: local inventory level, local work in progress level, downstream incoming orders, safety stock factor and lead time. To conclude, the classical order-up-to is composed by three terms: (1) forecast on the order from the subsequent echelon i (2) work in progress gap (3) inventory gap.

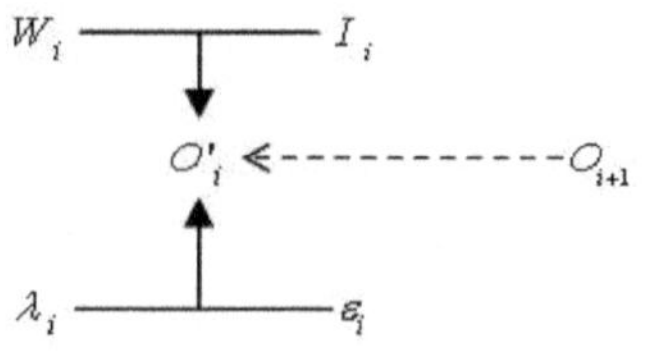

Fig. 2. Operational information for the classical (S, R)

Figure 2 summarises the information used by an echelon to generate the order quantity with the classical (S, R) policy.

In the following section we show how the three terms of the classical (S, R) modify when information is shared among partners and operations are coordinated.

3 Coordinated (S, R) Policy

The coordinated decision-making in the multi-echelon supply chain is modelled under the same conditions of the previous model. Furthermore, the following two relevant assumptions are considered:

- Market demand is visible to all echelons. All echelons adopt the exponential smoothing rule to forecast demand.
- A generic echelon i receives information about order quantity O'_{i+1} from the downstream adjacent echelon $i+1$, as in the classical (S, R) order-up-to, plus information on the up-to-date market demand d and on safety stock factors ε_j, lead times λ_j, inventory levels I_j, and work in progress levels W_j from all downstream echelons $j=i+1 \ldots K$.

First of all we underline that in the coordinated (S, R) policy, unlike the classical policy, the aim of a generic tier is not to serve the order generated by the subsequent adjacent stage, but the demand coming from market. This implies that the risk period (lead time plus review period) has to be referred to the entire time length needed to deliver the finished product from the generic tier to the final customer. According to Clark and Scarf's multi-echelon model [25], recognised as the seminal work in multi-echelon inventory analysis [26]-[29], in the determination of the replenishment quantity any echelon needs to consider its successors as a part of its inventory system.

Maintaining the same formalism as in the classical (S, R), we derive the formula for the coordinated order-up-to level S'. The S' level for a generic echelon i (eq. 7) is equal to the forecast of customer demand d during the review period R, plus the expected customer demand during the multi-echelon lead time λ_i', plus the multi-echelon safety stock to prevent shortages.

$$S_i' = \lambda_i'\hat{d} + R_i\hat{d} + \varepsilon_i'\hat{d} \ . \tag{7}$$

The multi-echelon lead time λ' for echelon i is the entire time period needed to deliver the finished product from the generic tier to the final customer (eq. 8).

$$\lambda_i' = \sum_{j=i}^{K} \lambda_j \,. \tag{8}$$

Analogously, the multi-echelon safety stock factor ε' from echelon i to customer K+1 is given by eq. 9.

$$\varepsilon_i' = \sum_{j=i}^{K} \varepsilon_j \,. \tag{9}$$

The inventory position at echelon i (eq. 10) is given by the multi-echelon inventory I_i' (eq. 11) plus the multi-echelon pipeline inventory or multi-echelon work in progress W_i' (eq. 12).

$$inventory\ position_i' = I_i' + W_i' \,. \tag{10}$$

$$I_i' = \sum_{j=i}^{K} I_j \,. \tag{11}$$

$$W_i' = \sum_{j=i}^{K} W_j \,. \tag{12}$$

The multi-echelon inventory I_i' is the sum of the inventory levels I_j from echelon i to echelon K (inventory level at echelon i plus inventories levels of all downstream echelons $j=i+1...K$). The multi-echelon work in progress W_i' is the sum of the work in progress levels W_j from echelon i to echelon K (work in progress at echelon i plus work in progress of all downstream echelons $j=i+1...K$).

Such as the classical (S, R), the order quantity O' is derived for R=1 (eq. 13, eq. 14 and eq. 15). Analogously to the previous configuration, the term $\varepsilon_i'\hat{d}$ can be viewed as a multi-echelon target net stock TI_i' and the term $\lambda_i'\hat{d}$ as a multi-echelon target pipeline stock or multi-echelon target work in progress TW_i'.

$$O_i' = \lambda_i'\hat{d} + \hat{d} + \varepsilon_i'\hat{d} - I_i' - W_i' \,. \tag{13}$$

$$O_i' = \hat{d} + (\varepsilon_i'\hat{d} - I_i') + (\lambda_i'\hat{d} - W_i') \,. \tag{14}$$

$$O_i' = \hat{d} + (TI_i' - I_i') + (TW_i' - W_i') \,. \tag{15}$$

The main difference between a classical (S, R) and a coordinated (S, R) can be inferred from the comparison of eq. 4-6 and eq. 13-15. In the classical (S, R) the determination of order quantity is based on the order quantity placed by downstream adjacent echelon. In the coordinated (S, R) the determination of order quantity is based on the market demand and the inventory levels and in-transit items from all downstream echelons. Figure 3 summarises the information used by a generic echelon to generate a coordinated order quantity.

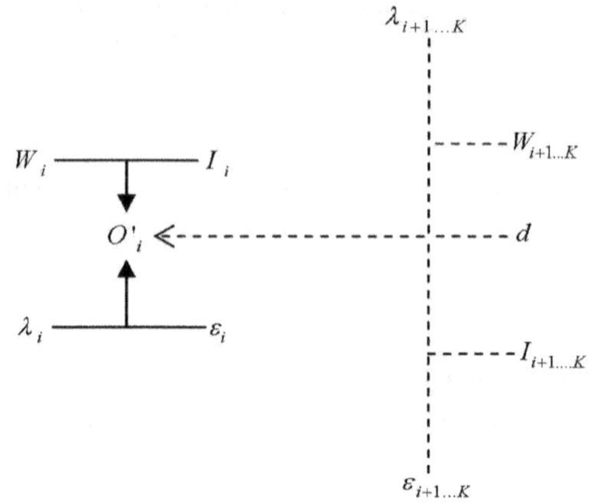

Fig. 3. Operational information for the coordinated (S, R)

4 Numerical Example

We now show a numerical example to highlight the different performance of the classical (S, R) policy and the coordinated (S, R) policy. In our example the two decision rules are adopted in a serial supply chain system.

The two decision rules are compared in terms of demand amplification and inventory stability, with two referenced metrics: Order Rate Variance Ratio [30] and Inventory Variance Ratio [10].

The presented model follows Sterman's traditional supply chain order of events [31]. Refer to [8] for details on the model. To set the numerical values for the experiment, we refer to contributions validated in the literature. The lead time and demand smoothing forecasting factor, the initial values of the state variables, safety stock factor, and the market demand pattern refer to the setting of Sterman's traditional supply chain model [31] and also used in Wikner et al. [32], Van Ackere et al. [33], John et al. [34], Crespo Márquez et al. [35], Machuca and Barajas [36], Jakšič and Rusjan [37], and Wright and Yuan [38].

The numerical experiments are performed under the following settings:

- The serial system is composed by K=4 echelons.
- Supply chain member adopt exponential smoothing to forecast customer demand and incoming orders.
- The initial values of the state variables are: $[W_i(0), I_i(0)]=[\lambda_i d(0), \varepsilon_i d(0)]$
- The lead time is equal to 3, demand smoothing forecasting factor is equal to 0.33 and safety stock factor is equal to 3.
- Numerical experiments are performed for a time length T=52.
- The solutions for the initial-value problem are approximated through Vensim PLE. The Euler-Cauchy method with order of accuracy $\Delta t = 0.25$ is adopted.
- The market demand *d* is initialised at 4 units per time unit, until there is a pulse at t=5, increasing the demand value up to 8 units per time unit.

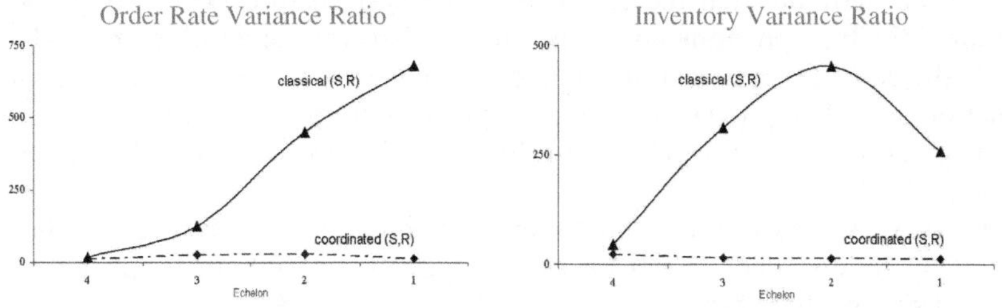

Fig. 4. Order Rate Variance Ratio and Inventory Variance Ratio for classical (S, R) and coordinated (S, R)

Figure 4 reports Order Rate Variance Ratio and Inventory Variance Ratio for classical (S, R) and coordinated (S, R) according to Dejonckheere et al.'s notation [39]. Echelon i=1 stands for the manufacturer and i=4 for the retailer.

The numerical example shows how the classical (S, R) exhibits a geometrical increase of order variance in up-stream direction (from retailer to manufacturer) and high inventory variability. On the contrary, under the same parameter setting, the coordinated (S, R) policy stabilises inventory and significantly avoids demand amplification phenomenon. On the basis of the adopted metrics, it is shown that the coordinated (S, R) largely outperforms the classical order-up-to policy. More specifically, the significant reduction of Order Rate Variance Ratio shows how shifting from classical (S, R) to a coordinated (S, R) results in a reduction of several production and distribution costs [37], [39]. Analogously, the reduction of Inventory Variance Ratio reflects lower holding and backlog costs, drastically diminishing the average inventory cost per period [17], [38].

5 Managerial Implications of Coordinated (S, R)

During the last decade, several companies adopted large-scale collaboration projects such as VMI. The implantation of this typology of collaboration practices between supplier and retailer have often exclusively modified the ownership of the inventory, without changing how the replenishment orders are generated with respect to a traditional system.

As reported by Holweg et al. [14], the retailer can and does pass sell-through data and inventory levels to the manufacturers' plants. The supplier exploits this information implicitly in strategic planning issues, such as capacity planning and manning levels in the factory, but very often she does not exploit the consumption information at a tactical planning level. In other words, supply chain members do not fully benefit from the shared operational information for the generation of a coordinated decision order. Case studies show how this VMI arrangement can degenerate in five-to-one increase in the bullwhip effect at each level of the supply chain [14]. Counter to the common perception, several VMI supply chains remain, from the operational perspective, mere traditional structures [14]. As a consequence, several companies did not succeed in eliminating inefficiencies such as demand amplification. It was largely shown in the literature that the classical (S, R) is by nature extremely prone to demand amplification phenomenon generation, whatever forecasting method is used [40]. On the contrary, it was shown that coordinated order decision making can allow elimination of bullwhip [41]-[43], reduction of inventory levels [44], better utilisation on transportation resources [45], controlling the risk for constrained components of material [46], [47].

In this paper we showed how adopting a collaboration system based on the coordination of operations leads to performance improvement and creates a win-win solution for all supply chain partners.

To conclude, the coordinated (S, R) represents a relatively simple solution to several inefficiencies in supply chain. Its implementation in the large-scale collaboration practices represents a concrete mean to achieve the recently advocated improvements in supply chain operations.

6 Conclusion and New Directions in Coordinated Policies

This paper presented the mathematical derivation of a new generation of the most largely used periodic review policy in supply chain: the coordinated (S, R) replenishment rule. We first derived the classical order-up-to model and then we modified it to generate the coordinated decision policy. We concluded on the managerial implications related to the implementation of the coordinated (S, R) order-up-to.

Even more supply chain collaboration models and practices appear in literature and practices. One of the novel examples is the global inventory policy for a two echelon supply chain discussed by Gaalman and Disney [48]: go-policy. This policy is like a VMI supply chain in reverse. The innovation of this policy stands in a feedback mechanism that corrects some of the manufacturer's forecast errors with the support of retailer. The realisation of new coordination order policies converts day-by-day in a tangible possibility. This is due to the recent advances of ICT and the related decreasing of the controlling inventory costs. However, in the future decade, researchers and practitioners should focus and overcome one of the real problems (rarely addressed in operation management literature) to concretise the operational collaboration: the reluctance of the companies to share with their partners/competitors operational information regarding their own core business.

Acknowledgements. This research was supported by the doctoral fellowship programme of the Italian Ministry of University and Research. We would like to express our gratitude to our PhD thesis supervisors, Mario Enea, Jose M. Framinan and David Canca Ortiz, and all colleagues and professors that have contributed to our education. We would like to dedicate this article to Laura Oreto.

References

1. McKinsey & Co.: Evaluating the impact of alternative store formats. Supermarket Industry Convention, Food Marketing Institute, Chicago (1992)
2. Cachon, G., Fisher, M.: Campbell Soup's continuous replenishment program: Evaluation and enhanced inventory decision rules. Productions and Operations Management 6, 266–276 (1997)

3. Disney, S.M., Towill, D.R.: The effect of vendor managed inventory (VMI) dynamics on the Bullwhip Effect in supply chains. International Journal of Production Economics 85, 199–215 (2003)
4. Andraski, J.C.: Foundations for a successful continuous replenishment programme. International Journal of Logistics Management 5, 1–8 (1994)
5. VICS Merchandising Issue Committee, CPFR Committee: Collaborative Planning, Forecasting, and Replenishment Guidelines.Voluntary Interindustry Commerce Standards Associations, Lawrenceville (1998)
6. Holmström, J., Främling, K., Kaipia, R., Saranen, J.: Collaborative planning forecasting and replenishment: new solutions needed for mass collaboration. Supply Chain Management: An International Journal 7, 136–145 (2002)
7. Lee, H.L., Padmanabhan, V., Whang, S.: Information distortion in a supply chain: the bullwhip effect. Management Science 43, 546–558 (1997)
8. Cannella, S., Ciancimino, E.: On the Bullwhip Avoidance Phase: Supply Chain Collaboration and Order Smoothing. International Journal of Production Research (2009)
9. Cachon, G., Fisher, M.: Supply chain inventory management and the value of shared information. Management Science 46, 1032–1048 (2000)
10. Disney, S.M., Towill, D.R.: A discrete transfer function model to determine the dynamic stability of a vendor managed inventory supply chain. International Journal of Production Research 40, 179–204 (2002)
11. Chatfield, D.C., Kim, J.G., Harrison, T.P., Hayya, J.C.: The bullwhip effect - Impact of stochastic lead time, information quality, and information sharing: A simulation study. Production and Operations Management 13, 340–353 (2004)
12. Shang, J.S., Li, S., Tadikamalla, P.: Operational design of a supply chain system using the Taguchi method, response surface methodology, simulation, and optimization. International Journal of Production Research 42, 3823–3849 (2004)
13. Kelepouris, T., Miliotis, P., Pramatari, K.: The impact of replenishment parameters and information sharing on the Bullwhip effect: a computational study. Computers and Operations Research 35, 3657–3670 (2008)
14. Holweg, M., Disney, S.M., Holmström, J., Småros, J.: Supply chain collaboration: making sense of the strategy continuum. European Management Journal 23, 170–181 (2005)
15. Disney, S.M.: The Dynamics of Material Flows in Supply Chains, `http://ssrn.com/abstract=1013326`
16. Wang, Y.C.: Evaluating flexibility on order quantity and delivery lead time for a supply chain system. International Journal of Systems Science 39, 1193–1202 (2008)
17. Boute, R.N., Disney, S.M., Lambrecht, M.R., Van Houdt, B.: Designing replenishment rules in a two-echelon supply chain with a flexible or an inflexible capacity strategy. International Journal of Production Economics 119, 187–198 (2009)
18. Kulp, S.C., Ofek, E., Whitaker, J.: Supply-Chain Coordination: How Companies Leverage Information Flows to Generate Value. In: Harrison, T.P., Lee, H.L., Neale, J. (eds.) The practice of supply chain management. Where theory and practice converge, pp. 91–107. Springer, New York (2004)
19. Hax, A.C., Candea, D.: Production and Inventory Management. Prentice-Hall, Englewood Cliffs (1984)
20. Disney, S.M., Lambrecht, M.R.: On Replenishment Rules, Forecasting, and the Bullwhip Effect in Supply Chains. Foundations and Trends in Technology, Information and Operations Management 2, 1–80 (2008)
21. Boute, R.N., Disney, S.M., Lambrecht, M.R., Van Houdt, B.: A win-win solution for the bullwhip problem. Production Planning and Control 19, 702–711 (2008)

22. Simon, H.A.: On the application of servomechanism theory to the study of production control. Econometrica 20, 247–268 (1952)
23. Holt, C.C., Modigliani, F., Muth, J.F., Simon, H.A.: Planning Production, Inventories and Work Force. Prentice-Hall, Englewood Cliffs (1960)
24. Forrester, J.: Industrial Dynamics. MIT Press, Cambridge (1961)
25. Clark, A.J., Scarf, H.: Optimal Policies for a Multi-Echelon Inventory Problem. Management Science 6, 475–490 (1960)
26. Whang, S.: Coordination in operations: a taxonomy. Journal of Operations Management 12, 413–422 (1995)
27. Dong, L., Lee, H.L.: Optimal policies and approximations for a serial multiechelon inventory system with time-correlated demand. Operations Research 51, 969–980 (2003)
28. Swaminathan, J.M., Tayur, S.R.: Models for supply chains in e-Business. Management Science 49, 1387–1406 (2003)
29. DeCroix, G.A.: Optimal policy for a multiechelon inventory system with remanufacturing. Operations Research 54, 532–543 (2006)
30. Chen, F., Drezner, Z., Ryan, J.K., Simchi-Levi, D.: Quantifying the bullwhip effect in a simple Supply Chain: the impact of forecasting, lead-times and information. Management Science 46, 436–443 (2000)
31. Sterman, J.D.: Modelling managerial behaviour: misperceptions of feedback in a dynamic decision-making experiment. Management Science 35, 321–339 (1989)
32. Wikner, J., Towill, D.R., Naim, M.M.: Smoothing supply chain dynamics. International Journal of Production Economics 22, 231–248 (1991)
33. Van Ackere, A., Larsen, E.R., Morecroft, J.D.W.: Systems thinking and business process redesign: An application to the beer game. European Management Journal 11, 412–423 (1993)
34. John, S., Naim, M.M., Towill, D.R.: Dynamic analysis of a WIP compensated decision support system. International Journal of Management Systems Design 1, 283–297 (1994)
35. Crespo Márquez, A., Bianchi, C., Gupta, J.N.D.: Operational and financial effectiveness of e-collaboration tools in supply chain integration. European Journal of Operational Research 159, 348–363 (2004)
36. Machuca, J.A.D., Del Pozo Barajas, R.: A computerized network version of the Beer Game via the Internet. System Dynamics Review 13, 323–340 (1997)
37. Jakšič, M., Rusjan, B.: The effect of replenishment policies on the bullwhip effect: a transfer function approach. European Journal of Operational Research 184, 946–961 (2008)
38. Wright, D., Yuan, X.: Mitigating the bullwhip effect by ordering policies and forecasting methods. International Journal of Production Economics 113, 587–597 (2008)
39. Dejonckheere, J., Disney, S.M., Lambrecht, M.R., Towill, D.R.: The impact of information enrichment on the bullwhip effect in Supply Chains: A control engineering perspective. European Journal of Operational Research 153, 727–750 (2004)
40. Dejonckheere, J., Disney, S.M., Lambrecht, M.R., Towill, D.R.: Measuring and avoiding the bullwhip effect: A control theoretic approach. European Journal of Operational Research 147, 567–590 (2003)
41. Kim, J.G., Chatfield, D., Harrison, T.P., Hayya, J.C.: Quantifying the bullwhip effect in a supply chain with stochastic lead time. European Journal of Operational Research 173, 617–636 (2006)
42. Agrawal, S., Sengupta, R.N., Shanker, K.: Impact of information sharing and lead time on bullwhip effect and on-hand inventory. European Journal of Operational Research 192, 576–593 (2009)

43. Ciancimino, E., Cannella, S., Canca Ortiz, J.D., Framiñán, J.M.: Supply chain multi-level analysis: two bullwhip dampening approaches [Análisis multinivel de cadenas de suministros: Dos técnicas de resolución del efecto bullwhip]. Revista de Métodos Cuantitativos para la Economía y la Empresa 8, 7–28 (2009)
44. Hosoda, T., Naim, M.M., Disney, S.M., Potter, A.: Is there a benefit to sharing market sales information? Linking theory and practice. Computers and Industrial Engineering 54, 315–326 (2008)
45. Mason, R., Lalwani, C., Boughton, R.: Combining vertical and horizontal collaboration for transport optimization. Supply Chain Management 12, 187–199 (2007)
46. Cannella, S., Ciancimino, E., Márquez, A.C.: Capacity constrained supply chains: A simulation study. International Journal of Simulation and Process Modelling 4, 139–147 (2008)
47. Ciancimino, E., Cannella, S.: Modelling the bullwhip effect dampening practices in a limited capacity production network. LNBIP, vol. 20, pp. 475–486. Springer, Heidelberg (2009)
48. Gaalman, G., Disney, S.M.: A new inventory policy for a two echelon supply chain. In: Euroma conference, Göteborg (2009)

Towards an Integrated Virtual Value Creation Chain in Sheet Metal Forming

Manfred Grauer, Daniel Metz, Ulf Müller, Sachin Karadgi,
Walter Schäfer, and Thomas Barth

Information Systems Institute, University of Siegen,
57076 Siegen, Germany
{grauer,metz,mueller,karadgi,jonas,barth}@fb5.uni-siegen.de

Abstract. OEMs are focusing more on manufacturing critical components whereas non-critical components are outsourced to suppliers. At supplier's side, production planning process is intervened between sales and production processes. Past experiences can be adapted to support production planning activities. Hence, it is crucial to identify knowledge from previously solved cases on demand and assimilate it. At the Information Systems Institute, various methodologies (e.g., similarity search in product and process data) are developed to support aforesaid knowledge intensive activities in sheet metal forming. However, these methodologies are used as stand-alone applications. Therefore, the current contribution introduces an integrated virtual value creation chain in sheet metal forming which utilizes previously developed methodologies. Also, these methodologies are elaborated as part of the integrated virtual value creation chain. The developed methodologies are heterogeneous and are coupled through a shared knowledge base.

Keywords: collision detection, cost calculation, knowledge reuse, sheet metal forming, similarity search.

1 Introduction

Sheet metal forming, comprising methods like stamping, blanking, (deep) drawing, and bending, is an important production method for several industries, especially automotive [1]. In accordance to the importance of sheet metal forming, metal forming machine tools have a share of around 20% of the total amount of produced machine tools in Germany [2]. Due to increased complexity of components and ongoing trend of OEMs to focus on core competencies, an increasing number of sheet metal components are outsourced to suppliers for design and production.

At any given time, a customer is catered by different suppliers and, likewise, a supplier offers product development and production capabilities to different customers. As a consequence, customer and supplier have an intricate relationship manifested in a plethora of complex and interdependent activities in the course of developing and producing complex metal forming parts. Before production, a supplier performs various production planning activities supported by sophisticated tools from information technology. These activities are knowledge intensive and

W. Dangelmaier et al. (Eds.): IHNS 2010, LNBIP 46, pp. 186–197, 2010.

time consuming, and can include cost calculation, design, and the application of methods and tools from virtual prototyping. To execute these activities, customer and supplier employ 'request-offer-order' processes as depicted in Fig. 1.

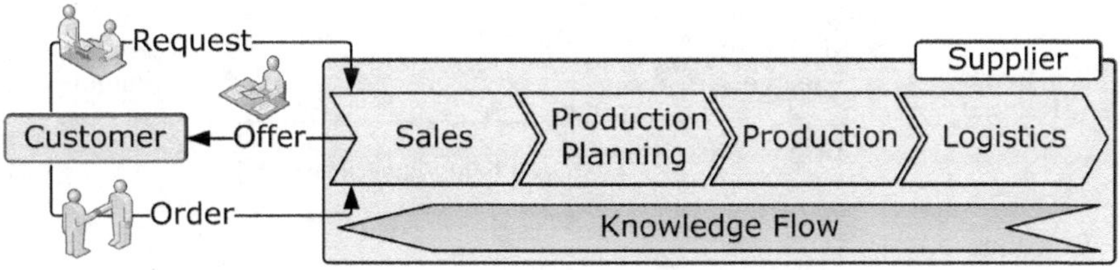

Fig. 1. Customer supplier interaction and value creation process at supplier

Production planning process is intervened between sales and production processes (s. Fig. 1) and collaborates with these processes to achieve production management goals [3]. Production planning process encompasses knowledge intensive and time consuming activities like process planning, scheduling and coordination of production processes. Hence for efficient execution of its activities, it should be equipped with knowledge about upstream and downstream processes. This knowledge can be assimilated to enhance value creation processes (e.g., resource configuration). Unfortunately, knowledge is not available for these processes as there are almost no efficient tools to identify it in the enormous amount of data generated during previously executed processes. In addition to knowledge, production planning activities should be supported with integrated virtual prototyping methodologies that optimize and ensure the required quality of value creation processes, in particular during production.

Due to these observations, the approach presented in this contribution aims at integrating knowledge-reuse by virtual prototyping methods. The remainder of this contribution is organized as follows. Section 2 elaborates an integrated virtual value creation chain in sheet metal forming. At each step of this value creation chain, knowledge can be identified on demand and applied, as elaborated from Section 3 to Section 6. Problem description, related work and methodology are elaborated for each process step. The contribution is concluded with a discussion on future work in Section 7.

2 Integrated Virtual Value Creation Chain

Customer-supplier interaction starts with a customer requesting information for capabilities provided by the supplier. The supplier should be able to respond to the customer's request quickly with reliable information on time and cost utilizing information technology which supports knowledge intensive and time consuming activities. In this context, it is crucial to identify knowledge from previously solved cases on demand and assimilate it. These knowledge identification and assimilation activities at various stages in production planning are depicted in Fig. 2. The principle approach of each of these stages can be summarized as follows:

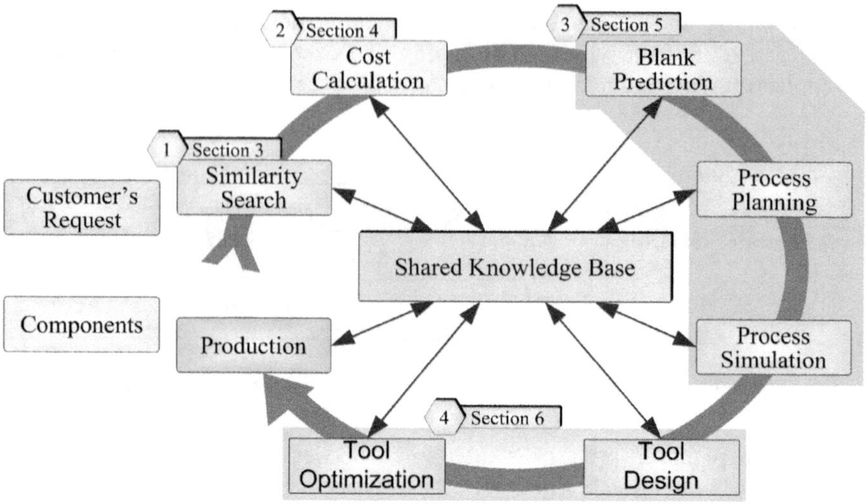

Fig. 2. Integrated virtual value creation chain in sheet metal forming

1. For a product requested by the customer, a similarity search system can be employed to search for similar products based on product's specification (e.g., material) stored in enterprise's product databases (e.g., Product Lifecycle Management (PLM) systems). The retrieved similar case and the associated product data (e.g., process plans) can then be utilized to enhance the production planning process [4] and also, act as a template which can be adapted by the domain experts to respond to customers' requests.
2. Manufacturing cost of a component includes tool-, material- and processing- cost. Information requested by the customer is derived using data gathered from different departments. This information should be sufficiently precise to increase the probability of getting an order and subsequently to avoid financial loss during order execution. However, manually obtaining data from different departments will delay the response to the customer. To speed up this task, a methodology has been developed which utilizes various systems (e.g., similarity search) and applies techniques from knowledge-based engineering (KBE) and case-based reasoning (CBR) [5].
3. Sheet metal forming process encompasses a sequence of metal forming operations like blanking, deep drawing, piercing, and so forth. Blanking is often the initial operation to provide a blank that can be further processed with the aforementioned metal forming operations. The prediction of an optimal blank shape (e.g., with minimal consumption of material) has already attained considerable attention in research [6]. The predicted blank shape is input for process planning to define the number and sequence of metal forming operations. Also, it will be used during cost calculation to estimate material utilization. Further, process planning is integrated with process simulation to ensure reliability of planned production process [7].
4. Based on a process plan, forming dies have to be designed and assembled using geometrical modeling process (i.e., CAD systems). Due to their complexity, sheet metal components have to be produced with multi-stage metal forming processes,

and subsequently, transportation of semi-finished components from one forming stage to the next has to be considered. As a consequence, collisions might occur between dynamic components (e.g., between transfer equipment and transfer dies). In addition, the configuration of transportation systems influences the achievable stroke rate of the forming press [8]. Hence, a methodology for collision analysis of metal forming processes using CATIA® V5 and a digital mock-up (DMU) software OpTiX® KINSIM has been developed [9]. The methodology is integrated with geometrical modeling of tool structure and aims at avoiding collisions during production as well as increasing the production performance in terms of achievable stroke rate. The stroke rate has to be (at least) equal to the value assumed during cost calculation process.

Each of the virtual prototyping methodologies mentioned above are utilized to address certain production planning activities, which are heterogeneous. However, these methodologies are coupled through a shared knowledge base. For instance, cost calculation utilizes similarity search to identify previous documents from a supplier's database, and blank prediction to determine the blank shape and material utilization.

After processing the integrated virtual value creation chain, the start of production is initiated. Executing the aforementioned virtual value creation chain mitigates production disturbances and increases production efficiency. The individual components of the virtual value creation chain are elaborated in the subsequent sections. The presented components (except for blank prediction) have been developed by the Information Systems Institute at the University of Siegen. For completeness, already existing tools for blank prediction, process planning and process simulation have been integrated into the outlined virtual value creation chain. Each section starts with a brief problem description, proceeds with presenting related work, and ends with elaboration of a methodology.

3 Similarity Search for Identifying and Reusing Knowledge on Demand

"In modern product development, as the complexity and variety of products increase to satisfy increasingly sophisticated customers, so does the need for knowledge and expertise for developing products" [10]. In addition, the necessity to shorten the time-to-market forces enterprises to accelerate and improve their product development by reusing knowledge. Knowledge related to development activities is tacit and embedded in the products. Products are described by product data (e.g., geometrical model, process data) which can be managed in enterprise's product databases (e.g., PLM systems). Enterprise members can access these product data and reuse it to shorten the time-to-market of new products. However, capabilities to efficiently search for various kinds of product data are not available.

3.1 Related Work on Similarity Search

Available search systems related to product data are mainly based on CBR. CBR can be explained as a cycle consisting of four steps [11]: First, RETRIEVE problem descriptions which are similar to a given (new) problem from a case base; second,

REUSE the assigned similar solution; third, REVISE this solution, and finally, RETAIN the current problem and the new solution in the case base.

There are two approaches for assessing the similarity between products. First, assessing the similarity based on the products' geometries represented by means of 3D models [4]. A survey of possible methods and techniques to assess and retrieve similar 3D models is available in [12]. Second, assessing the similarity based on products' assembly sequences (e.g., bill of material (BOM)). A comprehensive overview of methods to determine differences between BOMs and measure their similarity is presented in [13]. Several retrieval systems and search engines have been developed for searching in enterprise databases. Most of these systems offer similarity metrics for numerical or alphanumerical data, and only a few of them are capable to search for products' 2D or 3D geometry [14, 15, 16]. However, there is no integrated search system which is capable to search for all of these different types of data simultaneously.

3.2 Methodology

A methodology has been developed to set up an integrated search system which is capable to incorporate similarity metrics for numerical, alphanumerical and geometrical product data [17]. This search system is based on CBR, and an architectural overview of the search system is illustrated in Fig. 3. The architecture consists of three main components: case base, retrieval system, and search views.

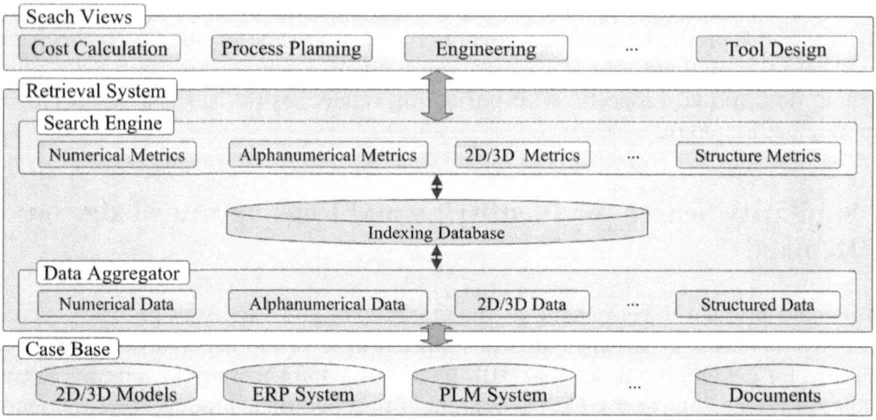

Fig. 3. Overview of search system architecture (adapted from [17])

In the context of product data, the case base consists of geometrical data (e.g., 2D/3D CAD models), bill of materials (BOM), cost calculations, process plans). Product-related documents along the product lifecycle (PLC) have to be analyzed to identify attributes that can be utilized to define similarity metrics. These attributes approximate the experts' idea of searching similar products, and are used to generate sufficient case descriptions. The case descriptions are managed in an indexing database and are linked to the cases in the case base. A data aggregator is in charge of deriving case descriptions to be stored or modified in the indexing database.

The core module of the retrieval system is a search engine encompassing various similarity metrics (s. "Retrieval System" in Fig 3). These similarity metrics combine techniques from different research areas to derive a similarity value. This value is used to retrieve cases from the case base. The search engine is capable of incorporating various data types (i.e., numerical, alphanumerical, and geometrical data) relevant in the context of PLC. View- and image- based methods are employed to measure the products' visual similarity [18]. A density-based shape descriptor is utilized to evaluate the similarity based on local surface features [19]. To assess structural similarities (e.g., between BOMs), a graph-based distance measure is applied utilizing an extended Jaccard similarity coefficient. Edit-distance and Euclidean-distance metrics are used for comparing simple data types (e.g., strings). Different search views (s. "Search Views" in Fig. 3) are generated to grant enterprise members access to the retrieval system. The search views depend on the user's profile and his privileges.

4 Cost Calculation for Quick and Precise Responses to Customers

Cost calculation is knowledge intensive and time consuming process, and the elaborated concepts and actual values depend upon a vast number of interdependent details within calculation, calculation depth, and the granularity of calculation. In many cases, a customer's request will be waiting for processing, increasing the lead time to respond to customer's request. Around 10% of the total offers are converted into orders [20]. Due to this reason, the cost calculation process is not performed in-depth in every case by domain experts from different departments. Instead, sales personnel often perform the cost calculation activities of domain experts. As a consequence, cost calculation will be based on sales personnel's limited knowledge and experience with the actual manufacturing technology. In addition, parameters used to derive the offer will act as inputs or constraints to downstream processes (e.g., process planning) and might cause problems during order execution. To overcome these problems and reduce unrealistic cost calculation, knowledge adopted by various departments has to be made available to sales personnel (s. knowledge flow in Fig. 1).

4.1 Related Work on Cost Calculation

The major components of an automotive body and chassis are complex sheet metal components and require metal forming operations to manufacture them [1]. Manufacturing cost of a component consists of tool-, material- and processing- cost [21, 22]. Cost calculation application is available with the sales department for metal forming processes [20]. Procedures to calculate amortization, processing costs, transportation and material utilization are embedded into this application and necessary information (e.g., material cost) is retrieved from databases. However, tool cost and forming press details should be provided as inputs which sales personnel is not in the position to determine these values preciously. In addition, tool cost depends upon the number of stages in a process plan and the actual metal forming operations performed at each stage. Research has been carried out in isolation to determine process plans in combination of feature-based engineering (FBE) and KBE [23, 24, 25], and sometimes using CBR [23, 26].

4.2 Methodology

To overcome aforementioned problems and support sales personnel with knowledge about downstream processes, a methodology to integrate various systems and techniques has been proposed as illustrated in Fig. 4 [5]. Information about enterprise resources, material and components along with previous offers/orders are stored in enterprise databases which form a case base (s. Fig. 3). A supplier will be collaborating with many customers and for a component in focus there is a high probability that a similar offer/order was created previously. Search system (s. Section 3) can be utilized to retrieve similar offers or orders and the result can be used as a template to create a new offer resulting in accelerated offer creation process and increasing reliability of the new offer.

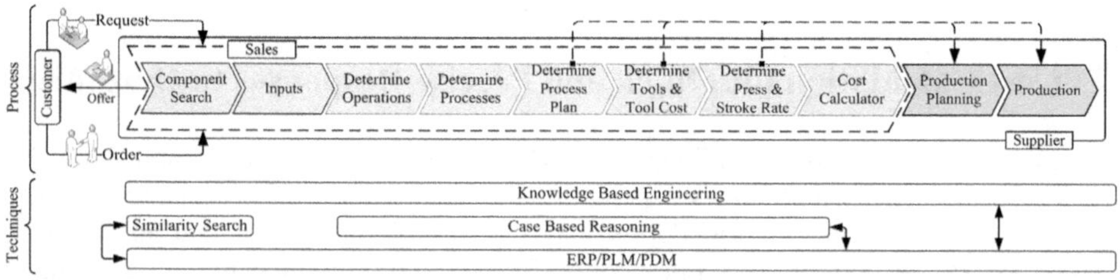

Fig. 4. Cost calculation process (adapted from [5])

Accuracy of the estimated values depends upon input data. Hence, data model of typical sheet metal forming features and operations can be hierarchically classified [23, 25, 27] and this classification can be further enhanced to include process plans and manufacturing resources. KBE is employed to capture individuals' expertise within an enterprise and manage it as rules. Rules are modeled based on the data model and assist sales personnel to validate inputs and determine unknown process parameters (e.g., operations). In many instances, design practices are based on past experiences and these experiences are adapted to solve new problems [26]. Knowledge and design experiences are stored in documents (e.g., 3D CAD models) which are part of the case base, and case base is enriched and enlarged with each step of the offer process. Rules are utilized to create an initial process plan which act as input to search the case base for old process plans and iteratively adapt the initial process plan resulting in an optimized process plan. This can be achieved using CBR technique, especially focussing RETRIEVE and REUSE steps. Once a process plan is determined and with assistance of rules, it is possible to calculate tool's size and cost, and select appropriate forming press and determine achievable stroke rate. The determined values can be provided as inputs to a cost calculation system [20] and subsequent production planning activities (s. Section 6).

The cost calculation methodology presented above is based on different techniques (e.g., KBE and CBR) and utilizes different systems (e.g., search system presented in Section 3) that support the reuse of existing knowledge. This can help to reduce mistakes in the early phases of PLC, and results in a quicker response to a customer's request comprising more precise information.

5 Blank Prediction, Process Planning and Process Simulation

For completeness, available third-party solutions for blank prediction, process planning and process simulation can be integrated into the virtual value creation chain. Process planning in sheet metal forming often starts with the determination of blank shape. A blank shape is an important input for calculation of material cost and has to ensure the feasibility of the forming process required to achieve the final shape. Research exists to provide methodologies for prediction of optimal blank shape [6, 28]. Further, commercial applications are available to optimize blank shapes [29].

Based on determined blank shape, further process planning activities are performed to derive the total number and sequence of metal forming operations. After design of every die face (metal forming operation) an incremental process simulation verifies the feasibility of the (partial) process plan [7]. Nowadays, commercial process simulation packages have been established in industry like Indeed® [30]. Problems (e.g., thinning of component) which would occur during production can be minimized or even eliminated before manufacturing the dies. Required corrections and redesign of blank shape and dies are done immediately, and subsequently, lead to smoother die try-out and shorter lead times [7].

6 Collision Detection and Optimization of Tool Structures

To realize production processes in sheet metal forming, progressive compound dies and transfer dies are used. In case of transfer dies, a peripheral transportation system (e.g., transfer system) has to be engaged to transport semi-finished components from one forming stage to the next. A tri-axis transfer system performs trajectories in three dimensions: in longitudinal direction to transport components, in transverse direction in the form of a closing movement for gripping parts and in vertical direction to lift out drawn parts [1].

A domain expert has to define the trajectories of the transportation system and forming press in the press control system. Collisions might occur during production due to inaccurate configuration of the transfer system and/or design faults in tool structures. Collisions are interferences between transfer equipment and transfer dies, but also between semi-finished components and the transfer dies. In addition, the transfer system configuration influences the achievable stroke rate of the forming press. Hence, main challenges during production planning are to design transfer dies and to configure the transfer system in order to avoid collisions and optimize the stroke rate.

6.1 Related Work on Collision Detection

Traditionally, collision curves are derived by the combination or overlay of motion curves (e.g., motion of opening/closing of the gripper rail). These collision curves are used to analyze if a critical point of the transfer equipment (e.g., gripper tip) is or will be in collision with other components during production [31]. The aforementioned approach is time consuming, and only includes a few parts of the geometrical model which are perceived by domain experts as critical. In addition, physical simulation is also carried out using manufactured transfer dies, transfer equipments and lower press [1]. Unfortunately, upper dies cannot be incorporated in such a collision analysis, and collisions detected after manufacturing of transfer dies result in costly elimination of errors [32].

Kinematics simulation has already been used to solve several engineering related problems. Extensive research has been carried out to simulate robots using different simulation tools and packages like MATLAB®, Open Dynamic Engine (ODE) and ADAMS™ (e.g., [33, 34, 35]). Also, validation of machining setup for selected NC machines can be done (e.g., using Tecnomatix RealNC [36]). Similarly to OpTiX® KINSIM, Tecnomatix Press Line Simulation enables the kinematics simulation of forming presses and handling devices [36]. Aforementioned approaches are expensive and non-optimal to detect collisions between transfer equipment and transfer dies.

6.2 Methodology

To overcome aforementioned drawbacks, a novel methodology for automatic collision analysis using DMU software OpTiX® KINSIM and CATIA® V5 has been developed [9]. An architectural overview of OpTiX® KINSIM is illustrated in Fig. 5. After analyzing various tri-axis transfer systems along with designed transfer dies, it can be deduced that the geometric shape of the dies and their assemblies varies whilst the trajectories of most dynamic components (e.g., grippers) are similar. Therefore, definition of kinematics mechanisms is separated from geometrical modeling of transfer dies by means of a kinematics reference model constructed in CATIA® V5.

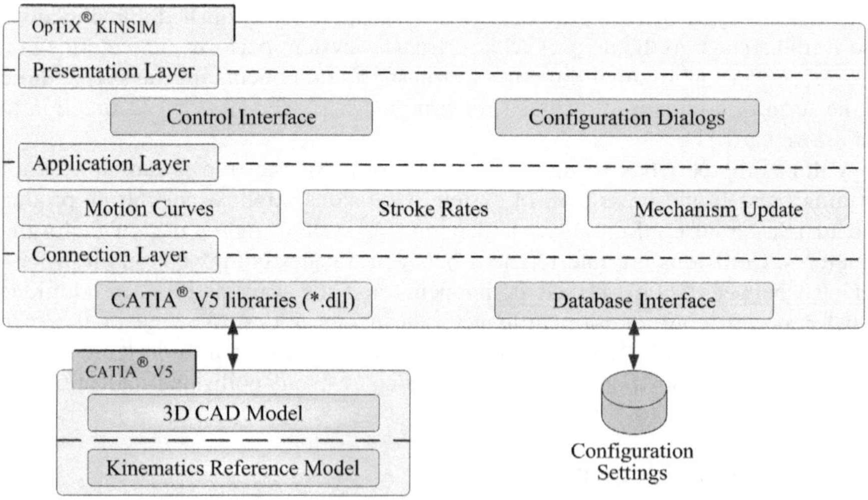

Fig. 5. Architecture of OpTiX® KINSIM and its interfaces to CATIA® V5 [9]

An interface between OpTiX® KINSIM and CATIA® V5 has been developed using CATIA® V5 APIs (connection layer). In the application layer, motion curves and corresponding stroke rates are calculated based on a given configuration of transfer equipment and forming press. This configuration is defined by a domain expert via presentation layer, and the resulting motion curves are transferred to the kinematics reference model using the mechanism update module (application layer). Configuration settings can be stored in a database which also contains vendor specific information about transfer systems and forming presses (e.g., accelerations).

After setting up the kinematics reference model within CATIA® V5, a collision analysis can be performed. As all dynamic components of tri-axis transfer are incorporated in the collision analysis, all collisions are detected automatically during simulation of production process. Although achieved results are adequate, the presented simulation approach could be improved by including dynamic effects (e.g., bending), which is now impossible with the standard functionality of CATIA® V5.

7 Conclusions and Future Work

In manufacturing scenarios, customers and vendors have an intrinsic relationship as well as interactions. At supplier's side, production planning is a critical process encompassing knowledge intensive and time consuming activities, and is intervened between sales and production processes. These activities have to be supported with knowledge to enhance value creation process. Information technology and virtual prototyping methodologies are required to identify and assimilate knowledge from previous experiences on demand. In addition, these methodologies need to be integrated to optimize and ensure required quality of value creation processes.

Current contribution has addressed the aforementioned requirements of production planning process. Integrated virtual value creation chain consisting of different methodologies has been elaborated from Section 3 to Section 6. Each of the methodologies supports different stages of production planning process by identifying knowledge on demand and assimilating it. Further, these methodologies interact with each other through shared knowledge base. Several methodologies of the outlined integrated virtual value creation chain have been put into practice by Co.Com Concurrent Computing GmbH, Siegen. The practical experiences gathered while using these methodologies can be utilized for further development activities.

In future, it is envisaged to enhance the available virtual prototyping methodologies and expand the integrated virtual value creation chain by including methodologies to support other stages (e.g., simulation-based feasibility study of forming components before process planning [7]) of the production planning process. Furthermore, research and development will also focus on the calculation of design space wherein engineers can design transfer dies without causing collisions during transportation and production.

Acknowledgment. Parts of the work presented here have been supported by German Federal Ministry of Science and Research within German InGrid initiative (#01AK806C) and FP6 EU-funded IST project BEinGRID (IST5-034702).

References

1. Schuler GmbH: Metal Forming Handbook. Springer, Heidelberg (1998)
2. VDW (ed.): The German Machine Tool Industry 2009 (2010)
3. Halevi, G., Weill, R.: Principles of Process Planning: A Logical Approach. Chapman, Boca Raton (1995)
4. Cardone, A., Gupta, S.K., Karnik, M.: A survey of shape similarity assessment algorithms for product design and manufacturing applications. J. of Computing and Information Science in Engineering 3(2), 109–118 (2003)
5. Karadgi, S., Müller, U., Metz, D., Schäfer, W., Grauer, M.: Cost estimation of automotive sheet metal components using knowledge-based engineering and case-based reasoning. In: IEEE Int. Conf. on Ind. Eng. and Eng. Management, Hong Kong, pp. 1518–1522 (2009)
6. Lin, C.-T., Kwan, C.-T.: Application of abductive network and FEM to predict the optimal blank contour of an elliptic cylindrical cup from deep drawing. J. of Materials Processing Tech. 209(3), 1351–1361 (2009)
7. Tisza, M., Lukács, Z., Gál, G.G.: Integrated process simulation and die-design in sheet metal forming. Int. J. of Material Forming 1(1), 185–188 (2008)
8. Roll, K., Bogon, P., Ziebert, C.: Innovative methods for the design of forming tools in the automotive industry (in German). wt Werkstattstechnik online 97(10), 753-759 (2007)
9. Metz, D., Grauer, M., Schäfer, W., Reichert, O.: Collision analysis for virtual production planning in sheet metal forming. In: Int. Conf. on Changeable, Agile, Reconfigurable and Virtual Production, pp. 367–376 (2009)
10. Ameri, F., Dutta, D.: Product lifecycle management: Closing the knowledge loops. Computer-Aided Design and Applications 2(5), 577–590 (2005)
11. Aamodt, A., Plaza, E.: Case based reasoning: Foundational issues, methodological variations, and system approaches. J. AI Communications 7, 39–59 (1994)
12. Tangelder, J.W., Veltkamp, R.C.: A survey of content based 3D shape retrieval methods. J. Multimedia Tools and Applications 39(3), 441–471 (2008)
13. Romanowski, C.J., Nagi, R.: On comparing bills of materials: a similarity/distance measure for unordered trees. IEEE Transactions on Systems, Man and Cybernetics, Part A: Systems and Humans 35(2), 249–260 (2005)
14. 3D model search engine, `http://shape.cs.princeton.edu`
15. 3D model similarity search engine, `http://merkur01.inf.uni-konstanz.de/CCCC`
16. Victory Project, `http://www.victory-eu.org:8080/victory/`
17. Lütke Entrup, C., Barth, T., Schäfer, W.: Towards a process model for identifying knowledge-related structures in product data. In: Reimer, U., Karagiannis, D. (eds.) PAKM 2006. LNCS (LNAI), vol. 4333, pp. 189–200. Springer, Heidelberg (2006)
18. Müller, U., Barth, T., Seeger, B.: Accelerating the retrieval of 3D shapes in geometrical similarity search using M-tree-based indexing. In: Workshop of Uncertainty, Knowledge Discovery and Similarity in Case-base Reasoning (UKDS'09), Seattle (2009)
19. Akgul, C.B., Sankur, B., Yemez, Y., Schmitt, F.: Density-based 3D shape descriptors. EURASIP J. on Advances in Signal Processing (2007)
20. Lütke Entrup, C., Barth, T.: Cost calculation system (in German). Tech. Report (2008)
21. Poli, C., Mahajan, P., Dastidar, P., Graves, R.J.: Design for stamping: A group technology-based approach. Concurrent Engineering 1(4), 203–212 (1993)
22. Poli, C., Dastidar, P., Mahajan, P., Graves, R.J.: Design for stamping - analysis of part attributes that impact die construction costs for metal stampings. J. of Mech. Design 115(4), 735–743 (1993)

23. Zheng, J., Wang, Y., Li, Z.: KBE-based stamping process paths generated for automobile panels. Int. J. Adv. Manuf. Tech. 31(7-8), 663–672 (2007)
24. Tor, S.B., Britton, G.A., Zhang, W.Y.: A knowledge-based blackboard framework for stamping process planning in progressive die design. Int. J. Adv. Manuf. Tech. 26(7-8), 774–783 (2005)
25. Ramana, K.V., Rao, P.V.M.: Data and knowledge modeling for design-process planning integration of sheet metal components. Journal of Intel. Manuf. 15(5), 607–623 (2004)
26. Zhang, W.Y., Tor, S.B., Britton, G.A.: Indexing and retrieval in case-based process planning for multi-stage non-axisymmetric deep drawing. Int. J. Adv. Manuf. Tech. 28(1-2), 12–22 (2006)
27. DIN 8582: DIN 8582 Manufacturing processes forming - Classification, subdivision, terms and definitions, alphabetical index. Ref. Nr. DIN 8582:2003-09 (2003)
28. Wang, J., Goel, A., Yang, F., Gau, J.-T.: Blank optimization for sheet metal forming using multi-step finite element simulations. Int. J. of Adv. Manuf. Tech. 40(7-8), 709–720 (2009)
29. Forming Technologies Inc.: FASTBLANK® `http://www.forming.com/fastblank.htm`
30. GNS mbH: Indeed, `http://www.gns-mbh.com/8.html`
31. Sander Automation GmbH: Seminar on progressive and transfer tools and their trajectories in forming press (in German) (2007)
32. Ou-Yang, C., Lin, T.S.: Developing an integrated framework for feature-based early manufacturing cost estimation. Int. J. of Adv. Manuf. Tech. 13(9), 618–629 (1997)
33. Samanka, M.: Robot task-level programming language and simulation. Proc. of World Academy of Science, Engineering and Technology 9, 99–103 (2005)
34. Shen, W., Gu, J., Ma, Y.: 3D kinematic simulation for PA10-7C robot arm based on VRML. In: Proc. of the IEEE Int. Conf. on Automation and Logistics, pp. 614–619 (2007)
35. Awa, Y., Kobayashi, K.: Kinematics simulation by using ODE and MATLAB. In: SICE Annual Conference, pp. 3090–3094 (2007)
36. Siemens Technomatix, `http://www.plm.automation.siemens.com/en_gb/products/tecnomatix`

Using ISO 10303-224 for 3D Visualization of Manufacturing Features

Tobias Teich, Jörg Militzer, Franziska Jahn, Daniel Kretz, and Tim Neumann

University of Applied Sciences, D-08012 Zwickau, Germany
{tobias.teich,joerg.militzer,franziska.jahn,daniel.kretz,tim.neumann}
@fh-zwickau.de
http://www.fh-zwickau.de

Abstract. Globalization leads to increasing competition for companies. Next to this, customer seek for individual products, leading to a boost in costs for product development. We try to tackle both aspects in our current research. We are trying to develop a solution for the automatic extraction of routings from a three-dimensional product description. This leads to two advantages. First, costs for product development are decreasing because labour intensive planning will be eliminated. Second, the time between customer inquiry and offer can be reduced, which is an additional advantage to get ahead of competition. To develop such a solution, different aspects have to be considered. This paper deals with the product representation, which should allow us to automatically assign process steps to design-elements.

Feature-based design is a very promising concept for optimizing the entire product development by enhancing the product model with significant information. An approach for its implementation is the ISO-standard 10303 "Standard for the Exchange of Product model data". ISO-10303 provides a formal reference model for a consistent, standardized and complete product description which is fundamentally required for automated process planning.

This paper summarizes the reasons for choosing the formally specified feature objects of the ISO-10303 application protocol 224 for 3D-visualization and how we implemented the concept "geometry from feature". Therefore we have developed a modular "viewer-concept" for connecting suitable geometric modeling cores like ACIS with our AP 224 feature library which will be presented in the paper. Additionally, a short overview of automated process planning will be given.

Keywords: Feature-based design, 3D visualization, ISO 10303, Product model data exchange, STEP.

1 Introduction

International companies have to face extreme pressure caused by globalization. Accompanied by this, customers seek individual solutions to a competitive price. To obtain potential orders, corporations have to react fast and exact to customer inquiries [1]. However, innovative and individual customer demands

W. Dangelmaier et al. (Eds.): IHNS 2010, LNBIP 46, pp. 198–209, 2010.

require cost- and time-intensive feasibility analysis and cost estimates in the preliminary stages of the product lifecycle, especially in the design and process development phase. To stay competitive in the future, a completely automated solution is required, which links design and calculation. This is already possible for similar orders, which were already executed and are well documented in an ERP- or PLM-System like SAP or PARTSolutions.

An efficient product development is prevented by heterogeneous system landscapes, their different kinds of data exchange formats and incomplete product models. Additionally we have recognized that *product data management* is extremely difficult to handle, because *product lifecycle management* is frequently confused with implementing product data management as an ordinary IT-system.

To solve this problem, it is fundamental to exchange a complete product model that includes functional, geometrical and technological aspects of the product. In fact, the digital product model is primarily created by using CAD-systems. These systems are not able to represent extended product model data. For example, complete high level objects for the description of forms, dimensions, materials, tolerances or development rules and associations between them cannot be provided. From our point of view, these are crucial prerequisites for automated process planning which leads to automated cost estimation derived from the product model. Incomplete product models require manual extraction and integration of necessary parameters during the later stages of product development which can be a very error-prone process.

Feature-based modeling is an approach for enhancing the virtual product model with extended and context-depended product data. This enables direct information exchange between particular stages of the product lifecycle. The ISO-standard 10303 focuses on the exchange between different application systems and software implementations within different stages of the product life cycle. Today's market-dominating CAD-systems are supporting feature-based design, but are only able to use geometric form-features. These low-level objects are neither standardized nor do the systems support a neutral product data exchange of those feature objects. Therefore the data exchange and the transformation of different product data models between different application systems are tremendous hampered and in many cases impossible. However, both aspects are mandatory for an efficient and automated interpretation of the digital product model.

The ISO-standard 10303 is titled "Standard for the Exchange of Product model data". This standard provides a solution for creating a complete, consistent, standardized and platform-independent product definition across the entire product lifecycle. The scope of STEP lies on an unambiguous digital data exchange format. It consists of different application protocols (AP's) to cover a wide range of application areas like shipbuilding industry, electrical engineering or the automobile industry. Surprisingly, there is a lack of implementation of STEP in existing CAD-Systems. CATIA for example only supports the storage of AP 203 and 214 Files. Next to this, the exported files only include a boundary representation of the designed product.

This paper focuses on AP 224 of the ISO-standard 10303 with the title "Mechanical product definition for process planning using machining features". It orientates on milling and turning operations and is a building block for automated information exchange between process planning and manufacturing. Each manufacturing feature can be associated with a corresponding manufacturing process. Using this AP as the basis of a CAD-system, the designer is able to develop a part corresponding to its manufacturing. The designer has to describe a minimum required volume as a base shape. Afterwards he applies volumes that should be removed or shapes that result from machining operations.

During our researches we have realized that there is a surprising lack of implementations of the specialized context-depended application protocol 224. That is why we have implemented manufacturing features of AP 224 as an object-oriented application programming interface within a C++ feature-library. A modular viewer-concept connects this feature-library with the geometric ACIS 3D modeling core and can be used for the stepwise simulation of manufacturing processes and design. This results in the visualization of the part-model with corresponding three-dimensional objects based on the AP 224 manufacturing features. This model can be used to simulate the manufacturing of the part stepwise. Therefore we derive significant geometric parameters from feature instances and construct the corresponding geometry by destructive solid geometry. This process is also called "geometry from feature" [2].

2 Geometric Product Models

Retrieving required product information the right time and at the correct place is essential for the success and efficiency of each product lifecycle stage. Digital product models like two-dimensional technical drawings or three-dimensional CAD-models represent the product shape informative and geometric exact. These methods of information exchange and representation also replace expensive physical prototypes or design studies [3]. Furthermore, virtual product models are combined with modern simulation technologies like digital fabrics or virtual reality. This way, production and assembly faults can be recognized and avoided in the forefront. Digital product models increase the efficiency of the product lifecycle significantly [4].

Geometric product shape information is very important in most stages of the product lifecycle, especially for product design, process planning and manufacturing. This was the reason why CAD-systems were invented for modeling, detailing, visualizing and analyzing digital product models [3]. The data structures of CAD-models primarily consist of geometric primitives like points, lines, faces and solids. Especially the export to neutral data exchange formats like STEP is limited to pure boundary representation. This is not satisfying our preconditions of detailed high level objects.

In fact, incompatible data exchange formats of different computer-aided systems (CAx-systems) complicate the information flow and create delays. Every type of a CAx-system supports a corresponding stage of the product lifecycle [5].

E.g., product design is supported by CAD-systems (computer aided design), process planning by CAPP-systems (computer-aided process planning) and manufacturing by CAM-systems (computer-aided manufacturing). Accordingly, most developed product lifecycle environments possess a heterogenious system landscape with different requirements and digital product data representations.

The product data exchange and model transformations are not standardized and cause information losses or disastrous misinterpretations. In worst case's, the collaboration can also be hampered within one stage of the product lifecycle. An example is the usage of different software versions of one vendor which can lead to incompatibility.

Digital product models represent the product shape with geometric information and the model topology. Extended product information like design intentions, functional requirements or controlling and resource management information are not represented. They must be documented additionally.

Figure 1 illustrates a further problem for adapting models that is caused by the model topology and the used CAD-system. There are non-parametric and parametric CAD-systems. In a non-parametric CAD-system, every new geometric element is a single instance. Parametric CAD-system refer to the entire CAD-model. Assume that diameter D of the entire groove in figure 1 has to be reduced to $D' < D$. In a non-parametric CAD-model, every geometric element has to be changed. The parametric CAD-model can solve this problem but allows no change of a single geometric element. This is only possible by destructing the entire model [6].

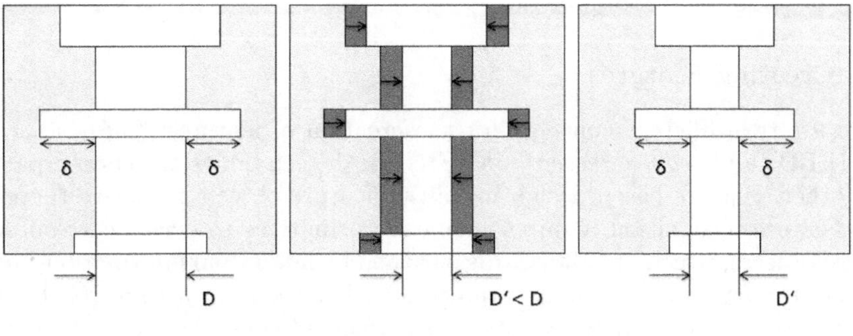

Fig. 1. CAD-model adaptation [6]

3 Feature-Based Modeling

An easier model adaptation and also product data exchange can be achieved with *feature-based modeling*. This approach enhances the digital product model with functional, geometric and technological product descriptions. Additionally, products can be designed faster because the designer only has to depose significant parameters instead of constructing difficult geometric product aspects like threads or gears. The concept of feature-based design is also used in the ISO-standard 10303 [7], especially in the application protocol 224 with a destructive design approach [8]. This application protocol offers a mechanical product definition and a standardized data exchange structure for process planning by using machining features. *Features* are an opportunity for creating interpretable product models, consisting of high-level objects.

3.1 Definition and Classification of Features

A *feature* defines physical and geometrical aspects of a part or assembly and contains context-depended information like shape, dimensions, material attributes, production time or tolerances. Additionally, a feature is an obvious entity with semantic significance in one or more engineering points of view. In conclusion Shah and Mantyla defined a feature as following [2]:

"A feature is a physical part of a building component or an assembly. It can be represented in a general way and owns engineering specific and predictable attributes. Each feature is an identifiable entity in a *feature model* and has some explicit representations."

The different engineering points of view require several specialized *feature-types* such as design-, manufacturing-, deburring- or inspection-features. In our work we focused on *manufacturing features*. The reason for this is the need for representation of important design- and manufacturing-information for process planning in high-level objects. Each manufacturing feature can be associated with a real milling or turning operation and with manufacturing operations in the resulting process plan.

3.2 Creating Features

There are two different concepts for the creation of features: *feature-based design* (FBD) and *feature-recognition* (FR). Both techniques are counterparts to each other. Feature-based design initializes feature-objects to derive the geometry. Feature-recognition requires geometric primitives to derive corresponding feature-objects, which are afterwards enriched by additional information. In general, feature-based design is also known as *geometry from feature* (GFF) while feature-recognition refers to *feature from geometry* (FFG) [2]. Figure 2 illustrates the contrary methods.

In both concepts the feature model is in relationship with the geometric model. The feature model is a functional and technical description of the building component or assembly. Derived from (or generated by) the design approach, the geometric model represents the corresponding geometry. Geometrical information can be used to visualize the building component or assembly as a two-dimensional graphic or a three-dimensional solid. We preferred the feature-based design approach cause of easier implementation and independence of proprietary data formats. In addition, feature recognition requires enrichment of derived objects which is an additional step.

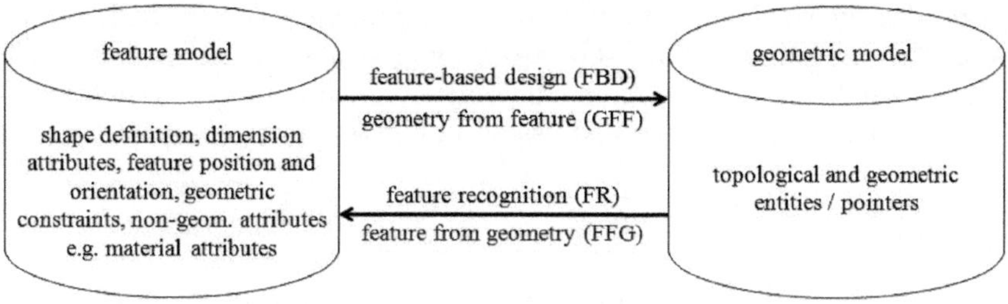

Fig. 2. Relationship between features and geometry [2]

3.3 Feature-Based Design

Feature-based design describes the part or assembly by initializing high-level feature-objects. The resulting semantics, associated parameters and relationships include information about form, position, dimension and geometric constraints. Feature-attributes can be used for derivation of the entire topology and geometry of the part and its visualization. Feature-based design can be divided into two different approaches for creating a feature: *destruction by machining-features* and *synthesis by design-features* [6], [9].

Destruction by machining-features orientates on destructive machining operations like turning, milling, drilling, planing and thread cutting. The part model of a building component or assembly is created by subtracting volume from its initial stock. This way the attributes of the subtracted volumes correspond to features.

Synthesis by design-features doesn't require an initial base shape volume. The building component or assembly can be built by adding and subtracting features. Each feature implies semantics about corresponding manufacturing operations like joining, welding and soldering or the operations named above.

Deriving the geometric model from the feature model requires a *feature library* for interpreting the feature-attributes like feature-position, -dimension or -orientation [2]. In our approach the destruction by machining-feature approach is used. The focus lies on feature descriptions of the application reference model in application protocol 224.

4 ISO-Standard 10303

The ISO-standard 10303 (mostly informally called STEP - "Standard for the Exchange of Product model data") pursues the superior goal to offer a mechanism for describing complete, obvious and platform-independent product definitions within the entire product lifecycle and the reuse or extension of existing product model data within each stage. The title of this standard implies that the data exchange is in the center of attention. This is not the only intention of the standard. It supports a continuous standardized product data representation and therefore also contributes computer integrated manufacturing. STEP connects the heterogenous computer aided systems by creating consistent data exchange standards for efficient and economical product development. As mentioned before, implementations are poor or missing in existing systems.

4.1 Architecture

STEP is an international standard for the product data exchange and was founded by worldwide experts under the patronage of the *International Organization for Standardization* (ISO). It includes a series of sub-standards. They are divided into several *application protocols* (AP's) which are listed for example in [10] and published separately.

Each application protocol is a standardized reference model for a specific application domain like machining industry, shipbuilding, electrical engineering or automobile industry. Different description-, implementation- and testing methods can be used for accessing common resources and creating a complex product data structure (see figure 3).

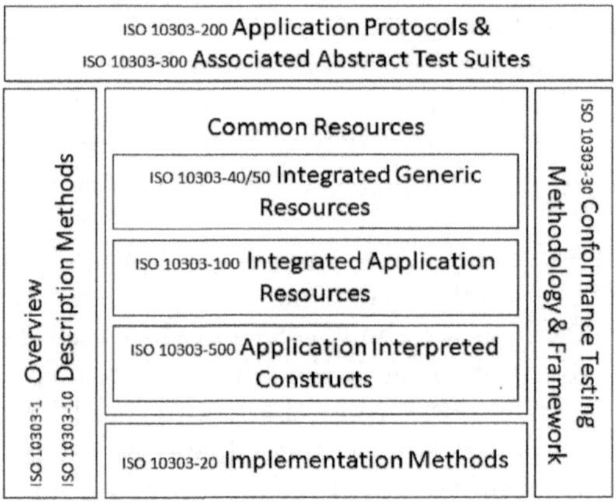

Fig. 3. STEP-Architecture [11]

Every application protocol includes three description models for its documentation and implementation. The *application activity model* considers the user's point of view by describing the scope of the AP with a special kind of activity diagrams (defined in IDEF-0 notation). Second, the *application reference model* presents possible objects (in AP 224 for example features) from an engineer's point of view by defining application specific product data. Third, there is an *application interpreted model* which supports the mapping of context-depend information in a neutral data exchange structure for the exchange between different application systems.

4.2 Application Protocol 224

Application protocol 224 (AP 224) supports feature-based design by using manufacturing features for the digital representation of product model data. It was standardized in 1999 [8], almost 30 years after the feature concept was invented. Recently, the 3rd edition was published by the ISO. AP 224 orientates on destructive machining operations and is suited for a feature-based design approach *"destruction by machining features"*.

Therefore it supports the application of manufacturing features and the digital representation of product data. AP 224 consists of thirteen different *units of functionality* and includes 262 object definitions for managing geometrical information, product- and production data.

Overall, AP 224 describes textually 48 manufacturing features with their specific attributes (see figure 4) which we have implemented object-oriented within a feature-library. For a visual representation some chosen manufacturing features are illustrated in addition.

The relationships between the manufacturing features are defined in the *application reference model* of AP 224. In the ISO-document it is represented textually with illustrations. Relations are mapped and visualized in several EXPRESS-G diagrams. Additionally, the relationships between feature attributes are documented in the *application interpreted model* which is provided by different EXPRESS reference paths. EXPRESS is a specific data model description language defined in ISO 10303-11 [12]. Its graphical pan don is EXPRESS-G which is useful for the interpretation of those complex data models [13]. The following figure 4 illustrates the example of an open slot, defined by the application reference- and the application interpreted model.

4.3 Automated Process Planning with AP 224

Process planning is a fundamental step in the creation of a new product or its optimization. The *process plan* (or routing) is a concrete guideline how to manufacture a part. To stay competitive it is important to generate a *process plan* under respect of the manufacturing environment and the requirements of the product in an economic way and a timely manner. Therefore the entire manufacturing environment with its available machines, their capacities and other resources should be considered. This information can be provided by today's enterprise resource planning systems (like SAP or Microsoft Dynamics).

Additionally, process planning requires knowledge about the product, especially its geometry, dimensions, material properties, tolerances and many other extended information. With a product model based on AP 224 manufacturing features, this information can be extracted automatically by an AP 224-based process planning system. Our primary approach of automated process planning can be described as a *generative planning* [14]. To give small and medium enterprises (SME's) a possibility to be ahead over competitors, we want to generate process plans for new components automatically.

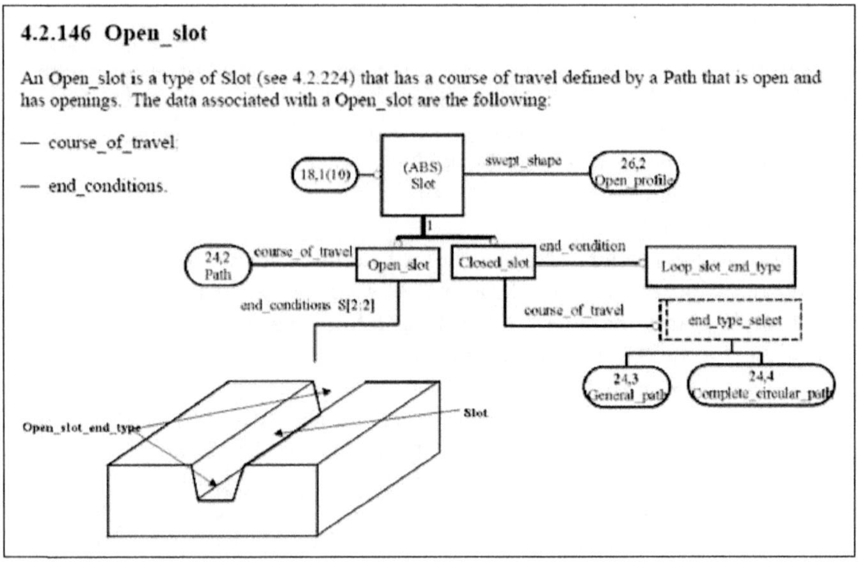

Fig. 4. Description of an Open_slot [8]

Another great advantage of AP 224 is that we can conclude corresponding manufacturing operations to each manufacturing feature used in the product model because of their semantics. First, based on the derived manufacturing operations it is possible to check the manufacturing environment if the required resources are available. Second, we can generate a dependency graph based on the features of this application protocol. Because of their semantics, significant attributes, relationships and rules, we can determine dependencies between single manufacturing features. Afterwards we divide these features into *dependent* and *independent* ones. Independent features can be employed directly. Dependent features require others to be employed in the forefront. Taking these two rules, we are able to build up an *dependency graph.* Finally we are going to find an optimized sequence of employing AP 224 features automatically by using genetic algorithms to produce the final part. The genetic algorithm is used to assess a process variant by scheduling it within a given factory environment. Further description of our approach of automated process planning and cost-evaluation can be found in [15] and [16].

5 Implementing STEP

In our work we have modeled the manufacturing feature-objects of AP 224's *application reference model* in UML-class diagrams. Afterwards we generated the relating source code with a model driven software architecture framework *GeneSEZ* in a C++ feature-library. This library is able to generate e.g. the building component in figure 5 and enables the definition of shapes, dimensions attributes, feature positions and orientations.

For three-dimensional visualization of complex building components we have developed a modular viewer-concept. It connects the geometric modeler ACIS with the generated AP 224 feature-library. The geometric ACIS modeler visualizes the geometric model in figure 5. Therefore the modeler obtains the necessary geometric information for the three-dimensional visualization from the AP 224 feature-library (feature model) from the viewer component.

With this solution we can visualize and create a product defined by AP 224 manufacturing features. During our researches we have used a roll axis-demonstrator (figure 5) which can be designed by using our implemented *destruction by machining-features* approach at runtime.

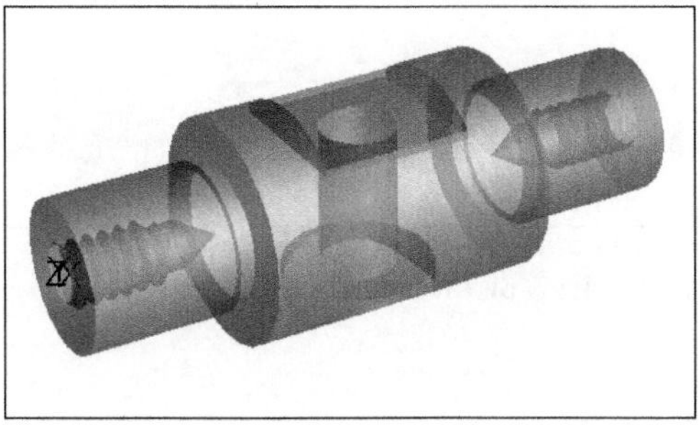

Fig. 5. Roll Axis-Demonstrator

First, the attributes of the roll axis-demonstrator are retrieved from the instantiated feature objects of the product model. This instantiation is done by the designer in the product design phase. The initialization starts with the cylindrical base shape (the initial stock) and subtracts all instantiated manufacturing features to create the final shape of the roll axis-demonstrator. The order of destruction is given by the derived routing.

We define the minimal geometric information for three-dimensional visualization of the cylindrical base shape with additional attributes like *base_shape_length* and *diameter*. Each manufacturing feature includes different shape attributes for retrieving its geometry. A *round_hole* for example can own a *taper* or a special *bottom_condition*.

After the initialization of an AP 224 library-object it will be transferred to its special viewer. This viewer extracts the required object attributes and constructs the geometric representation with combined ACIS method calls. The ACIS method `api_solid_cylinder_cone` can be used for drawing the cylindrical base shape. Because of the complexity of several manufacturing features we have created flexible geometric auxiliary constructions for their three-dimensional visualization.

Future work will lead to a graphical user interface for designing the building component or assembly. Afterwards, our own, FBD-based CAD-system will be used to design example parts more easier and will be fed by our automated process planning CAPP/CAM system.

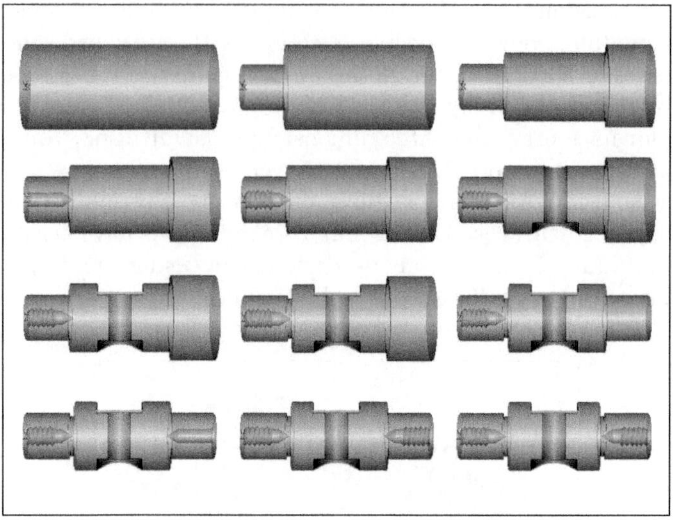

Fig. 6. Intermediate Variations

6 Conclusion

During our researches we have realized that there is a huge potential of STEP to optimize the product data exchange of the whole product lifecycle. An implementation is essential for fast and effective product development. There is still a surprising lack of implementations of the specialized context-depended application protocol. Although AP 224 and other protocols have great potential for implementing feature-based solutions and developing integrated CAx-systems for continuous product data integration.

Our presented viewer-concept realizes the approach "destruction by machining features" by subtracting different features from a base shape at runtime. The modular structure of our viewer concept separates the feature model from the geometric model clearly and practically demonstrates the approach of "geometry from feature". Every component of the viewer-concept can be exchanged, reused and extended.

References

1. Brosze, T.: Seminar Anwendungen der Kontrolltheorie im Supply Chain Management. GriN - Verlag für akademische Texte, Karlsruhe (2004)
2. Shah Jami, J., Mantyla, M.: Parametric and Feature Based CAD/Cam. In: Concepts, Techniques, and Applications, John Wiley & Sons Inc., New York (1995)
3. Meißner, M.: Methoden zur qualitätsgerechten CAD-Modellerzeugung für die virtuelle Produktentwicklung am Beispiel der Automobilindustrie. Shaker Verlag GmbH, Aachen (2000)

4. Scheer, A.W., Boczanski, M., Muth, M., Schmitz, W.G., Segelbacher, U.: Prozessorientiertes Product Lifecycle Management. Springer, Heidelberg (2006)
5. Vajna, S., Weber, C., Bley, H., Zeman, K.: CAx für Ingenieure: Eine praxisbezogene Einführung. Springer, Heidelberg (2007)
6. Xu, X.: Integrating Advanced Computer-Aided Design, Manufacturing, and Numerical Control - Principles and Implementation. Idea Group Publishing (2009)
7. ISO - International Organisation for Standardization: ISO 10303 Industrial automation systems and integration - Product data representation and exchange - Part 1: Overview and fundamental principles. Geneva (1994)
8. ISO - International Organisation for Standardization: ISO 10303 Industrial automation systems and integration - Product data representation and exchange - Part 224: Application Protocol: Mechanical product definition for process planning using machining feature. Geneva (2006)
9. Kusiak, A.: Intelligent Design and Manufacturing. John Wiley & Sons, Inc., Chichester (1992)
10. ISO - International Organisation for Standardization: STEP application handbook. SCRA, North Charleston (2006)
11. Nell, J.G.: STEP on a Page. National Institute of Standards and Technology (2003)
12. ISO - International Organisation for Standardization: ISO 10303 Industrial automation systems and integration - Product data representation and exchange - Part 11: Description methods: The EXPRESS language reference manual. Geneva (2004)
13. Schenck, D.A., Wilson, P.R.: Information modeling: the EXPRESS way. Oxford University Press, New York (1994)
14. Nasr, E.A., Kamrani, A.K.: Computer-Based Design and Manufacturing. Springer Verlag, Berlin (2006)
15. Teich, T., Militzer, J., Jahn, F., Neumann, T., Kretz, D.: Step Standardized Product Data Representation and Exchange for Optimized Product Development and Automated Process Planning. In: Proceedings of the 6th CIRP-Sponsored International Conference on Digital Enter-prise Technology - DET, Hongkong, pp. 41–56 (2009)
16. Teich, C., Kretz, D., Richter, M., Militzer, J., Neumann, T.: Interaction of natural inspired methods in process planning. In: Proceedings of the 19th International Conference on Flexible Automation & Intelligent Manufacturing (FAIM '09), pp. 105–111. Gemini International Limited (2009)

Combined Working Time Model Generation and Personnel Scheduling

Maik Günther and Volker Nissen

Ilmenau University of Technology, Information Systems in Services, D-98684 Ilmenau
maik.guenther@gmx.de, volker.nissen@tu-ilmenau.de

Abstract. Workforce management is comprised of several phases, such as working time model generation and personnel scheduling. The combination of these phases has significant potential, especially for volatile personnel demand. This article shows that the concepts for the automatic generation of working time models already used in retail can be transferred to personnel scheduling in the logistics industry. Through this, the assignment of personnel can be accurately adapted to personnel demand. The results suggest the use of heuristics, especially meta-heuristics such as the evolution strategy or constructive methods which are adapted to the problem at hand.

Keywords: working time model, workforce scheduling, workforce management, evolution strategy, constructive heuristic.

1 Introduction

Despite many advantages in automation, logistics is still very labour intensive. At the same time logisticians are under considerable pressure, among other things due to legal regulations, stronger customer orientation and increasingly tough international competition. Above all, though, the current economic situation forces logisticians to take measures in order to remain competitive. An important parameter for cost-cutting is demand-oriented workforce management.

Employees spend up to 36% of their working time unproductively, depending on the branch [8]. Major reasons for this include a lack of planning and controlling. The problem can be dealt with using demand-oriented workforce management. Key planning goals are increased productivity, reduction of personnel costs, prevention of overtime and better motivation of employees [13].

In order to achieve these goals, accurate personnel assignment adapted to demand is required. A multi-level approach to workforce management in seperate steps can be very inefficient, especially for volatile demand. In retail, where personnel demand generally depends on customer frequency, which is very volatile, other methods must be used. For several years working time models have been created automatically during personnel scheduling, meaning these models are not preset. Rather, there are rules which are applied directly during personnel assignment planning (e.g., the minimum and maximum length of the models). The success of this concept in retail suggests its use in other fields in which

W. Dangelmaier et al. (Eds.): IHNS 2010, LNBIP 46, pp. 210–221, 2010.

volatile personnel demand also occurs. Therefore, the present work attempts to investigate the automatic generation of working time models as a direct part of personnel scheduling using practical examples.

First, the four phases of workforce management for volatile demand are explained. Possibilites for combining individual phases are shown at the end of section 2, in particular the combination of working time model generation and personnel scheduling. Section 3 discusses the necessity of accurate working time models using an example from logistics. For this, the limitations of the current practice (two-phase approach) are highlighted. The remedy is the automatic generation of suitably adapted working time models directly in personnel scheduling, as has already been done in retail. This concept is presented in section 4. Also, corresponding practical problems from retail, which have already been worked on, are discussed as well as solution methods used to integrate model generation and personnel scheduling. The transfer of these approaches from retail to logistics is done in section 5 using a practical application. Finally, the paper concludes with a summary and some indications for future work.

2 The Four Phases of Demand-Driven Workforce Management

Workforce management is a central component of business action and is intricately linked to many other processes in the organisation. Several software manufacturers follow an idealised approach, which is shown in figure 1 [12]. This method is quite practical and useful; however, the combination of phases can bring about further potential for usage. The individual phases are explained below and the possibilities for combining them are then discussed.

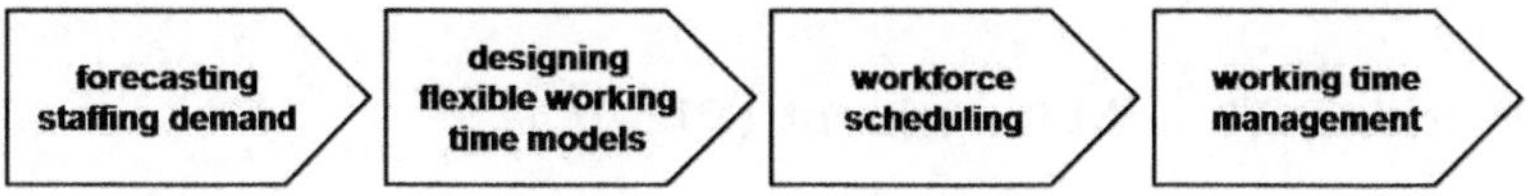

Fig. 1. Four phases of demand-driven workforce management

2.1 Staffing Demand Forecasts (Phase 1)

It is necessary for efficient personnel assignment to determine personnel demand as exactly as possible. According to a study by Miebach Consulting [7], errors in this phase alone can cause up to 15% higher personnel costs. In practice, various methods are employed depending on the planning horizon, with simple approximation dominating for short-range estimation of demand.

Demand determination based solely on experience is often suboptimal. However, the planner frequently lacks high-performance support tools. Modern solutions are able to predict the expected work volume using past data. Another form of demand determination is event-oriented. In logistics, for instance, if it is known ahead of time when a lorry (or van, aeroplane or train) of a specific type will arrive or depart again, the personnel demand can be ascertained.

2.2 Designing Flexible Working Time Models (Phase 2)

A working time model defines work beginning and ending time on a particular day, among other things. Starting with personnel demand, working time models are generated which cover that demand as well as possible. Legal and contractual regulations must be taken into account. The amount of effort required to generate models is usually quite large. Sometimes, they must also be authorised by the works council. A set of working time models, onces generated, are usually rarely changed and are used for later planning. An extreme negative example for an ill-suited working time model would be a plan with only one model (work begins at 7:00, ends at 15:00, variable break of 30 minutes, no flexitime). If all employees are scheduled using this model, sub-daily, weekly and even seasonal variations in personnel demand cannot be compensated for. Unnecessary costs due to idle time and overtime as well as bad service and low employee motivation would be just some of the effects.

2.3 Workforce Scheduling (Phase 3)

This phase involves the actual plan generation. It is determined which employee is assigned to which workstation at what time or rather which task is to be completed. The size of the planning horizon can be variable, as is the point in time at which the staff can view a new plan. Various constraints are considered during the planning, e.g., qualifications, desires of the employees, absences, balance of flexitime hours, legal and contractual regulations, aspects of fairness and so on.

2.4 Working Time Management (Phase 4)

Time management involves more than just recording the entering and exiting times of employees. In this phase, the flexitime balances of each employee are calculated based on their working times. These calculations and the regulations on which they are based can be quite complex. Absences are also planned and tracked. Additionally, as part of time management all data are gathered which are necessary for input into the payroll system.

2.5 Combination of Phases

Generally the four phases are carried out in sequence but there are sensible exceptions. The integration of personnel demand determination (phase 1) into personnel scheduling (phase 3) can be found, for example, in call centres, where personnel demand cannot be separated from scheduling. If the service level sinks sufficiently due to a lack of manpower, many customers will phone again in subsequent periods, affecting the personnel demand of that period as well.

The combination of working time model generation (phase 2) and personnel scheduling (phase 3) is also of interest. These approaches are sometimes used in retail when working time models are dynamically created during each planning session (phases 2 and 3 in one step). This combination significantly increases the flexibility of assignment planning and makes accurate scheduling adapted to demand possible. Sections 4 and 5 discuss the combination of these phases in more detail.

Even more optimisation potential lies in a combination of all three phases 1 to 3, as one could not only fit personnel supply to demand, but also shift demand in a favourable way for planning. However, this is only possible when personnel demand is fairly deterministic and the precise moment of occurence is not completely externally determined. Such flexibility is available, for instance, when customer orders can be individually planned ahead and to some extent be moved between different time slots.

Various approaches can be found with respect to phase 4; it can be arranged before or parallel to other phases. In this way, personnel scheduling can influence time management. On the other hand, if the time management data flow into the next planning session, flexitime balances and absences can be taken into account. Planning for days off alone would require a parallel approach because providing days off reduces personnel capacity, possibly requires adjusted working time models and even influences personnel demand (refusal/postpone of customer orders due to reduced personnel capacity).

3 Demand-Oriented Working Time Models

This section discusses the effects from the generation of demand-oriented working time models. The limits of this approach are also examined. The differences between fixed and demand-oriented working time models are illustrated using a case from logistics.

Figure 2 shows real data from a logistics service provider, who mainly carries out loading and unloading activities. The personnel demand fluctuates throughout the five days. However, it is not possible with only three available working time models to cover the demand well. The result indicates large phases of over- and understaffing due to the volatile demand with unadapted personnel assignments. This situation is critical for the logistics company because a high service level is contractually obligated to the customers.

This problem can be countered with the introduction of further working time models. Figure 3 clearly shows how the personnel assignment is oriented toward demand. By better utilising the normal working times of all employees, over- and understaffing can be greatly reduced while maintaining the same number of employees and the same level of effort. The effects which result from this are obvious: cost reduction through improved utilisation of employee time and fewer temporary workers, increase of turnover through a higher level of service and a rise in employee motivation.

Through an increase in the number of working time models, even short-term fluctuations in personnel demand can be covered economically. This, of course,

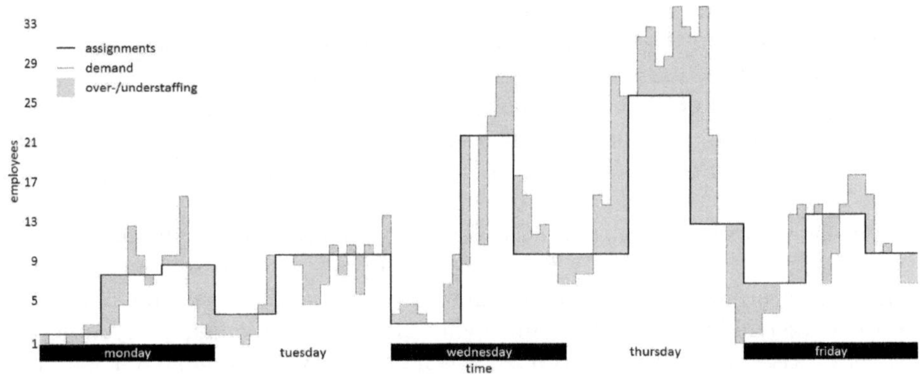

Fig. 2. Personnel demand and personnel assignment with three working time models

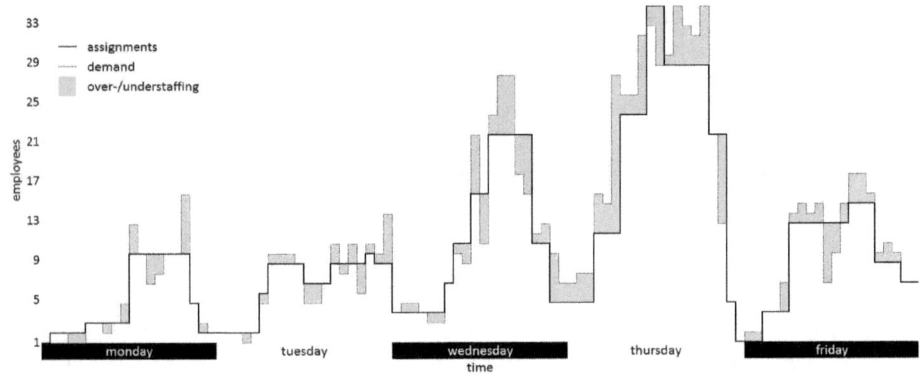

Fig. 3. Personnel demand and personnel assignment using more working time models

means a rise in complexity and effort needed for the generation and maintenance of the working time models. It is impractical to do this manually or by using spreadsheets, but a powerful software solution is needed.

On the basis of the working time models generated in this phase, employees are subsequently assigned working time models and workstations or orders for a defined time period in the assignment planning phase (phase 3). During this phase, qualifications, flexitime balances, teams, sub-daily workstation switches, planned absences and so on are taken into account. Using an approach in which phases 2 and 3 are not integrated, the latter phase must necessarily leave some constraints and some data unaccounted for. An otherwise optimal solution for phase 2 and subsequently for phase 3 could be determined to be sub-optimal when the result is regarded as a whole. Therefore, a combination of both phases is desirable.

Such as combination, however, requires that personnel demand can be forecasted accurately, using, for instance, past demand data, fixed customer events or information from returns data. However, this already applies, when the phases

2 and 3 are not integrated, but only the number of working time models is increased. Any attempt for demand-oriented personnel planning must ultimatively be based on reliable demand figures. Moreover, stochastic influence (dynamic demand) should be fairly low. Otherwise, there are more effective measures to capture stochastic demand peaks, for instance employing temporary workers, who can be ordered to work or sent home at short notice (but at extra cost).

4 Integrating the Phases 2 and 3: Lessons from Retail

Through the automatic generation of working time models directly within workforce scheduling it possible to generate a plan without fixed working time models. More specifically, automatic working time model generation refers to separating oneself from the idea of using only a few rigid working time models, because they do not always provide optimum coverage of the actual personnel demand. Instead, the planner only needs to provide some rules for the generation of working time models suitably adapted to demand. These include the minimum and maximum duration of the models as well as the limits for their beginning and end. Individual working time models are then generated for each employee for each day. Naturally, aspects as employee availability and planned weekly working hours are accounted for.

This apporach has been successfully implemented at many retailers and several software manufacturers provide specialized methods for this task. This stresses that the combination of phases 2 and 3 in workforce management is both possible and indeed successful. However, only a few scientific sources have dealt with this topic, yet. Below, we discuss this related work.

4.1 Work of Others

In 2007 Sauer and Schumann [11] introduced a constructive method for a personnel planning problem with automatically generated working time models in retail. The procedure was implemented in the workforce management system of a software manufacturer. Demand is viewed as deterministic and forecasted based on historical data. No demand dynamics are considered at the time of planning. Due to the fact that the scheduling algorithm was designed for interactive scheduling, greedy heuristics were chosen for two reasons. Expert knowledge can easily be integrated into the algorithm to generate acceptable plans for the human planner. And greedy heuristics generate a solution quickly. Unfortunately, this solution method cannot take into account more than one workstation or sub-daily workstation switches. Additionally, the planning horizon is limited to a maximum of one week.

Prüm [9] also addresses different variants of a personnel planning problem from retail. He uses 20 test problems with 4 to 168 timeslots as well as 2 to 576 employees. Prüm creates working time models parallel to assignment planning with varying demand in one-hour intervals. However, only one workstation is present in the scenarios of Prüm, and sub-daily workstation rotations are not included. He experiments with simplex, branch & bound as well as a hybrid method. His results indicate that problems of realistic size with constraints can in general not be successfully solved with exact methods.

4.2 Own Work

A practical retail problem with two workstations and also including planned sub-daily job rotations is considered in our own, previously unpublished work. Here, working time models are generated directly as part of personnel assignment planning. Fifteen employees work in the department for ladies' wear at a department store. The store is open Monday to Saturday from 10:00 to 20:00 and closed on Sunday and holidays. Six employment contracts exist, differing in the amount of planned weekly working time from between 10 and 40 hours. Employees are assigned to two different workstations (till and sales), with all employees trained for both stations. Qualifications are therefore not required to be taken into account. Many other factors influence planning, such as regulations, employee availability, promotions and time sheets. The personnel demand can be forecasted fairly accurately. It is given in one-hour intervals and centrally determined based on past data.

In order to find a solution to the problem described, automatic working time model generation is performed. Because of high fluctuations in demand, sub-daily workstation changes are allowed. Several hard and soft constraints must be considered during the planning process. They include rules supplied by the personnel planner of the company to ensure reasonable working time models. One year is planned ahead based on available historical data, resulting in a very complex search space with 131,400 dimensions (decision variables). If the assigment plan violates soft constraints, this is punished with error points that reflect the companies requirements as inquired through interviews. The planning problem can now be modeled as an optimisation problem where the sum of error points must be minimised. For real data sets and benchmarks see [14].

Staff scheduling is a hard optimisation problem. Garey and Johnson [3] demonstrate that even simple versions of staff scheduling problems are NP-complete. Kragelund and Kabel [6] show the NP-hardness of the general employee timetabling problem. Thus, heuristic approaches appear justified for the combined tasks of automatically generating working time models and scheduling the workforce in a company. In particular, we compare heuristics based on the evolution strategy (ES) [1], particle swarm optimisation (PSO) [5], multi-start local search (MLS) and a commercially successful constructive heuristic.

Our solution approaches based on adapted versions of ES, PSO and MLS create working time models automatically in the following way: The problem is represented as a two-dimensional matrix of employees and time periods, where the cells are filled with workstation assignments. A dummy workstation is used to mark times where an employee is not available. The heuristics change these workstation assignments, implicitly generating working-time models that are evaluated through the penalties in the objective function. Faulty models (wrong length, gaps etc.) are additionally corrected by a repair heuristic within ES and

PSO. For each heuristic, different strategy parameter settings were tested. 30 runs were performed for each setting using 400,000 evaluated solutions as the uniform termination criterion.

We compare these results to a commercial workforce management software package that delivers an adequate constructive method, capable of solving the problem at hand. All restrictions are supported and the associated error points can be entered into the software. Unfortunately, no code is available and no information is given in the documentation as to how the working time models are generated, so it must be considered a black box. However, our application of the constructive heuristic was supported by the software manufacturer, so errors in software handling can be excluded.

Table 1 presents results for the different solution approaches to the retail problem. For ES and PSO only the best strategy parameter sets are given, i.e. a swarm size of 20 particles in a gBest neighbourhood structure for PSO, and a (1, 5)-selection strategy for the evolution strategy. The constructive heuristic can be regarded as a benchmark, since it is part of a commercially successful software package that is actually used at around 300 companies for personnel planning.

ES with classical Gaussian mutation (Type G) yields the best mean error points. ES with a mutation adapted to combinatorial search spaces using the concept of maximum entropy (Type E) [10] finds the best of all solutions. If one inspects the assignment plans generated with the ES, one sees that they can hardly be improved upon, even with highly complex manual changes. For this reason, and because of the superior performance over the other solution methods, these plans can be regarded as very usable.

ES and PSO significantly outperform the commercial constructive heuristic. The constructive method delivers unsatisfactory results, because it is unable to cope with the various employee contracts. However, one run requires only some 10 minutes as opposed to 6 hours for a single run with the ES or PSO. Multi-start local search, as some form of 'brute-force' solution approach, is the worst heuristic and not able to generate reasonable solutions at all.

5 Transfer to an Application from Logistics

5.1 Description of the Problem

It can be seen in table 1 that several solution methods were able to almost eliminate over- and understaffing using automatically generated working time models. While this form of personnel planning has already gained significance in retail, practical applications in logistics are unknown to us. Very small personnel budgets, a tendency for understaffing and intense competition have promoted automatic generation of working time models in retailing.

However, logistics in many areas is also characterized by tough competition, cost pressure and volatile demand. Therefore, a transfer of workforce management approaches originally developed in a retail context appear useful. At this stage, the applicability of individual solution methods is not our focus. Rather,

Table 1. Comparison (minimisation of error points) of the different approaches, based on 30 independent runs each. Best results are bold and underlined.

heuristic	error		number of job-changes	under-staffing in (h/yr)	over-staffing in (h/yr)	too much weekly working time in (h/yr)	more than one working time model per employee per day
	mean	min					
constructive heuristic	84,690.0	84,690	0.0	25.0	0.0	1,386.5	0.0
multistart local search	4,265,931.9	3,807,375	2,434.5	209.5	1,420.8	955.8	410.8
PSO (20)	37,117.9	14,385	389.9	13.9	0.3	597.9	0.0
ES (1,5) Type G	**8,267.1**	5,924	214.3	13.9	0.1	120.2	0.0
ES (1,5) Type E	8,464.8	**4,911**	248.0	14.0	0.1	122.8	0.0

the objective is to check whether this type of integrated work force planning can be reasonably transferred from retail to a given logistics service provider scenario, which is typical for many companies in that sector. Again, the idea is to automatically generate working time models as part of workforce scheduling. For this reason, three variations with diverse preset working time models will be compared to the combination of model generation and assignment planning. The fixed working time models were directly taken from the human resources department of the respective company.

The problem shown below exhibits many similarites to the logistics problem discussed in [4]. Personnel demand varies greatly at sub-daily intervals and for different workstations. Sub-daily workstation rotations are to be planned as well. Actual data are available for an entire year. The logistics provider operates between 6:00 and 19:00 with 480 employees. Their contracts are the same in most cases, so that the contractual weekly amount of work varies only slightly. In order to keep personnel assignments flexible, flexitime balances are kept and workers may maintain the balance within the range of -30 to +80 hours. The work volume is subject to economic and seasonal fluctuations, but can be regarded as fairly deterministic. Sub-daily jumps in personnel demand occur as well. In order to automatically generate working time models, the regulations found below are in effect and violations are assigned an amount of error points that reflects the companies requirements as inquired through interviews:

- compliance with the given allocations (no over- or understaffing)
- employees should not be assigned to workstations with no demand
- compliance with the contractually arranged maximum number of hours
- working time models should not be shorter than 3 hours
- working time models should not be longer than 9 hours
- working time models must not be split up during the course of a working day

- working time models must not begin or end outside operating hours
- working time models begin and end on the hour
- work must not be done on Sundays and holidays
- absent employees (holiday etc.) may not be planned for the corresponding day
- sub-daily workstation changes are only allowed on the hour
- no unnecessary workstation changes (no improvement of assignment plan)

5.2 Results and Discussion

Table 2 shows the effects of increasing the number of working time models and adding automatic model generation. The calculations were done for an entire calendar year, with the focus of the optimisation in each case on reducing over- and understaffing. All of the working time models operate on the period between 6:00 and 19:00. As a rule, working time models were generated whenever possible with a length of 7.5 hours. Stronger consideration of employee desires or unplanned absences would worsen the results. However, the main tendency seen in table 2 remains even then.

Personnel demand is viewed as deterministic and planning is based on historical data. In practice, dynamic changes (late truck arrival etc.) occur, but only rarely. Nevertheless, this points to the limits of personnel planning. Personnel demand must be known in advance and stochastic influence should be low to allow for demand-oriented, integrated planning as suggested here.

It is clear from the table that over- and understaffing in personnel assignment are reduced through the corresponding increase in the number of models. The fluctuating demand can be covered significantly better with more models available. In principle, a maximum of 62 different models could be generated given the existing restrictions. The generation, administration and possible later adjustment of these models, however, is significantly time consuming and error-prone. These problems can be avoided through the use of automatic model generation in parallel to workforce scheduling. Moreover, the problem of over- and understaffing is completely eliminated with this method, as can be seen in table 2. This demonstrates that an integrated approach to automatically generate working time models and schedule personnel to work at given workstations can indeed be very beneficial in logistics. It must be pointed out that the resulting working time models respect the constraints (rules) on model generation as provided by the planner. Moreover, the resulting plan can be made available to employees long before the relevant working periods actually begin, so that also contractual regulations regarding plan stability are covered.

The results in table 2 were generated with the constructive heuristic also shown in table 1. While this approach performed poorly for the retail problem, it proved to be very useful for the logistics problem. The constructive method is more suited to staff whose contractual amount of hours are similar, as is the case in the logistics problem. The retail problem, by contrast, presented highly variable contracts (full-time, student helpers, so-called 400€ workers etc.). With this diversity in retail, the constructive approach had significant difficulties and, consequently, was unable to produce adequate sub-daily staffing changes.

Table 2. Effects of increasing the number of (preset) working time models and introducing automatic working time model generation on over- and understaffing in a representative logistics personnel planning application

working time models	overstaffing (in h/yr)	understaffing (in h/yr)	total (in h/yr)
5 different working time models	26,094	528	26,622
6 different working time models	16,488	4,734	21,222
15 different working time models	3,426	438	3,864
automatically generated working time models	0	0	0

As a rule, constructive heuristics are suitable for automatic generation of working time models when the method is specifically designed to solve the particular problem at hand. Small changes of the application area can be enough to severely limit the quality of the results, though. Because the constructive heuristic fits well to the logistics problem and produces very satisfactory results in a short time, the implementation of further and potentially more time-consuming approaches, such as metaheuristics, was dispensed with for now.

6 Conclusion and Future Work

It was shown that the demand-oriented generation of working time models can contribute significantly to accurate workforce scheduling adapted to personnel demand. This enables the minimisation of overtime and idle time, among other things, which positively affects personnel costs, workload on individual employees and service level. The limits of a two-phase approach for working time model generation and subsequent personnel assignment planning were shown. The remedy is the combination of both phases into one planning step. This approach has already gained significance in retail, where personnel demand is extremely volatile. The concept was successfully transferred from the retail domain to a logistics personnel planning application that is representative for many similar problems in the logistics sector. This opens up the possibility to use a practically tested instrument for the flexibilisation of workforce scheduling under volatile personnel demand in logistics. The corresponding heuristic solution methods should be integrated in todays workforce management software to fully leverage the benefits.

Future work will involve the implementation and comparison of further solution methods for the mentioned logistics application as well as other scheduling applications from the logistics and retail domains.

References

1. Beyer, H.G., Schwefel, H.P.: Evolution strategies: a comprehensive introduction. Natural Computing 1, 3–52 (2002)
2. Ernst, A.T., Jiang, H., Krishnamoorthy, M., Owens, B., Sier, D.: An Annotated Bibliography of Personnel Scheduling and Rostering. Annals of OR 127, 21–144 (2002)

3. Garey, M.R., Johnson, D.S.: Computers and Intractability. A Guide to the Theory of NP-Completeness. Freeman, New York (1979)
4. Günther, M., Nissen, V.: Sub-Daily Staff Scheduling for a Logistics Service Provider. KI Künstliche Intelligenz 25 (2010) (accepted)
5. Kennedy, J., Eberhart, R.C., Shi, Y.: Swarm Intelligence. Kaufmann, San Francisco (2001)
6. Kragelund, L., Kabel, T.: Employee Timetabling. An Empirical Study. Master's Thesis, Dep. of Computer Science, Univ. of Aarhus (1998)
7. Miebach Consulting GmbH (ed.): Personalbedarfsplanung. Personalplanungs- und Flexibilisierungsmethoden in der logistischen Praxis. Frankfurt am Main (2008)
8. Proudfoot Consulting (ed.): Produktivitätsbericht 2007. Company Report (2007)
9. Prüm, H.: Entwicklung von Algorithmen zur Personaleinsatzplanung mittels ganzzahliger linearer Optimierung. Untersuchung eines bedarfsorientierten Modells mit flexibler Schichtkonfiguration. Master's Thesis, FH Trier, FB Informatik (2006)
10. Rudolph, G.: An Evolutionary Algorithm for Integer Programming. In: Davidor, Y., Männer, R., Schwefel, H.-P. (eds.) PPSN 1994. LNCS, vol. 866, pp. 139–148. Springer, Heidelberg (1994)
11. Sauer, J., Schumann, R.: Modelling and Solving Workforce Scheduling Problems. In: Sauer, J., Edelkamp, S., Schattenberg, B. (eds.) Proc. of PUK 2007, pp. 93–101 (2007)
12. Scherf, B.: Ein Vorgehensmodell zur Personaleinsatzplanung. In: Fank, M., Scherf, B. (eds.) Handbuch Personaleinsatzplanung, Datakontext, pp. 15–53 (2005)
13. Scherf, B.: Wirtschaftliche Nutzenaspekte der Personaleinsatzplanung. In: Fank, M., Scherf, B. (eds.) Handbuch Personaleinsatzplanung, Datakontext, pp. 55–83 (2005)
14. Sub-Daily Staff Scheduling Data Sets and Benchmarks, `http://www.tu-ilmenau.de/fakww/2608+M54099f70862.0.html`

Knowledge Oriented Implementation of Collaborative Supply Chain Management

Priscilla Heinze and Marcus Hake

University of Potsdam,
Chair of Business Information Systems and Electronic Government,
August-Bebel-Str. 89, 14482 Potsdam, Germany
{priscilla.heinze,marcus.hake}@wi.uni-potsdam.de
http://wi.uni-potsdam.de

Abstract. Knowledge and information flow within supply chain has gained more and more significance. A timely and accurate information exchange among the actors of the supply chain prevents unnecessary loss due to failed planning and missed forecasting. The method of collaborative supply chain management should reduce uncertainties in supply chain. Nevertheless, because of its complexity, its implementation has not been fully optimized. Knowledge and information flows within the supply chain has been unattended. To achieve transparency and trust among partners in the supply chain, a new approach, which allows the visualization of these knowledge and information flows and their participants, is required. Knowledge Modeling and Description Language (KMDL) offers a means to meet this requirement.

Keywords: knowledge intensive process, collaborative supply chain, information flow, supply chain management, knowledge management.

1 Introduction

Due to the growing competition, companies are forced to perform their best in order to survive in the market. Every supply chain management approach aims to produce the right goods at the right time for the right customers for the right price [1]. Any companies unable to stand up to this expectation are most likely to fail [2]. The importance of an efficient supply chain management should not be underestimated if one tends to accomplish alignment to new markets [3].

All decisions made in the supply chain are based on certain information. However, this information is not always available. Therefore, decisions made in the supply chain are always connected to risks and an insignificant misstep at the beginning may cause a huge damage during the further execution of the chain (bullwhip effect) [4], [5]. In a collaborative supply chain all partners in the supply chain are integrated to facilitate an accurate and timely information exchange [6].

Collaborative supply chain management (CSCM) has nowadays been widely used by most companies. However, due to the complexity of a collaborative supply chain, it is often difficult to control and comprehend its information flow [7].

W. Dangelmaier et al. (Eds.): IHNS 2010, LNBIP 46, pp. 222–233, 2010.

The main reason is because the knowledge intensive activities within this flow have been merely individually and randomly performed. CSCM is in need of a method that provides a person oriented representation of the knowledge and information flow in order to enforce transparency of the knowledge intensive activities. This paper introduces the application of Knowledge Modeling and Description Language (KMDL), a tool that is developed to visualize and analyze the flow of knowledge and information of knowledge intensive processes [8]. In this paper we will demonstrate how KMDL solves the knowledge oriented shortcomings of the implementation of CSCM.

In the next section (section 2) we will briefly recapitulate the definition of supply chain management. Within this section the implementation and potentials of CSCM will be explained in a thorough manner. Theoretical basis of KMDL follows in the consecutive section (section 3), different types of knowledge and the definition of a knowledge intensive process will be described. This is followed by the procedural model and view concepts of KMDL. In section 4 we will demonstrate the integration of KMDL into CSCM. Section 5 will cover the conclusion of the paper and the needs for further research.

2 Supply Chain Management

Supply chain is a term that emerges from the practice. Back then, it was a chain of events that occurs naturally within a value creation process. Researchers in management science have been developing methods and structures to manage these events. The divergent objectives of the practitioners and the various concepts of supply chain management have led to inconsistency in the definition of supply chain management [9]. In this paper we agree with Scholz-Reiter, who defines supply chain management as "an inter-organizational coordination of material and information flows through the whole value creation process from the production of raw materials [...] up to the end-customer, with the aim to organize the entire process at optimal costs and time" [10].

In planning a supply chain, one should master the coordination of each participant within the phases of value creation, otherwise it would lead to increase in variability in the supply chain. This phenomenon is the so-called bullwhip effect [9], [1], [5]. The bullwhip effect highlights the importance of information flow in supply chain. Timely and accurate information from each company participating in the chain and from the customers reduces the bullwhip effect significantly because it reduces unnecessary uncertainties. The next subsection will briefly describe the collaborative planning, forecasting and replenishment (CPFR) method as a concrete implementation of CSCM.

2.1 Collaborative Supply Chain Management (CSCM)

According to Anthony, "supply chain collaboration occurs when two or more companies share the responsibility of exchanging common planning, management, execution, and performance measurement information" [6]. It is crucial that the collaboration is also based on a partnership, which is a "tailored business relationship based on mutual trust, openness, shared risk, and shared rewards that yields a competitive advantage, resulting in business performance greater than would be achieved by the firms individually" [11]. In other words, a CSCM is based on mutual trust by two or more companies, who share risks, responsibilities and rewards, in order to reach competitive advantage. CPFR is a business model of multiple companies participating in a supply chain, which require a collective preparation and organization of the supply chain. Fig. 1 shows the CPFR process model.

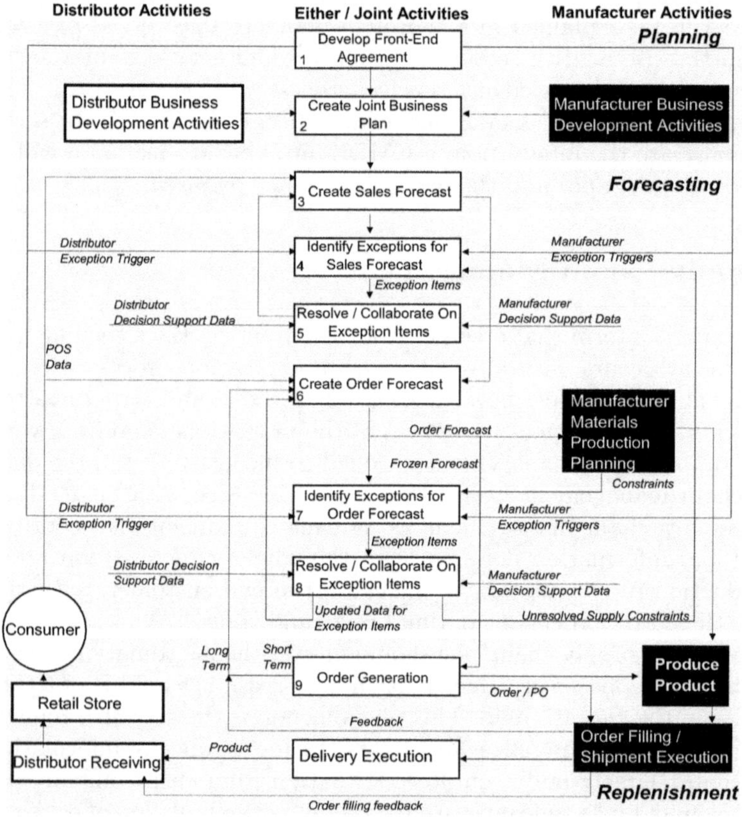

Fig. 1. CPFR Model [12]

Collaborative planning marks the first step of CPFR. The participating companies develop an agreement containing their expectations and the required resources [9]. Based on this agreement a joint business plan is created, in which the ground rules are defined. In the collaborative forecasting, the partners predict sales based on the exchanged information. After the sales forecast is created, the partners predict the exceptions that might appear during the supply chain and determine the solutions to the problems arising from the expectations.

The next step in the collaborative forecasting is generating the customer order forecast. The obtained prognosis from the previous steps are brought together with data regarding the stock of inventory. This step aims to assure the capacity and disposition of each partner in terms of the amount of order within a certain period of time. Afterwards, the exceptions that might occur after the receipt of order are identified and their solutions are figured out. In the collaborative replenishment, a binding order is generated, which then results in the production of the goods by the manufacturers.

2.2 Knowledge Oriented Prerequisites of Collaborative Supply Chain Management

The success of collaborative supply chain management depends significantly on the fulfillment of some substantial prerequisites. Adequate communication is one of them [7]. From each partner a person responsible for supply chain communication should be selected. Having to communicate with different persons every time specific information should be provided or a collective decision should be made results in communication delay and content loss. Furthermore, the assignment of competence in the decision making should be defined. Otherwise, it might happen that some decisions are made without incorporating the partners. This would result in a deteriorating relationship in the collaboration.

It is imperative that the partners clarify their expectations [7]. According to Ireland, it is even a requirement to have partners that have a "synchronous collaborative vision" [13]. Vague expectations causes misunderstanding and unfair conditions. The partners should understand what is being expected from each other. Another prerequisite of CSCM is the availability of relevant technologies [7]. At the beginning of the collaboration process, the technical requirements between partners should be adjusted for a timely and effective information flow. It is important to determine which technology is required in order to avoid incompatibility during the collaboration process.

The success of CSCM depends highly on reliable information from the partners. Trusting each other is a critical factor but the information being provided is not always as accurate as it should be. This might result from a human error or a partner betrayal. A possibility to track the information as in from whom or where it came and where it is being transferred would help to pursue transparency of the information flow.

As implied by Barratt and Oliveira, there are only few sources found regarding which information should be shared to create the visibility offered by CPFR and which business function should be involved in the CPFR process [14]. Lack of documentation on these matters results in a confusing and an unclear mess of information exchange. Recording the information needed, its origin and its use would save the time needed to figure them out in case a new employee should substitute the person in charge.

3 Knowledge Modeling and Description Language (KMDL)

Attempts to integrate the knowledge management approaches into aspects of supply chain management have been made [15],[16], [17]. However, none of the existing tools and methods consider the person-bound aspect of knowledge and how it affects the flow and exchange in the supply chain. KMDL was chosen over other modeling languages because of the method's capability to model how knowledge is originated, how it develops and how it is being shared among the participants of the process [18]. It also enables a potential analysis of the model for the process improvement. In this section we will briefly describe the theoretical foundation of the method, followed by the procedural model and the view concepts of KMDL.

3.1 Knowledge Intensive Processes

Davenport and Prusak state that a process is knowledge intensive when its input and output are uncertain [19]. Hoffmann defines the large varieties of sources and medias as well as high variance and dynamic development of the process organization as indicators for a knowledge intensive process [20]. These definitions are coherent to the characteristics of a supply chain management process. As such, a supply chain management process can be categorized as a knowledge intensive process.

3.2 Procedural Model

Fig. 2 illustrates the procedure of KMDL implementation in the practice. The project begins with phase 0, in which the KMDL method is introduced to the project partner. In phase 1, the project partner and the analysts agree on a project aim and the relating objectives. Based on this, they decide on the knowledge intensive process, which will be modelled and analyzed. The next phase (phase 2) begins with the analysts gathering information from the employees participating in the process (process participants). Commonly they use a semi-formal interview. Based on this information, the analysts model the process view, which represents the chain of events in the process. This view is presented to the process participants and they provide inputs to improve the model if necessary. This phase ends when the process participants state their final validation.

Not all tasks within the process are knowledge intensive. Although the selected process may fulfill the characteristics of a knowledge intensive process, some administrative tasks needing less creativity or innovation ability within this process should nevertheless be conducted in order to provide a structured framework for the process. However, since knowledge non-intensive tasks are not the focus of KMDL, in phase 3, these tasks are identified and filtered out. The knowledge intensive tasks are used as basis for phase 4, where the process participants share their activities within each task in detail. The analysts use the obtained information to model the activity view, which shows the knowledge oriented activities of the process and the process participants validate the model.

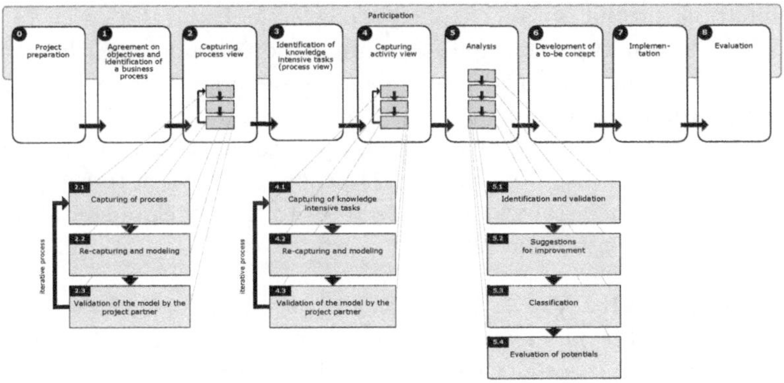

Fig. 2. KMDL procedural model [8]

By means of the generated process and activity views, the analysts then identify and evaluate the weak points of the captured process in phase 5. The analysts then provide recommendations for improvement based on this finding. The recommendations are confronted with the company's vision and aim as well as available resources. In phase 6, a to-be concept is developed under the consideration of the found weak points and the suitable and agreed improvement recommendations. In phase 7 and 8, this to-be concept is implemented and evaluated.

3.3 KMDL Layer Concept

Tacit and Explicit Knowledge. There are numerous ways to categorize knowledge. KMDL uses the categorization proposed by Polanyi, who classifies knowledge into tacit and explicit knowledge [21]. Tacit knowledge is produced by personal experiences, individual perceptions and value systems. It is difficult to articulate and formalize [22]. The most important characteristic of tacit knowledge is its person-orientation, meaning that it is originated in the mind of its holders [19].

Explicit knowledge can be found in documents, repositories, processes and other transferable and verbalizable forms. It is a type of knowledge, which is independent from the context and the knowledge holder [8]. In this case the utilization of information and communication technologies plays an important role. Both knowledge types are connected to each other. Nonaka and Takeuchi define the interdependency of both knowledge types as "knowledge conversion" [22], which is explained as follows:

Socialization is an exchange of experiences between one tacit knowledge holder and another. This can occur through informal conversation during coffee breaks or by expert observation.

Externalization marks the conversion of tacit into explicit knowledge. Knowledge sharing is enabled through articulating, documenting and formalizing tacit knowledge, so that it does not remain only in mind of the holder.

Combination produces new explicit knowledge from the existing explicit knowledge. Using information and communication technologies one can combine, categorize, complement, delete and revise documents in order to generate new context of information.

Internalization converts explicit knowledge into tacit knowledge. Reading a book, watching a film, listening to a presentation, etc. belong to the practices of this conversion. Explicit knowledge contributes to the development of a person's tacit knowledge through internalization.

Process and Activity View. KMDL is a semi-formal and graphic-based modeling language. In this paper we refer to version 2.1, which offers two alternative views: process view and activity view. These views are interdependent and should be taken into account in their entirety.

The process view describes the operational flow of tasks and their alternative routes as explained in section 3.2. Using this view, it is also possible to identify the units performing the tasks in the process and the information system used.

The activity view enables a detailed description of these tasks. The main objects of this view consist of tacit knowledge and explicit knowledge. Tacit knowledge is symbolized as *knowledge object (KO)* and explicit knowledge is symbolized as *information object (IO)*. As explained in section 3.3, tacit knowledge is always person oriented. This view enables the identification of each knowledge crucial for the process and its bearers as well as its shortfalls.

This view also shows the knowledge conversions along with their frequencies within each task. These conversions can be refined into *conversion methods*, in which the type of activity performed is explicitly mentioned. For example: the method used for the knowledge conversion *internalization* is *reading a manual.*

Some conversions can only be completed under the circumstances that certain conditions are fulfilled. These conditions are symbolized as *requirements.* A *function* defines the type of action enabled by the information system (of the process view). Fig. 3 shows the objects of the process and activity view.

4 Integration of Knowledge Management into Supply Chain Management

4.1 Modeling CPFR

In this section we present the potentials of KMDL in helping to improve the implementation of the collaborative supply chain. The following example is applied to the CPFR method. Fig. 4 and 5 shows an exemplary KMDL model of the knowledge flow within the first sub-process in CPFR: collaborative planning[1].

[1] As observed by Barratt and Oliviera and by the authors themselves, there exists no step-by-step procedure in the literature on how to implement CPFR [14]. The presented KMDL model is only a showcase and is intended to be advisory and not regulatory. The development of this model is based on many years of practice experiences in the field.

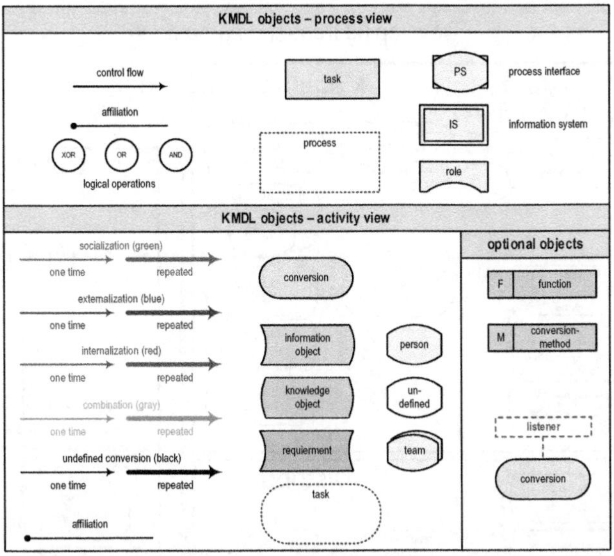

Fig. 3. Objects of the KMDL process and activity view [8]

Collaborative planning consists of the tasks: *developing front-end agreement* and *creating a joint business plan*. Within the first task an ERP system is used. Participants of the process are the manufacturer and the distributor.

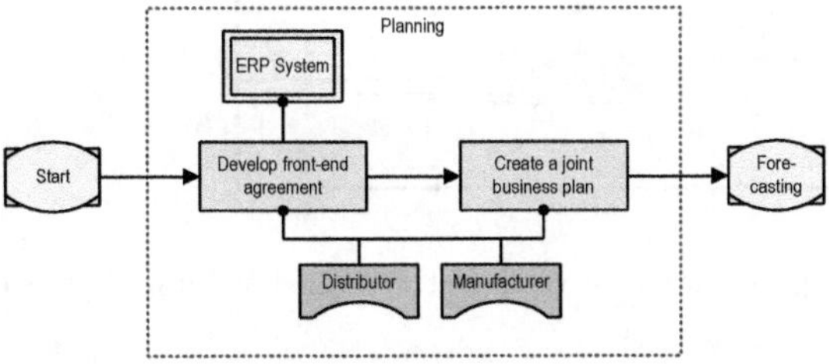

Fig. 4. Exemplary process view of the collaborative planning

Fig. 5 shows the activity view of the task: *developing front-end agreement.* In this case, the KMDL notation is adjusted in order to enable the visualization of an interorganizational knowledge flow. The roles presenting the business functions, which usually belong to the process view, are dragged down to the activity view so they can be assigned to the person.

Within this view, person A from the production department of the manufacturer company discusses the company's resource and product availability with person G, the supply chain manager, using his knowledge about the company's production capacity and spectrum. Person D from the legal department, together

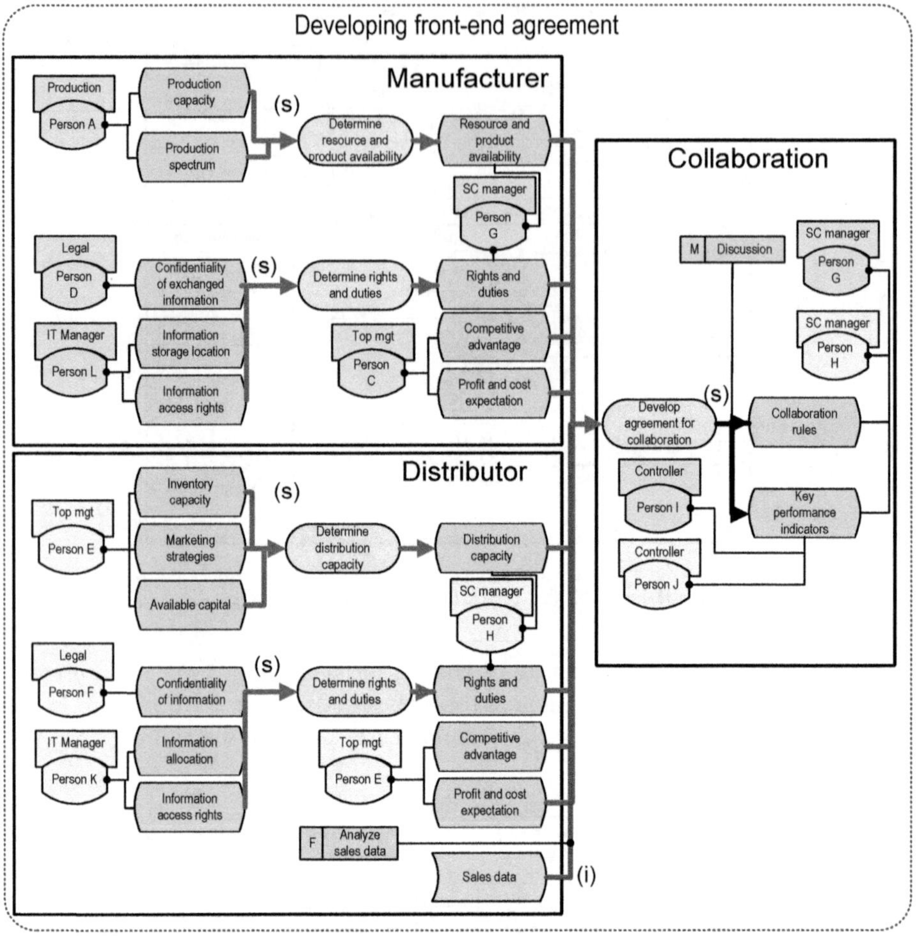

Fig. 5. Exemplary activity view of the task developing front-end agreement

with person L, the IT manager, contribute their knowledge about the confidentiality of the potentially exchanged information as well as its storage location and access rights to the discussion with person G (the SC manager) to determine the company's preferred rights and duties.

At the distributor side of this view, person E as a representative of the top management discusses about the company's distribution capacity with person H, the supply chain manager of the distribution company, by contributing his knowledge about the company's inventory capacity, marketing strategies and the available capital. In determining the companies' rights and duties, the same knowledge activity at the distributor as well as manufacturer side was performed.

In order to collaborate, an agreement between both sides has to be developed. Person G, the supply chain manager of the manufacturer company contributes his knowledge about the resource and product availability and the preferred rights and duties of his company. Person H, the supply chain manager from the distributor company shares his knowledge about his company's distribution capacity and the preferred rights and duties. Both person C and person E from the top management of both companies take into account the expected competitive advantages as well as the profits and costs emerging from the collaboration. Collaboration rules are the result of this discussion. Person I and person J, controllers from each company, analyze the sales data, which is accessible from the CRM system, and decide on the key performance indicators for later investigations.

The given example shows that mostly tacit knowledge is being exchanged and transformed while explicit knowledge, such as documents, is hardly used in this part of the CPFR process.

4.2 Potentials of KMDL

KMDL not only enables the visualization of knowledge and information flow within a CPFR process. Since it is a process oriented knowledge management method it also describes the operationalization of the CPFR model (see Fig. 1). The activities happening within each task can be presented and adjusted by the process participants with a high level of detail along with the knowledge and information flow.

KMDL also enhances the performance of CPFR as it offers solutions to improve its implementation by helping to fulfill the prerequisites (see section 2.2). A transparent view of the corresponding persons (and their business functions) reduces the confusion caused by the unclear assignment of person being responsible for providing information. Knowing who to contact is a crucial condition to establish an adequate communication. In case of personal or hierarchical change, the model can be adjusted as well as expanded. This serves as a prerequisite of an effective knowledge transfer if a new employee enters the process. Decisions can be made faster when knowing who has the competence to make them. The companies can identify their employees' knowledge and competence shortfalls and effectively acquire them, which in turn gives them the opportunity to strategically plan their employment and qualification policies.

The process participants can also make early planning regarding the technology used to facilitate the information exchange. The process view shows the type of the information system and the activity view presents the type of action conducted using the technology. This way, both companies can be prepared if the need to adjust or upgrade their technologies arises, especially when more documents are exchanged at the later stages of CPFR.

Companies are able to identify the origin of the knowledge, its holders and development. In case of flawed information it is possible to track down the provider of the knowledge at the first place for clarification. The model allows the identification of key persons, which are the ones holding core knowledge and handling important information, and the impacts that might occur if they leave the company. One of the conditions mentioned in section 2.2 is the clear expectation of companies. This is not easy to fulfill since tacit knowledge is hardly articulable. The KMDL model also provides a framework showing which knowledge and information is needed, what it is needed for, the interrelating information that adds the value of the information, etc. The framework serves as a basis of requirements to be fulfilled when submitting an information or requesting it.

Last but not least, due to its adaptability the KMDL model can be used as a guideline to operationalize the CPFR process. If a failure is discovered during the implementation, improvement measures can be planned and modelled first before they are being analyzed for implementation. It is also possible to build a to-be process based on the actual process, which serves as a target process. This way, a continuous process improvement can be performed.

5 Conclusion and Outlook

This paper proposes KMDL as a method to operationalize and visualize the CPFR steps by integrating the aspects of knowledge management in order to enhance the knowledge and information flow. Due to the space limitation and complexity of the process this model has been significantly simplified. Many potential aspects of KMDL have not yet been covered in this paper, for example the reporting possibility and the pattern recognition. The next paper should include these aspects. Further research can be conducted in many areas. Due to the growing trend of outsourcing decision making processes in a collaboration, the cultural differences and their impact on the collaboration are interesting aspects to cover. The use of suitable information and communication technologies to facilitate the collaboration process also needs closer investigation, especially in the Web 2.0 era. Special attention should be given to applications such as social software and social network concerning their implementation within the collaborative supply chain management.

References

1. Simchi-Levi, D., Kaminsky, P., Simchi-Levi, E.: Managing the Supply Chain - The Definitive Guide for the Business Professional. McGraw-Hill, New York (2004)
2. Tan, K.C., Kannan, V.R., Handfield, R.B., Ghosh, S.: Supply chain management: an empirical study of its impact on performance. International Journal of Operations & Production Management 19(10), 1034–1052 (1999)
3. Min, S., Roath, A.S., Daugherty, P.J., Genchev, S.E., Chen, H., Arndt, A.D., Richey, R.G.: Supply chain collaboration: what's happening? The International Journal of Logistics Management 16(2), 237–256 (2005)
4. Forrester, J.: Industrial dynamics – a major breakthrough for decision makers. Harvard Business Review 36(4), 37–66 (1958)
5. Lee, H., Padmanabhan, V., Whang, S.: The bullwhip effect in supply chains. Sloan Management Review 38(3), 93–102 (1997)
6. Anthony, T.: Supply chain collaboration: success in the new internet economy. Achieving Supply Chain Excellence through Technology, vol. 2. Montgomery Research Inc., San Francisco (2000)
7. Mentzer, J., Foggin, J., Golicic, C.: Collaboration: the enablers, impediments, and benefits. Supply Chain Management Review 4(4), 52–58 (2000)

8. Gronau, N.: Process Oriented Management of Knowledge: Methods and Tools for the Employment of Knowledge as a Competitive Factor in Organizations (Wissen prozessorientiert managen: Methode und Werkzeuge für die Nutzung des Wettbewerbsfaktors Wissen in Unternehmen). Oldenbourg Verlag, München (2009)
9. Gronau, N.: Enterprise Resource Planning and Supply Chain Management - Architecture and Funktions (Enterprise Resource Planning und Supply Chain Management - Architektur und Funktionen). Oldenbourg Verlag, München (2004)
10. Scholz-Reiter, B., Jakobza, J.: Suppply Chain Management – Überblick und Konzeption. HMD 207, 7–15 (1999)
11. Lambert, D., Emmelhainz, M., Gardner, J.: So you think you want a partner? Marketing Management 5(2), 24–41 (1996)
12. Knolmayer, G., Mertens, P., Zeier, A.: Supply Chain Management based on SAP-Systems. Springer, Berlin (2002)
13. Ireland, R., Bruce, R.: CPFR: only the beginning of collaboration. Supply Chain Management Review, 80–88 (September/October 2000)
14. Barratt, M., Oliveira, A.: Exploring the experiences of collaborative planning initiatives. International Journal of Physical Distribution & Logistics Management 31(4), 266–289 (2001)
15. Wadhwa, S., Saxena, A.: Knowledge management based supply chain: An evolution perspective. Global Journal of e-Business and Knowledge Management 2(2), 13–29 (2005)
16. Voigt, G.: Supply chain coordination with information sharing in the presence of trust and trustworthiness: a behavioral model. Femm working paper no. 06, Otto-von-Guericke-Universität Magdeburg (March 2009)
17. Yuva, J.: Knowledge management: The supply chain nerve center. Inside Supply Management, 34–43 (July 2002)
18. Gronau, N.: Modelling knowledge intensive engineering processes with the Knowledge Modeler Description Language KMDL. In: Weber, F., Kulwant, S., Thoben, K.D. (eds.) Enterprise Engineering in the Networked Economy, pp. 195–202. University of Nottingham (2003)
19. Davenport, T., Prusak, L.: Working Knowledge. How Organizations manage what they know. Harvard Business School Press, Boston (1998)
20. Hoffmann, M.: Analysis and Support of Knowledge Processes as Requirement for Successful Knowledge Management (Analyse und Unterstützung von Wissensprozessen als Voraussetzung für erfolgreiches Wissensmanagement). In: Abecker, A., Hinkelmann, K., Maus, H., Müller, H. (eds.) Geschäftsprozessorientiertes Wissensmanagement, Springer, Heidelberg (2002)
21. Polanyi, M.: The tacit dimension. Peter Smith, Gloecester, Mass (1966)
22. Nonaka, I., Takeuchi, H.: The Knowledge-Creating Company – How Japanese Companies Create the Dynamics of Innovation. Oxford University Press, New York (1995)

Reference Modeling of an IT-Based Logistics System

Iris Hausladen

Heinz Nixdorf Chair of IT-based Logistics, Leipzig Graduate School of Management (HHL), Jahnallee 59, Leipzig, Germany
iris.hausladen@hhl.de

Abstract. Both the logistics as well as the production management have to face challenging requirements nowadays. Not only evolutions in markets and new organizational structures induce important both systems and methods adaptations, but especially the enhanced application of information and communication technologies along the worldwide distributed value creation chain results in a sustainable paradigmatic change. While in the field of production an already field-tested concept, the production system approach exists to meet the various demands, a similar concept is lacking in the area of IT-based Logistics. The given resemblances between the two disciplines favor a transfer of concept by reference modeling of an IT-based Logistics System depending on the production system framework.

Keywords: IT-based Logistics, production system, reference modeling.

1 Introduction

IT-based Logistics, often also known as E-Logistics, becomes more and more important within a globalized and complex networked value creation chain to coordinate the material and information flows in an efficient manner. Researchers as well as practitioners have outlined the benefits of intelligent information and communication technologies for the logistics area in recent years [1]. In addition to the emphasizing of their benefits on the underlying business processes (see Fig. 1.), meanwhile also the specific impact of IT-Logistics tools on a company's competitiveness has been analyzed by an empirical study [2]. However, a clear conceptual both strategic as well as methodical and operational system framework is missing compared to other management areas, such as for example the production or marketing management. Predominantly, only isolated tools and strategic guidelines are found in literature and praxis, considering and discussing aspects related to the design, realization and controlling of IT-based Logistics applications and the respective business processes. Regarding the challenges (IT-based) Logistics has to face nowadays as a multifarious and networked field of action, practitioners have to answer a lot of questions, like: "How to handle the diversity of IT-based Logistics, how to define the right strategy, at the right time, at the right location supposing the right action guidelines for the people involved in the IT-supported Logistics function?". For that reason, complexity in theory and praxis often "calls for" an overarching concept to structure the related decision matters and design variables, thus to induce transparency, to balance targets and finally to allow a target-oriented allocation of resources.

W. Dangelmaier et al. (Eds.): IHNS 2010, LNBIP 46, pp. 234–244, 2010.

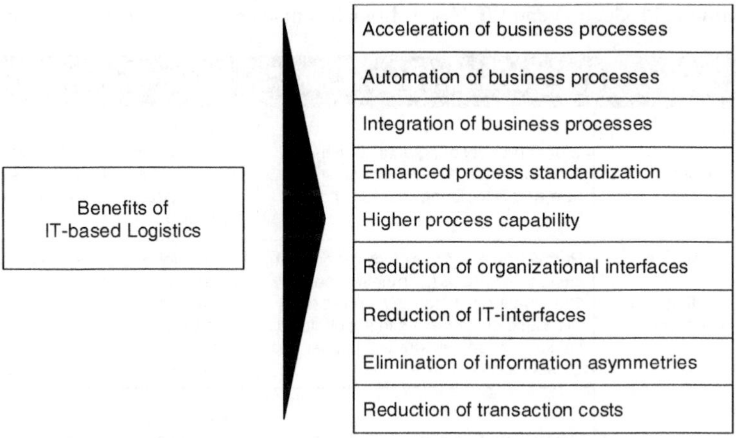

Fig. 1. Benefits of IT-based Logistics – Overview

One possible way to resolve this problem consists in the reference modeling approach, i.e. looking at other disciplines, outlining the given structural and content related similarities and finally transfer of concept by adjusting the reference model to the paradigm related characteristics.

In search of an adequate reference concept to model IT-based Logistics, the production system approach seems to be a promising scheme from my point of view. Over the past two decades numerous methods and tools evolved from innovative production management concepts that pushed companies to think about an integration concept to structure systematically all activities in the manufacturing area: the production system approach was born. For me, the situation turns out to be quite similar to the IT-Logistics development. Especially big companies started to build up individual production systems, in most cases based on the well-known benchmark the Toyota Production System [3]. The automotive industry acted as a first mover in applying the Toyota Production System as a role model to their business concept [4]. Meanwhile also small and medium sized enterprises (SME) are more and more interested in specific aspects of the production system approach and the concept has been transferred to different industries, also to the Health Care Management sector [5]. So why not using it as a reference model for the IT-based Logistics?

2 Production and Logistics as Interdependent Paradigms

The deduction of an IT-based Logistics System approach depending on the production reference model does at first not only require the availability of a content related but also of a structural equivalence of both paradigms.

By looking at the crucial trends in the field of production and (IT-based) Logistics management, as shown in Table 1, the interdependency and coexistence of both disciplines should be examined.

Table 1. Production and (IT-based) Logistics management – A trend comparison

Trend	Production Management	(IT-based) Logistics Management
Internationalization	Global production, global sourcing and worldwide distribution require new and flexible logistics structures.	Establishment of logistics networks and collaboration with logistics service providers due to the changes of value creation structures. Enhanced IT-application.
Cost reduction and Value orientation	Permanent need to reduce production costs to safeguard competitiveness. Outsourcing/Offshoring as one option to reduce costs results in a globalized value creation network with higher demands on logistics.	Higher logistics costs induced by the Outsourcing trend in industry. Cost reduction and Value orientation as basic targets of logistics management as a consequence. Logistics costs share depends e.g. on industry [6].
Time and Speed Management	Reduction of processing and lead times as a core function of nowadays production management and essential part of organizational development programs.	Speed is one essential pillar of logistics management and thus part of every service level agreement with logistics service providers. Reduction of e.g. transportation and storage times as a mandatory tasks. Speed can be enhanced by using IT.
Quality Management	Improvement of product, process and of service quality. Application of Quality Management Systems like the Total Quality Management (TQM) or the (Lean) Six Sigma approach.	Process and service quality as decisive factors for customer satisfaction. The product quality is influenced e.g. by the handling and transportation process. TQM and (Lean) Six Sigma concept are gradually transferred to the field of (IT-based) Logistics [7].
Mass Customization	Combination of the advantages resulting both from Mass Production and from Customization, i.e. fulfilling customer requirements [8]. Higher demands on (IT-based) Logistics.	Mass Customization sets higher demands on (IT-based) Logistics processes, leads to a higher complexity and requires a broader task range from the logistics service provider's point of view [9].

Conclusion. Although the multiple aspects demonstrated above represent no complete list both the content and structural related analogousness of the paradigms production and (IT-based) Logistics management become evident from my point of view. Therefore the production *management* concept forms an adequate reference approach for the IT-Logistics topic. In the next step it has to be outlined how the production *system*, as background of a successful production management, is structured to prepare the following reference modeling of an IT-based Logistics System.

3 Production Systems as a Reference Modeling Approach

Today production systems are characterized by important design factors that should act as guiding structural principles for a planned reference modeling in the context of IT-based Logistics later on (see paragraph 4).

- Strategic focus: Companies have to concentrate their activities on the present and future core competences. Therefore a long-term planning approach, i.e. a clear strategy has to be captured. Furthermore a market orientation of the production process as well as a deduction of long- and medium-term objectives has to be realized.
- Operational focus: The famous "Lean" approach serves as a guiding concept in the sense of an operational improvement program. Lean Production implies that all business processes connected to the field of manufacturing are to hundred percent value-adding, i.e. free of any wastage (e.g. idle times, unnecessary storage times, times for getting information) [10]. Recently the "Lean" philosophy has been transferred to nearly all business functions, thus also to the cross-sectional function logistics [11]. The initiation of a companywide Continuous Improvement Process (CIP) helps to realize cost reductions in manufacturing, logistics and further business process areas.
- Integration focus: In the sense of a holistic organizational design it is not useful to develop and apply isolated methods and tools, with respectively perhaps conflicting targets. In order to realize existing synergies it is indispensable that a consistent range of concepts is outlined during the design and implementation phase. Hence those structures are named Integrated Production Systems.

Integrated Production Systems consist of different subsystems and elements. On the upper level production strategies are formulated that represent the future pathway for all activities in this area (see Fig. 2). Total Quality, Just-in-Time, Asset-Light, Flexibility & Agility, and Efficiency are considered as the relevant strategic course of action for a successful production management to meet the challenges in a global value creation network.

Design guidelines on the second level act as a kind of rule for all participants of the Integrated Production System. They concretize the production strategies in order to facilitate the respective implementation on the shop floor level. Thus Standardization, the realization of Synergies, Transparency, Self-responsibility, Consistency as well as a Continuous Improvement should guide the staff members through their day-to-day business.

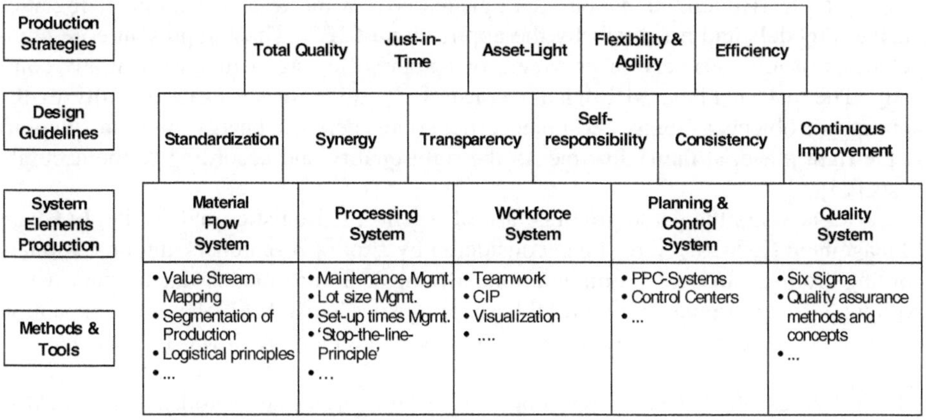

Fig. 2. System Structure of an Integrated Production System (modified; according to [12])

Production strategies and the related design guidelines are defined companywide to establish a common understanding of what production management is about and to enhance a creative corporate culture.

Methods and tools are structured in subsystems on level three. Material Flow, Processing, Workforce, Production Planning & Control, and the Quality System constitute the pillars of an Integrated Production System. Each subsystem contains methods as system elements that are assigned to the respective subsystem according to their main functional and process related impact. As a consequence the tool allocation is not free of overlap, but reality is too complex to consider the interdependent aspects all at once. In some examples nearly one hundred different methods are part of a production system [13]. Regarding the local supply infrastructure the elements of the individual subsystems have to be designed company or site specific.

From an IT point of view the link of the physical production with information technologies represents an important milestone in the evolution of production management concepts. Therefore the above mentioned modeling framework has to be modified to overcome the shortcoming in that field of action, especially if the transfer of concept should take place for the IT-based Logistics.

The definition of the underlying conception of a *system* emerges as a necessary task in order to realize such a structural organization in praxis anyway. A system represents a network of elements that influence each other – and if open systems are considered – it interacts with the surrounding environment. A system itself consists of several subsystems and is furthermore part of a meta or super system. The paradigm, in that sense the production system (later on: the IT-based Logistics System) manifests itself furthermore as a social system [14]. People interact in a workshop for example and their behavior is characterized by their former socialization, qualification as well as by the corporate culture of the company they are working for.

4 Reference Modeling of an IT-Based Logistics System

Before developing a suitable concept and system framework the paradigm considered has to be clearly defined. IT-based Logistics covers the transformation of logistics business models and processes by the application of ICT. Those represent core competencies of logistics service providers or comprise logistical functions in any company. The task of IT-based Logistics consists in providing the exact quantity of the appropriate objects as logistical subjects (i.e. goods, persons, energy, and information) at the right place, at the right time, in the right quality and according to the accurate costs [15].

In recent years the paradigms IT-based Logistics, E-Logistics and E-Supply Chain Management (E-SCM)[1] have been considered by a lot of researchers and practitioners coming from a business administration, technical or informatics disciplinary view. Although some frameworks for IT-based Logistics exist [16], the implementation stage of a corresponding system approach is located on a more or less basic level.

[1] In this context the terms IT-based Logistics and E-Logistics are considered as coincident concepts. E-SCM represents from the author's point of view a more comprehensive concept that includes all cooperating partners along the complete value creation chain.

High flexible and dynamic structures like they are found in the IT-based Logistics emerge both from a process, functional, and ICT related point of view as risky management topics. Therefore the specific system elements have to be continuously monitored, controlled and coordinated in order to secure the fulfillment of the ambitious targets set by the management. The overarching concept (i.e. in the sense of a meta system) of a respective framework requires as a consequence – in addition to the already introduced production system reference model - an adequate anchorage in the company's management system. From my point of view a suitable approach in this context is represented by the coordination oriented controlling concept developed by Küpper.

Following this understanding each company consist of a *leadership* system on the one hand and a *performance* system on the other hand [17]. The controlling is responsible for the coordination of all *leadership* subsystems, i.e. the Planning, Control, Information, Organization, and Human Resource System. Controlling itself can be seen as an individual subsystem that is marked as a cross-functional task in every company, regardless the related industrial sector or business model. From the main coordination function also the adaptation and innovation function, the target orientation function as well as the service function can be deducted as additional tasks [18]. Controlling is responsible for the coordination within one subsystem and between the single subsystems on the *leadership* system level.

Logistics and therefore also IT-based Logistics are conceived as management topics with their own strategies, targets, resources etc. IT-based Logistics then represents one specific element in the *performance* system that is coined by the cross-functional integration of the IT-supported diversified logistics and supply chain, i.e. mainly of the procurement, production, distribution and redistribution logistics. Following the coordination oriented controlling approach each individual subsystem (here: the IT-based Logistics System) within the *performance* system (further subsystems are e.g. R&D, Procurement, Production, Sales) disposes over its own *decentralized leadership* system. Consequently the controlling function includes (a) the coordination of the leadership subsystems within the IT-based Logistics System (decentralized controlling), (b) the coordination with the centralized company controlling and (c) the coordination with the decentralized controlling of other functional or business process fields, i.e. subsystems within the performance system [19].

Coming from a logistics service provider company, IT-based Logistics represents a core competence and defines the business model. Thus the coordination oriented controlling approach locates the IT-Logistics paradigm on the upper *leadership* system level, i.e. centralized (a). Considering an industrial company or a retailer IT-based Logistics is processed on the *performance* system level, as part of the decentralized controlling function. Implementing the controlling task on this level IT-based Logistics emerges as a cross-company linking pin to all available decentralized leadership subsystems (b). The perspective assumed in this article, coming from the production management, will be the one of an industrial company; therefore option (b) is followed up.

The next reference modeling level of an IT-based Logistics System consists in the definition as well as allocation of strategies and design guidelines (see the conceptualization of Integrated Production Systems). Coming from the system and controlling approach as described above the *decentralized leadership* system (as part of the performance system of a company) is now constructed. The respective inherent subsystems are basically structured as shown in Fig. 3.

- The planning subsystem includes the IT-based Logistics management strategies, i.e. Total Quality (e.g. TQM, Six Sigma), Just-in-Time (production and demand synchronous supply), Asset Light (concerning warehouse stocks, WIP, logistics assets like trucks, loading equipment etc.), and Flexibility/Agility (i.e. response time to changes in the logistics environment). Besides the modeled production management strategies, also Value Orientation/Efficiency (i.e. IT-based Logistics affects a company's profit in a positive way by relying on optimized material and information flows), Sustainability (i.e. balancing of economic, ecological and social dimensions of an IT-based Logistics) and Technology/Innovation Orientation (i.e. ICT management, technology transfer in the field of IT-based Logistics and invention of IT-based Logistics applications, tools, concepts and methods).

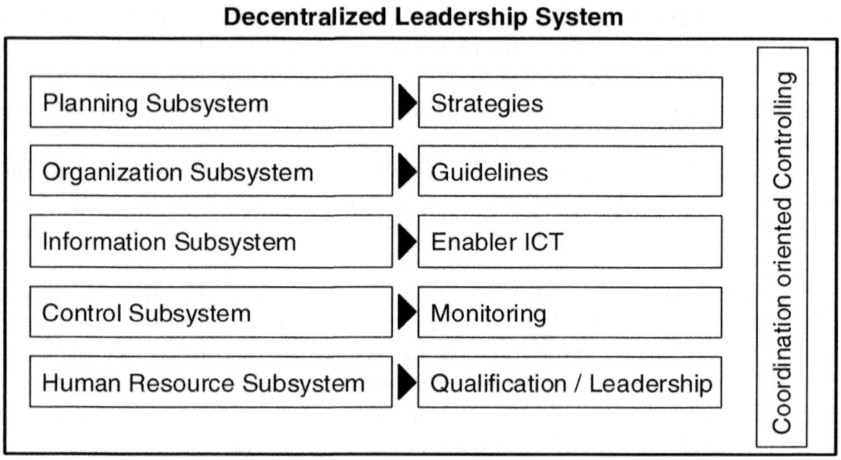

Fig. 3. Decentralized IT-based Leadership System – Basic Structure

- The organization subsystem includes all relevant design guidelines of an IT-based Logistics System, i.e. (in analogy to the reference modeling approach Integrated Production Systems) Standardization (standardized IT-supported processes, standardized organizational and IT-interfaces), Synergies (realization of the inherent economies of scale and scope), Transparency (as regards e.g. strategies, design guidelines, philosophy & culture, targets, processes, responsibilities etc.), Self-responsibility (i.e. employees act as co-leaders), Consistency (harmonization of tools and methods application both from a business and ICT related point of view; a fact that leads to a kind of "doubled complexity"), and Continuous Improvement (i.e. CIP as an important task of each system member).
- The information subsystem covers the enabling information and communication technologies of an IT-based Logistics System, i.e. for locating purposes (e.g. GPS, Galileo), for identification purposes (e.g. Barcode, RFID), for mobile communication (e.g. GSM/GPRS, UMTS, Satellite communication, WLAN), mobile

ICT-devices (e.g. On-board vehicle computer, Handheld device), and for electronic data exchange purposes (e.g. Data networks, XML, EDI, Telematics).

- The control subsystem is dedicated to the monitoring of the IT-based Logistics strategies and guidelines. Therefore concepts like for example risk management (focused both on the logistics and IT paradigm), event management (pro-active control of the decentralized leadership system) as well as approaches to safeguard the IT-based Logistics chain (by integrating safety, i.e. a continuous supply is guaranteed and security, i.e. protection against criminal or natural occurrences) constitute the related framework.
- The Human Resource subsystem is responsible for the development of responsible and effective managers and workers which are part of the IT-based Logistics System. Therefore training and qualification activities are tailor-made regarding the age, pre-qualification, and career plan of all system members (target group orientation). Requirement specification, personnel and leadership audit as well as organizational development, and the evolution of a sustainability oriented corporate culture represent basic elements of this respective subsystem.
- The controlling subsystem predominantly incorporates on this level the coordination and service function, especially by providing an (IT-based) Logistics System controlling (e.g. Key Performance Indicator Management, Balanced Scorecard Reporting, Pro-active Notification and Counter Measurement Tracking).

In comparison to the Integrated Production System approach the IT-based Logistics System inherent planning (strategies), organization (design guidelines) and control (safety & security aspect) subsystems have to be designed in a company-wide standard. Regarding the information and Human Resource subsystem the basic rules have to be done company-wide, the related tools and methodological concepts must be defined by consideration of the site specific situation.

Continuing the reference modeling of an IT-based Logistics System, the production system approach requires the definition and structuring of the methodical subsystem level, i.e. the *performance* system content resulting from the coordination oriented controlling point of view. IT-based Logistics emerges as a cross-function and thus as a logistics process spanning task along the whole value creation chain. Consequently, regarding the specific logistics stage the related ICT applications are primarily dedicated to [20], the subsystem structure as illustrated in Table 2 can be defined. The selection and description of the single methods and tools per subsystem have to be done company or site specific regarding the local conditions. Similar to a production system, the methods level of an IT-based Logistics System is also not free of overlap and the responsible persons designing such a framework are asked to solve the resulting conflict, i.e. to keep primarily the guidelines Synergy and Transparency.

The integration of the two core modeling steps leads to a holistic structure of an IT-based Logistics System developed on the reference basis Integrated Production System and related to the coordination oriented controlling approach (see Fig. 4).

From the viewpoint of a practitioner one could ask how such a more theoretical framework actually contributes to realize synergies along the IT-based logistics chain in praxis. The answer is as follows: some enterprises (regardless the company type and the related industrial sector) already dispose over single elements of such a kind of system, albeit the basic philosophies, strategies, guidelines and methods are not clearly structured and documented in a written way. As a consequence the elaborated

Table 2. Decentralized IT-based Performance System – Overview

Subsystem	IT-based Logistics Applications (Exemplary)
Supply Chain Subsystem	• Production Planning & Control Systems (PPC) • Enterprise Resource Planning Systems (ERP) • Advanced Planning Scheduling Systems (APS) • Supply Chain Management Systems (SCM Systems) • Order Processing Tools
Procurement Logistics Subsystem	• Supplier/Demand catalogs • Supplier E-Kanban • Online Auctions • Virtual Marketplaces
Production Logistics Subsystem	• Just-in-Time (JIT) • Just-in-Sequence (JIS) • Production E-Kanban • Digital Factory/Virtual Logistics
Maintenance Logistics Subsystem	• Computerized Maintenance Management Systems (CMMS) • Supervisory Control And Data Acquisition Systems (SCADA) • Condition Monitoring Systems • Maintenance Platforms
Warehouse Logistics Subsystem	• Vendor Managed Inventory (VMI) • Merchandise Information Systems (MIS) • Cross Docking • eConsignment • Robogistics • Warehouse Management Systems (WMS)
(Re-)Distribution Logistics Subsystem	• Efficient Consumer Response (ECR) • Customer Relationship Management (CRM) • Logistics platforms • Tour planning & Route optimization • Tracking & Tracing • Telematics • Milkrun • ePayment • Last Mile Logistics

framework has to be applied to the practical company situation, i.e. design and visualize the integrated model and communicate it to all involved staff members. Thus the benefits of integrated, harmonized and pro-actively controlled company activities can be realized. If the topic IT-based Logistics itself hasn't been discussed yet, it is important, regarding the success relevance of the paradigm, to start up with the concept development and to go through all the defined modeling steps.

The system framework designed in this article serves as a future pathway both to meet the challenges of an intelligent logistics management nowadays and to cope with the complexity resulting from IT-supported logistics processes in a global, networked value creation chain.

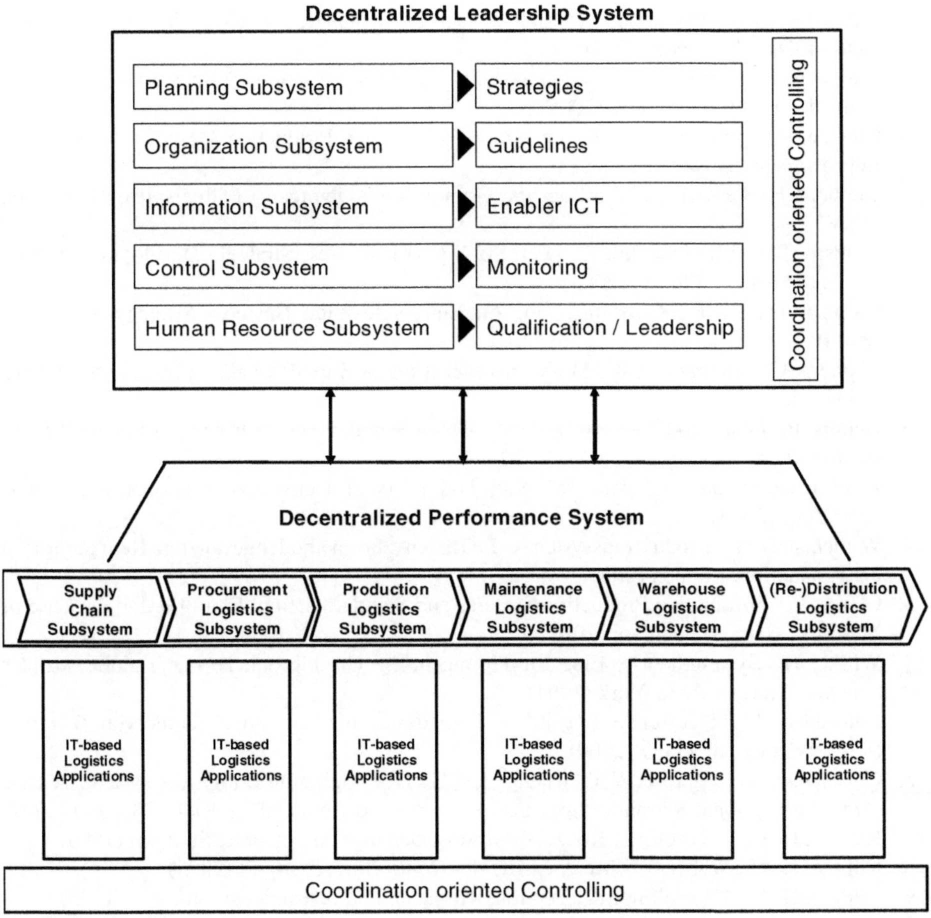

Fig. 4. Structure of an IT-based Logistics System

From a scientific point of view, the next steps would be to implement the IT-based Logistics System in praxis in order to get a feedback via different case studies, thus to proof the concept regarding its practicability and finally if necessary to modify it.

Additionally, the reference modeling can be done for logistics service provider whose business model is characterized by the core competence “production of logistics services”.

References

1. Atkinson, W.: How E-Logistics Changes Shipper-Carrier Relationships. Logistics management and distribution report April, E10–E14 (2001)
2. Azevedo, S.G., Ferreira, J., Leitão, J.: The Role of Logistics Information and Communication Technologies in Promoting Competitive Advantages of the Firm. The Icfai University Journal of Managerial Economics VI(3), 7–21 (2008)
3. Liker, J.K.: The Toyota Way: 14 Management Principles from the Worlds’ Greatest Manufacturer, New York et al (2004)
4. Clarke, C.: Automotive Production Systems and Standardisation: From Ford to the Case of Mercedes-Benz, Heidelberg (2005)
5. Chalice, R.: Improving Healthcare Using Toyota Lean Production Methods: 46 Steps for Improvement, Milwaukee (2007)
6. Gleißner, H., Femerling, J.C.: Logistik: Grundlagen – Übungen – Fallbeispiele, Wiesbaden (2008)
7. Goldsby, T.J., Martichenko, R.: Lean Six Sigma Logistics: Strategic Development to Operational Success, Florida (2005)
8. Gardner, D.J.: Mass Customization: An Enterprise-Wide Business Strategy, Cupertino (2009)
9. Chandra, C., Kamrani, A.K.: Mass Customization: A Supply Chain Approach, New York (2004)
10. Dennis, P.: Lean Production Simplified: A plain language guide to the world’s most powerful production system, New York (2007)
11. Baudin, M.: Lean Logistics: The Nuts and Bolts of Delivering Materials and Goods, New York (2005)
12. Wildemann, H.: Produktionssysteme: Leitfaden zur methodengestützten Reorganisation der Produktion, Munich (2009)
13. Clarke, C.: Automotive Production Systems and Standardisation: From Ford to the Case of Mercedes-Benz, Heidelberg (2005)
14. Willke, H.: Systemtheorie: Eine Einführung in die Grundprobleme der Theorie sozialer Systeme, Stuttgart/New York (1991)
15. Hausladen, I.: IT-gestützte Logistik – Logistikmanagement im Zeitalter von Bits und Bytes. Wist (9), 452–457 (2009)
16. Gunasekara, A., Ngai, E.W.T., Cheng, T.C.E.: Developing an e-logistics system: A case study. International Journal of Logistics: Research and Applications 10(4), 333–349 (2007)
17. Küpper, H.-U.: Controlling: Konzeption, Aufgaben und Instrumente, Stuttgart (1995)
18. Küpper, H.-U., Weber, J.: Grundbegriffe des Controlling, Stuttgart (1995)
19. Küpper, H.-U.: Controlling: Konzeption, Aufgaben und Instrumente, Stuttgart (1995)
20. Hausladen, I.: IT-gestützte Logistik – Logistikmanagement im Zeitalter von Bits und Bytes. WiSt (9), 452–457 (2009)

An Autonomous Control Concept for Production Logistics

Henning Rekersbrink, Bernd Scholz-Reiter, and Christian Zabel

BIBA - Bremer Institut für Produktion und Logistik GmbH, Hochschulring 20, 28359 Bremen, Germany
{rek,bsr,zbl}@biba.uni-bremen.de

Abstract. The German Collaborative Research Centre 637 'Autonomous Cooperating Logistic Processes' tries to make a paradigm shift from central planning to autonomous control in the field of logistics. Among other things, autonomous routing algorithms based on internet routing protocols are developed. The Distributed Logistics Routing Protocol (DLRP) was originally designed for transport networks to match goods and vehicles and to continuously make route decisions. Now the protocol was transferred to production logistics as a promising autonomous control method. The DLRP enables the abilities for logistic objects, orders and machines, to make own decisions with the information actually and locally available. In contrast to common scheduling algorithms, the DLRP is not a planning, but a control method with the capability for multiple, user defined optimization goals. The new autonomous control concept for production logistics will be presented in this paper and a first evaluation with common scheduling heuristics will be given.

Keywords: Flexible Flowshop, Scheduling, Dynamics, Autonomous Control.

1 Introduction

Due to growing dynamics and complexity of logistics systems, common concepts of hierarchical planning and control are questioned. A possible alternative is a shift from central planning to decentralized, autonomous control strategies (see e.g. [1], [2], [3], [4]). This concept of autonomous control is the main research area of the German Collaborative Research Centre 637 „Autonomous Cooperating Logistic Processes – A Paradigm Shift and its Limitations".

One possibility to implement an autonomous control strategy is to transfer existing routing protocols from data communication to similar routing problems in transport logistics. This idea lead to the development of a new autonomous control concept called *Distributed Logistics Routing Protocol* (DLRP). The DLRP was originally designed for transport logistics. The general idea was to make a paradigm shift from central planning to decentralized, autonomous control where each logistic entity is able to interact and decide autonomously (see [5], [6], [7]).

After the suitability and performance of the DLRP concept was shown for the field of transport logistics ([6]), it was obvious to transfer the basic concept to other logistic fields, like production control. Due to this transfer, two different names are

W. Dangelmaier et al. (Eds.): IHNS 2010, LNBIP 46, pp. 245–256, 2010.

introduced: DLRPt and DLRPp for transport and production respectively. The long term vision is here to connect all concepts to one system where a product order is able to route its way through a production environment, between production sites (global production networks) and after its completion through a transport environment to reach the costumer at the right time.

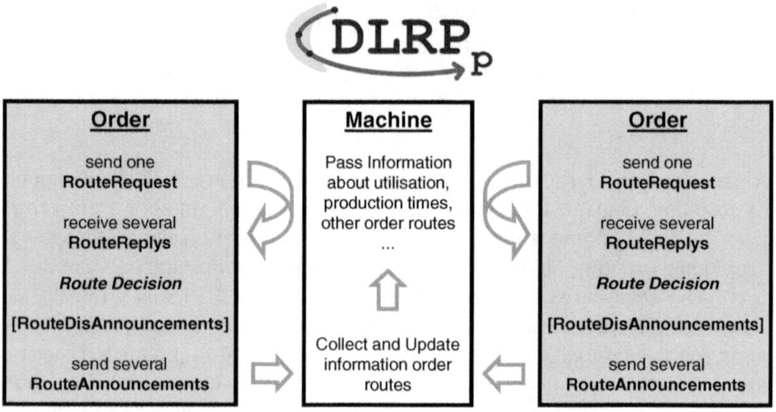

Fig. 1. Basic scheme for the Distributed Logistics Routing Protocol for Production

A brief overview about the DLRPp can be illustrated on the basis of fig. 1. When an order enters the system, it needs a route through the production environment. It sends a RouteRequest object to the next machine which fills it with necessary information. The machine now manifolds the RouteRequest object and sends ahead the objects to successional machines (not to all possible machines; this point is discussed later). These do the same until the last production step is reached. By this scheme each of the many RouteRequest objects represents a possible routes through the production set. At the last production step all the collected information are sent back to the order as a RouteReply object. This RouteReply object is basically a RouteRequest object on its way back to the order. The order receives several RouteReply objects as route alternatives by this discovery scheme. After its route decision, the order announces the new routes and possibly disannounces old routes from previous routing processes by sending RouteAnnouncement and RouteDisAnnouncement objects respectively. Together with the RouteAnnouncement, plenty of information can be passed to the relevant machines, like production times, setup times, probabilities, urgencies etc. Because of the ongoing processes, this scheme leads to a continuous cooperative structure - at any time there is enough information for any decision. Orders decide their preferred routes, machines decide their setup plan and the dispatching.

In contrast to classical scheduling algorithms, the DLRPp is designed for controlling an ongoing process. It does not need all information in advance – although it would be possible to announce roughly planned future orders to the system. The structure of the DLRP leads to some crucial advantages like adaptivity, robustness, possible manual interventions, estimation of future net states and an arbitrary decision process (see [7]).

In the next section, the DLRPp concept is presented in detail. Section 3 then gives a brief evaluation of the DLRPp by taking common scheduling algorithms as reference. The problem used for this evaluation is a dynamic flexible flowshop problem with unrelated parallel machines and sequence-dependent setup times. The paper closes with a conclusion of the evaluation and an outlook on further research and application possibilities.

2 Distributed Logistics Routing Protocol for Production

The core function of this autonomous control concept is to route orders through a production logistics scenario. Figure 2 exemplarily shows a simple flexible flowshop scenario. At this, some optimization criteria may be defined such as minimization of the throughput time or the maximization of the machine utilization.

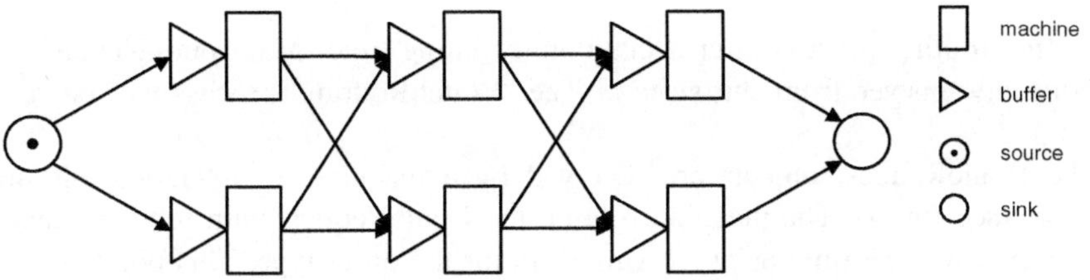

Fig. 2. Example of a production logistics scenario

In order to deal with all possible optimization criteria, all logistic objects involved have to be defined within the protocol. For this case, the DLRPp defines two logistic object types: machines and orders. The main attributes for these objects are shown in Table 1, where italic names are representatives for larger data structures.

In addition to logistic objects, there are four different data objects defined by the DLRP: RouteRequest, RouteReply, RouteAnnouncement and RouteDisAnnouncement. These objects are data packages, which are sent from one logistic object to another. Every single instance of these data objects belongs to one routing process of one order. To identify this process, every data object has the two attributes 'OwnerID' and 'RouteRequestID' according to the order attributes. The routing information itself is stored within these data objects as a special data structure, which may be user-defined according to the application.

To give a detailed description of the DLRPp, let's follow an order object through the system. Figures 3 and 4 show the procedures of the logistic objects, which are described in the following. Firstly, the order has to be initialized. This means that the order object does its first routing process. It gets a 'StartRouting'-message from the system (see fig. 3). The order is in 'no Routing'-state, so it creates a new RouteRequest object and sends it to all possible machines for the first production step (this information has to be defined within the 'Production Steps'-attribute). After that, it sends two delayed messages to itself. One is to ensure the next routing process after a maximum time interval (MaxRoutingInterval). The other one is to ensure the end of

Table 1. Main attributes of the logistic objects

Logistic Object: Order	
Name	*Type*
ID	string
Type	string
ReleaseTime	double
DueTime	double
Production Steps	
MaxRoutingInterval	double
MaxRoutingTime	double
MaxCountReceivedRouteReplys	integer
MaxAnnouncedRoutes	integer
Position	
State ('routing'/'no routing')	string
RouteRequestID	integer
CountReceivedRouteReplys	integer
Data from routing process	
...	

Logistic Object: Machine	
Name	*Type*
ID	string
OrderTypes	strings
RetoolingTimes	double
ProcessingTimes	double
TransportTimes	double
Neighbour-IDs	strings
Data for buffer content	
Data for announced orders	
Data for server	
...	

the actual routing process after a maximum routing time (MaxRoutingTime) – even without any answer from the system. The 'WatchRouting'-process can be retraced with fig. 3.

The RouteRequest objects are received by a machine as a 'GetRouteRequest'-message (see fig. 4). The machine fills up the RouteRequest with relevant data such as expected waiting time or processing time for the order type. This point (see '1' in fig. 4) is discussed below. The crucial point here is, that the machine writes its own ID, expected arrival and departure time into a list of the RouteRequest. By this procedure, the way of a RouteRequest object is collected and creates a possible route for the order.

If this machine is not the last production step for the order, it generates a recipient list of possible following machines for the order (this information must be placed within the RouteRequest). For the performance of the protocol, this is a very crucial point (see '2' in fig. 4) and is discussed below. Multiple copies are made of the RouteRequest object and they are sent to all recipients. These recipients follow the same scheme. When a machine is the last production step, it sends back the RouteRequest as a RouteReply to the order. This RouteReply object has now all data from all machines on its way.

The order now receives the RouteReply object as a 'GetRouteReply'-message. If the routing process is still active (state = 'Routing'), the data from the RouteReply is saved and the counter is counted up. When the order has received a maximum number of RouteReplys (MaxCountRouteReplys), the order sends an 'EndRouting'-message to itself. With this 'EndRouting'-message, the RouteReuestID is counted up and the order decides for a route (see '3' in fig. 3). The pool of received RouteReplys defines all route alternatives for this decision. This point is described below. After the routing decision, the order disannounces older route decisions and announces the actual routes at the corresponding machines by sending 'RouteDisAnnouncement'- and 'RouteAnnouncement'-objects.

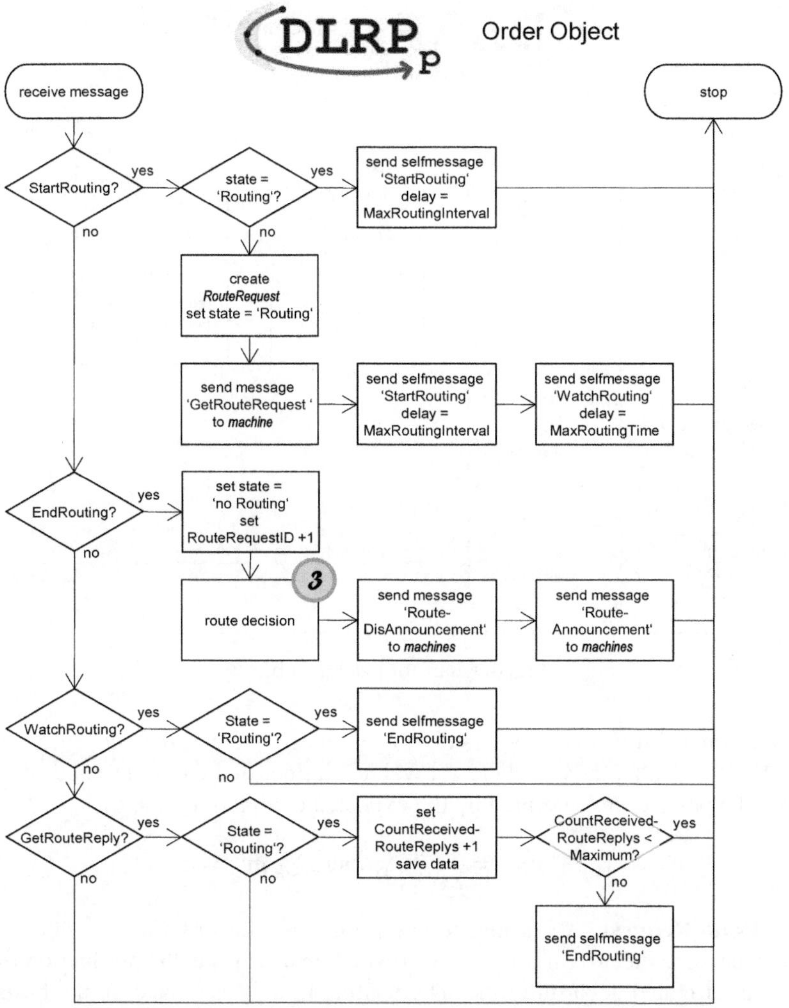

Fig. 3. Chart for order objects

In the following, some important points and parameters for the autonomous control are described. There are many different possibilities to configure the DLRPp at these points. These possibilities are the focus of actual and future work concerning the protocol.

Route Decision. The decision process itself is not a part of the DLRP. It may be user defined depending on the application. The route decision of an order object is based on the received RouteReplys, so the machines have to put the necessary information into the RouteRequest objects. For the evaluation presented below, the decision was made very simple for this first step. The smallest expected completion time C'_k is the only criterion for the decision. The completion time of an order is directly connected to its throughput time and the makespan (see below).

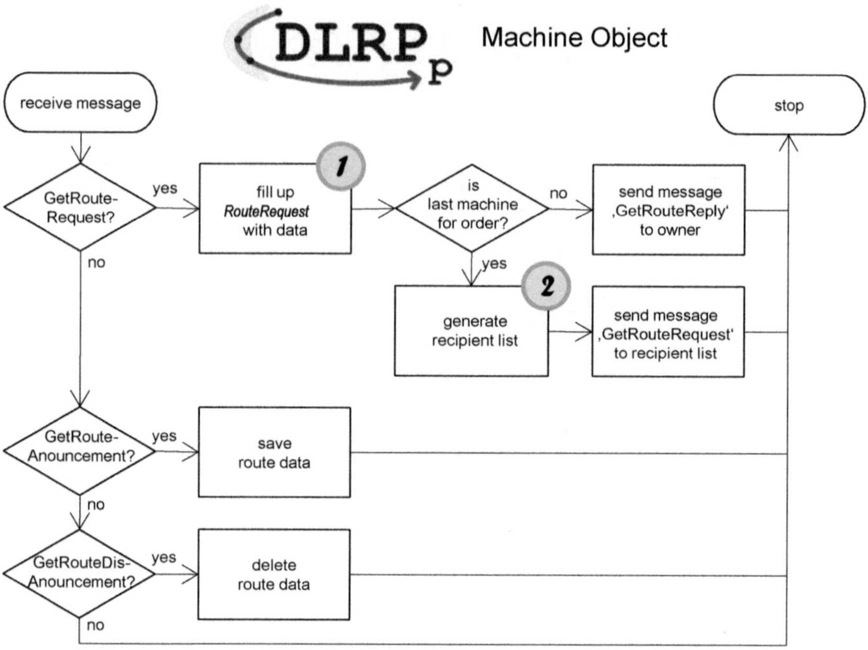

Fig. 4. Chart for machine objects

MaxAnnouncedRoutes. The orders do not announce one definite but several possible routes. For these alternatives, there are values defined for the preference pr_k for the route k. The preference is defined by the expected completion time C'_k:

$$pr_k = \frac{1/C'_k}{\sum_k 1/C'_k}, \text{ where } \sum_k pr_k = 1$$

Fill up RouteRequest. According to the route decision criteria, the values for the RouteRequest are the arrival time, the estimated waiting time, the production time and finally the estimated departure time. The waiting time is estimated on the basis of the actual orders in the machine buffer and the announced orders. The workload from the announced routes are allocated with the preference of the route.

MaxCountRouteReplys. This value is important for the route decision quality. When this value is too small, there may be only bad RouteReplys as alternatives.

Generate Recipient List. This is a deciding point for the DLRP. In small scenarios, it is the best to sent the RouteRequest to every possible next machine. But this would lead to a combinatory explosion. Therefore, an intelligent method for the reduction of the recipient list is needed. For the actual DLRP, this reduction is realized in a complex way and is still under investigation.

Order Path Decision. This decision process is not part of the DLRP. It may be user defined depending on the application. When an order is finished at a machine, it has to decide where to go next. Because of the announcement of several routes, there possibly are alternative machines as next step. This decision is made randomly, while the probability for the alternatives are set to the according preference of the route. At this point, it has to be ensured that all announced routes have a chance to be chosen. Otherwise the announcement of more than one route has no effect. The random decision is a simple possibility here and it also ensures that the route preference of an order has an effect. Other possibilities which include machine states, for example, are still under investigation.

Dispatching. This decision process is not part of the DLRP. It may be user defined depending on the application. The order with the shortest retooling time is taken as next order.

3 Evaluation

In order to evaluate this new autonomous control concept, a *dynamic flexible flowshop scenario with unrelated parallel machines* was chosen. Figure 2 sketches an exemplary scenario with three stages and two parallel machines. On the basis of this production scenario, a brief evaluation study for the DLRPp is presented below.

A crucial advantage of the DLRP is the ability to optimize multiple goals. The decision points can be user defined depending on the application. For a complete evaluation of the logistic performance of a system many key figures are relevant. Work in process, utilization, throughput time and due date reliability are some of these values (see e.g. [8]). The reference heuristics for this evaluation study try to minimise the makespan of the system as a single goal. In order to keep the comparability between the systems, the DLRP had to be simplified to minimise the makespan only.

3.1 Problem Description and Instances

In order to obtain comparability for this evaluation, an existing problem formulation from Jungwattanakit, Reodecha, Chaovalitwongse and Werner [9] is chosen. It defines a flexible flowshop problem with unrelated parallel machines and sequence-dependent setup times. The term dynamic denotes the nature of the job arrivals defined by Conway (see [10]). In a static problem a certain number of jobs arrive simultaneously. In a dynamic problem the shop is a continuous process - Jobs arrive intermittently. In addition to this problem formulation, order types were defined here. Order types assign orders into groups with equal production and setup times.

Consider a flexible flowshop system like sketched in fig. 2. There are T stages and M^t unrelated machines at every stage t. All J orders have to pass every stage from $t = 1$ to $t = T$, whereas the machine choice in every stage is free. There are S different order types defined for all orders. All order types have different processing times $p_{m,s}^t$ on the different machines and they have different sequence-dependent setup times $s_{m,u,s}^t$. The completion time C_j and the throughput time T_j are defined for the whole network at the end of last stage $t = T$.

Parameters:

J	number of orders
S	number of order types (indices s and u)
T	number of stages
M^t	number of parallel machines at stage t
r_j	release time of order j
$s^t_{m,u,s}$	setup time from order type u to s at machine m at stage t
ps^t_s	standard processing time of order type s at stage t
v^t_m	relative speed of machine m at stage
$p^t_{m,s}$	processing time of order type s on machine m at stage t where $p^t_{m,s} = ps^t_s / v^t_m$

Variables:

C_j	completion time of order j
C_{max}	makespan, where $C_{max} = \max(C_j)$
T_j	throughput time of order j; where $T_j = C_j - r_j$
TPT	mean throughput time; where $TPT = mean(T_j)$

The problem instances for this evaluation were chosen very close to the so called 'large size problems' defined by Jungwattanakit et al in [9].

- the number of orders J is 250
- the number of order types S is set to 10, they are uniformly distributed
- the number of stages T and the number of parallel machines M^t are set to the variants 3x5, 5x3, 5x10 and 10x5
- the standard processing times ps^t_s are integers uniformly distributed in the interval [1 50]
- the relative speeds v^t_m are uniformly distributed in [0.7 1.3]
- the setup times $s^t_{m,u,s}$ are integers uniformly distributed in [1 10]
- the release times r_j are all 0 (st, static) or a poisson process with λ orders per time; λ can be st, 0.5, 0.1, 0.075, 0.05, 0.025 for the small and 0.5, 0.1, 0.075, 0.05 for the larger scenarios (due to computation time). Lambda can be considered as a parameter for the dynamic or the workload of the scenario.

3.2 Reference Heuristics

The flexible flowshop scheduling problem itself is NP-hard (see e. g. [11]). Therefore there are no algorithms known so far for finding an optimal solution for a larger *dynamic* flowshop scheduling problem in feasible time, not to mention unrelated parallel machines. In order to find approximate solutions, many heuristic methods were developed for the classical flexible flowshop problem (see [12], [13] or [14] for an overview for the more general job shop scheduling). These heuristics were taken as basis for the development of heuristics for the dynamic flowshop problem.

Jungwattanakit et al made large simulation studies to evaluate some heuristics for the flexible flowshop problem mentioned (see [9]). They used the following sequencing algorithms as basis for their heuristics:

- PAL: a slope index heuristic by Palmer, see [15]
- CDS: a best choice heuristic by Campbell, Dudek and Smith, see [16]
- GUP: a slope index heuristic by Gupta, see [17]
- DAN: a heuristic by Dannenbring, see [18]
- NEH: a constructive heuristic by Nawaz, Enscore and Ham, see [19]

All heuristics try to minimise the makespan of a given problem instance. They are taken as reference algorithms to evaluate the new developed method.

3.3 Results

The problem instances were solved by all five reference algorithms and by a DLRPp simulation. The results of these calculations are shown in figures 5 and 6. Each point in the graphs represents an average value for five instances. To have an estimation for a lower bound for the makespan, an average of a minimum makespan is calculated:

$$\min C_{max} = \max \begin{pmatrix} \max(r_j) + T \cdot \text{mean}(p_{m,s}^t) + T \cdot \text{mean}(s_{m,u,s}^t) \\ J \cdot (\text{mean}(p_{m,s}^t) + \text{mean}(s_{m,u,s}^t))/M^t \end{pmatrix}$$

Similar to the makespan, a lower bound for the throughput time is estimated:

$$\min \text{TPT} = T \cdot \text{mean}(p_{m,s}^t) + T \cdot \text{mean}(s_{m,u,s}^t)$$

Concerning the makespan, fig. 5 shows that NEH is the best of the reference algorithms. For more static scenarios (larger λ or $\lambda = st$) and those with less stages than parallel machines, the DLRPp and the NEH are both close to the lower bound. One can say that they are more or less equal for these problem instances. For instances with more stages than parallel machines, the DLRPp gets better than the NEH heuristic only with more dynamic workload (smaller λ). This is quite reasonable, because the scheduling algorithms are originally made for static scenarios and they are best with smaller scenarios, because of the extraordinary increase of scheduling alternatives with larger scenarios. Around a reasonable system workload of $\lambda = 0.1$, the DLRPp gets better than the reference and it even gets near to the estimated minimum.

In fig. 6, one can see the main disadvantage of scheduling algorithms in a dynamic environment. They are made to optimize a schedule for a given set of orders, but they don not look on throughput times. For an ongoing production process, where orders continuously enter and leave the system, they cause high throughput times. The DLRPp in contrast is made for a continuous process - therefore the results are best for dynamic instances.

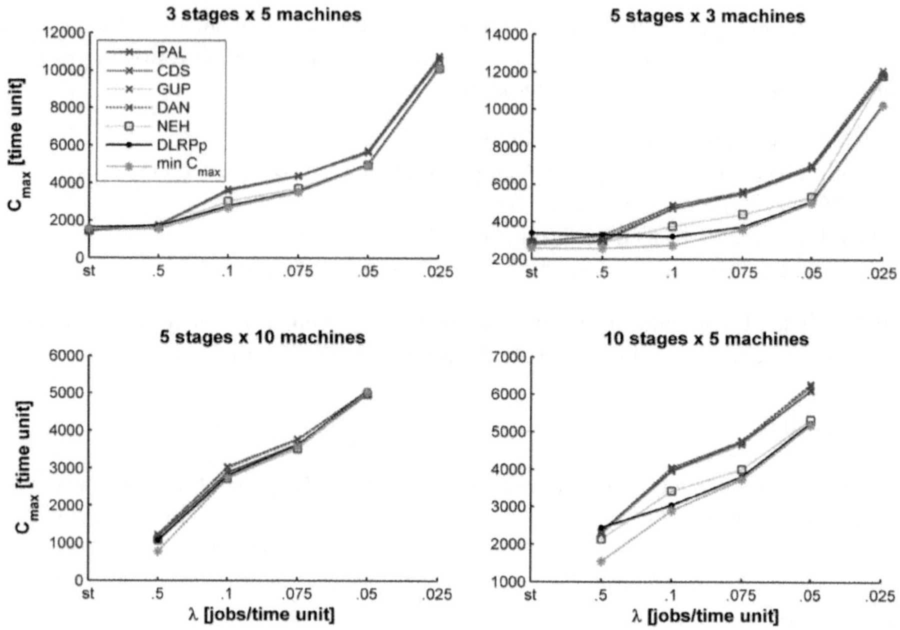

Fig. 5. Makespan against scenario size and workload

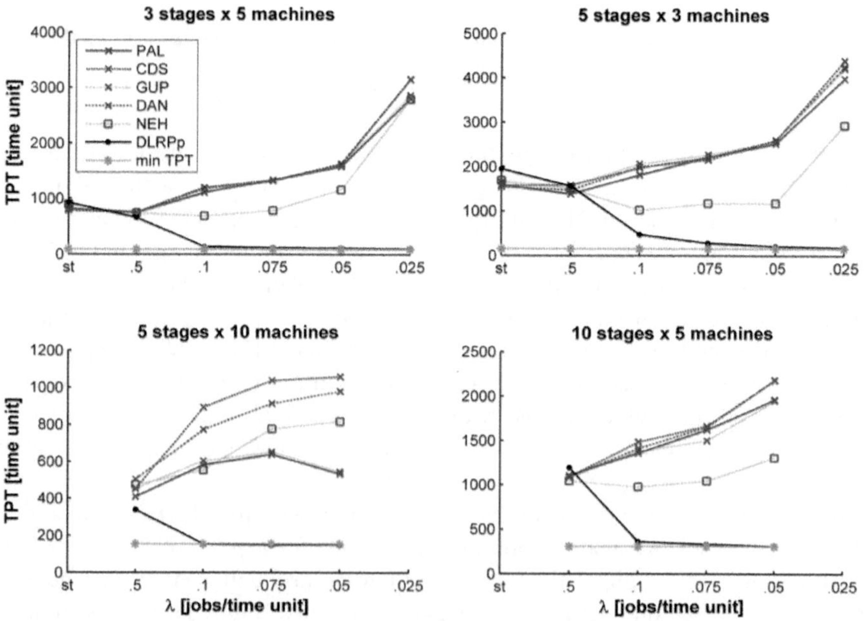

Fig. 6. Throughput time against scenario size and workload

4 Conclusion

A new autonomous control approach for production logistics was presented and the Distributed Logistics Routing Protocol for production was described in detail. The evaluation study showed, that the DLRPp is a promising alternative to common scheduling algorithms.

For the scenarios chosen, it is shown, that the DLRPp is nearly equal the reference algorithms for small and more static scenarios. For larger and more dynamic scenarios, the DLRPp shows better results. These results are rooted in the basic difference of the concepts: On the one hand prior static and centralized planning, which leads to good results in static environments. On the other hand the ongoing and decentralized concept of autonomous control, which leads to good results in large and dynamic environments. Additionally, it is shown that the DLRPp has crucial advantages with multiple optimization goals. Additional key figures and goals like tardiness or machine utilization could easily be implemented. As an ongoing autonomous control, it is plainly able to cope with dynamic environments and process disturbances. And in contrast to scheduling algorithms, the DLRP does not need all information in advance but is able to adapt to new incoming orders and new situations.

Future research will extend these investigations to other production topologies and different DLRPp variants. The deciding points of the DLRPp mentioned above are all under current investigation. Additionally an extended evaluation study will be made, e. g. with additional reference algorithms with rolling horizon planning concepts.

The autonomous control concept DLRP is not only useful with the special scenarios presented, but it can cope with many different logistic tasks. After transportation logistics and now production logistics, it is suggestive to transfer the protocol to more tasks like assembly or warehousing. The long term vision is here to connect all concepts to one integrative autonomous control system.

Acknowledgments. This research is founded by the German Research Foundation within the Collaborative Research Centre 637 'Autonomous Cooperating Logistic Processes – A Paradigm Shift and Its Limitations'.

References

1. Scholz-Reiter, B., Görges, M., Philipp, T.: Autonomously controlled production systems-Influence of autonomous control level on logistic performance. CIRP Annals – Manufacturing Technology 58(1), 395–398 (2009)
2. Windt, K., Jeken, O.: Allocation Flexibility – A New Flexibility Type as an Enabler for Autonomous Control in Production Logistics. In: 42nd CIRP Conference on Manufacturing Systems (2009)
3. Scholz-Reiter, B., Teucke, M., Özsahin, M.-E., Sowade, S.: Smart Label – supported Autonomous Supply Chain Control in the Apparel Industry. In: 5th International Congress on Logistics and SCM Systems (ICLS 2009), pp. 44–52 (2009)
4. Böse, F., Piotrowski, J., Scholz-Reiter, B.: Autonomously controlled Storage Management in Vehicle Logistics - Applications of RFID and Mobile Computing Systems. International Journal of RF Technologies: Research and Applications 1(1), 57–76 (2009)

5. Freitag, M., Herzog, O., Scholz-Reiter, B.: Selbststeuerung logistischer Prozesse – Ein Paradigmenwechsel und seine Grenzen. Industrie Management 20(1), 23–27 (2004)
6. Rekersbrink, H., Makuschewitz, T., Scholz-Reiter, B.: A distributed routing concept for vehicle routing problems. Logistics Research 1(1), 45–52 (2008)
7. Scholz-Reiter, B., Rekersbrink, H., Freitag, M.: Internet routing protocols as an autonomous control approach for transport networks. In: 5th CIRP International Seminar on Intelligent Computation in Manufacturing Engineering, pp. 341–345 (2006)
8. Philipp, T., Böse, F., Windt, K.: Evaluation of Autonomously Controlled Logistic Processes. In: 5th CIRP International Seminar on Intelligent Computation in Manufacturing Engineering, pp. 347–352 (2006)
9. Jungwattanakit, J., Reodecha, M., Chaovalitwongse, P., Werner, F.: An evaluation of sequencing heuristics for flexible flowshop scheduling problems with unrelated parallel machines and dual criteria. Otto-von-Guericke-Universität Magdeburg, 1-23 (2005) (Preprint May 28)
10. Conway, R.W., Maxwell, W.L., Miller, L.W.: Theory of scheduling. Addison-Wesley, Reading (1967)
11. Garey, M.R., Johnson, D.S.: Computers and Intractability: A Guide to the Theory of NP-Completeness. Freeman, San Francisco (1979)
12. Park, Y.B., Pegden, C.D., Enscore, E.E.: Survey and evaluation of static flowshop scheduling heuristics. International Journal of Production Research 22, 127–141 (1984)
13. Ruiz, R., Maroto, C.: A comprehensive review and evaluation of permutation flowshop heuristics. European Journal of Operational Research 165, 479–494 (2005)
14. Jones, A., Rabelo, L.C.: Survey of Job Shop Scheduling Techniques. National Institute of Standards and Technology (2000)
15. Palmer, D.S.: Sequencing jobs through a multi-stage process in the minimum total time - a quick method of obtaining a near optimum. Operational Research Quarterly 16(1), 101–107 (1965)
16. Campbell, H.G., Dudek, R.A., Smith, M.L.: A heuristic algorithm for the n-job m-machine sequencing problem. Management Science 16(10), 630–637 (1970)
17. Gupta, J.N.D.: A functional heuristic algorithm for the flow-shop scheduling problem. Operations Research Quarterly 22(1), 39–47 (1971)
18. Dannenbring, D.G.: An evaluation of flow shop sequencing heuristics. Management Science 23(11), 1174–1182 (1977)
19. Nawaz, M., Enscore Jr., E., Ham, I.: A heuristic algorithm for the m-machine, n-job flowshop sequencing problem. OMEGA 11(1), 91–95 (1983)

Towards Agile Business Processes Based on the Internet of Things

Benedikt Schmidt and Markus Schief

SAP Research Darmstadt, Bleichstrasse 8,
64283 Darmstadt, Germany
{benedikt.schmidt,markus.schief}@sap.com

Abstract. Though agility is a core demand for firms, software based process modeling and execution approaches realize strict and inflexible formalization. We propose an extension of process modeling and execution based on the integration of real world information as trigger for process variance. As decisions are knowledge-intensive they are propagated to knowledge workers to support the decision process. These activities are again tracked, to identify demand for process redesign. Thus, processes become basis for proactive support and a representation of reality, even in change-ridden domains.

Keywords: agile business processes, internet of things, knowledge-intensive.

1 Introduction

In today's world, speed of business environment changes is rapidly increasing. So, companies need to enable agile business processes. One key prerequisite for agility is transparency. The integration of valuable information is hence a key success factor. Agile business processes can be defined by tackling two information blind-spots that will be identified in this paper: a lack of real-world object information integration in process-driven business environments (blind spot 1) and detection of actual compared to planned process execution (blind spot 2). Firstly, the evolution of the Internet of Things (IoT) for reality based object information increases transparency. Secondly, transparency is needed in terms of process execution deviations as an indicator that current processes are too static. According to our research, both blind spots are particularly relevant to operational, complex and knowledge-intensive processes.

2 Initial Example

Motivation and Status quo. In production logistics, just-in-time (JIT) arrangements as an interface between production processes and JIT-supplier logistics are important for process efficiency. By leveraging the information quality at this interface, processes such as Receiving Inspection can be improved. An exemplary smart phone producer relying on a JIT strategy supposes displays to arrive in time and in quality. Today, employees at receiving inspection run checks according to their best judgment or based on stochastic approaches such as parts per million checks. If quality

W. Dangelmaier et al. (Eds.): IHNS 2010, LNBIP 46, pp. 257–262, 2010.

deviations, for instance due to vibrations during transport, are not revealed they hamper succeeding production processes. If they are revealed, follow-up measures have to be defined. A production manager in charge has to decide if acceptance of delivery has to be denied and if production processes have to be adapted. In any case, information has to be aggregated in order to adjust supplier classifications for future contract negotiations such as suppliers' lot sizes. The impact on various business processes is hence tremendous.

Opportunities. Transparency is a key prerequisite for any proactive measure as well as reactive follow-up activities. The earlier a potential deviation can be identified and communicated, the better process scheduling can be adjusted, particularly in a JIT-environment. At best, quality deviations are avoided at all based on a continuous root-cause analysis. For instance, transport containers could be improved. Secondly, quality deviations are indicated by respective logistic partners, proactively. Smart tags on transport containers can monitor vibrations, for example. At the latest, deviations in quality should be revealed at receiving inspection. In any case, instant and valuable information is needed on an information granularity that is as specific as possible while still economically reasonable. By that, flexibility of business processes can be improved accordingly. With that respect, knowledge workers being in charge of process adjustments need to be supported. Information retrieval, handling, distribution, and integration are hence in the sweet spot of any potential solution.

3 Big Picture

Business process management (BPM) means supporting business processes by using methods, techniques, and software to design, enact, control, and analyze operational processes involving humans, organizations, applications, documents and other sources of information [1]. Workflow engines are used to model and run business processes. Workflows are used to automate business processes in whole or parts according to a set of procedural rules [2]. Many attempts to add dynamics and flexibility to processes and workflows have been introduced, always in the light of automation: e.g. change patterns, change support techniques and business rule integration [3]. Still, only few idealized process traces integrated in static models. These are foundation for process instances to coordinate and control reality. Two blind spots of current approaches are of specific importance (see Fig. 1.):

Blind-spot 1. "Lack of transparency and timely delay with respect to real world information in a firm's respective environments". Environmental factors that have a tight relationship to company processes are only partly tracked and often with a delay in time. Likewise, the broader environment of physical states of things, regulations etc is generally completely out of standard business processes' scope.

Blind-spot 2. "Lack of transparency with respect to actual process execution on functional level". The actual execution process of human tasks is neither modeled, nor is a feedback mechanism of actual execution integrated. The resulting intransparency of the execution process might conceal a discrepancy between designed process activities and actual process execution.

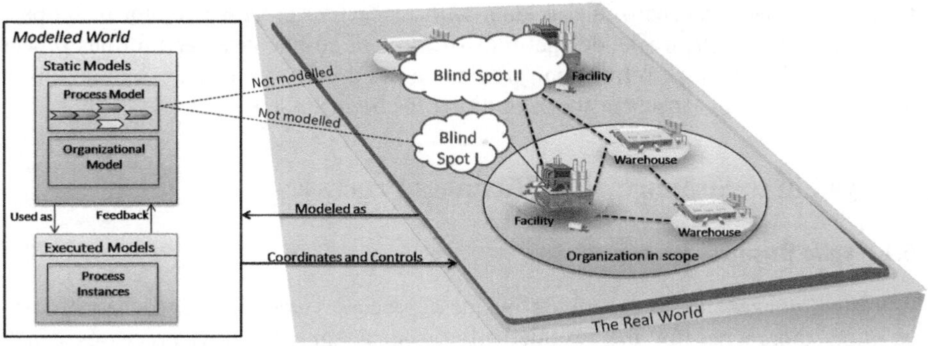

Fig. 1. Transition in business process model to reflect blind-spot information

Processes shall become a reflection of real world execution which also hold true in dynamic domains beyond pure production workflows. Therefore, we introduce an extension of process modeling, which integrates awareness of the environment and awareness of the actual process execution to include triggers for process variance. This information, captured in static models generates executed models of processes which coordinate and control processes based on actual data.

4 State of the Art

4.1 Blind Spot 1 – Awareness of the Environment

The term IoT encapsulates a variety of sensor based technologies which provide event and status information to connect information systems and the physical-world-based sensor data. Such real world integration still lacks means to connect such devices and integrate them in a standardized way into existing business process management systems. We see two demands: standardized collection and business oriented processing of the accessible information. [4] proposes smart tags to create a general object memory, standardizing the storage and distribution of information collected during the lifetime of a product. The IoT generates high value event streams of data, which need to be organized and processed. [5] describes complex event processing (CEP) as useful approach to identify patterns of interest in streams of real world information, enabling the aggregation of such information to events. Such events need to be contextualized with respect to the business. [6] proposes an extended modeling approach integrating event and process modeling.

4.2 Blind Spot 2 - Awareness of the Actual Process Execution

The second identified blind-spot reflects the transparency of human-process interaction. The discovery of such execution processes demands the externalization of activities. [7] generate Task Patterns which mediate weakly structured processes by means of abstracted and personalized information based on ad-hoc task objects. [8] uses sensors to track user activity on the operating system. Thereby, the data is used to generate task models and provide execution support once a known task is detected.

Connecting such externalized task data with respective processes is not in scope of these approaches. [9] tracks delegation structures of ad-hoc tasks and refines process models. Although such refinement provides a good basis for process extraction, the identification of all aspects of the execution process is missed.

5 System to Support Agile Business Processes

5.1 Agile Business Processes

A system concept to support agile business processes was developed based on the solution approaches for the revealed blind-spots. Agile business processes can be realized in business process design, planning, and execution.

During business process instance planning and execution, employees need support in various dimensions. Firstly, instance specific information is needed. Speed and relevance are core quality properties that have to be met by information inputs. With that respect, IoT is expected to be the information input channel that will leverage existing process transparency, significantly. Likewise, the design of business processes can be improved. For instance, information over certain time intervals is very important to identify continuous deviations between designed business processes and actual business process instances. Information revealed from either of both blind-spots can significantly support the business process design.

In any case, dynamic design, planning, and execution need to be supported. The combination of information, supporting tools such as modeling environments, and business context enables knowledge-workers to implement agile business processes. The definition of different business process variants should be as easy and fast as possible. User context detection is one of the key elements in this area. In order to ensure consistency of the modeled business processes, certain business constraints should be indicated as well. Finally, IoT-based pro-active solutions can be achieved by defining leading indicators.

5.2 Human Process Interaction Platform

As detection and handling of process variance cannot be fully automated, supporting human process interaction is a core demand for agile business processes. We propose the integration of proactive delivery of process related IoT information in a controlling and task execution platform. Current human process interaction focuses on task lists. Delivery of necessary information and support of execution process externalization are usually not in focus. We realize it by a process linked dashboard with whiteboard functionalities for task-specific service and information composition. Organization of working tasks, adaptation of processes, and problem identification are supported by the composition of information objects, KPIs, webservices, and widgets. General whiteboard capabilities are provided such as spatial arrangement of objects, annotations, and collaboration by whiteboard sharing. The whiteboards persist execution knowledge in objects arrangements that can be re-used in future process instances. User activities are externalized based on desktop sensors to identify the users' plans and support them by recommendations.

5.3 Real World Integration

Real world transparency is one of the key prerequisites in our solution concept. Relevant information sources as well as information retrieval rules have to be identified. All kinds of events that are retrieved via the IoT have to be defined, analyzed, structured, and aggregated. In the context of the ADiWa project (www.adiwa.net) a general event classification was developed as shown in Fig. 2.

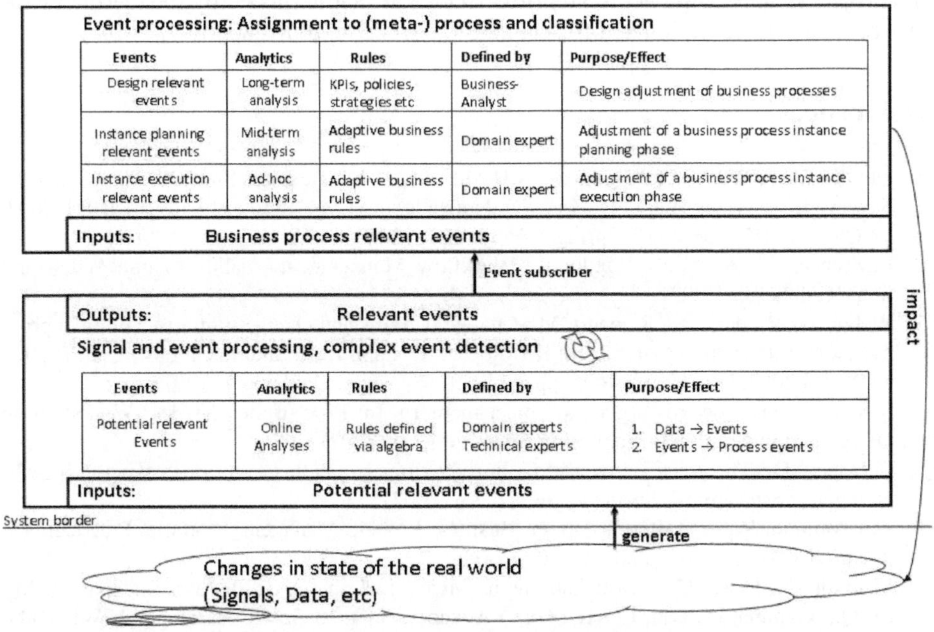

Fig. 2. Classification of events and the impact thereof

Potential relevant events are those events that are generated out of the real world along a pre-defined set of input information sources. Via technology-based algorithms such as CEP relevant events will be identified, structured and aggregated. Relevant events, then, present the information source for event subscribers. The retrieved events are input parameters for the agile business process adjustments, which can be performed on a business process design or a business process instance level. Referring to the initial example, smart tags transmit vibration events and hence production planning can be adjusted on an instance and meta level. The environmental blind spot is hence transformed into event based inputs for agile business processes.

6 Outlook

In the present paper we have described an approach for enabling IoT-based agile business processes. Therefore, we have extended business process models by information on variance and triggers for variance. To integrate these aspects we have shown that two blind-spot need to be brightened: the environment and the actual process activities. We have described concepts how to realize these challenges in a system, using internet of things integration and a dedicated process execution environment. Within the ADiWa project, the described ideas are currently refined and will be realized in a prototype system. This will be used to evaluate the applicability of the described approach.

Acknowledgement. The work presented in this paper was developed in the context of the project Allianz Digitaler Warenfluss (ADiWa) that is funded by the German Federal Ministry of Education and Research. Support code: 01IA08006.

References

1. van der Aalst, W.M.P., ter Hofstede, A.H.M., Weske, M.: Business Process Management: A Survey. In: van der Aalst, W.M.P., ter Hofstede, A.H.M., Weske, M. (eds.) BPM 2003. LNCS, vol. 2678, pp. 1–12. Springer, Heidelberg (2003)
2. Lawrence, P.: Workflow Handbook, Workflow Management Coalition. John Wiley and Sons, Chichester (1997)
3. Weber, B., Rinderle, S., Reichert, M.: Change patterns and change support features in process-aware information systems. In: Krogstie, J., Opdahl, A.L., Sindre, G. (eds.) CAiSE 2007 and WES 2007. LNCS, vol. 4495, pp. 574–588. Springer, Heidelberg (2007)
4. Schneider, M.: Towards a general object memory. In: Proc of the First Int. Workshop on Design and Integration Principles for Smart Objects (2007)
5. Luckham, D.: The power of events: an introduction to complex event processing in distributed enterprise systems. Springer, Heidelberg (2002)
6. von Ammon, R., et al.: Event-Driven Business Process Management and its Practical Application Taking the Example of DHL (2008)
7. Schmidt, B., Riss, U.V.: Task Patterns as Means to Experience Sharing. In: Spaniol, M., Li, Q., Klamma, R., Lau, R.W.H. (eds.) Advances in Web Based Learning – ICWL 2009. LNCS, vol. 5686, pp. 353–362. Springer, Heidelberg (2009)
8. Lokaiczyk, R., Faatz, A., Beckhaus, A., Goertz, M.: Enhancing just-in-time e-learning through machine learning on desktop context sensors. In: Kokinov, B., Richardson, D.C., Roth-Berghofer, T.R., Vieu, L., et al. (eds.) CONTEXT 2007. LNCS (LNAI), vol. 4635, pp. 330–341. Springer, Heidelberg (2007)
9. Stoitsev, T., et al.: Enabling End Users to Proactively Tailor Underspecified, Human-centric Business Processes. In: Filipe, J., Cordeiro, J. (eds.) ICEIS 2008. LNBIP, vol. 19, pp. 307–320. Springer, Heidelberg (2008)

Methods for the Calculation of CO_2 Emissions in Logistics Activities

Hartmut Zadek and Robert Schulz

University Magdeburg "Otto-von-Guericke", Chair of Logistics, Institute of Logistics and Material Handling Systems, Universitaetsplatz 2, 39106 Magdeburg, Germany
Zadek@ovgu.de, Robert.Schulz@ovgu.de

Abstract. This paper outlines general methods, calculation models and some available software for the calculation of carbon dioxide (CO_2) emissions in logistics activities. General calculation formulas are presented mostly for transportation by truck, but by train as well. Listed are CO_2 emission factors (EFs) for the different types of fuels and partially average numbers for the various transport carriers. The relevance of the calculation of CO_2 emissions in the logistics sector is described to show the growing importance of sustainability in the world of logistics.

Keywords: sustainability, Carbon dioxide (CO_2), emission factors (EFs), CO_2 equivalents, CO_2 emissions, CO_2 balance, life cycle assessment (LCA), fuel-based method, distance-based method.

1 Introduction

According to the International Energy Agency (IEA) the transportation sector is responsible for approximately 23% of the global carbon dioxide (CO_2) emissions [1]. However, there is quite a big energy savings potential in those transportation activities, too. In many cases these potentials cannot be used because of economical reasons. Another problem is the identification of energy savings potentials which could often times lead to a cost reduction.

A lot of companies calculate the carbon footprint of their products and their supply chains (SCs) to identify these potentials. This paper describes which methods, calculation models or software tools are used to calculate CO_2 emissions in the SCs of industrial companies or logistics service providers (LSP). For those firms intending to calculate their CO_2 emissions this paper gives an overview of how the possible calculations.

2 Existing Methods for the Calculation of CO_2 Emissions

2.1 The Greenhouse Gas Protocol

The Greenhouse Gas Protocol (GHGP) is one of the most important standards for the calculation of CO_2 emissions. It offers the general instructions to accomplish a CO_2 balance. The GHGP distinguishes between direct and indirect emissions (see Fig. 1).

W. Dangelmaier et al. (Eds.): IHNS 2010, LNBIP 46, pp. 263–268, 2010.

Scope 1 covers all direct emissions occurring from sources owned or controlled by the company. Scope 1 does not include direct CO_2 emissions from the combustion of biomass. This kind of emissions shall be reported separately. Greenhouse Gas (GHG) emissions resulting from the generation of purchased electricity consumed by equipment or operations owned or controlled by the company is included in scope 2. Thus, these emissions account for the largest part of GHG emissions in many companies and are the best chance to reduce these emissions. Last, scope 3 covers all other indirect GHG emissions. According to the GHGP scope 3 emissions are optional and not caused directly by the company itself, but indirectly by its activities [2]. However, these are the most important activities in the logistics sector. A standard for the calculation of scope 3 emissions is still missing, but its development is in progress and will be finished in December 2010.

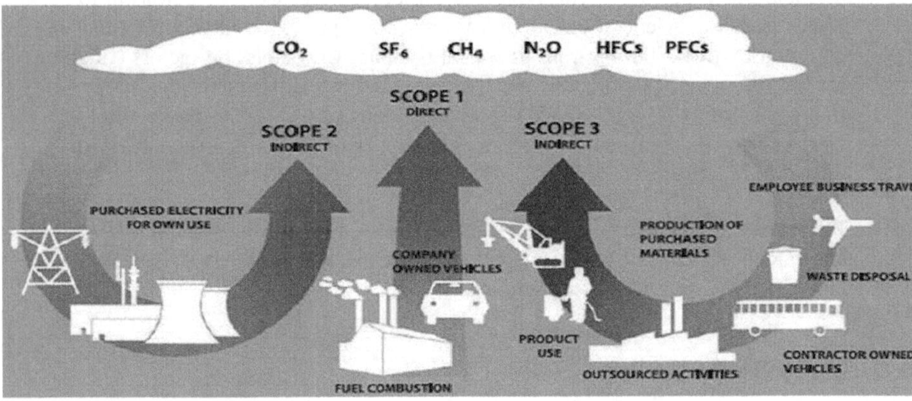

Fig. 1. Overview of scopes and emissions across a value chain [2], Fig. 3, p. 26

There are two possibilities to calculate the CO_2 emissions of mobile sources: the fuel-based method and the distance-based method. Both methods use CO_2 or CO_2 equivalent emission factors (EFs) for the calculation of CO_2 emissions. These factors express the calculated ratio between GHG emissions and activity data, for example fuel consumption and distance traveled. EFs can also include the emissions of the previous SC, for example emissions occurring during the production of diesel or its transport to the gas station.

If the absolute fuel consumption or the distance traveled and the average fuel consumption (AFC) are available the fuel-based method can be applied. The data for the distance-based method is usually available from records of odometer logs or company fleet records [3]. To calculate the absolute fuel consumption the following formula is to be deployed:

$$\text{Absolute fuel consumption} = \text{distance traveled} \cdot \text{average fuel consumption}\,. \tag{1}$$

The CO_2 emissions in the fuel-based method depend on the heating value of the specific fuel, the share of the oxidized carbon in the combustion (usually between 99 %

and 100 %, thus simplified to 100 %), and the carbon content of the fuel. This correlation is expressed by formula 2. Examples for different fuel types are shown in Table 1.

$$CO_2 \text{ emissions} = \text{absolute fuel consumption} \cdot \text{heating value} \cdot \text{emission factor} . \quad (2)$$

Table 1. Examples of heating values and emission factors for different fuel types

Fuel type	**Heating value [GJ / l]**	**Emission Factor [kg CO_2 / GJ]**	**Emission Factor [kg CO_2 / l]**
Gasoline / Petrol	0.0344	69.25	2.3822
Diesel	0.0371	74.01	2.7458
Propan	0.0240	62.99	1.5118
Kerosene	0.0357	71.45	2.5508

The distance-based method is used when the distance traveled is available but the AFC is missing. In this case the distance-based EFs (in ton kilometer - tkm, person km - pkm or vehicle - vkm) are required for the calculation of CO_2 emissions. Formula 3 expresses this correlation. Table 2 shows examples for EFs of different vehicle types.

$$CO_2 \text{ emissions} = \text{distance traveled} \cdot \text{emission factor} . \quad (3)$$

Table 2. Examples of emission factors for different vehicle types in the distance-based method

Vehicle Type	**Average Fuel Consumption [l / 100 km]**	**Emission Factor [kg CO_2 / km]**
Gasoline light truck	16.8	0.4002
Gasoline heavy truck	39.2	0.9338
Diesel light truck	15.7	0.4311
Diesel heavy truck	33.6	0.9226

Another method for the calculation of transports will be introduced in the standard for scope 3 emissions: the activity-based method which uses cargo transport activity based EFs (e.g. g / tkm). Formula 4 shows the calculation. In addition, the new standard will introduce the calculation of emissions from storage (see formula 5) and terminal activities.

$$CO_2 \text{ emissions} = \text{transported weight} \cdot \text{distance traveled} \cdot \text{emission factor} . \quad (4)$$

$$CO_2 \text{ emissions} = \text{storage days} \cdot \text{emission factor} . \quad (5)$$

2.2 Databases for Emission Factors

One main database for EFs in the road transportation sector is the Handbook Emission Factor for road transport (HBEFA). HBEFA provides data related to the emissions of several air pollutants such as carbon monoxide (CO), CO_2, hydrocarbons (HC), nitrogen oxides (NO_x) and particles based on g per vkm. The results of the calculation depend on the vehicle category (car, light-duty vehicle, heavy-duty vehicle, public service bus, coach, and motorcycle), the emission category (hot, start, etc.), the fuel type (gas, diesel), and the base year (1990 to 2020). In addition, the professional version of HBEFA distinguishes between different traffic situations, pitch attitudes, the driving style of the driver, and engine categories. HBEFA does not calculate the amount of emissions, but the EFs. An example for 2005 is shown in Table 3 [4].

Table 3. Sample emission factors from HBEFA

Year	Vehicle category	Pollutant	Fuel	Emission factor	Unit
2005	Heavy-duty vehicle	CO_2	diesel	685.63	[g/vkm]
2005	Light-duty vehicle	CO_2	diesel	245.997	[g/vkm]
2005	Light-duty vehicle	CO_2	gas	250.959	[g/vkm]

Another important database especially for the life cycle assessment (LCA) is the Global Emission Model for Integrated Systems (GEMIS). GEMIS is a tool for the assessment of emissions and resource consumption over the life cycle of products and processes. The database offers information on energy sources (fossil fuels, renewable energies, nuclear energy, biomass and hydrogen), processes for electricity and heat production, sourcing of materials (especially raw materials), and transportation processes (cars, trucks, trains, ships, airplanes, etc.). All relevant processes are included in the calculation, e.g. from raw material sourcing to the waste treatment. For each process the following data is available [5]: efficiency, power, capacity factor, lifetime, direct air pollutants (SO_2, NO_x, CO, etc.), GHG emissions (CO_2, CH_4, N_2O, etc.), solid wastes (ashes, overburdens, process wastes and others), liquid pollutants, and land use. In addition, a calculation of costs is possible with GEMIS since key figures for fuels, energy and transportation systems (investment and operating costs) are available.

2.3 CO_2 Calculation Tools

At this point some tools for the calculation of CO_2 emissions shall be introduced. The first one is the Transport Emission Model (TREMOD). TREMOD was developed by the Institute For Energy and Environmental Research (IFEU), Heidelberg, Germany. For the calculation of CO_2 emissions arising from road traffic TREMOD relates to the HBEFA. TREMOD takes all traffic performances on road, rail, water and air into account for different scenarios between 1960 and 2030. The results for the assessed traffic performances are the primary energy consumption (EC) (in MJ), the end EC (in t or kWh), and emissions (for example CO_2). The model distinguishes between direct and indirect emissions. Direct emissions result from the operation of the vehicle, whereas indirect emissions arise during the upstream energy and SC [6].

Another tool for the calculation of CO_2 emissions is the Ecological Transport Information Tool (EcoTransIT) which was developed by the IFEU, the Rail Management

Consultants GmbH and by several European rail companies. EcoTransIT uses data from HBEFA and TREMOD and is available online [7]. It is possible to calculate the ecological impacts of a transport for a given origin and destination in Europe. The basis for the calculation are the flow direction of rivers, the fuel type, the energy mix in each country, the topography, the infrastructure, and the transportation distance. For the transports by airplane the straight-line distance is multiplied with 1.1 to take the start and landing processes into account. The mass and volume of the goods determine the needed quantity of means of transportation. Thus the more vehicles are required the more energy is used.

The underlying calculation formulas are described in the German logistics magazine "VerkehrsRundschau" [8]. For each mode of transportation formula 6 applies. The specific EC can be given as l per 100 km, Wh per tkm, g per tkm, etc. Formula 7 is required for the calculation of emissions from road transportation caused by additional empty trips.

$$CO_2 \text{ emissions} = \text{energy consumption}_{\text{specific}} \cdot \text{emission factor} . \quad (6)$$

$$EC_{\text{net load}} = EC_{\text{empty}} + (EC_{\text{empty}} - EC_{\text{full}}) \cdot (\text{weight load} / \text{load capacity}) . \quad (7)$$

3 Conclusion

A number of methods, calculation models and software tools for the measurement of CO_2 emissions already exist, especially for traffic and transportation activities. But first, there is no standard companies can rely on. Second, the existing tools are not completely elaborated yet. Last, the attribution to the causer is difficult since most transports are executed by LSPs and not by the company itself. For the LSP it is complicated to attribute the CO_2 emissions to a specific client because the cargo of one truck consists of goods of multiple clients in many cases. Vice versa it is difficult to provide the clients with the relevant data. Hence standards for the calculation of CO_2 emissions must be developed and it must be assured that the emissions can be attributed to the causer. Not only LSPs can be made accountable for CO_2 emissions and their reduction.

Since most transports are realized by LSPs these companies must be proactive in developing a standard approach for each means of transportation, e.g. truck, train, ship, and airplane because the calculation varies between these means of transportation. A good example is the train. The CO_2 emissions depend on the length of the train, the type of transported goods, the topography, and the electricity mix in the specific country. At the moment the CO_2 calculation formulas do not distinguish between the means of transportation. The basis for an approach to do so should be an international recognized standard like the Greenhouse Gas Protocol. The development of a GHGP standard for scope 3 emissions is in progress. The EFs of this future standard should depend on the specific means of transportation and whether a LSP operates nationally or internationally.

Some companies have started to calculate the CO_2 footprint of their products but this is not target-oriented. Logistics managers must be able to calculate the CO_2

emissions of the SC network or the CO_2 emissions of the transports allocated to their customers. Once the basis for the calculation by customer is established a monitoring tool can be implemented.

In the future it will be very important for the LSP to be ecologically efficient because this will be one judgment criteria of their customers. Furthermore, it is relevant to take CO_2 emissions into account during the design of a SC network, in addition to the costs and service levels. Knowing about the calculation of CO_2 emissions will become necessary due to future worldwide legislative decisions. Every company must calculate the company's overall CO_2 emissions and will be judged by its efforts to reduce them.

References

1. International Energy Agency: Transport, Energy and CO_2: Moving Toward Sustainability. OECD Publishing (2009)
2. World Business Council for Sustainable Development, World Resource Institute: The Greenhouse Gas Protocol - A Corporate Accounting and Reporting Standard. Washington (2004)
3. Greenhouse Gas Protocol Initiative: Calculating CO_2 Emissions from Mobile Sources, guidance, http://www.ghgprotocol.org/calculation-tools/all-tools
4. Departments for Environment of Germany, Austria, and Switzerland: Handbook emission factors for road transport (HBEFA), http://www.hbefa.net/
5. Oekoinstitut Darmstadt, Germany, http://www.oeko.de/service/gemis/en/index.htm
6. Knörr, W., Reuter, C., et al.: Daten- und Rechenmodell: Schadstoffemissionen aus dem motorisierten Verkehr in Deutschland 1980 - 2020. Heidelberg/Dessau (2005)
7. EcoTransIT, http://www.ecotransit.org/railion/index.en.phtml
8. Rundschau, V.: CO2 Berechnung (42-48), Verlag Heinrich Vogel. Springer, Munich (2009)

A p-Robust Capacitated Network Design Model with Facility Disruptions

Zuli Liu[1], Songshan Guo[1], Lawrence V. Snyder[2], Andrew Lim[3], and Peng Peng[3]

[1] Sun Yat-sen University
Guangzhou, China, 510275, P.R. China
liuzuli@gmail.com
[2] Lehigh University
200 West Packer Avenue, Mohler Lab Bethlehem, PA 18015, USA
larry.snyder@lehigh.edu
[3] City University of Hong Kong, Tat Chee Avenue, Kowloon, Hong Kong SAR
adxypeng@cityu.edu.hk

Abstract. This paper studies a strategic supply chain management problem of designing robust networks which perform well under both normal condition and disruptions. A mix-integer programming model which incorporates p-robust measure is presented. The objective is to minimize the total nominal cost, while setting upperbounds on relative regrets in disruption scenarios. A GA-based hybrid metaheuristic algorithm is proposed and tested. Computational results demonstrate that system robustness can be substantially improved with little increase in cost. Our solution is also less conservative compared with common robustness measures.

Keywords: Network Design, Facility Disruption, p-Robust, Metaheuristics.

1 Introduction

Recent events have highlighted supply chain risks arising from facility disruptions. Hurricanes Katrina and Rita trembled U.S. oil production and refining industry in 2005 [13]. Ericsson lost 400 million euros after an eight-minute fire at a supplier's semiconductor plant, and eventually gave up the mobile phone market in 2000 [12]. Facilities are forced to be shut down temporarily during disruptions, and it may take a long time to recover and resume operation. Failing to consider disruption in the design phase, and thus lack of countermeasures may result in huge costs during a disruption event and serious ripple consequences. The popularization of "lean" concept which allows minimum redundancy and the development of global supply chains has exacerbated the problem. Empirical studies [7][8] have shown significant decline in sales growth and stock returns for companies after major disruption events.

An emerging research topic is to design supply chain systems which attain continuity and efficiency in presence of all sorts of disruptions. Tang [19] discussed

W. Dangelmaier et al. (Eds.): IHNS 2010, LNBIP 46, pp. 269–280, 2010.

the need and strategies for mitigating supply chain disruptions. Drezner [4] studied unreliable p-median and (p, q)-center location problems, assuming that a facility has a given probability of becoming inactive. Snyder and Daskin [15] formulated the reliable models for facility location problems, in which supply facilities may fail with certain probabilities and customers are assigned to suppliers in a hierarchical fashion. These location problems are special cases of network design problems with two echelons, while designing general networks will result in much complex network structures and larger problem sizes. Bunschuh et al. [3] studied multistage network design problems, in which they tried to improve network robustness by creating supplier redundancy. Snyder et al. [18] summarized a variety of analytical models developed for supply chain network design problems under threat of disruptions.

The requirement for accurate estimation of the probability distribution of disruptions has restricted the application of above stochastic programming approach in practice. Robust optimization provides an alternative approach by which no probability has to be estimated. Kouvelis and Yu [11] proposed a framework for robust discrete optimization and introduced two widely considered robustness measures: *minimax cost* and *minimax regret*, which minimize maximum cost or regret across all scenarios. A major drawback of this approach is being overly pessimistic by planning against worst-case scenarios which rarely occur. To enable the companies to create a flexible capability to design the supply chain around both long-term goals and short-term unexpected events, we try to minimize the nominal cost, which is the operational costs under normal conditions, while adopting the p-robust measure to set bounds on the relative regret during disruption. The p-robust measure was introduced in a facility layout problem [10] and later in uncapacitated network design problems [6]. Snyder and Daskin [17] first adopted the term "p-robust" and formulated stochastic models for two classical facility location problems.

To the best of our knowledge, we are the first to take this approach in designing general robust supply chain network to insure high-level performance during disruption. We follow the definition of p-robust by Snyder and Daskin [17]. For a given set of scenarios S, let P_s be the deterministic minimization problem for scenario s, which is a fixed-charge capacitated network design problem and c_s^* be the optimal objective value for P_s. Let $\mathbf{X}$ be a feasible network design and ,$\mathbf{Y}$ be the corresponding optimal flows, and $c_s(\mathbf{X}, \mathbf{Y})$ be the corresponding objective value. $(\mathbf{X}, \mathbf{Y})$ is called p-robust if for all $s \in S$,

$$\frac{c_s(\mathbf{X}, \mathbf{Y}) - c_s^*}{c_s^*} \leq p_s \tag{1}$$

or equivalently,

$$c_s(\mathbf{X}, \mathbf{Y}) \leq (1 + p_s) c_s^* \tag{2}$$

where $p_s \geq 0$ is a given constant, indicating the desired robustness level for scenario s. The left-hand side of (1) is the relative regret for scenario s. The p-robust measure is to set bounds on maximum allowable relative regret for each scenario. For the sake of convenience, we assume that $p_s = p$ for all $s \in S$. All of our results can be extended to the more general case.

The rest of the paper is organized as follows. In Section 2, the p-robust network design model is formulated. A GA-based hybrid metaheuristic algorithm is proposed in Section 3. We report the numerical results in Section 4, as well as management insights drawn from these results. Section 5 concludes our studies.

2 The p-Robust Network Design Model

We propose the following mix-integer programming model for the p-robust network design problem. Suppose that we have a general network $(\mathcal{V}, \mathcal{A})$. Let $\mathcal{V}_S, \mathcal{V}_T, \mathcal{V}_D$ denote the sets of supply, transshipment and demand nodes, respectively. For notational convenience, define $\mathcal{V}_0 = \mathcal{V}_S \cup \mathcal{V}_T$ as the set of all supply and transshipment nodes. Let S be the set of scenarios in which different sets of facilities fail simultaneously. The scenarios can be identified by examining the potential threats to the supply chain, and how the facilities will be affected. For example, facilities which are located in the same area have a higher probability of becoming inactive together during disruption events such as earthquake, electricity shortage and so on. Then we have a good reason to consider a scenario in which all facilities in the region are disrupted. Specifically, denote $s = 0$ as the nominal scenario, i.e., the scenario with no disruption. We assume that all unmet demands will be satisfied by an emergency facility that is always open, with unlimited capacity but much higher costs. This can be interpreted as outsourcing from another supplier, strategic emergency buffers or lost-sale costs.

Let f_j be the fixed cost to open node $j \in \mathcal{V}_0$; q_{ij} be the unit transportation cost on arc $(i, j) \in \mathcal{A}$. Let s_j denote the supply $\forall j \in \mathcal{V}_S$, d_j be the demand $\forall j \in \mathcal{V}_S$ and k_j be the capacity $\forall j \in \mathcal{V}_0$. We use a_{js} to indicate whether facility j is disrupted in scenario s ("1") or not ("0"). Note that in general, a_{js} may take any value in $[0, 1]$, which represents that the facility only suffers partial loss of capacity during disruption. The robustness coefficient p is the maximum allowable relative regret. c_s^* is the optimal scenario cost for scenario s.

For each $j \in \mathcal{V}_0$, define a binary variable X_j, which is 1 when facility j is opened and 0 otherwise. Finally, for every $(i, j) \in \mathcal{A}$ and $s \in S$, define a non-negative variable Y_{ijs} to represent the flow from node i to node j. The p-robust network design model is then formulated as follows:

$$\text{Minimize} \quad \sum_{j \in \mathcal{V}_0} f_j X_j + \sum_{(i,j) \in \mathcal{A}} q_{ij} Y_{ij0} \tag{3}$$

Subject to

$$\sum_{j \in \mathcal{V}_0} f_j X_j + \sum_{(i,j) \in \mathcal{A}} q_{ij} Y_{ijs} \leq (1 + p) c_s^* \quad \forall s \tag{4}$$

$$\sum_{(j,i) \in \mathcal{A}} Y_{jis} - \sum_{(i,j) \in \mathcal{A}} Y_{ijs} = b_j \quad \forall j, s \tag{5}$$

$$\sum_{(j,i)\in\mathcal{A}} Y_{jis} \leq (1 - a_{js})k_j X_j \quad \forall j, s \tag{6}$$

$$X_j \in \{0,1\} \quad \forall j \tag{7}$$

$$Y_{ijs} \geq 0 \quad \forall i, j, s \tag{8}$$

where b_j equals s_j for all supply nodes, $-d_j$ for all demand nodes and 0 for all transshipment nodes.

The objective function (3) minimizes the nominal cost, including fixed location costs and transportation costs. Constraints (4) enforce the p-robust criterion, such that the scenario cost should not be more than $(1+p)\%$ of the optimal scenario cost c_s^*. Constraints (5) are the flow conservation constraints. Constraints (6) ensure that the total flow through a node can not exceed its capacity when it is opened and fully functional in that scenario, while no flow is permitted when it is closed or disrupted. Constraints (7) and (8) are the standard binary and nonnegativity constraints.

3 GA-Based Hybrid Metaheuristic Algorithm

Algorithms based on Lagrangian relaxation and Benders decomposition have been proposed to solve the p-robust models [6,10,17]. However, as the problem size grows exponentially with the network size and the number of scenarios, the previous methods will not be as efficient. Practitioners are more interested in obtaining quality near-optimal solutions in relatively short time. Genetic Algorithm(GA) is a powerful global search heuristics inspired by evolution theory(Holland [9]). GA has gained much popularity for ease of implementation and successful application in a wide variety of optimization and search problems [5][16]. Unlike other heuristics such as tabu search, which work on improving one single solution, GA keeps a pool of solutions and perform comparatively little work on each one.

We propose a hybrid metaheuristic algorithm which combines genetic algorithm, local improvement and shortest augmenting path method. A major improvement to the basic GA scheme is that for each generation, we apply a local search procedure called *learning* to improve the average population fitness. Numerical experiments have shown that the algorithm outperforms CPLEX in terms of CPU time and solution quality for most data sets. The algorithm can be easily adapted to solving extensions of the problem, such as additional constraints or defining different objective functions.

3.1 Representation Scheme

We use a n-digit binary string chromosome structure to represent a solution $\mathbf{X}$, where n is the number of candidate sites. The i-th digit on the string indicates whether the ith facility is open ("1") or not ("0").

In order to expedite the later local improvement procedure, the digits on a chromosome are sorted by two criteria: (1) *Cost-saving of each facility.* We first assume all candidate sites $j \in \mathcal{V}_0$ are opened and set to their original capacities k_j, then solve the min-cost flow problem(MCFP) and obtain the optimal flow cost c_f^*. For each facility j, we increase its capacity by 1 unit with all other facilities remain unchanged, and solve the resulting MCFP again for the optimal flow cost $c_f^*(j)$. The cost-saving for facility j is defined as $c_f^* - c_f^*(j)$. We solve the MCFP $|\mathcal{V}_0| + 1$ times to calculate cost-saving for all candidate facilities. The more cost a facility can save, the higher probability it will be opened in the optimal solution. (2) *Total times of opening.* We solve for optimal scenario solutions at the beginning to get c_s^*, for all $s \in S$. For each candidate sites, we count the number of scenarios in which it is opened in the optimal solution. The more times a facility is chosen to be opened, the it will be more likely to be open in the optimal master problem solution. We sort the facility indexes on the binary string non-increasingly first by (1) and then by (2).

3.2 Fitness Function

For a specific solution $\mathbf{X}$, we solve the MCFP to get optimal flow assignments $\mathbf{Y}$ by shortest augmenting path method [1] for each scenario, and obtain optimal objective value $c_0(\mathbf{X}, \mathbf{Y})$, as well as the minimum scenario costs $c_s(\mathbf{X}, \mathbf{Y})$. The object function is a commonly used fitness function to justify the quality of a solution $\mathbf{X}$. However, it cannot determine the fitness of the solutions which are infeasible, i.e., have violated some p-robust constraints. Therefore, we introduce a modified fitness function as follows:

$$f(\mathbf{X}) = c_s(\mathbf{X}, \mathbf{Y}) + \omega \cdot \sqrt{\sum_{s \in S} [\mathbf{1}_A(c_s(\mathbf{X}, \mathbf{Y}) - (1+p)c_s^*)]^2} \qquad (9)$$

where c_s^* is the optimal scenario cost, $\mathbf{1}_A(x)$ is a indicator function defined as

$$\mathbf{1}_A(x) = \begin{cases} x & \text{If } x > 0 \\ 0 & \text{Otherwise} \end{cases} \qquad (10)$$

A smaller value indicates better fitness. If a solution $\mathbf{X}$ is feasible, the second part in the equation is 0 and the fitness value is exactly the optimal objective value; if a solution violates some p-robust constraints, its fitness will be penalized by the regret from the scenario optimum. Finally, the value of ω is a factor that the decision-maker can adjust to weigh the penalty of violating the p-robust constraints in different scenarios. We set $\omega = 50$ in numerical tests.

3.3 Initial Population

The experimental work by Anlander [2] suggests that a value between n and $2n$ is optimal for the population size, where n is the length of a chromosome ($\mathcal{V}_0$) in our case). We use 100 as population size (*pop_size*), which is within $[n, 2n]$ for most of the data sets we have tested.

The initial population is generated by first selecting the optimal solutions of all scenarios by solving the scenario subproblems P_s. If we have more than *pop_size* scenarios, choose the first *pop_size* best solutions. If the number of scenarios is smaller than *pop_size*, the rest of individuals are generated by crossover between randomly drawn pairs from the scenario optimal solutions.

3.4 GA Operators

We perform the *crossover* and *mutation* operations after the learning process. Crossover starts by randomly pairing up the individuals from current population. For each pair, two new individuals are generated by switching some randomly chosen digits of the same position on the strings. A probability is drawn uniformly from set $\{0, 10\%, ..., 90\%, 1\}$ to decide whether these digits are to be switched or not. Mutation serves as a mechanism to prevent solutions from being trapped in local optima and to ensure a diverse population. Each individual has a probability of 20% of being chosen for mutation. If selected, one digit will be randomly selected and changed from 1 to 0 or 0 to 1.

3.5 The Learning Process

Before the crossover and mutation operation, we perform a *learning* process, which is a local improvement procedure for all individuals in each iteration. The philosophy behind this process is that we want to "train" the individuals to be more "stronger", thus increasing the probability of producing better quality offsprings. Local improvement heuristics have added a great deal of power to GAs shown by various researchers(e.g., [16]). The process has greatly shortened the time needed for convergence in GA, therefore, improved the solution speed.

Constructing Search Tree. We construct a *search tree* with the individual to take the learning process as the root node. The children nodes are obtained by setting the "1"s to "0"s one by one, starting from the end of the chromosome. Therefore, the number of children for a node is the number of "1"s in it. A sample of a complete search tree with "11100" as root node is illustrated in Fig. 1. Note that there are duplicated branches if a search tree is constructed in this way, so that we will check and discard the duplicated branches during the search.

Local Search. We apply the *depth-first-search(DFS)* algorithm to traverse the search tree. The search starts from the root node, and explores as far as possible along each branch before backtracking. To accelerate the search procedure, We adopt Deterministic Annealing(DA) [14] with slightly modification as the

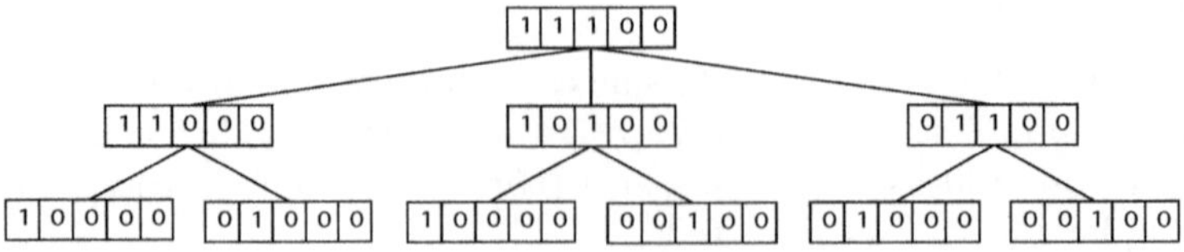

Fig. 1. Search Tree

pruning criterion. Let the current node be $\mathbf{X}$ with fitness $f(\mathbf{X})$. We calculate fitness $f(\mathbf{X}')$ for its first child $\mathbf{X}'$, then compare $f(\mathbf{X}')$ with $f(\mathbf{X})$. In traditional DA, we will only search $\mathbf{X}'$ and its children if $f(\mathbf{X}') < \delta f(\mathbf{X})$, where $\delta > 1$ is a given gap. In our algorithm, δ is initially set to 1 and kept being updated during the search. We will search $\mathbf{X}'$ if $f(\mathbf{X}') < f(\mathbf{X})$. δ will then be updated to $f(\mathbf{X}')/f(\mathbf{X})$. After searching node $\mathbf{X}'$ and its children, we then consider the second child $\mathbf{X}''$ of $\mathbf{X}$. Similarly, we will search $\mathbf{X}''$ only if $f(\mathbf{X}'') < \delta f(\mathbf{X})$, and update δ if needed.

One potential problem is that it is possible to prune good descents of node $\mathbf{X}$, which may even contain the optimal solution. However, we use the crossover and mutation operations to create new individuals as a compromise. Numerical test does not show significant effects on performance, i.e. in most of the test instances, we are able to find the optimal or high quality near-optimal ($gap < 5\%$) solutions. Furthermore, We have set a time limit of 5 seconds for the learning process, and replace root node solution with the solution of best fitness at termination. We have tried longer time, i.e., 20 seconds, but the improvement was not obvious.

3.6 Termination Criteria

In numerical test, we have observed that the best objective value converges in 20 generations most of the time. Therefore, we terminate the algorithm after 30 iterations, or the improvement to the best solution is less than 10^{-5} for 10 consecutive iterations. We also set a time limit of 1200 seconds and optimality gap of 5% to terminate the algorithm. Note that we do not require the best solution to be kept in population to avoid being trapped in local optima, but we always keep record of the best solution during one execution. At termination, we compare the best solution in current population with the best solution kept in record, and return the better one.

4 Computational Results

Numerical experiments have been performed to evaluate the performance of our algorithm. The algorithm was coded in C++ and executed on a computer with Intel Xeon E5430 2.66GHz processor and 2 gigabyte of RAM, operating under Microsoft Windows XP Professional SP2. We benchmark our results to solution by branch-and-bound algorithm in CPLEX 11.0, which runs in the same environment. Computing times are reported in seconds.

4.1 Experimental Design

We have generated 46 data sets of different problem sizes based on total number of facilities, scenarios and the edge density. In detail, we first choose the number of supply, transshipment and demand nodes, respectively; then construct arcs between nodes based on probability specified by the edge density.

The fixed costs f_j and unit transportation costs q_{ij} are drawn uniformly from $[5000, 15000]$ and $[1, 500]$. The emergency cost is 1500. The demands d_j are drawn uniformly from [50, 110]. Supplies s_j are determined as follows. We first calculate the average supply as $\overline{s} = (|\mathcal{V}_D|/|\mathcal{V}_S|) \cdot \overline{d}$, where $\overline{d}$ is the average demand. Each s_j is then drawn uniformly from $[1.5\overline{s}, 2.5\overline{s}]$. The capacity of a supply node is the same as its supply, while the capacity of each transshipment node is determined similarly by first calculating the average capacity $\overline{c} = (|\mathcal{V}_D|/|\mathcal{V}_T|) \cdot \overline{d}$, and then drawn uniformly from $[1.5\overline{c}, 2.5\overline{c}]$. We generate disruption scenarios randomly with facility failure rate of 10%. We set $p = 0.15$ for all instances.

4.2 Algorithm Performance

We compare the performance of our algorithm with that of CPLEX in Table 1. The "Test Problem" column gives the problem setting, in which the "Density" column gives the density of the arcs, the "Problem Size" column describes problem sizes. The four numbers linked by " – " are the numbers of supply, transshipment and demand nodes and number of scenarios, respectively. Test results including the running time("Time"), objective value("Cost") and Optimality gap("Gap"). Results are compared in column "Comparison", where "Time" column gives the fraction of time that our algorithm requires to obtain solution at least as good as that given by CPLEX, and the "Gap" column gives the difference between the optimality gap given by our algorithm and CPLEX. A negative value indicates that our algorithm has found a better solution.

From the results we can see that when compared with CPLEX, our algorithm is able to find the same or better solutions for 34(or 73.9%) of the 46 data sets, by taking only a small fraction of CPLEX's time(26.8% on average). There are three instances for which CPLEX could not return a 5% solution in 1200 seconds, while our algorithm has found near optimal solutions in much shorter time.

4.3 Comparison with Other Robustness Criteria

Widely studied robustness measures minimax cost and minimax regret [11] are usually too conservative from a managerial stand-point, by protecting against merely the worst-case scenarios. Our modeling approach is to think more from the business side. Minimizing the nominal cost is more suitable goal for the companies in long-term, while limiting the scenario costs greatly reduces the short-term risk when exposed to disruptions. This approach is less conservative compared with other robustness measures, as demonstrated in numerical tests(See Table 2). We have chosen a set of randomly generated data of size 20%-10-20-40-20. The optimal objective values by the p-robust model, minimax cost and regret models are compared and the differences(in %) are given in parentheses. We can see that the minimax cost model only protects the worst-case scenario, which is scenario 13 in this case, and has much higher cost in other scenarios. Compared with the minimax regret model, the p-robust model results in lower costs in all but four scenarios. Another advantage of the p-robust measure is that it provides decision makers more flexibility by allowing them to adjust the p values, while the minimax regret approach only gives one solution which minimizes the regret over all scenarios. We will next discuss this in detail.

Table 1. Algorithm Performance

Test Problem		CPLEX			Heuristic			Comparison	
Density	Problem Size	Time	Cost	Gap	Time	Cost	Gap	Time	Gap
20%	10-20-30-10	2.7	1089309	0.0213	0.6	1083225	0.0156	22.22%	-0.0057
	10-20-30-20	4.4	1083397	0.0158	1.1	1100347	0.0316	25.00%	0.0159
	10-20-30-30	7.3	1083225	0.0120	0.8	1098120	0.0260	10.96%	0.0139
	10-20-40-10	2.1	1301360	0.0082	0.6	1297906	0.0055	28.57%	-0.0027
	10-20-40-20	4.0	1308196	0.0112	0.9	1297906	0.0033	22.50%	-0.0080
	10-20-40-30	6.6	1307370	0.0072	1.9	1307370	0.0072	28.79%	0
	20-30-50-10	4.1	1445399	0.0736	1.9	1356548	0.0076	46.34%	-0.0660
	20-30-50-20	11.6	1352886	0.0048	3.7	1356548	0.0076	31.90%	0.0027
	20-30-50-30	48.0	1357762	0.0084	5.1	1356548	0.0075	10.63%	-0.0009
	20-30-50-40	31.3	1363529	0.0042	3.5	1367487	0.0072	11.18%	0.0029
	20-40-60-10	10.2	1265404	0.0082	2.8	1265404	0.0082	27.45%	0
	20-40-60-20	22.8	1286958	0.0248	5.2	1267788	0.0095	22.81%	-0.0153
	20-40-60-30	41.9	1290052	0.0188	5.7	1277615	0.0090	13.60%	-0.0098
	20-40-60-40	159.1	1306235	0.0308	7.1	1277615	0.0082	4.46%	-0.0226
	40-50-60-10	207.8	1061264	0.0154	175.4	1069501	0.0233	84.41%	0.0079
	40-50-60-20	1200	N/A	N/A	176.7	1116626	0.0585	14.73%	N/A
	40-50-60-30	976.7	1084595	0.0180	807.4	1097533	0.0301	82.67%	0.0121
	40-50-60-40	1200	N/A	N/A	359.1	1115546	0.0465	29.93%	N/A
30%	10-20-30-10	2.3	775589	0.0223	0.7	772216	0.0178	30.43%	-0.0044
	10-20-30-20	4.9	786124	0.0362	1.4	772216	0.0179	28.69%	-0.0183
	10-20-30-30	9.8	777349	0.0232	2.3	774351	0.0193	23.47%	-0.0039
	10-20-40-10	3.6	837035	0.0185	0.5	836723	0.0181	13.89%	-0.0004
	10-20-40-20	6.9	837585	0.0136	1.1	836723	0.0126	15.94%	-0.0010
	10-20-40-30	14.2	848359	0.0263	1.6	836723	0.0123	11.27%	-0.0141
	20-30-50-10	22.4	937162	0.0197	2.4	929912	0.0118	10.71%	-0.0079
	20-30-50-20	79.5	947762	0.0213	74.5	944494	0.0177	93.71%	-0.0035
	20-30-50-30	62.5	952897	0.0311	7.6	951886	0.0300	12.16%	-0.0011
	20-30-50-40	664.0	972631	0.0302	223.3	972183	0.0297	33.63%	-0.0005
	20-40-60-10	32.7	946954	0.0047	4.5	944588	0.0022	13.76%	-0.0025
	20-40-60-20	46.6	980612	0.0226	8.9	985606	0.0278	19.10%	0.0052
	20-40-60-30	170.4	983037	0.0205	140.1	987159	0.0247	82.22%	0.0043
	20-40-60-40	1200	N/A	N/A	17.7	1020961	0.0547	1.48%	N/A
	40-50-60-10	89.1	908885	0.0679	44.4	894260	0.0507	49.83%	-0.0172
	40-50-60-20	733.2	915288	0.0335	432.3	943158	0.0649	58.96%	0.0315
50%	10-20-30-10	3.8	540019	0.0278	0.6	528375	0.0057	15.79%	-0.0222
	10-20-30-20	7.2	548073	0.0247	1.1	541350	0.0121	15.28%	-0.0126
	10-20-30-30	22.9	547223	0.0173	1.9	545787	0.0146	8.30%	-0.0027
	10-20-40-10	5.8	666770	0.0244	1.1	665422	0.0223	18.97%	-0.0021
	10-20-40-20	10.8	696778	0.0698	2.3	667780	0.0253	21.30%	-0.0445
	10-20-40-30	24.4	672728	0.0309	3.0	670613	0.0277	12.30%	-0.0032
	20-30-50-10	46.4	700737	0.0251	19.5	697477	0.0203	42.03%	-0.0048
	20-30-50-20	170.8	697477	0.0225	13.1	697477	0.0225	7.67%	0
	20-30-50-30	258.9	706586	0.0315	15.8	707212	0.0324	6.10%	0.0009
	20-30-50-40	227.4	706654	0.0334	17.7	709919	0.0382	7.78%	0.0048
	20-40-60-10	89.5	694156	0.0173	26.0	701026	0.0274	29.05%	0.0101
	20-40-60-20	570.5	707029	0.0321	177.4	705661	0.0301	31.10%	-0.0020

Table 2. Comparison with other Robust Criteria

S	*p*-robust	Minmax Cost	Minmax Reg.	S	*p*-robust	Minmax Cost	Minmax Reg.
0	1304742	1380486(5.81)	1340636(2.75)				
1	1417387	1536837(8.43)	1435828(1.30)	11	1304742	1381811(5.91)	1340636(2.75)
2	1536831	1718892(11.85)	1539185(0.15)	12	1342785	1471419(9.58)	1381522(2.88)
3	1424161	1616208(13.48)	1420871(-0.23)	13	1843482	1763248(-4.35)	1816693(-1.45)
4	1395719	1471926(5.46)	1408501(0.92)	14	1488826	1511722(1.54)	1502212(0.90)
5	1520669	1681322(10.56)	1558724(2.50)	15	1394089	1502552(7.78)	1430055(2.58)
6	1574351	1530685(-2.77)	1562348(-0.76)	16	1414395	1477127(4.44)	1441102(1.89)
7	1526260	1575124(3.20)	1555332(1.90)	17	1336351	1492141(11.66)	1373204(2.76)
8	1368789	1477890(7.97)	1401844(2.41)	18	1304742	1380486(5.81)	1340636(2.75)
9	1338619	1424519(6.42)	1374447(2.68)	19	1513518	1713979(13.24)	1476344(-2.46)
10	1338619	1424519(6.42)	1374447(2.68)	20	1417273	1502552(6.02)	1430055(0.90)

Table 3. Nominal Cost vs. Maximum Regret

Problem $P1$(20-40-60-20-20%)					Problem $P2$(30-40-50-20-20%)				
p	Cost	% Inc.	Max Reg.	% Dec.	p	Cost	% Inc.	Max Reg.	% Dec.
∞	1081873	N/A	0.6926	N/A	∞	1240547	N/A	0.6856	N/A
0.6925	1084742	0.27%	0.5972	13.75%	0.6855	1240634	0.01%	0.5497	19.79%
0.5972	1087107	0.22%	0.5100	14.60%	0.5497	1241184	0.04%	0.5463	0.60%
0.5099	1089461	0.22%	0.4325	15.17%	0.5463	1244359	0.26%	0.5344	2.16%
0.4325	1089760	0.03%	0.4213	2.57%	0.5344	1244441	0.01%	0.5212	2.46%
0.4213	1089967	0.02%	0.4197	0.35%	0.5212	1245636	0.10%	0.5122	1.70%
0.4197	1091741	0.16%	0.4183	0.31%	0.5122	1245737	0.01%	0.3942	23.01%
0.4183	1093949	0.20%	0.4145	0.88%	0.3942	1247396	0.13%	0.3908	0.84%
0.4145	1095008	0.10%	0.4139	0.13%	0.3908	1249465	0.17%	0.3241	17.04%
0.4139	1094801	-0.02%	0.4106	0.76%	0.3241	1250898	0.11%	0.2961	8.61%

4.4 Interpret p-Robustness

One of the main purposes of this research is to show the "*price of robustness*", in other words, how much it costs to design a more robust system. We generate a tradeoff curve between the maximal relative regret and nominal costs as follows. We first set $p = \infty$ and solve the problem. We calculate the maximal relative regret over all scenarios. Note that when $p = \infty$, the p-robust constraints become inactive and the problem becomes a fixed-charge capacitated network design problem. Then we set p to the maximal relative regret less 0.000001 and re-solve the problem, find the maximal relative regret again and reset p. By doing this, the objective value will increase because we have tightened the p-robust constraints. Continue this process until no feasible solution can be found. Note that the smallest p value with a feasible solution also provides a lowerbound for the minimax regret problem.

This experiment was performed on two data sets $P1$ and $P2$. The result is summarized in Table 3. The column "p" gives the value of p-robust level. The column "Cost" returns the optimal objective value. "% Inc." is the percentage by with the objective value has increased compared to the value obtained by previous p value. The "Max Reg." is the maximal relative regret over all scenarios, while "% Dec." is the percentage by which the maximal relative regret has decreased. We only include the left 10 points on the curves because they will be of more interests to the managers. The tradeoff curve between objective value and maximal relative regret of both $P1$ and $P2$ are plotted in Fig. 2. We can see that left portions of the curves are quite steep, which means that the maximal relative regret can be greatly reduced with small increase in the objective value. For example, in $P1$, we are able to reduce the maximal relative regret from 0.6925 by 26.3% to 0.5099, with a slightly increase in cost from 1084742 to 1089461, or 0.4%. In other words, with little additional investment, one can obtain robust supply chain networks which are more resilient to disruptions.

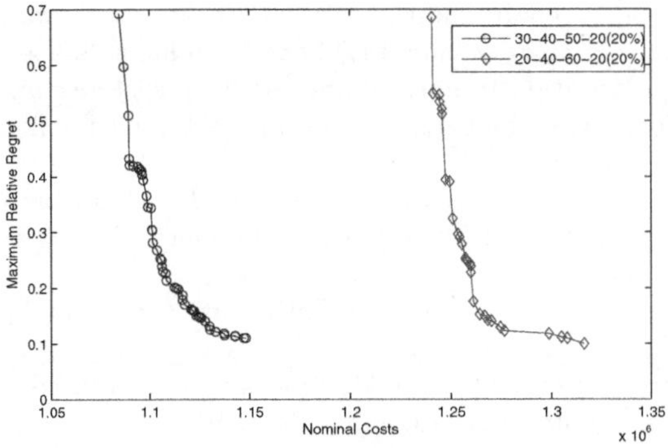

Fig. 2. Tradeoff between maximal relative regret and nominal cost

5 Conclusion

In this paper, we present a p-robust network design model to protect against risk of disruption. A GA-based metaheuristic is proposed and computational experiments have shown that quality solutions can be obtained in a relative short time compared with CPLEX, which makes it attractive in practise where numerous data tests are required. Although GA has been successfully applied to many combinatorial problems and solves our problem well, it is still interesting to test whether other metaheuristic methods, such as Tabu Search to see if they can provide better alternative solution options in future research. We also demonstrate that this approach is less conservative when compared with traditional robustness measures. From managerial point of view, we have shown that system robustness can be improved greatly without large increase in investment and creating financial constraints to companies.

References

1. Ahuja, R.V., Magnanti, T., Orlin, J.: Network Flows: Theory, Algorithms, and Applications. Prentice Hall, New York (1993)
2. Alander, J.T.: On optimal population size of genetic algorithms. In: Proceedings of CompEuro., vol. 92, pp. 65–70. IEEE Computer Society Press, Silverspring (1992)
3. Bunschuh, M., Klabjan, D., Thurston, D.L.: Modeling Robust and Reliable Supply Chains. Working Paper (2006)
4. Drezner, Z.: Heuristic Solution Methods for Two Location Problems with Unreliable Facilities. J. Oper. Res. 38(6), 509–514 (1987)
5. Gen, M., Cheng, R.: Genetic Algorithms and Engineering Optimizations. Wiley, New York (2000)
6. Gutiérrez, G.J., Kouvelis, P., Kurawarwala, A.A.: A Robustness Approach to Uncapacitated Network Design Problems. Eur. J. Oper. Res. 94, 362–376 (1996)
7. Hendrickand, K.B., Singhal, V.R.: An Empirical Analysis of the Effect of Supply Chain Disruptions on Long-Run Stock Price Performance and Equity Risk of the Firm. POMS 14(1), 35–52 (2005)
8. Hicks, M.: When the chain snaps, eWeek - Enterprise News & Reviews (2002), `http://www.eweek.com/c/a/Channel/When-the-Chain-Snaps/`
9. Holland, J.H.: Adaptation in natural and artificial systems: an introductory analysis with applications to biology, control, and artificial intelligence. University of Michigan Press, Michigan (1975)
10. Kouvelis, P., Kurawarwala, A.A., Gutiérrez, G.J.: Algorithms for Robust Single and Multiple Period Layout Planning for Manufacturing Systems. Eur. J. Oper. Res. 63(2), 287–303 (1992)
11. Kouvelis, P., Yu, G.: Robust Discrete Optimization and Its Applications. Kluwer Academic Publishers, Nowell (1997)
12. Latour, A.: Trail by fire: A blaze in Albuquerque sets off major crisis for cell-phone giants. Wall Street Journal A1 (January 29, 2001)
13. Mouawad, J.: Katrina's Shock to the System. In: NY Times, September 4 (2005)
14. Rose, K.: Deterministic Annealing for Clustering, Compression, Classification, Regression, and Related Optimization Problems. In: Proceedings of the IEEE, vol. 86(11), pp. 2211–2239 (1998)
15. Snyder, L.V., Daskin, M.S.: Reliability Models for Facility Location: The Expected Failure Cost Case. Trans. Sci. 39(3), 400–416 (2005)
16. Snyder, L.V., Daskin, M.S.: A Random-Key Genetic Algorithm for the Generalized Traveling Salesman Problem. Eur. J. Oper. Res. 174(1), 38–53 (2006)
17. Snyder, L.V., Daskin, M.S.: Stochastic p-robust Location Problems. IIE Trans. 38(11), 971–985 (2006)
18. Snyder, L.V., Daskin, M.S.: Models for Reliable Supply Chain Network Design. In: Murray, A.T., Grubesic, T.H. (eds.) Critical Infrastructure: Reliability and Vulnerability, vol. B, pp. 257–289. Springer, Berlin (2007)
19. Tang, C.: Robust Strategies for Mitigating Supply Chain Disruptions. International J. of Logistics 9(1), 33–45 (2006)

A Resource Based Mixed Integer Modelling Approach for Integrated Operational Logistics Planning

Jens Peter Kempkes[1], Achim Koberstein[2], and Leena Suhl[2]

[1] ORCONOMY GmbH Paderborn
jens.peter.kempkes@orconomy.de
[2] Decision Support and Operations Research Lab (DS&OR-Lab), University of Paderborn
{koberstein,suhl}@dsor.de

Abstract. The paper considers the operational planning task in the supply network of an Original Equipment Manufacturer (OEM), including the logistics from first tier suppliers to the assembly areas of the OEM. We propose an integrated planning approach both for the external and internal part of the network. The approach is based on a mixed-integer optimization model with a multi-commodity network design formulation. We present a generic modelling construct, which is capable of representing various tariff structures and discount schemes. These tariffs and discounts are parameters to the model and can be configured without altering the model formulation. Finally, we present a case study including numerical results for instances of practically relevant magnitude using a standard solver as well as a specially tailored heuristic algorithm.

1 Introduction

In this paper we consider the operational planning task in the supply network of an Original Equipment Manufacturer (OEM), including the logistics from first tier suppliers to the assembly areas of the OEM. Figure 1 shows an example with four suppliers, two consolidation points, i.e., intermediate storage facilities where freight is repacked and rerouted, and two plants. The left hand side represents the external part and the right hand side the internal part of the network. The transportation tasks of the external part are taken over by freight forwarding services over maritime, air, highway or railway connections, whereas operations in the internal part are carried out by employees of the OEM. The internal transportation requires consideration of various alternative connections that are capacitated by one or more resources such as volume or weight. Connections between different plants or locations of a plant are given in the network, so that incoming goods may pass several plant locations before arriving at the final location. Thus the material flow of numerous articles between the supplier and the demand location can be considered. Within the time frame considered here

W. Dangelmaier et al. (Eds.): IHNS 2010, LNBIP 46, pp. 281–294, 2010.

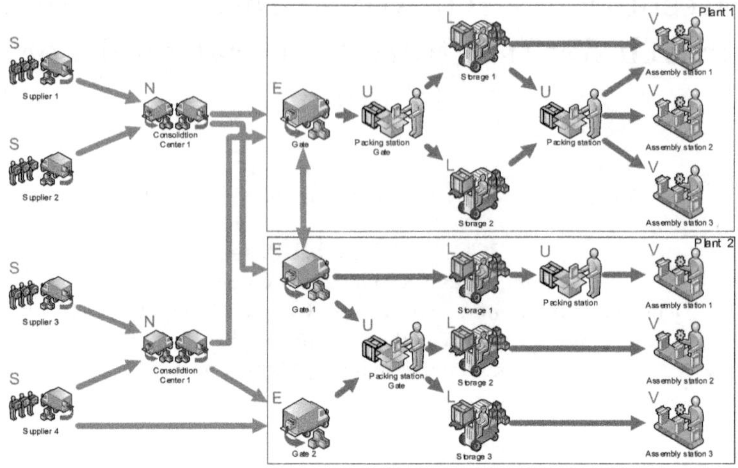

Fig. 1. Supply network with four suppliers, two consolidation points and two plants

the network structure remains unchanged: in each time period identical arcs and nodes are available. However, the capacities and/or costs have a dynamic character and can change between the time periods.

In the operational time scope the production plan is considered as fixed, and it covers a given time period, typically 52 weeks. The dominant planning tasks in the external part of this network, i.e., from the first tier supplier to the stock receipt of the OEM are supplier selection, order sizing and replenishment planning. In the internal part from stock receipts to assembly work stations optimal material flows, inventories and safety stocks have to be determined, crates have to be assigned and workforce and other resources have to be allocated and deployed efficiently.

In the supply chain literature, these different planning tasks have mostly been treated independently. [1] showed in the early 1970's that capacitated lot sizing models can be used for order sizing and procurement planning. [2,3] revised and extended these models in the early 1990's. In these times, discounts were not considered. More recently, [4] and [5] proposed extended formulations for the consideration of quantity discount schemes. [6] and [7] investigate special cases of inventory lot sizing with multiple transportation modes. [6] describes a model which includes transport modes with individual container sizes and set-up costs, discounts are not used though. This is a single product model and therefore container sizes can be determined during a preprocessing phase. The model shown in this paper allows variable amounts of different products to be packed in containers. [7] shows another single product model considering transport modes and even time windows. A comprehensive overview over (joint) replenishment models is given in [8]. The DJRP basis model stated in this publication does not include the choice of supply paths and any discounts. [9] integrated quantity discounts into replenishment models and suggest efficient heuristics. The model uses individual discount levels for product prices, which in contrast to the model shown in this paper, do not interfere with each other. The work of [10] is one of the few, in which network design models are used for operational logistics planning. The authors consider multi-commodity, multi-mode transportation and time-windows. With deterministic and dynamic demands as well as prohibited backlogging the model shares some assumptions with the one shown in this paper. Nevertheless discounts are not considered in this model.

In this paper, we propose an integrated planning approach that integrates the external and internal logistic operations. The approach is based on a mixed-integer optimization model with a multi-commodity network design formulation. The model is able to consider different tariff structures and discount schemes. We present a generic modeling construct to handle complex tariff systems including quantity discount schemes, multi-mode transportation and other internal and external cost drivers.

The new model considers time divided into periods so that the arcs between nodes may start and end in different periods. All logistic operations utilize resources within their capacities. The overall usage of resources leads to total cost. Resource costs may be discounted based on their total usage in an unlimited number of steps. The actual discount step is based on the usage of an unlimited number of resources. This discount system is utilized to represent discounts on transports as well as product prices. Additionally the presented model representation covers different product crates (carriers) for each product. Operations on crates utilize resources as well, based on the number of crates required to pack a certain amount of items.

The remainder of the paper is structured as follows. In section 2 we describe and define the operational logistics planning problem. Furthermore, we highlight the importance of complex tariff systems and transportation modes in the external and resource utilization in the internal part. In this section we also present and illustrate the two modeling concepts resource pools and resource groups. In section 3 we develop a mixed integer programming formulation for these concepts and embed it into a multi-commodity network flow formulation. The formulation is improved through symmetry breaking constraints. The practicability of the whole approach is demonstrated in a case study in section 5 including a computational study. Finally, we finish with conclusions and further research directions in section 6.

2 A Resource Based View on Operational Logistics Planning

The main goal of operational logistics planning is to fulfill production demands at the lowest possible costs. [11] shows how costs incurred in logistics networks can be represented by the utilization of independent resources. Resources can be capacitated such as the working time of an employee, available storage space, number of transportation vehicles etc., or uncapacitated such as stock holding cost. The costs of utilizing a resource can be linear, piece-wise constant or piece-wise linear. An example for linear costs are interest expenses, piece-wise constant costs could be incurred by wages which are paid shift-wise. Examples for piece-wise linear costs are given below. In the internal part main cost drivers are linear stock holding costs and piece-wise constant handling costs. These steps reflect shift costs. The usage of these resources is linear in the number of operations, though. In the external part of the logistics network the costs are highly influenced by long-term contracts the OEM establishes with several logistics service providers such as freight forwarding businesses. In the context of operational planning each of such contracts constitutes a *transportation mode*, which can differ in the type of carrier (truck, rail, ship, aircraft, etc.), the type of shipping space (container, parcel, etc.) and several other conditions, which determine its capacity and tariff system (see [12] for a detailed study and overview of transportation modes). Transportation tariffs are usually differentiated by both volume and weight. Figure 2 shows a selection of seven typical tariff systems.

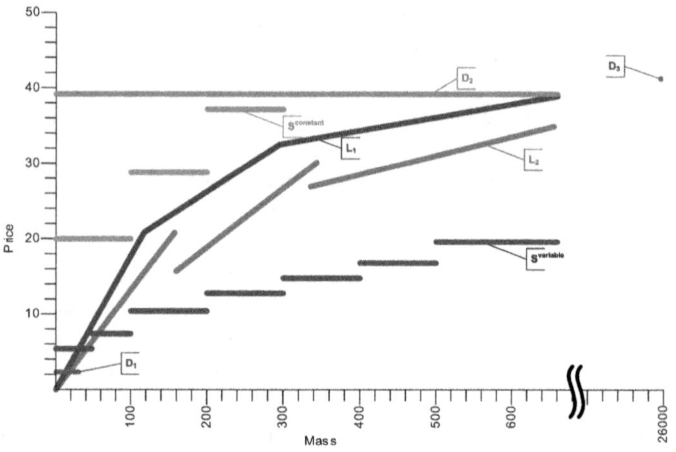

Fig. 2. Cost structure of tariff systems based on a single resource

D_1, D_2 and D_3: In all of the three systems a tariff cap is given, which can only be valid for a certain weight or volume as D_3, e.g., a full container, within certain intervals as D_1 or can be totally independent of the amount of freight as D_2.

$S^{variabel}$, $S^{constant}$: Costs of transportation depend on weight and are differentiated into several levels, which can be valid within intervals of constant ($S^{constant}$) or variable ($S^{variable}$) length. For each level a constant price applies, which cannot be charged against other levels.

L_1 and L_2: In system L_1 quantity discounts lead to costs which depend piece-wise linearly on weight. Additional level discounts in L_2 result in overlapping pieces in terms of price.

Transport agreements with courier, express mail and parcel services are often similar to tariff D_1, those with exclusive shipping storage to tariff D_2. Freight forwarders price their pre-leg transport using freely negotiable weight levels as in $S^{variable}$ and $S^{constant}$. A variety of further combinations of tariff levels with arbitrary lower and upper limits, base prices and linear component are possible. A comprehensive overview of tariff systems and their application areas is given by [13].

Complex tariff systems often consider several freight characteristics at a time, e.g. weight and volume. These characteristics can be regarded as resources which are utilized by shipping a certain amount of freight. We call the subset of resources which are cost drivers of a specific tariff system a *resource group*. A resource group can be used to represent a tariff system with piece-wise linear or piece-wise constant discount levels. This generic formulation leads to a reusable model formulation. The model adopts to different tariff structures and discounts by changed parameters only, altering the model formulation is unnecessary. Furthermore, resource groups can be used to add logical *or* and *and* expressions on resource usages. Table 1 shows an example of a resource group with two resources, weight and volume, representing a pre-leg transport tariff system. Up to a volume of 108 units the discount level is determined by weight only (within steps of 50, 100 and 200 units). The discount levels $n-1$ and n were defined to cover the special cases of utilizations of over 90% of weight (25200) or volume, respectively. In between, all kinds of discount levels are possible. Also, fixed and variable costs can be associated with each discount level as described above.

Table 1. Resource group representing a tariff system for a shipping space, RS: discount level, R: resource, LB and UB: lower and upper limit of a validity interval

RS	R	LB	UB
1	weight	0	50
1	volume	0	108
...	...	...	...
n-1	weight	25200	28000
n-1	volume	0	120
n	weight	0	28000
n	volume	108	120

Of course, the concept of resource groups can also be used to represent complex cost systems in the internal part of the network, such as costs for capacity expansion etc, non-linear handling costs etc. Furthermore, the concept is capable of representing logic expressions on resource usage.

3 A Multi-period Multi-commodity Network Design Formulation

Let $\mathcal{T}$ be the set of time periods. The planning problem is represented by a time-space network with nodes $\mathcal{N}$, supplier nodes $\mathcal{N}^{\mathcal{S}} \subset \mathcal{N}$ and arcs $\mathcal{A} \subseteq \mathcal{N} \times \mathcal{N} \times \mathcal{T} \times \mathcal{T}$. A set $\mathcal{P}$ of products is routed through the network such that demands $d_{i,p,t}$ of products $p \in \mathcal{P}$ at nodes $i \in \mathcal{N}$ in time periods $t_2 \in \mathcal{T}$ are fulfilled. All products are packed in a carrier chosen from a set of given carriers $c \in \mathcal{C}$ with a maximum amount of $q_{p,c}$ per carrier. This results in the finally requested supply $s_{j,p,t'}$ from supplier $j \in \mathcal{N}$ in time periods $t' \leq t$.

Given a set $\mathcal{R}$ of the independent resources of a supply network, each resource $r \in \mathcal{R}$ can be characterized by lower and upper bounds LB_r^R and UB_r^R and unit usage costs g_r. A material flow of product p on arc $(i, j, t, t') \in \mathcal{A}$ which starts in time period t at node i and arrives in period t' $(t \leq t')$ at node j is represented by a flow variable $x_{i,j,p,t,t'}$ and the integer number of required carriers $y_{i,j,p,t,t'}$. It can utilize a resource $r \in \mathcal{R}$ in time period t by a rate of $u_{i,j,p,r}$ per product item and $\widetilde{u}_{i,j,c,r}$ per carrier. The utilization is represented by a variable $k_{r,t}$. Let $\mathcal{RG}$ be the set of all resource groups and $\mathcal{R}_{rg}^{RG} \subseteq \mathcal{R}$ be the subset of resources constituting resource group $rg \in \mathcal{RG}$. A resource can only be part of at most one resource group, which means that the sets $\mathcal{R}_{rg}^{RG}$ must be pairwise disjunctive.

Complex cost systems, e.g., costs of transportation given by complex tariff systems, can be completely represented by the cost values associated with resources and resource groups as indicated in the preceding section. The cost system associated with a resource group $rg \in \mathcal{RG}$ is represented by a set of discount levels $\mathcal{RS}_{rg}$. Each discount level $rs \in \mathcal{RS}_{rg}$ is characterized by a fixed cost value $f_{rg,rs}^{RS}$ and variable costs $c_{r,rg,rs}^{RS}$ as well as a validity interval $[LB_{rg,rs,r}^{RS}, UB_{rg,rs,r}^{RS}]$ for each resource $r \in \mathcal{R}_{rg}^{RG}$. A discount level is active if the utilizations of all the resources in the associated resource group fall within their respective validity interval. Only one discount level of a resource group can be active at the same time. The definition of a resource group is valid if there is an active discount level for all possible resource utilizations.

In the following, we present a basic model formulation. Note, that, for the sake of simplicity, we do not consider storages, crates and transportation vehicles for simplicity. Storages are considered implicitly in the sense that the set A can contain *storage arcs* $(i, i, p, t, t + 1)$ to represent a storage at a node i. Storage costs can be accounted for by associate resources to storage arcs. To consider crates and transportation vehicles, further dimensions can be added to the set of arcs. Furthermore, the model can be enhanced by discrete utilizations of resources, e.g. starting with constant offsets, which requires further binary variables. An elaborated version of the model including all these extensions is presented in [14].

$$
\begin{aligned}
\min Z = & \sum_{t \in \mathcal{T}} \sum_{r \in \mathcal{R}} g_r \cdot k_{r,t} \\
& + \sum_{t \in \mathcal{T}} \sum_{rg \in \mathcal{RG}} \sum_{rs \in \mathcal{RS}_{rg}^{RG}} \sum_{r \in \mathcal{R}_{rg}^{RG}} (c_{r,rg,rs}^{RS} \cdot (k_{r,t,rg,rs}^{RG} - LB_{rg,rs,r}^{RG} \cdot \omega_{rg,rs,t}^{RG})) \\
& + \sum_{t \in \mathcal{T}} \sum_{rg \in \mathcal{RG}} \sum_{rs \in \mathcal{RS}^{RG}} f_{rg,rs}^{RS} \cdot \omega_{rg,rs,t}^{RS} \qquad (1)
\end{aligned}
$$

subject to

$$\sum_{j,t':(j,i,t,t')\in\mathcal{A},c\in\mathcal{C}} x_{j,i,p,c,t,t'} - \sum_{j,t':(i,j,t',t)\in\mathcal{A},c\in\mathcal{C}} x_{i,j,p,c,t',t} = d_{i,p,t} \qquad \forall t\in\mathcal{T}, i\in\mathcal{N}\setminus\mathcal{N}^{\mathcal{S}}, p\in\mathcal{P} \quad (2)$$

$$\sum_{j,t':(i,j,t',t)\in\mathcal{A},c\in\mathcal{C}} x_{i,j,p,c,t',t} - \sum_{j,t':(j,i,t,t')\in\mathcal{A},c\in\mathcal{C}} x_{j,i,p,c,t,t'} = s_{i,p,t} \qquad \forall t\in\mathcal{T}, i\in\mathcal{N}^{\mathcal{S}}, p\in\mathcal{P} \quad (3)$$

$$k_{r,t} = \sum_{\substack{i,j,t':(i,j,t,t')\in\mathcal{A},\\ p\in\mathcal{P},c\in\mathcal{C}}} \begin{pmatrix} u_{i,j,p,r}\cdot x_{i,j,p,c,t,t'} \\ +\widetilde{u}_{i,j,c,r}\cdot y_{i,j,p,c,t,t'} \end{pmatrix} \qquad \forall r\in\mathcal{R} \quad (4)$$

$$k_{r,t} = \sum_{rs\in\mathcal{RS}^{RG}_{rg}} k^{RG}_{r,t,rg,rs} \qquad \forall rg\in\mathcal{RG}, r\in\mathcal{R}^{RG}_{rg} \quad (5)$$

$$y_{i,j,p,c,t,t'} \geq \frac{x_{i,j,p,c,t,t'}}{q_{p,c}} \qquad \forall (i,j,t,t')\in\mathcal{A}, p\in\mathcal{P}, c\in\mathcal{C} \quad (6)$$

$$\sum_{rs\in\mathcal{RS}^{RG}_{rg}} \omega^{RS}_{rg,rs,t} \leq 1 \qquad \forall rg\in\mathcal{RG} \quad (7)$$

$$k^{RG}_{r,t,rg,rs} \leq UB^{RG}_{rg,rs,r}\cdot\omega^{RG}_{rg,rs} \qquad \forall t\in\mathcal{T}, rg\in\mathcal{RG}, rs\in\mathcal{RS}^{RG}_{rg}, r\in\mathcal{R}^{RG}_{rg} \quad (8)$$

$$k^{RG}_{r,t,rg,rs} \geq LB^{RG}_{rg,rs,r}\cdot\omega^{RG}_{rg,rs} \qquad \forall t\in\mathcal{T}, rg\in\mathcal{RG}, rs\in\mathcal{RS}^{RG}_{rg}, r\in\mathcal{R}^{RG}_{rg} \quad (9)$$

$$LB^{R}_{r} \leq k_{r,t} \leq UB^{R}_{r} \qquad \forall t\in\mathcal{T}, r\in\mathcal{R} \quad (10)$$

$$x_{i,j,p,c,t,t'} \geq 0 \qquad \forall (i,j,t,t')\in\mathcal{A}, p\in\mathcal{P}, c\in\mathcal{C} \quad (11)$$

$$y_{i,j,p,c,t,t'} \geq 0 \qquad \forall (i,j,t,t')\in\mathcal{A}, p\in\mathcal{P}, c\in\mathcal{C} \quad (12)$$

$$\omega^{RG}_{rg,rs,t} \in \{0,1\} \qquad \forall rg\in\mathcal{RG}, rs\in\mathcal{RS}^{RG}_{rg}, t\in\mathcal{T} \quad (13)$$

In our formulation the objective function (1) consists merely of resource driven cost components, i.e., a component linear in the resource utilizations, one that is linear in certain discount levels of resource groups and a fixed cost component associated with discount levels. As shown above all kinds of linear, piece-wise linear and piece-wise constant cost systems can be represented by defining proper resources and resource groups. Constraint set (2) assures material flow balance and demand satisfaction for each product in each node except supplier nodes. Flow balance for supplier nodes is shown in constraint set (3). We assume that total supplies are free of choice and demands are given. The concept of resource groups is established by constraint sets (4) to (10). Resource utilizations are derived from material flows in constraints (4). If a resource is part of a resource group, constraint set (5) assigns the utilization to a valid discount level. Note, that we assumed that all discount levels of a resource group are pair-wise disjunctive. Constraints (7) assures that only one discount level per resource group can be active. Active discount levels are indicated by binary variables $\omega^{RG}_{rg,rs}$ and indicator constraints (8) which establish also the upper limits of the level validity intervals. Lower limits of discount levels are modeled in constraints (9).

Symmetry breaking. Practical instances of the model presented above can comprise symmetries, e.g., by offering alternative transportation modes which have the same cost structure. Symmetries are regions of the set of feasible solutions which contain virtually equivalent solutions from the viewpoint of the planner. More specifically, if one of the symmetric regions contains a solution then each of the other regions contains an equivalent one. If one region does not contain a solution then neither do the other regions. Therefore, it suffices to search one of the regions and discard the other regions from the set of feasible solutions (s. [15]). Variants of the branch-and-bound solution method that cope with symmetries have been developed by [16], [17] and [18]. We use the following idea developed by [16], which eliminates symmetries on trucks by adding valid inequalities.

The shown model uses arcs and resources in a generic manner. In the following, symmetries created by alternative arcs representing several trucks between two locations, are broken by additional constraints on resources. Let $\mathcal{SYM}$ denote the set of symmetry groups $\mathcal{R}^{SYM}_{sym}$. Each group is an ordered set of affected resources of a symmetry group $sym \in \mathcal{SYM}$ and $item(\mathcal{R}^{SYM}_{sym}, n)$ the n-th element of an ordered set. Within this order, the preceding resource must have a greater or equal usage than its successor. A group is added for each group of alternative arcs, i.e., for a set of trucks with identical limitations and tariff systems. Then, symmetries can be eliminated by adding the following constraint:

$$k_{item(\mathcal{R}^{SYM}_{sym},n),t} \geq k_{item(\mathcal{R}^{SYM}_{sym},n+1),t} \forall sym \in \mathcal{SYM}, n = 1..|\mathcal{R}^{SYM}_{sym}| - 1, t \in \mathcal{T} \tag{14}$$

Extending the generic model formulation by this constraint enables the model to use any resource for symmetry breaking, instead of the usual limitation to flow variables. A formulation with these constraints is used in the numerical tests below.

4 A Construction Heuristic

A heuristic approach is required to find a first valid solution to the problem in reasonable time. Every heuristically generated solution may be transferred to the MIP solver as starting value and can reduce the branch and bound search space. The general procedure of the heuristic algorithm is shown in Figure 3. It consists of two phases. Phase 1 calculates a valid solution from scratch. Without considering the objective, it constructs flows ignoring resource constraints. If any

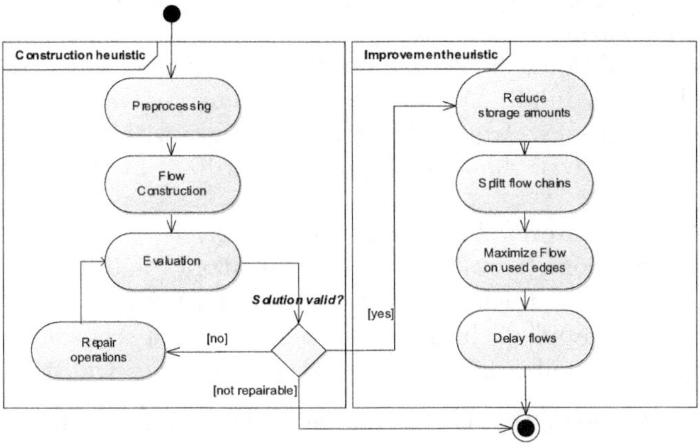

Fig. 3. Heuristic activity diagram

constraint is violated certain repair mechanisms are applied to fix the solution. Phase 2 improves the valid solution of phase 1 in the sense of the objective function.

The internal functions are designed to operate on flow chains. A flow chain is considered as a flow from a single supplier node to a single demand. The amount of flow on all path elements of this flow chain is equal. A flow chain never splits or joins. If a part of the amount shall be routed on a different path, the flow chain amount is reduced and a new flow chain, following the different path, is added to the solution.

Phase 1: Construction. During Phase 1 a new flow chain is constructed for each demand. For a depth first search from the demand node to some supplier node, the network is considered with reversed arcs. The flow is constructed from the first supplier found to the demand node (considering the network with original arcs). These flow chains are established in the solution without considering resource constraints.

If some constraint is violated, a repair cycle starts. It iterates over all broken constraints and applies a repair strategy. The strategy is chosen based on the type of the broken constraint and the direction of violation. The usual repair strategies are lower resource usage if a resource is above its upper bound and shift of flows to different edges. If echelon storage is allowed, additional flows may be generated. If a resource is below its lower bound flows are shifted to different edges. A simple search strategy for finding and evaluating appropriate edges is used in both cases. If a resource group is violated, several levels are selected as repair goals. The resource repair methods shown before are applied to all affected resources. Symmetry groups are repaired by a minimal number of swaps of flows. The cycle round number determines the depth of the neighborhood search applied to find an alternative flow chain to fix the issue. The strategies are:

- Reroute: Reroute a flow chain over a different path or add storage arcs in order to use earlier arcs.
- Add: Add additional flow amounts to the network, which results in final storage but may fulfill lower bounds on resource usages.
- Fix resource groups: Adjust the resource usages within a resource group to match a single level. *Add* and *Reroute* are utilized by this strategy.

Phase 2: Improvement. Phase 2 improves a given valid solution in terms of the objective function. It evaluates several strategies and keeps chains only if the objective improves. Each improvement step must end with valid solution. A repair operation is applied, if an improvement step breaks some constraint. If the repair operation fails, this particular step is undone. The strategies are:

- Minimize storage: Delay flow chains in order to reduced stored amounts.
- Split flow chains: Flow chains can only be rerouted completely. In order to allow for a more diverse network usage the existing flow chains are split into individual ones.
- Maximize flow on used edges: Scale effects on resources may be gained by moving flow chains on used edges and thereby freeing other edges.
- Delay flows: Group of flow chains are stored and delivered in earlier periods to maximize resource utilization on earlier arcs.

5 Case Study

In this section, the effect of optimization is evaluated in a case study with realistic data. We consider inbound logistics of an original equipment manufacturer (OEM) where an area forwarding network is used. The network contains seven areas each with 1 to 20 suppliers. Within a pre-leg the deliveries are first transported into a consolidation point of an area, and then within a main leg into the OEM. The basic structure is shown in Fig. 4. The optimization balances internal and external costs generated by transporting and storing products and crates. Storage costs are considered as interest and per crate cost.

In practice, the delivery times are irregular and given through the orders which are placed according to the employed order policy. The transportation is fixed as area forwarding mode. The optimization model presented in this work was then used to evaluate other possible delivery modi. The tariff structure and discounts used are D_2, D_3, $S^{constant}$ and $S^{variable}$ (cmp. Fig. 2). In particular, we evaluated the use of different transportation modi between a supplier and the OEM, not necessarily fixed to area forwarding. As an alternative it is possible to use direct transportation from supplier to customer or, on the other hand, Courier, Express and Parcel (KEP) services can be provided. As another aspect a fixed delivery profile for each supplier was evaluated. This means that the delivery of a given supplier arrives regularly with a given frequency, for example on Mondays, Tuesdays etc. or on two given weekdays. We implemented and tested 12 different delivery profiles for each supplier within the optimization. The following three models were tested and compared:

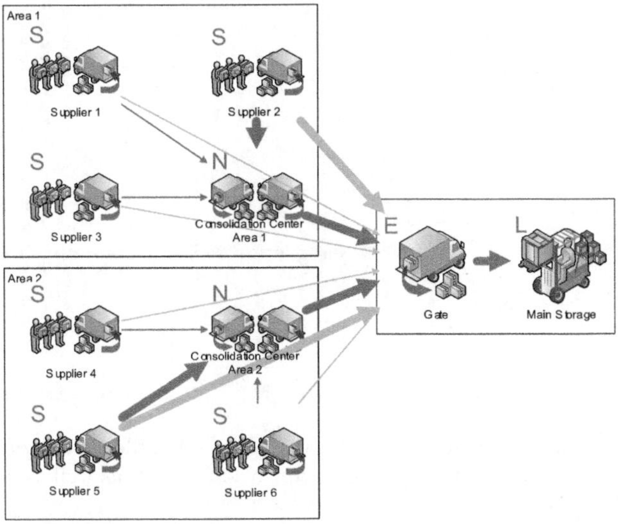

Fig. 4. Network structure of case study Transportation Mode and Delivery Profile Optimization: dark arrows show selected transportation connections, grey arrows represent alternative routes, line thickness corresponds to the amount transported

- TM (Transportation mode): A transportation mode is assigned to each supplier, it can be area forwarding, direct transportation, or KEP services. The amounts transported remain as in the original model.
- DP (Delivery profile): One of the 12 different delivery profiles is assigned to each supplier.
- TM&DP: Both TM and DP are combined within one single model: both transportation mode and delivery profile are assigned to each supplier.

The required quantities were derived from the original model and expressed as aggregated daily amounts of each item and supplier. When the delivery dates change because of a given delivery profile, additional storing of items is usually required. This happens especially when it becomes necessary to deliver earlier than needed because of the fixed delivery frequency. The quantities stored and their costs were considered when generating the delivery quantities within the model. Thus upper bounds for the storage capacity were included into the model and realized as a resource. Because the deliveries from different areas are independent from each other, it is possible to decompose the optimization model into smaller models, each considering one area.

Results of the optimization with the original model compared with the TM, DP, and TM&DP models are illustrated in Fig. 5. The costs are shown as sum of the transportation costs with KEP services, direct transportation as well as area forwarding including costs of pre-leg, main leg and storage. The percentage of the costs with TM&DP model of the original costs of the area forwarding model are shown for each area.

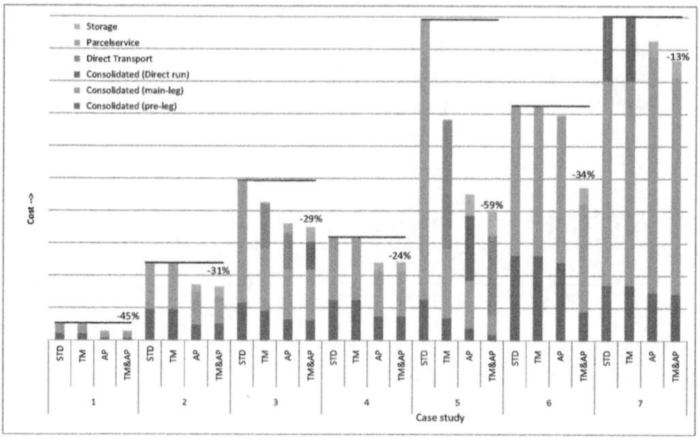

Fig. 5. Case study Transportation Mode and Delivery Profile Optimization: Results of the seven areas with total costs without optimization (STD) as well as with the models TM, DP and TM&DP

Figure 5 shows that optimization results in high cost reduction for all areas. The amount of stored items is rather small as expected. The relative cost reduction is significant up to 59%. It tends to be higher in areas where the trucks are not fully loaded compared to areas with full trucks. In such cases it is advantageous to combine deliveries and to deliver on certain given days. Overall, the optimization implies a significant cost reduction. The results can be integrated into planning and dispatching systems in a simple way.

Further numerical tests have been carried out with data of test networks having a structure similar to real logistics networks in practice. In particular the quantities number of suppliers, products, crates, tariff discount levels, and nodes are equals. Product mass, create mass and sizes, demands and all prices are equal in average value and variance to the original instances. The numerical study involves 14 different networks with up to 102 products, six time periods and 696 rebate levels. Table 2 shows the properties of the optimized instances.

Table 2 also states the solution quality and computing time for all instances. Computational results are shown for three model formulations: the original formulation, an extended formulation with symmetry breaking as well as extended formulation with explicit definition of semi-continuous variables and special ordered sets (SOS) where appropriate. The maximum optimization time is 14400 seconds. An integer gap smaller than 0.1% is considered as solved to optimality. The solver ILOG CPLEX 11.0 was used on a 2.9 GHz Intel Core 2 Duo. 7 GB of RAM were available for optimization. The heuristic algorithm provided a valid solution to CPLEX as mip starting values.

On average symmetry breaking speeds up optimizations by 42.1% compared to the original formulation. Using semi-continuous variables gives an average time advantage of 7.4%. When using symmetry breaking only three, instead of five, instances could not be solved to optimality within time. In only one case (instance 10) symmetry breaking did not gain the best result. However, this

Table 2. Instance properties: only unique elements are counted, edges are distinct not considering given product and carrier combinations. Remaining gap in percent and optimization time (sec.) are shown for each instance.

Instance	Products	Periods	Carriers	Demands	Nodes	Edges	Resources	Resourcegroups	Rebate Levels	Original	Symmetry Breaking	SemiCont, SOS
1	24	6	12	11	11	150	123	21	250	0,1% 80	0,1% 290	0,1% 687
2	61	5	23	64	18	205	262	35	424	17,74% 14432	17,21% 14459	19,58% 14410
3	48	5	18	50	11	150	205	26	270	0,25% 14404	0,1% 77	1,88% 14416
4	39	5	18	24	14	165	184	29	354	0,1% 2915	0,1% 202	0,1% 318
5	1	5	1	1	4	30	19	4	38	0% 0	0% 0	0% 0
6	102	5	21	87	25	340	439	62	696	4,77% 14410	0,1% 1841	3,73% 14410
7	61	5	23	64	18	205	262	35	424	0,2% 14412	0,18% 14407	0,85% 14406
8	33	5	15	33	12	145	158	25	290	0,09% 24	0,04% 17	0,09% 24
9	8	5	7	4	9	95	62	15	178	0% 0	0% 0	0% 0
10	42	5	15	43	20	290	243	54	598	2,1% 14424	2,35% 14409	2,23% 14412
11	12	5	8	8	9	95	75	15	178	0% 1	0% 1	0% 4
12	30	5	12	19	13	145	149	25	314	0,1% 141	0,09% 52	0,1% 172
13	11	5	5	10	9	95	72	15	178	0% 2	0% 1	0% 1
14	8	5	5	8	5	45	46	7	74	0% 0	0,1% 1155	0,1% 1709

instance could not be solved to optimality in time and the shown gap may only be considered as an approximation of the best model formulation.

6 Conclusions

We have presented a mixed integer optimization model to minimize total logistics costs of an integrated internal and external supply network. The model comprise the generic concepts of resource groups which is an elegant and flexible tool to represent several piece-wise linear tariff systems. This can be done by adjusting the model parameters only, no modifications of the model formulation is required. Our numerical experiments show that models of practical size can be solved to optimality after a first feasible solution is generated by a heuristic procedure. In a practical case study significant savings of up to 59 per cent could be realised.

References

1. Glaser, H.: Zur Bestimmung kostenoptimaler Bestellmengen bei deterministisch gleichbleibendem und deterministisch schwankendem Bedarf. PhD thesis, Universität zu Köln (1973)
2. Günther, H.: Bestellmengenplanung aus logistischer Sicht. Zeitschrift für Betriebswirtschaft 61(5), 641–666 (1991)
3. Günther, H.: Ein Verfahren zur Bestellmengenplanung bei beschränkter Lagerkapazität. In: Operations Research Proceedings 1990, pp. 491–498. Springer, Heidelberg (1991)
4. Reith-Ahlemeier, G.: Ressourcenorientierte Bestellmengenplanung und Lieferantenauswahl: Modelle und Algorithmen für Supply-Chain-Optimierung und E-Commerce. Reith-Ahlemeier, G., Leichlingen (2002)
5. Stadtler, H.: A general quantity discount and supplier selection mixed integer programming model. OR Spectrum 29(4), 723–744 (2007)
6. Jaruphongsa, W., ÇEtinkaya, S., Lee, C.: A dynamic lot-sizing model with multi-mode replenishments: polynomial algorithms for special cases with dual and multiple modes. IIE Transactions 37(5), 453–467 (2005)
7. Jaruphongsa, W., Lee, C.: Dynamic lot-sizing problem with demand time windows and container-based transportation cost. Optimization Letters 2, 39–51 (2008)
8. Boctor, F., Laporte, G., Renaud, J.: Models and algorithms for the dynamic-demand joint replenishment problem. International Journal of Production Research 42(13), 2667–2678 (2004)
9. Cha, B., Moon, I.: The joint replenishment problem with quantity discounts under constant demand. OR Spectrum 27(4), 569–581 (2005)
10. Chen, P.S.: Cost Minimization in Multi-Commodity, Multi-Mode generalized Networks with Time Windows. PhD thesis, Texas A&M University (2005)
11. Schönsleben, P.: Integrales Logistikmanagement: Operations and Supply Chain Management in umfassenden Wertschöpfungsnetzwerken. Springer, Heidelberg (2007)
12. Sahin, B., Yilmaz, H., Ust, Y., Guneri, A.F., Gulsun, B.: An approach for analysing transportation costs and a case study. European Journal of Operational Research (2007)
13. Pooler, V., Pooler, D.: Purchasing and Supply Management: Creating the Vision. Kluwer Academic Publishers, Dordrecht (1997)
14. Kempkes, J.P.: Kostenoptimale Materialflüsse in der operativen Zulieferungslogistik der Nutzfahrzeugindustrie. PhD thesis, Universität Paderborn (2009)
15. Fahle, T., Schamberger, S., Sellmann, M.: Symmetry breaking. In: Goos, G., Hartmanis, J., van Leeuwen, J. (eds.) CP 2001. Fahle, T., Schamberger, S., Sellmann, M, vol. 2239, pp. 93–107. Springer, Heidelberg (2001)
16. Sherali, H., Smith, J.: Improving Discrete Model Representations via Symmetry Considerations. Management Science 47(10), 1396–1407 (2001)
17. Sellmann, M.: Reduction Techniques in Constraint Programming and Combinatorial Optimization. PhD thesis, Universität Paderborn (2002)
18. Stickel, M.: Planung und Steuerung von Crossdocking-Zentren. PhD thesis, Universität Karlsruhe (2006)

Job Shop Scheduling with Buffer Constraints and Jobs Consuming Variable Buffer Space

Andreas Witt[1] and Stefan Voß[2]

[1] GESIS Gesellschaft für Informationssysteme mbH, Eisenhüttenstraße 99, 38239 Salzgitter
a.witt.mail@web.de
[2] Universität Hamburg, IWI, Von-Melle-Park 5, 20146 Hamburg
stefan.voss@uni-hamburg.de

Abstract. Job shop scheduling is among the most well studied problems in production planning. Yet, there is still a lack of comprehensive treatment of intermediate buffers. This paper deals with limited intermediate buffers with the important extension that not all jobs need to be handled in the same way regarding buffer space consumption. A heuristic to find feasible solutions is described and computational results are presented. While a reference procedure from literature devoted to case of unit buffer space consumption does not guarantee feasibility in all cases, the approach developed in this paper does. Moreover, it allows for variable space consumption, a feature which is new to literature paying better attention to more practical requirements than other models.

Keywords: job shop, buffer, generation scheme.

1 Introduction

In traditional job shops every job j of a set $J = \{j_1, \ldots, j_n\}$ traverses a set $M = \{\mu_1, \ldots, \mu_m\}$ of machines on a specific route $(\mu_{1j}, \ldots, \mu_{k_j j})$ where k_j is the number of operations of j. That is, each j can be sub-divided into k_j operations forming a set $\{o_{1j}, \ldots, o_{k_j j}\}$. Associated with each operation o_{ij} is a machine $\mu_{ij} \in M$. Usually the number k_j of operations equals the number m of machines. Operation o_{ij} has to be processed on $\mu_{ij} \in M$ for p_{ij} time units. Contrary to a flow shop the routings may arbitrarily differ from each other but are fixed in advance. Furthermore, every machine is only visited once by a job and no preemptions are allowed. A feasible schedule S is generated by assigning starting times s_{ij} and completion times c_{ij} to each o_{ij} on μ_{ij} in a way that the predefined routing for each job is met. Besides, the conditions $s_{ij} + p_{ij} = c_{ij}$ must hold and neither the time intervals $[s_{ij}, c_{ij}]$ on the same machine μ_{ij} nor the time intervals assigned to the operations of a job j may overlap. A common objective in job shop scheduling is to minimize the makespan, i.e., the time from start to finish for completing all jobs. Another important objective is the minimization of the sum of flow times of all jobs defined for each job by the time span from its start up to its completion.

In a classical job shop as defined so far every job leaves a machine immediately after its processing. However, in a modified job shop with buffer constraints this is only

W. Dangelmaier et al. (Eds.): IHNS 2010, LNBIP 46, pp. 295–307, 2010.

possible if enough buffer capacity is available to store a processed job. If this is not the case the job must either wait on the machine until the relevant buffer offers enough free capacity or it is not allowed to schedule the job in the respective time interval. The first alternative reflects a blocking of the machine where a job has not to leave a machine after its processing. The second alternative implies that the machine must be immediately unblocked at c_{ij}. A third alternative arises if a *no-wait* problem is considered where every job j must immediately be manufactured on the next machine $\mu_{i+1,j}$ after its completion on μ_{ij}. Both the blocking case and the *no-wait* case imply buffer capacities $\kappa_i = 0$ in front of every machine μ_i (behind a machine or between two machines, respectively). In the scheduling literature it is usually assumed that buffers with capacities $\kappa_i > 0$ offer exactly one buffer place for one job. However, particularly in real-world problems the jobs might consume different space in a buffer. Such a modified situation can be modeled by introducing an additional parameter q_{ij} for each operation o_{ij} reflecting the buffer space that is consumed by an operation. For instance, a material quantity that is associated with an operation could be represented by q_{ij}. By setting $q_{ij} = 1$ such a considered problem reduces to the case in which every job (operation, respectively) consumes one place in a buffer.

In a job shop with limited buffer capacities specific *deadlock* situations might occur that are characterized as follows: One operation o_{ij} of job j is processed on machine μ_{ij} and another operation o_{ab} of job b is processed on machine μ_{ab}. If the conditions $i + 1 = a$, $a + 1 = i$ and $c_{ij} = \mathrm{c}_{\mathrm{ab}}$ hold and if the buffers in front of both machines i and a are filled, a deadlock situation arises because none of the two jobs is able to leave its current machine. Such deadlocks cannot occur in flow shops because all jobs traverse the machines in the same direction and cannot meet or "collide" face-to-face as in a job shop. In an effective scheduling procedure such deadlocks have to be avoided systematically in order to generate feasible schedules.

This paper considers a job shop problem where limited buffer capacities have to be observed. It takes the cases into account where every job consumes exactly one buffer place and the other where different jobs consume different buffer space. The blocking case is not considered. That is, in this paper every job has to leave a machine immediately after its processing and must not block it. After a literature review in Section 2 a heuristic that guarantees to generate feasible schedules is presented in Section 3. Computational results are given in Section 4. The paper concludes with some final remarks in Section 5.

2 Literature Review

For the classical job shop problem with a makespan objective a well known benchmark problem with 10 jobs and 10 machines was introduced in [1] and could be solved only 25 years later in [2] based on a branch-and-bound algorithm. Other exact methods are presented, e.g., in [3], [4], and [5]. In [6] a 15 × 15 instance is solved which seems to be among the largest instances solved exactly up to now. Apparently larger problems must be solved by heuristics so far; see for a recent approach, e.g., [7]. Further solution approaches and general descriptions can be found, among others, in [8] and [9]. Most of the approaches make use of a disjunctive graph representation.

Buffer constraints are mostly considered in flow shop scheduling; see, e.g., [10] and [11]. In [10] intermediate storage (buffer, respectively) capacities κ_i for flow shops are defined based on the following classification:

1. Unlimited Intermediate Storage (UIS): Storage capacities κ_i are very large and may be considered as unlimited. That is, storage capacities are irrelevant.
2. No Intermediate Storage (NIS): There are no intermediate buffers between any two machines. That is, all storage capacities are zero. Consequently, all jobs have to be processed on the next machine after they were finished on a machine without any waiting time. Alternatively, a blocking might occur if a job is allowed to remain on a machine after its processing since the subsequent machine is not idle. These cases comply with *no-wait* and blocking problems mentioned in Section 1.
3. Limited Intermediate Storage (LIS): Between two machines there are buffers with limited capacities $\kappa_i \geq 0$. At least one buffer has a capacity $\kappa_i > 0$. Every job consumes exactly one place inside a buffer.

According to this classification the case LIS is considered in this paper, however, with the important extension that jobs may consume different space in a buffer.

The job shop problem with buffer constraints as a generalization of the classical problem received considerable interest only recently. In [12] an approach based on an alternative graph formulation which is a modification of disjunctive graphs is presented. They distinguish two cases of blocking: In a blocking problem with swaps allowed jobs can move simultaneously between machines and buffers whereas in a blocking problem without swaps jobs can only move strictly after the subsequent machine has become available. Furthermore, heuristics and a branch and bound algorithm are described. However, the heuristics to generate initial solutions do not guarantee to find feasible schedules.

Representations of solutions for job shops with limited buffers are investigated in [13]. Furthermore, a constructive heuristic for a job shop with output buffers is presented. In [14] the following types of buffers are defined:

1. A buffer is called *general* if any assignment of operations to buffers is possible.
2. If every job has its own buffer the buffers are called *job-dependent.*
3. If the assignment of operations to buffers depends on the machines μ_i the type is called *pairwise*. In this situation a separate buffer is associated with every pair (μ_i, μ_a) of machines μ_i and μ_a.
4. If the assignment of operation i to a buffer depends on the machine μ_{i+1} on which the successor operation $i + 1$ is processed this type is called *input buffer*. Symmetrically, another type *output buffer* can be defined.

According to this classification the type *input buffer* is considered in this paper.

The topic of systematic deadlock avoidance seems to be rather scarce in the literature on job shop scheduling with buffer constraints. Recently, in [15] an insertion heuristic based on matrices and "latin rectangles" was presented. This heuristic is capable to systematically avoid deadlocks and to take into account limited buffer capacities. However, the computation times appear to be too long especially when applying the procedure within a metaheuristic. Contrary to the job shop literature flexible manufacturing systems (FMS) are another area of scheduling research where deadlocks receive a large amount of interest. In [16] a neighborhood policy named RUN as a deadlock avoidance policy for FMS is developed. They apply a buffer capacity reservation scheme where high-capacity workstations are treated as central storage buffers for jobs. In [17] a deadlock policy for computer operating systems called "Banker's algorithm" (BA) was proposed. In [18] and [19] the interaction between these deadlock avoidance policies and dispatching rules is investigated.

Deadlock avoidance methods using Petri net models are developed and implemented for FMS among others in [20], [21], and [22]. In most cases reachability graphs have to be computed to determine if a marking of the Petri net could lead to a deadlock. As this might be very time consuming heuristics are applied in most cases.

All above mentioned works on limited buffer space except [11] assume that each job consumes exactly one place inside a buffer. However, it may be important to consider different measures related to different jobs. Therefore, the case where jobs consume different space inside a buffer is considered. This seems to be particularly interesting for real-world applications. Furthermore, related general solution approaches are needed in advanced planning systems to be able to handle various configurations that might be realized in such a system. That is, approaches with the assumption that every job consumes one buffer place do not seem to be completely sufficient for advanced planning systems.

3 A Heuristic to Generate Initial Feasible Solutions

In the sequel different heuristics are developed. Note that a disjunctive graph formulation or any other graph representation is not necessary as there appears to be no graph formulation for job shop problems where the jobs consume variable buffer space.

The basic approach of the heuristic to generate feasible solutions is to schedule one job after another whereas the operations of a job are scheduled recursively one after another by a procedure *ScheduleOperation( )*. This idea is adopted from [23] who apply a similar procedure to a traffic control problem and illustrate the analogy to job shop scheduling. After all operations of a job were scheduled all operations of the next job are scheduled and so on. The sequence for the scheduling of the jobs is arbitrary. At first for every operation o_{ij} that is currently considered a set Σ_{ij} of possible start times is calculated that consists – according to the definition determined here – of the earliest possible start time s_{ij}^* and all completion times c on the relevant machine μ_{ij} which are greater than s_{ij}^*. That is, $\Sigma_{ij} \leftarrow \{s_{ij}^*\} \cup \{c \in \Phi_{\mu_{ij}} \mid c > s_{ij}^*\}$ where $\Phi_{\mu_{ij}}$ is the set of all current completion times on machine μ_{ij}. The earliest possible start time s_{ij}^* equals the completion time of the preceding operation of the same job or zero if the first operation of a job is considered. All start times are sorted in an increasing order. Alternatively, all periods $t > s_{ij}^*$ into Σ_{ij} could be included what would imply a more detailed search for a starting time on the one hand but much longer computation times on the other hand. However, the preferred approach in this paper is to limit the computation times. Therefore, this alternative is neglected in the sequel. In the next step of the heuristic it has to be verified for every possible start time $s \in \Sigma_{ij}$ if o_{ij} can be inserted into the current schedule at s. For this verification a latest next start time $LNST_{i+1,j}$ for the next operation $o_{i+1,j}$ on the next machine $\mu_{i+1,j}$ is calculated in the following way: For every period t greater/equal than the current potential completion time $c = s + p_{ij}$ up to the planning horizon T it has to be examined if the condition $I_{i+1,t} + q_{i+1,j} \leq \kappa_{i+1}$ holds where I_{it} indicates the inventory in front of machine μ_i in period t. If it does not hold the procedure sets $LNST_{i+1,j} = t$. That is, the next operation $o_{i+1,j}$ has to start not later than $s_{i+1,j} = LNST_{i+1,j}$. Otherwise, the maximal buffer capacity κ_{i+1} would be exceeded. If $I_{i+1,t} + q_{i+1,j} \leq \kappa_{i+1}$ holds for every period t up to the planning horizon T the procedure sets $LNST_{i+1,j} = T$. Note that minimal and maximal waiting times for the operations could easily be considered by a modified calculation of s^*_{ij} and $LNST_{ij}$. However, in the sequel the paper continues without such an enhancement.

Consequently, an operation o_{ij} may be inserted at the current potential start time s if the following three conditions hold:

1. The current potential start time s has to be lower/equal $LNST_{ij}$.
2. The current potential completion time $c = s + p_{ij}$ has to be lower/equal $LNST_{i+1,j}$.
3. Machine μ_{ij} is idle within the time interval $[s, s + p_{ij}]$.

In the following algorithm a Boolean indicator variable *insertable* is used that is true if the defined three conditions are true. If an insertion of o_{ij} is allowed at s the current schedule is extended accordingly and the inventories in front of μ_{ij} and $\mu_{i+1,j}$ are updated. That is, for all $t = c$ up to $t < LNST_{i+1,j}$ the procedure has to set $I_{i+1,t} := I_{i+1,t} + q_{ij}$ and for all $t = s$ up to $t < LNST_{ij}$ it has to set $I_{it} := I_{it} - q_{ij}$. Furthermore, the earliest start time for the next operation must be set by $s^*_{i+1,j} := c_{ij}$ and the current potential start time s has to be removed from Σ_{ij}. After that the next operation $o_{i+1,j}$ is scheduled recursively in the same way etc.

It might happen that the variable *insertable* is not true for a succeeding operation. In such cases the recursively called procedure *ScheduleOperation( )* returns another Boolean variable *inserted* = false. Consequently, the scheduling of the preceding operation $o_{i-1,j}$ has to be reset, i.e., the earliest start time s^*_{ij} and the inventories I_{it} and $I_{i-1,t}$ have to be set to their former values. Then the next start time $s \in \Sigma_{i-1,j}$ is considered and the procedure continues as described above.

Unfortunately, it might happen that no further potential start time is available for an operation to be scheduled. That is, $\Sigma_{ij} = \varnothing$ and all attempts to schedule the operations returned *inserted* = false. How such a situation arises is illustrated by the following example. A *no-wait* job shop with two jobs, five machines and the data provided in Table 1 is considered.

Job j =1 is scheduled first. Due to the *no-wait* assumption the heuristic provides the start and completion times shown in Table 2. Afterwards job j =2 is scheduled. For operation o_{12} the set $\Sigma_{12} = \{0, 146\}$ is obtained where s=146 is chosen as machine μ_4 is not idle in the interval [0, 77]. For operation o_{22} the set $\Sigma_{22} = \{223\}$ is obtained where s=223 may be chosen. In conclusion Σ_{32}={302, 320} is calculated where s=302 must not be selected since μ_2 is not idle and s=320 would violate the *no-wait* constraint. Consequently, the procedure *ScheduleOperation( )* returns *inserted* = false for o_{32}. A reset of the preceding operation o_{22} does not solve the problem because of $\Sigma_{22} = \varnothing$. Hence, the heuristic so far does not provide a feasible solution.

Table 1. Instance of a *no-wait* job shop

p_{ij} / i	j = 1	2
1	69	77
2	77	79
3	87	43
4	87	75
5	93	96

μ_{ij} / i	j = 1	2
1	μ_3	μ_4
2	μ_4	μ_3
3	μ_1	μ_2
4	μ_2	μ_1
5	μ_0	μ_0

Table 2. Solution for job $j = 1$

	$j = 1$		
i	s_{ij}	c_{ij}	μ_{ij}
1	0	69	μ_3
2	69	146	μ_4
3	146	233	μ_1
4	233	320	μ_2
5	320	413	μ_0

Table 3. Solution for job $j = 2$

	$j = 2$		
i	s_{ij}	c_{ij}	μ_{ij}
1	164	241	μ_4
2	241	320	μ_3
3	320	363	μ_2
4	363	438	μ_1
5	438	534	μ_0

A very easy but time consuming solution could be realized by adding all periods t greater than the last completion time on the relevant machine to Σ_{ij} as mentioned above. One of these new potential start times would finally lead to a feasible solution. In the following an approach executed by an additional procedure *SetNewFirstStartTime( )* implying a more efficient alternative is proposed:

The current potential start time s that causes *insertable* = false under the condition $\Sigma_{ij} = \varnothing$ is used to calculate a new earliest start time for the first operation o_{0j} by subtracting the current processing time p_{ij} from s within every recursive call of *ScheduleOperation( )* until the first call, i.e., the first operation o_{0j} is reached. In the example illustrated above the procedure returns the new earliest start time $s^*_{0j} = 320 - p_{22} - p_{12} = 320 - 79 - 77 = 164$ for the first operation that results in the feasible solution for job $j = 2$ ($j = 1$ remains unchanged) shown in Table 3.

The heuristic described in this section guarantees to find feasible solutions for all input buffer scenarios except the blocking case because deadlocks are systematically avoided due to the recursive scheduling of all operations of a job. If an operation currently under consideration faces a deadlock the respective start time s is neglected and the next potential start time from Σ_{ij} is chosen. Figures 1, 2 and 3 depict a formal description of the heuristic.

```
Main algorithm:
S ← ∅
for all j ∈ {j1, …, jn} do
        inserted = false
        ScheduleOperation(S, o1j, inserted, T)
endfor
return S
```

Fig. 1. Main algorithm

```
Procedure SetNewFirstStartTime():
Parameter: insertable, current set of start times Σij, currently considered start time s
if insertable(S, oij, s, LNSTij, LNSTi+1,j) = false and Σij = ∅ and i > 0
    s0j^new = s
    for h = i – 1 down to h = 0 do
        s0j^new = s0j^new – phj
    endfor
endif
```

Fig. 2. Procedure *SetNewFirstStartTime( )*

4 Computational Results

To provide computational results the generation scheme described in Section 3 was implemented in C++ on a PC with a 2.2 GHz dual-core CPU and 1 GB RAM. As test data five instances abz05 – abz09 presented in [8], the three instances ft06, ft10, and ft20 from [24], 40 instances la001 – la040 from [25], ten instances orb01 – orb10 from [9], and ten instances swv11 – swv20 from [26] were used.* For every instance ten different scenarios established by different buffer capacities as shown in Table 4 were solved. Scenarios A – D reflect the case where every job consumes one place inside a buffer whereas scenarios E – J represent cases where the jobs consume variable buffer space.

Table 4. Buffer capacities and buffer consumptions

	A	B	C	D	E	F	G	H	I	J
κ_i	0	1	2	3	$1 \cdot p^{\max}$	$1.25 \cdot p^{\max}$	$1.5 \cdot p^{\max}$	$2 \cdot p^{\max}$	$3 \cdot p^{\max}$	∞
q_{ij}	1	1	1	1	Σp	Σp	Σp	Σp	Σp	Σp

* All instances are available at http://people.brunel.ac.uk/~mastjjb/jeb/orlib/files/jobshop1.txt

```
Procedure ScheduleOperation():
Parameter: current schedule S, operation o_ij, bool inserted, LNST_ij
if i < k_j
    Σ_ij ← {s*_ij} ∪ {c ∈ Φ_μij | c > s*_ij}          //calculate potential start times for o_ij
    while Σ_ij ≠ ∅ and inserted = false do
        s ← min Σ_ij                                   //choose next potential start time
        Σ_ij = Σ_ij \ {s}
        calculate LNST_i+1,j for c = s + p_ij
        calculate insertable(S, o_ij, s, LNST_ij, LNST_i+1,j)
        if insertable(S, o_ij, s, LNST_ij, LNST_i+1,j) = true
            s_ij ← s
            c_ij ← s_ij + p_ij
            S ← S ∪ {o_ij}
            s*_i+1,j ← c_ij
            for t = c_ij to t = LNST_i+1,j do
                I_κi+1,t = I_κi+1,t + q_ij             //update inventory in front of i + 1
            endfor
            if i > 0
                for t = s to t = LNST_ij do
                    I_κi,t = I_κi,t − q_ij             //update inventory in front of i
                endfor
            endif
            ScheduleOperation(S, o_i+1,j, inserted, LNST_i+1,j)  //continue recursion with o_i+1,j
            if inserted = false
                initialize s_ij and c_ij
                S ← S \ {o_ij}                         //reset current schedule
                reset s*_i+1,j
                for t = c_ij to t = LNST_i+1,j do
                    I_κi+1,t = I_κi+1,t − q_ij         //reset inventory in front of i + 1
                endfor
                if i > 0
                    for t = s to t = LNST_ij do
                        I_κi,t = I_κi,t − q_ij         //reset inventory in front of i
                    endfor
                endif
            endif
        endif
        SetNewFirstStartTime(insertible, Σ_ij, s)
    end while
else
    inserted ← true                                    //last operation of job j was inserted
endif
```

Fig. 3. Procedure *ScheduleOperation( )*

Table 5. Makespan for a descending and ascending sorting

Instance	Size	descending sorting										ascending sorting									
	(n, m)	A	B	C	D	E	F	G	H	I	J	A	B	C	D	E	F	G	H	I	J
abz05	(10, 10)	3601	1874	1890	1890	1874	1874	1874	1890	1890	1890	4113	2320	2137	2093	2320	2320	2320	2137	2093	2093
abz06	(10, 10)	3392	2060	1306	1306	2060	2060	1896	1306	1306	1306	3313	1690	1397	1397	1690	1690	1534	1397	1397	1397
abz07	(15, 20)	4382	1343	1142	1065	1343	1343	1343	1142	1065	964	4259	1527	1194	1058	1527	1527	1527	1194	1058	1058
abz08	(15, 20)	3584	1381	1205	990	1381	1381	1250	1205	990	990	4177	1511	1155	1202	1511	1511	1436	1155	1202	1110
abz09	(15, 20)	3635	1382	1244	1082	1382	1382	1361	1244	1059	947	3851	1612	1290	1249	1612	1612	1452	1290	1249	1224
ft006	(6, 6)	138	72	67	67	72	67	67	67	67	67	146	96	94	94	96	96	94	94	94	94
ft010	(10, 10)	2621	1541	1272	1272	1541	1541	1460	1272	1272	1272	2586	1477	1530	1530	1477	1477	1477	1530	1530	1530
ft020	(20, 5)	2778	1936	1617	1580	1936	1765	1666	1537	1544	1544	2500	1888	1745	1602	1872	1817	1799	1855	1602	1602
la001	(10, 5)	1653	1208	1033	1033	1208	1123	1208	1033	948	948	1964	1453	1210	1210	1453	1270	1270	1210	1210	1210
la002	(10, 5)	1525	1118	985	1084	1118	999	1009	1084	882	882	1701	1180	939	966	1072	1090	939	966	966	966
la003	(10, 5)	1744	1156	895	843	1156	1156	817	778	843	843	1669	1030	899	897	1030	972	940	899	897	897
la004	(10, 5)	1541	1042	892	1017	1042	955	943	892	968	968	1380	1185	1248	1055	1185	1333	1185	1248	1055	976
la005	(10, 5)	1435	814	754	699	770	754	754	699	699	699	1396	1067	905	905	1067	929	929	905	905	905
la006	(15, 5)	2304	1512	1157	1196	1419	1283	1219	1196	1196	1196	2218	1328	1303	1333	1328	1303	1303	1333	1333	1333
la007	(15, 5)	2421	1483	1153	1205	1475	1173	1176	1205	1123	1098	2106	1482	1279	1282	1482	1360	1300	1244	1282	1282
la008	(15, 5)	2541	1263	1070	1003	1125	1218	1070	1003	1003	1003	2146	1427	1393	1348	1427	1427	1383	1393	1348	1348
la009	(15, 5)	2316	1641	1552	1280	1493	1563	1446	1404	1280	1280	2362	1687	1465	1384	1687	1687	1687	1465	1384	1384
la010	(15, 5)	2358	1573	1223	1160	1550	1329	1213	1223	1151	1151	2272	1660	1509	1509	1660	1596	1509	1509	1509	1509
la011	(20, 5)	3226	1763	1458	1389	1657	1661	1541	1429	1498	1498	2892	1931	1689	1653	1894	1741	1741	1689	1653	1653
la012	(20, 5)	2795	1633	1409	1166	1383	1289	1379	1220	1143	1143	2807	1918	1499	1456	1747	1898	1843	1456	1456	1456
la013	(20, 5)	2832	1759	1618	1467	1759	1573	1541	1512	1467	1384	3360	1927	1644	1671	1944	1811	1811	1671	1671	1671
la014	(20, 5)	2845	1866	1519	1482	1621	1651	1477	1452	1470	1388	3189	1928	1888	1691	2285	2109	2064	1987	1669	1669
la015	(20, 5)	2582	1900	1695	1533	1859	1731	1651	1574	1533	1533	3245	1913	1904	1770	1913	2101	2020	1876	1900	1900
la016	(10, 10)	2752	1411	1262	1262	1411	1411	1533	1262	1262	1262	3284	1709	1644	1371	1709	1958	1644	1644	1371	1371
la017	(10, 10)	2695	1264	1154	1167	1264	1264	1343	1154	1167	1167	3054	1701	1560	1543	1701	1701	1764	1646	1439	1416
la018	(10, 10)	3431	1590	1117	1117	1590	1590	1487	1117	1117	1117	2919	1753	1384	1353	1753	1753	1753	1384	1353	1353
la019	(10, 10)	3182	1586	1146	1146	1586	1586	1586	1146	1146	1146	2726	1448	1302	1302	1448	1448	1448	1302	1302	1302
la020	(10, 10)	3562	1768	1245	1327	1768	1698	1321	1245	1327	1321	3375	1481	1447	1447	1481	1481	1460	1447	1447	1447
la021	(15, 10)	4129	2017	1623	1550	2017	1883	1586	1623	1367	1367	4662	2767	1918	1906	2767	2130	1985	1874	1906	1806
la022	(15, 10)	3960	1468	1687	1374	1468	1468	1468	1563	1374	1374	5210	2526	1677	1755	2526	2526	2077	1677	1778	1736
la023	(15, 10)	3977	1792	1523	1441	1792	1792	1763	1523	1441	1441	4847	2308	1645	1677	2308	2308	2130	1645	1677	1677
la024	(15, 10)	4644	1954	1643	1609	1954	1954	2019	1513	1609	1606	5045	2184	1856	1624	2184	2184	1826	1856	1624	1624
la025	(15, 10)	4401	1823	1533	1374	1823	1755	1529	1350	1465	1465	4101	2027	1849	1849	2027	2027	1982	1849	1849	1849
la026	(20, 10)	6289	2470	2044	1886	2470	2316	2212	2044	1717	1886	5415	2465	2355	2240	2465	3151	2315	2355	2240	2133
la027	(20, 10)	5885	2431	2213	2012	2431	2431	2016	2343	2012	1844	6238	2899	2616	2067	2899	2899	2887	2616	2160	2086
la028	(20, 10)	6040	2407	1962	1867	2407	2309	2087	1865	1867	1867	6527	2716	2162	2119	2716	2864	2607	2309	2119	2145
la029	(20, 10)	5436	2827	1841	1735	2827	2215	1771	1936	1735	1703	5718	2770	2118	2145	2770	2336	2747	2196	2154	2154
la030	(20, 10)	5000	2559	2186	2012	2559	2274	2238	2070	1858	1858	6241	2813	2333	2500	2813	2719	2806	2306	2500	2444
la031	(30, 10)	8074	3466	2865	2758	3466	3214	3027	2673	2532	2425	7819	3762	2969	3037	3762	3262	3346	2804	2934	2822
la032	(30, 10)	8232	4166	3094	2699	4166	3751	3020	2938	2722	2521	8788	4316	3260	3063	4316	4069	3616	3068	2880	2880
la033	(30, 10)	7343	3306	2568	2440	3306	2952	2566	2462	2361	2273	7746	3740	2833	2699	3740	3339	3432	2833	2723	2514
la034	(30, 10)	7506	3493	2736	2455	3493	3493	2926	2736	2455	2466	7658	4071	2673	2792	4071	4071	3232	2903	2943	2570
la035	(30, 10)	8148	3598	2894	2574	3598	3360	3183	2874	2574	2437	7971	3841	3136	3352	3841	3841	3795	2918	3352	2892
la036	(15, 15)	5678	2420	1995	1982	2420	2420	2232	1995	1982	1982	6739	2738	2280	2280	2738	2738	2304	2280	2280	2280
la037	(15, 15)	6669	2321	1907	1907	2321	2321	2321	1907	1907	1731	7118	2493	2204	2176	2493	2493	2493	2204	2176	2176
la038	(15, 15)	6214	1910	2235	1752	1910	1910	1871	2235	1752	1752	7117	2777	2081	2081	2777	2777	2472	2081	2081	2081
la039	(15, 15)	5912	2299	2263	2065	2299	2299	2129	2263	2065	2065	5573	2681	1998	2033	2681	2681	2681	1998	2033	2033
la040	(15, 15)	7111	2273	2023	1942	2273	2273	2058	2023	1916	1916	5367	2657	2007	2146	2657	2657	3049	2007	2146	2231
orb01	(10, 10)	2906	1558	1427	1391	1558	1558	1262	1427	1391	1391	2415	1945	1627	1489	1945	1945	1644	1627	1489	1489
orb02	(10, 10)	2985	1689	1500	1320	1689	1689	1453	1500	1320	1320	2744	1528	1609	1370	1528	1528	1528	1609	1370	1370
orb03	(10, 10)	2415	1518	1360	1360	1518	1518	1518	1360	1360	1360	2131	1689	1472	1463	1689	1689	1689	1472	1463	1463
orb04	(10, 10)	3292	1748	1674	1674	1748	1595	1674	1674	1674	1674	3260	1993	1604	1604	1993	1773	1773	1604	1604	1604
orb05	(10, 10)	3605	1497	1180	1180	1497	1497	1476	1180	1180	1180	2833	1283	1314	1296	1283	1283	1283	1314	1296	1296
orb06	(10, 10)	2852	1730	1713	1713	1730	1730	1723	1713	1564	1674	3101	1906	1562	1530	1906	1906	1478	1562	1530	1530
orb07	(10, 10)	1831	789	702	666	789	789	761	702	666	666	1427	984	618	618	984	984	984	618	618	618
orb08	(10, 10)	1997	1317	1207	1207	1317	1317	1375	1207	1207	1207	2188	1449	1386	1424	1449	1449	1558	1386	1424	1424
orb09	(10, 10)	2951	1787	1493	1450	1621	1621	1621	1450	1450	1450	3218	1852	1398	1398	1852	1852	1852	1398	1398	1398
orb10	(10, 10)	4092	1643	1437	1343	1643	1643	1355	1437	1343	1343	4694	1436	1501	1501	1436	1436	1436	1501	1501	1501
swv11	(50, 10)	10255	5671	4718	4640	5643	4897	4518	4430	4277	4195	10472	5279	4524	4495	5211	5050	4980	4321	4275	4279
swv12	(50, 10)	9597	5748	4656	4443	5748	5327	4759	4388	4194	4033	10980	5702	4566	4255	5702	5663	5001	4664	4201	4219
swv13	(50, 10)	10122	5722	4923	4439	5530	5134	4865	4882	4385	4231	10691	5936	4688	4131	5692	5796	5096	4723	4140	4314
swv14	(50, 10)	9885	5266	4639	4344	5155	4732	4422	4277	4318	3991	11128	5521	4368	4432	5473	5732	5293	4547	4231	4149
swv15	(50, 10)	10136	5471	4595	4237	5471	4904	4481	4553	4225	4194	10186	5297	4321	4172	5297	5441	5337	4722	4184	4229
swv16	(50, 10)	12047	5343	4050	3584	5343	5105	4397	4016	3505	3505	12312	5715	4210	4365	5715	5555	5040	4562	4158	4164
swv17	(50, 10)	10380	5173	4037	3643	5173	4627	4189	3985	3506	3425	11767	5614	4377	3964	5614	5415	4757	4343	4132	3765
swv18	(50, 10)	10797	5097	4274	3551	5097	4787	4307	4115	3551	3571	12012	5723	4126	4000	5723	5813	5015	4512	3932	3768
swv19	(50, 10)	11088	5495	4299	3782	5495	5079	4464	4191	3708	3570	12430	5570	4700	4236	5570	5359	5741	4669	4444	4190
swv20	(50, 10)	11263	5696	4127	3573	5696	5499	4106	3969	3457	3379	11639	5622	4419	4030	5622	5752	4874	4204	4040	3998

Scenario A corresponds to the *no-wait* case. For E – J the space that is consumed by a job was determined by setting $q_{ij} = \sum_{i=0}^{m} p_{ij} = \sum p$. That is, the buffer consumption of a job is proportional to its total processing time. The buffer capacities κ_i for E – J in front of each machine i of the same instance were defined by $\kappa_i = f \cdot \max\left(\sum_{i=0}^{m} p_{ij}\right) = f \cdot p^{\max}$. The values chosen for f are depicted in Table 4.

Table 6. Makespan for a mixed sorting

Instance	Size	M&P		mixed sorting									
	(n, m)	A	J	A	B	C	D	E	F	G	H	I	J
abz05	(10, 10)	2781	1318	4536	2096	1906	1906	2096	2096	2096	1906	1906	1906
abz06	(10, 10)	2743	985	3800	2081	1387	1527	2081	2081	1619	1387	1527	1527
abz07	(15, 20)	2538	753	4111	1474	1190	1050	1474	1474	1474	1190	1050	1050
abz08	(15, 20)	3145	783	4082	1442	1217	1271	1442	1442	1539	1217	1271	1085
abz09	(15, 20)	2350	777	3493	1733	1150	1129	1733	1619	1278	1150	1129	1129
ft006	(6, 6)	73	55	104	75	74	74	75	74	74	74	74	74
ft010	(10, 10)	2271	985	2095	2094	1426	1426	2094	1652	2066	1426	1426	1426
ft020	(20, 5)	1820	1338	2702	1940	1548	1510	2007	1748	1570	1510	1464	1489
la001	(10, 5)	1106	666	1879	1411	1212	919	1411	1058	1110	919	919	919
la002	(10, 5)	1296	694	1858	1060	840	840	952	952	840	840	840	840
la003	(10, 5)	955	735	1731	909	831	831	909	909	815	831	831	831
la004	(10, 5)	1039	679	1685	1309	991	991	1139	931	903	991	907	907
la005	(10, 5)	1094	593	1583	894	894	857	894	894	894	857	857	857
la006	(15, 5)	1758	926	2256	1765	1162	1139	1765	1576	1139	1067	1139	1067
la007	(15, 5)	1609	984	2464	1384	1193	1114	1371	1335	1325	1193	1114	1114
la008	(15, 5)	1580	873	2116	1535	1379	1141	1521	1469	1379	1141	1141	1141
la009	(15, 5)	2430	986	2789	1651	1172	1207	1872	1872	1409	1295	1207	1207
la010	(15, 5)	1506	1009	2500	1444	1380	1268	1359	1465	1380	1427	1276	1276
la011	(20, 5)	2226	1239	3141	1803	1743	1454	1837	1675	1623	1582	1387	1387
la012	(20, 5)	1935	1039	3316	1762	1345	1322	1742	1696	1345	1322	1322	1322
la013	(20, 5)	1867	1161	2704	1757	1553	1484	1813	1578	1556	1542	1516	1433
la014	(20, 5)	2381	1301	3096	1753	1505	1615	1787	1505	1500	1615	1615	1615
la015	(20, 5)	2134	1369	3177	2182	1700	1673	1816	1836	1610	1578	1554	1554
la016	(10, 10)	2337	979	2949	1720	1441	1382	1720	1456	1503	1382	1382	1382
la017	(10, 10)	1898	800	2630	1291	1159	1146	1291	1291	1231	1159	1146	1146
la018	(10, 10)	2197	916	2539	1565	1259	1249	1565	1565	1641	1259	1249	1249
la019	(10, 10)	2245	846	4478	1696	1358	1358	1696	1696	1696	1358	1358	1358
la020	(10, 10)	1971	930	2526	1796	1452	1372	1796	1664	1452	1452	1372	1372
la021	(15, 10)	3033	1241	4045	2500	1678	1589	2500	2385	1588	1599	1515	1515
la022	(15, 10)	2668	1032	3759	1948	1554	1554	1948	1948	1947	1554	1554	1453
la023	(15, 10)	3296	1131	5019	2119	1801	1801	2119	2119	2119	1801	1801	1801
la024	(15, 10)	2473	999	4154	2124	1703	1652	2124	2124	1649	1703	1686	1686
la025	(15, 10)	2750	1071	3926	1874	1625	1525	1874	2037	1594	1525	1525	1525
la026	(20, 10)	3687	1378	6932	2873	2287	1974	2873	2388	2253	2156	1815	1815
la027	(20, 10)	3488	1353	5639	3094	2081	1880	3094	2640	2737	2081	1880	1838
la028	(20, 10)	4806	1322	5314	2815	2268	1895	2815	2601	2001	2104	1833	1833
la029	(20, 10)	3086	1392	5260	2543	1977	2027	2508	2136	2090	2027	1912	1912
la030	(20, 10)	3392	1476	5390	2390	2124	2105	2963	2670	2407	2351	2025	2025
la031	(30, 10)	4838	1871	7907	3284	2561	2699	3284	3676	2790	2919	2617	2633
la032	(30, 10)	5469	1942	7859	3904	3382	2755	3904	3913	3122	3082	2601	2444
la033	(30, 10)	-	1897	7381	3165	2774	2611	3134	3718	2653	2606	2486	2486
la034	(30, 10)	5016	1942	7586	3667	2958	2641	3667	3667	3184	2958	2572	2543
la035	(30, 10)	5360	2017	8493	3781	2898	2705	3781	3607	3850	2898	2606	2637
la036	(15, 15)	3651	1347	7322	2644	2156	2116	2644	2644	2797	2156	2116	2116
la037	(15, 15)	4326	1547	6742	2623	2465	2420	2623	2623	2623	2465	2420	2164
la038	(15, 15)	3872	1342	6387	2233	1915	1915	2233	2233	2567	1915	1915	1915
la039	(15, 15)	-	1361	5846	2143	1974	2017	2143	2143	2412	1974	2017	2040
la040	(15, 15)	4422	1340	6571	2574	1893	1909	2574	2574	2318	1893	1909	1837
orb01	(10, 10)	1675	1213	2593	1732	1364	1364	1732	1732	1365	1364	1364	1364
orb02	(10, 10)	2324	924	2747	1736	1348	1348	1736	1736	1736	1348	1348	1348
orb03	(10, 10)	1647	1113	2054	1526	1471	1471	1526	1526	1709	1471	1471	1471
orb04	(10, 10)	1952	1108	3391	1961	1494	1418	1961	1961	1618	1494	1418	1418
orb05	(10, 10)	2261	924	2697	1543	1224	1109	1543	1543	1543	1224	1109	1109
orb06	(10, 10)	1788	1107	2653	1491	1470	1470	1491	1491	1691	1470	1470	1470
orb07	(10, 10)	922	440	1472	616	645	645	616	616	733	645	645	589
orb08	(10, 10)	1559	950	2114	1420	1344	1344	1420	1420	1468	1344	1344	1344
orb09	(10, 10)	1542	1015	2718	1598	1456	1456	1598	1598	1625	1456	1456	1456
orb10	(10, 10)	2467	1030	2782	2375	1473	1473	2375	2375	2075	1473	1473	1386
swv11	(50, 10)	-	-	10403	5512	4556	4244	5916	5430	4616	4481	4233	4126
swv12	(50, 10)	-	-	11123	5626	4523	4406	5626	5298	4748	4522	4181	4238
swv13	(50, 10)	-	-	10229	5627	4635	4714	5784	5501	4615	4986	4091	4079
swv14	(50, 10)	-	-	11269	5694	4527	4239	5569	5326	4599	4545	4016	4126
swv15	(50, 10)	-	-	10123	5334	4663	4198	5334	5483	4515	4456	4127	3960
swv16	(50, 10)	-	-	11576	6005	4432	3893	6008	5397	4846	4364	3855	3648
swv17	(50, 10)	-	-	11084	5206	4175	3650	5791	5469	4335	4325	3677	3564
swv18	(50, 10)	-	-	10872	5550	4310	3871	5550	5232	5042	4211	3771	3611
swv19	(50, 10)	-	-	12312	5775	4624	4157	5948	5393	5040	4427	3860	3749
swv20	(50, 10)	-	-	11204	5372	4514	3767	5662	5834	4506	4040	3503	3542

Furthermore, three different kinds of sorting for the job list were defined determining the sequence for the jobs to be scheduled by the generation scheme: In the first case the jobs were sorted according to their total processing time in a descending sequence and in the second case with an ascending sequence. The third kind of sorting was mixed in a way that it alternated between a job with longest total processing time first, a job with the shortest total processing time, followed by a job with second longest total processing time, one with second shortest total processing time and so on.

The computed results for the makespan for each instance and each defined scenario A – J are shown in Tables 5 and 6. For the scenarios A (*no-wait* case) and J (unlimited buffer capacities) the objective values heuristically obtained in [12] are depicted in column "M&P" in Table 6. Overall, 680 problem instances (68 instances in 10 scenarios) were solved for three times. Our results clearly indicate their advances and their weaknesses. Results of [12] are better than our results for those cases where they are applicable. However, [12] does not guarantee feasibility, and does not allow for solution of all scenarios. Figure 4 depicts the distribution of the makespans for the scenarios E to J based on the instances swv11 – swv20.

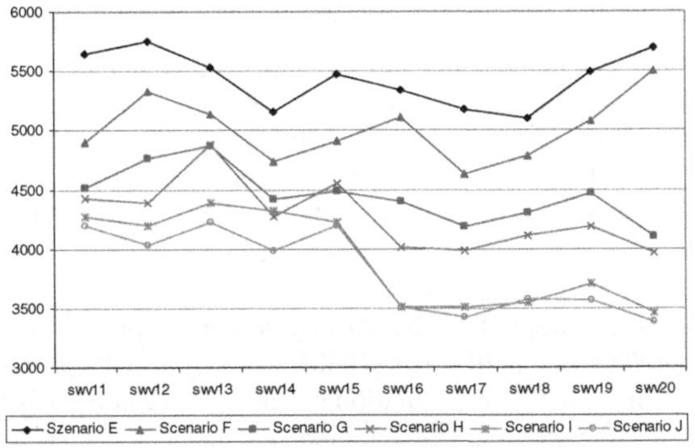

Fig. 4. Makespan distribution for swv11 – swv20

The best solution concerning the makespan was obtained with a descending sorting in 406, with an ascending sorting in 77 and with the mixed sorting in 197 cases. Computational times are almost neglectable for all instances but swv11–swv20, where the generation scheme might re-iterate due to deadlock prevention (this occurs for up to three minutes in the worst case swv14, scenario G, descending sorting).

5 Conclusions

In this paper a generation scheme to calculate feasible schedules for job shops with buffer constraints and jobs that consume different buffer space was presented. Computational results were obtained by applying the solution procedure to problem instances known from the scheduling literature.

The objective function values generated by a heuristic for the scenarios A and J in [12] outperform the results calculated by the heuristic presented here. However, the heuristic described in [12] does not guarantee feasibility in all cases and the job-wise scheduling within the generation scheme described above is essential to avoid deadlocks and guarantees to find feasible solutions in all cases. Furthermore, the generation scheme is capable to handle variable consumptions of buffer space what is new and seems to be particularly interesting for real-world applications and advanced planning systems. In conclusion, a generation scheme with an operation-wise scheduling in combination with deadlock avoidance and variable buffer consumption by the jobs seems to be an interesting topic for future research.

References

1. Muth, J.F., Thompson, G.L.: Industrial Scheduling. Prentice Hall, Englewood Cliffs (1963)
2. Carlier, J., Pinson, E.: An algorithm for solving the job-shop problem. Management Science 35, 164–176 (1989)
3. Pinson, E.: The job shop scheduling problem: A concise survey and some recent developments. In: Chretienne, P., Coffman, E.G., Lenstra, J.K., Liu, Z. (eds.) Scheduling Theory and Its Applications, pp. 277–294. Wiley, New York (1997)
4. Blazewicz, J., Domschke, W., Pesch, E.: The job shop scheduling problem: conventional and new solution techniques. European Journal of Operational Research 93, 1–33 (1996)
5. Jain, A.S., Meeran, S.: Deterministic job-shop scheduling: past, present and future. European Journal of Operational Research 113, 390–434 (1999)
6. Brinkkötter, W., Brucker, P.: Solving open benchmark instances for the job shop problem by parallel head-tail adjustments. Journal of Scheduling 4, 53–64 (2001)
7. Nowicki, E., Smutnicki, C.: An advanced tabu search algorithm for the job-shop problem. Journal of Scheduling 8, 145–159 (2005)
8. Adams, J., Balas, E., Zawack, D.: The shifting bottleneck procedure for job shop scheduling. Management Science 34, 391–401 (1988)
9. Applegate, D., Cook, W.: A computational study of the job shop scheduling problem. ORSA Journal on Computing 3, 149–156 (1991)
10. Leisten, R.: Flowshop sequencing problems with limited buffer storage. International Journal of Production Research 28, 2085–2100 (1990)
11. Witt, A., Voß, S.: Simple heuristics for scheduling with limited intermediate storage. Computers & Operations Research 34, 2293–2309 (2007)
12. Mascis, A., Pacciarelli, D.: Job-shop scheduling with blocking and no-wait constraints. European Journal of Operational Research 143, 498–517 (2002)
13. Brucker, P., Heitmann, S., Hurink, J., Nieberg, T.: Job-shop scheduling with limited capacity buffers. OR Spectrum 25, 151–176 (2006)
14. Heitmann, S.: Job-shop scheduling with limited buffer capacities. VDM Dr. Müller, Saarbrücken (2007)
15. Fahmy, S.A., ElMekkawy, T.Y., Balakrishnan, S.: Deadlock-free scheduling of flexible job shops with limited capacity buffers. European Journal of Industrial Engineering 2, 231–252 (2008)
16. Lawley, M., Reveliotis, S., Ferreira, P.: Structural control of flexible manufacturing systems and the neighborhood policy: Parts 1 & 2. IIE Transactions 29, 877–899 (1997)
17. Habermann, A.: Prevention of system deadlocks. Communications of the ACM 12, 373–377 (1969)
18. Lawley, M., Mittenthal, J.: Order release and deadlock avoidance interactions in counterflow system optimization. International Journal of Production Research 37, 3043–3062 (1999)
19. Aytug, H., Barua, A., Lawley, M., Uzsoy, R.: Observations on the interactions among deadlock avoidance policies and dispatching rules in automated manufacturing systems. International Journal of Production Research 41, 81–95 (2003)
20. Xing, K., Hu, B., Chen, H.: Deadlock avoidance policy for Petri net modeling of flexible manufacturing systems with shared resources. IEEE Transactions on Automatic Control 41, 289–295 (1996)

21. Zandong, H., Lee, G.: Application of Petri nets for deadlock analysis and avoidance in flexible manufacturing systems. International Journal of Advanced Manufacturing Technology 25, 735–742 (2005)
22. Kim, Y.W., Suzuki, T., Narikiyo, T.: FMS scheduling based on timed Petri Net model and reactive graph search. Applied Mathematical Modelling 31, 955–970 (2007)
23. Hatzack, W., Nebel, B.: The Operational Traffic Control Problem: Computational Complexity and Solutions. In: Cesta, A., Borrajo, D. (eds.) Recent Advances in AI Planning, 6th European Conference on Planning (ECP 2001), pp. 49–60. Springer, Berlin (2001)
24. Fisher, H., Thompson, G.L.: Probabilistic learning combinations of local job-shop scheduling rules. In: Muth, J.F., Thompson, G.L. (eds.) Industrial Scheduling, pp. 225–251. Prentice Hall, Englewood Cliffs (1963)
25. Lawrence, S.: Resource constrained project scheduling: an experimental investigation of heuristic scheduling techniques (Supplement). Graduate School of Industrial Administration. Carnegie-Mellon University, Pittsburgh (1984)
26. Storer, R.H., Wu, S.D., Vaccari, R.: New search spaces for sequencing instances with application to job shop scheduling. Management Science 38, 1495–1509 (1992)

Maturity Progression Model for Sustainable Supply Chains

Hendrik Reefke[1], David Sundaram[1], and M. Daud Ahmed[2]

[1] Department of Information Systems and Operations Management,
University of Auckland, Private Bag: 92019, Auckland 1142, New Zealand
{h.reefke,d.sundaram}@auckland.ac.nz
[2] School of Computing and Information Technology,
Manukau Institute of Technology, Private Bag: 94006 Auckland, New Zealand
Daud.Ahmed@manukau.ac.nz

Abstract. The three dimensions of sustainability have predominantly been addressed in a standalone fashion by academic researchers. Businesses and their supply chains often follow this trend, as environmental and social initiatives tend to be disconnected from overall strategies. Most sustainability frameworks are not directly applicable to the inherent complexity of supply chains. Supply chains need to address the sustainability dimensions in a strategic way that encompasses all stakeholders and aligns supply chain goals with the aims of individual members. Supply chains have to adjust their transformation strategy depending on the level of maturity. Our research proposes a model that integrates sustainability in supply chains with maturity considerations. Furthermore a sustainability modeling and reporting system is developed that facilitates a progression towards higher maturity levels. The proposed artifacts leverage the features from established frameworks and may be used in a prescriptive manner for design and implementation.

Keywords: Supply Chain Management, Strategy, Social, Environmental, Economic, Sustainability, Framework, Maturity, Modeling, Reporting.

1 Introduction

Corporate sustainability is about integrating social, environmental, and economic considerations into managerial decision making so that companies and their wider supply chains (SC) can survive on a long term basis [1-3]. These sustainability dimensions are highly interdependent and even though the concept of sustainability has been around for many years, businesses and their SC still struggle to implement more sustainable practices. This is partially caused by the contestable opinion that socially and economically responsible actions cannot be viable on an economic basis [1-2, 4]. Many firms have applied sustainability concepts into their processes, but often considered the sustainability dimensions independently rather than interdependently [5]. Organizations are increasingly faced with global competition and effective supply chain management (SCM) has become essential to all firms [6-8]. Often organizations no longer compete as single independent entities but rather as SC [9]. Hence, the

W. Dangelmaier et al. (Eds.): IHNS 2010, LNBIP 46, pp. 308–319, 2010.

concept of sustainability has to be extended from an organizational focus to entire SC. Decision makers designing SC have to consider the impact of their decisions on quality of life, safety, health, and welfare of the public to act in an ethical manner [10]. Compliance to management standards to improve environmental and social performance is commonplace nowadays [11], which pressures companies to adopt more sustainable operations and sustainability measures across the SC [12].

The interaction of sustainability and SCM is a timely and critical step in examining operations and sustainability [2, 13-14] and focuses on a product or service throughout the whole production and delivery process [14]. This research attends to both sustainability and SCM to provide results of interdisciplinary nature to benefit researchers and practitioners alike. There is an apparent lack of theory-building research in the field of sustainable supply chain management (SSCM) [14-15]. A conceptual theory building process provides a balance between inductive and deductive reasoning and gives researchers a possibility to develop valid theories and guide business practice. A review of existing literature and frameworks provides the foundation for this research by interpreting and synthesizing published research [16] and providing the reader with the major developments, questions, and issues in the field of SSCM [17]. Evaluation and comparison allows us to extract the common elements, contrast the differences, and extend the existing work with logical reasoning [18]. The emerging main research artifacts are a maturity model with a cyclical progression component presented in Figure 1 and Table 1, which connect and extend existing frameworks and studies. A sustainability modeling and reporting system that supports the cyclical progression model explicitly is presented in section 5.

2 Sustainable Supply Chains

SCM has become a method to effectively manage organizations and their business relationships, in particular to realize organizational goals when business partners are dispersed around the world [6, 9]. Successful integration and management of key business processes along the SC and the network of relationships between business partners can determine the success of the single firm [9], as it enhances competitiveness, customer service, and profitability [6]. SCM is a concept used by many organizations as an essential way of planning, designing, and controlling their business networks and tasks involved in the different stages of the SC [7]. The concept involves the integration of operations, technologies, information, processes, capital flows, and resources with business partners to ensure better coordination along the SC and to have a sustained competitive advantage [6-7, 19-20].

SC are evolving towards dynamic supply networks with every member influencing the finished product/service and the processes involved. The applicability of sustainability for the different aspects of SC has been pointed out by various studies. Sustainable development is accepted as a guiding principle, but actual application in business practices is still low [21]. The pressures on SC to operate sustainably on environmental, social and financial levels are steadily increasing. Often new legislations, customer expectations or competitive opportunity drive companies and their SC to align their operations to improve sustainability [22]. Practical implementation of sustainability measures is complex [23] and is often only applied to one's own SC and rarely extends towards second order SC. Research concerned with the connection between several SC remains also rather limited [13]. There is a gap in SC research when it comes to the effects of decisions on present and future ecosystems [24]. It is also not clear what constitutes an appropriate balance between individual tastes, the rights of others and the collective good [25]. The pressure to focus on the sustainability of SC is especially relevant for global market products [26]. In production or manufacturing the focus for more sustainable development is generally on supporting resource efficiency, as a very significant amount of today's production output is actually waste [4]. Furthermore ecological damage can be minimized by designing all production stages in an environmentally sustainable way [27]. A higher degree of social and ecological control throughout the SC can yield financial benefits and support long term relationships between business partners, which in turn support financial sustainability [3, 28]. Cooperation up- and downstream in a SC can improve environmental performance for the whole SC [29].

Many companies have embraced the concept of sustainability for their own company. However, sustainability management has to take the SC into account as often 50-70 percent of the value of a product is derived from suppliers. Thus, efforts in sustainability management that only concentrate on the corporate environment cannot achieve genuine sustainability [30]. SSCM is an extended approach to the traditional SCM based on the triple bottom line [5] to improve the long-term economic performance of the firms involved in the SC, and the SC itself [13]. The concept of sustainable SC refers to the transparent integration and achievement of the organization's economic, environmental and social goals by effectively coordinating the main inter-organizational business processes [5, 31]. There is no generally accepted definition of SSCM and it is also difficult to quantify the results that companies can achieve. Carter and Rogers [5] employ conceptual theory building in order to derive a number of potential benefits that companies can reap from SSCM. Though well justified by literature, overall validity has yet to be empirically verified. Key aspects of SSCM practices include the sustainability of the SC network and SC environment, application of environmental-friendly strategies, and embracing social responsibilities [13]. In order to ensure the sustainability of an entire SC, each of the individual companies has to be sustainable [1]. Thus, sustainable SC need to extend beyond the fundamental matters of SCM such as design, manufacturing, the creation and use of by-products, and product recovery and disposal [22].

Several internationally operating firms proposed initiatives to reduce waste and pollution in China's export-focused manufacturing industries, e.g. Wal-Mart, Procter & Gamble, FedEx, Lenovo, and Coca-Cola [32]. Wal-Mart acknowledges that its environmental footprint is primarily based on their SC as a company [32]. It is important to note that many companies seem to be interested in the sustainability of their SC for economic, ethical or marketing reasons. However, these efforts can only be seen as a starting point [33]. Such developments are often not holistic approaches to improve the organization's impacts on all sustainability dimensions and the inter-relations between the dimensions are only rarely considered [5, 20]. Some sustainability efforts can also easily be categorized as unverifiable claims in order to 'greenwash' the company's or SC's operational practices [34].

Overall the reviewed publications and the practical examples show that there is an urgent requirement for directed sustainability strategies within SC.

3 Maturity Models

A maturity level can be defined as an evolutionary plateau of process improvement [35]. SC maturity can be seen as a series of successive levels or stages that a SC goes through from early forming stages to a collaborative SC environment with common goals, measures and extensive information and resource sharing. Generally skipping maturity levels is described as counter-productive as the levels are interconnected and lower levels provide a foundation for subsequent levels [35-36]. Thus, a series of maturity levels or stages can also be assumed for SSCM including the assumption that a SC has to undergo each step in an evolving manner.

A well recognized capability model with five maturity layers has been developed by Carnegie Mellon [35]. At maturity level one, processes are ad hoc and chaotic. Business success depends solely on the competence of individuals. More visibility is evident in level two and requirements, processes, work products, and services are managed. Maturity level three is based on well documented standard processes. For the next level performance measures have to be established so that processes become quantitatively managed. Thus, the performance of processes becomes controlled and predictable. Maturity level five focuses on a continuous improvement process, so that all the goals for the preceding levels can be optimized further.

Yusuf et al. [37] propose a three stage model which shows agility and capability of a SC in relation to its SC maturity. Three interdependent dimensions of SC maturity are used, i.e. customer interaction, asset configuration, and knowledge leverage. This SC maturity model is based on a model developed by Venkatraman and Henderson [38] which focuses on virtual organizing. The three stages are meant to be used to "evaluate progress on each of the three dimensions of SC maturity" [37]. Venkatraman and Henderson [38] also add two more considerations in their work, i.e. target locus which refers to focus point of action and performance objectives for each of the three stages. The target locus would be on task units for stage one, on the organization for stage two and finally extends to inter-organizational relations in stage three. The performance objectives for the three stages change from improved operating efficiency for stage one to enhanced economic value added which evolves further to sustained innovation and growth in stage three. Yusuf et al (2004) point out that a SC aiming for agility should aim to reach the characteristics specified in stage three which according to their model contain the requirements for the highest level of SC maturity. These requirements are "ownership of customer communities or niche markets, membership of manufacturing resource coalitions, and possession of a workforce that operates within a community of professional experts" along with an inter-organizational target locus which drives "competitive strategies, plans and innovation" leading to sustained innovation and growth [37].

A five stage SC maturity model is proposed by Lockamy III and McCormack [36] which show "the progression of activities toward effective SCM and process maturity". All five levels are connected to maturity characteristics such as predictability, capability, control, effectiveness and efficiency. At the 'Ad hoc' level the SC is undefined and there are no established process measures. Process performance is therefore unpredictable and not on target. Functional cooperation between SC members is low which results in high SCM costs and unsatisfied customers. SC maturity increases from the 'ad hoc' level to the 'defined' level where processes and targets are defined and become more predictable. Companies are still not working together but are more concerned about local optimization. Hence, SCM costs are still relatively high and customer satisfaction is low. The 'linked' level is referred to as the breakthrough level as SCM is applied in a strategic fashion with structures in place that support SCM. Measures and goals are shared among SC members and problems are solved on a SC level which results in lower SC costs and higher customer satisfaction. Collaboration on a process level is common place in the 'integrated' level along with organizational structures which fully support the SCM strategy. SCM is a deeply integrated strategy and SC members work collaboratively with regard to planning and forecasting. SCM costs are low and customer satisfaction is high. In the 'extended' level finally competition is on a SC level and SC members work closely together with complete information sharing and deeply embedded collaboration in place. Risks and rewards are shared among SC members, e.g. investments are taken on together and benefits are given back to the entire SC. A high level of trust and mutual dependency encourages firms to develop common goals and processes to fulfill customer demand in a collaborative fashion [36].

Maturity of corporate sustainability has also been described in a five level model [39], focusing on required capabilities to reach the five maturity levels. Companies on level one are unaware and non-compliant to any regulations and undertake no sustainability efforts. Basic compliance with a limited awareness of sustainability issues and reactive compliance has been achieved in the next level. The company develops internal skills in level three, undertakes more directed voluntary efforts and exceeds regulations. In the fourth level sustainability becomes integrated into the business model including proactive sustainability programs and regular corporate reports on sustainability. The last level finally is based on sustainability leadership with sustainability values deeply integrated into the company culture and setting business standards.

4 SSCM Maturity Model

In section 2 the aspects and issues connected to SSCM were discussed and the concept of maturity levels was introduced in section 3. When looking at the described maturity models for SCM, corporate sustainability, and process capability, it becomes evident that there are many similarities. The lessons learned in the reviewed frameworks can be leveraged to achieve higher sustainability maturity in SC as well. A maturity model helps to establish a common structure to achieve a specific goal or a number of objectives. It is similar to a typical strategic roadmap and should provide a clear direction and shared vision by outlining the overall purpose of the business transformation, provide a common language by setting clear goals and objectives, provide guidelines, determine the roles of all parties involved, define action plans outlining the steps to fulfill the goals and objectives, and help users to communicate and evaluate the strategic decision [35, 40-41]. Furthermore the time dimension needs to be considered by outlining a progression strategy between the current state of the business and the long-term vision [41].

To address the mentioned requirements we propose the SSCM maturity model as depicted in Figure 1 which offers support for a long term sustainable SC strategy. The goals and requirements of each maturity level as well as a description are provided in

Table 1. There is a clear requirement for a sustainability maturity model specific to SC due to the unique requirements in SC outlined. There are few models which connect maturity considerations in SC with sustainability imperatives. However, such a model is needed in order to address the challenges that companies and their SC are faced with.

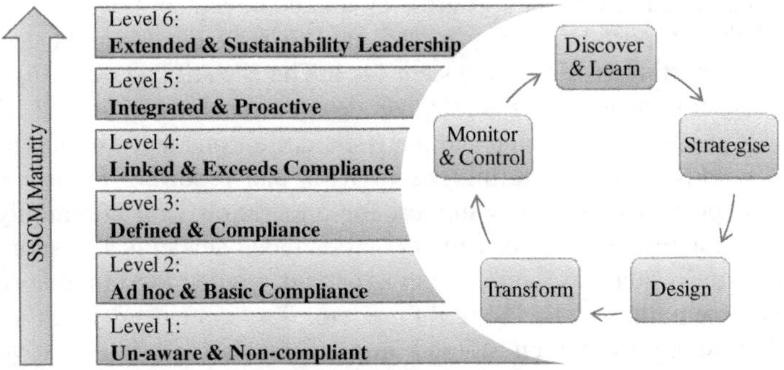

Fig. 1. SSCM Maturity Model

Table 1. Description, Goals, and Requirements of SSCM Maturity Model

Level	Description	Goals and Requirements
6	Processes are systematically managed through continuous improvement. Full SC collaboration embracing sustainability leadership role	Keep optimizing processes and ensure future leadership role
5	Sustainability has become a fully integrated concept and SC has moved towards proactive measures.	Make strategic concepts available to others and move towards leadership role
4	SC is linked and includes a comprehensive sustainability performance measurement system.	Move from compliance level towards proactive sustainability efforts.
3	Sustainability goals/standards have been defined and SC members are compliant with regulations.	Establish key indices to measure sustainability performance within SC.
2	Sustainability measures are disconnected from strategic direction. Compliance on a basic level.	Align sustainability goals and efforts with defined processes. Establish consistency.
1	SC are unaware and non-compliant to any regulations and undertake no sustainability efforts.	Raise sustainability awareness. Introduce sustainability initiatives.

The time dimension is clearly addressed by outlining the strategic progression between the potential current state of a SC and the long term vision. This strategic progression between the levels of maturity is supported through an iterative roadmap made up of clearly defined stages of discovery and learning, strategizing, design, transformation, and monitoring and controlling [40]. After the last phase a new cycle can be performed to progress from one maturity level to the next. This cyclical multi-step approach for maturity progression is a common approach for business process transformation as shown by e.g. the ARIS framework or the Natural Step [42-43].

In the following section we introduce a system that supports these stages explicitly allowing organizations to progress from being unaware and non-compliant to sustainability leadership.

5 Sustainability Modeling and Reporting System

The SSCM maturity model embraces the sustainable business transformation roadmap [40] for discovery, learning, strategic planning and alignment through incremental design, application and adaptation, and management of the process through continuous monitoring, controlling, communication through reporting and iteration of the process again and again. Gradual progression and achievement towards staged maturity levels of the proposed maturity model can be realized using the sustainability modeling and reporting (SMART) system [44]. The SMART system supports every aspect of the discovery and learn, strategize, design, transformation, and implementation phases. It supports data entry and analysis, definitions and execution of models (simulation, workflow, data), definitions of KPIs and reporting for monitoring and control, and it provides extensive support for integration with commonly available databases and systems. A group of experts that included academics, system architects, system and business analysts, and domain experts evaluated the framework, architecture and implementation of the SMART system. The system provides a number of functions that relate to the end-to-end implementation of the iterative roadmap discussed above and thereby the progression from one maturity level to another.

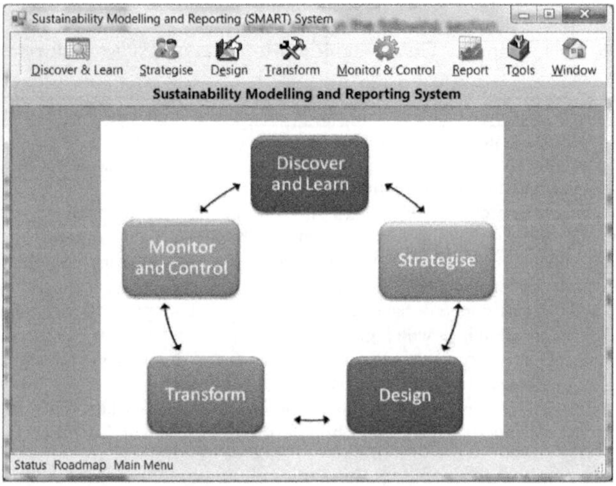

Fig. 2. Main Screen of the SMART System

The SMART system's graphical user interface directly maps to the iterative roadmap (Fig. 2). The five high-level phases of the roadmap are shown as five main menu items on the menu bar or five interactive buttons for user interaction. Each of these menu items or graphical buttons incorporates several sub-menus or context-sensitive menus for representing various steps for each of the SC transformation roadmap phases. The Discover and Learn menu supports discovery and learning activities. The Strategize menu supports import, selection, instantiation, simulation and analysis of the sustainability models. It also supports strategy documentation and domain model-based sustainability development and simulation. The Design menu supports flow-through of the roadmap processes as well as integration of business process models and their simulation and analysis for the business process design. This design phase includes development of the scenario model and management, development of domain object, data models, organization model and functional model, design documentation and business blueprint. The Transform menu supports project planning, and actual execution of the business process models, information systems, and organizational structure. The Monitor and Control menu supports continuous monitoring of the newly implemented system using a dashboard. It also supports the use of report and data models for sustainability modeling and reporting. As shown in Fig. 2, we have also included several other menu items such as Report, Tools and Window for supporting activities. The Report menu pulls all the reports from various phases, (such as discover and learn, strategize, design, monitoring and controlling) under one heading. It also includes sustainability and ad-hoc reports. Tools and Window menus contain various supportability and usability features such as configuration tool, connection to third party systems and modeling environments.

The SMART system enables the decision makers to undertake the required decision support activities ranging from strategic to tactical to operational towards the accomplishment of the tasks at various maturity levels of the proposed model. For example, the SMART system supports maturity level 1 by enhancing sustainability awareness and enabling sustainability initiatives through its directed and undirected knowledge discovery process (Fig. 3). It facilitates organization and presentation of various compliance requirements, leading corporate practices and innovations within the similar industries and sustainability business and SC leaders.

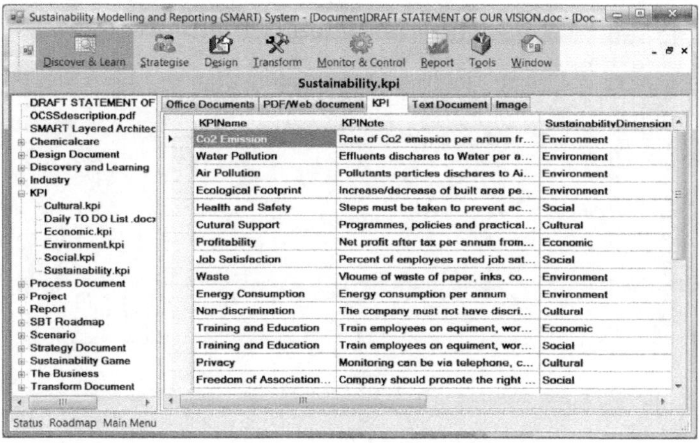

Fig. 3. Discovery and Learning

At maturity level 2, sustainability goals are disconnected from the strategic directions. The SMART system however supports measuring performance indicators of the existing SC processes and operations for establishing consistency and achieving basic compliance at level 2 of the maturity model. The system furthermore supports aligning sustainability vision, goals and strategic objectives through system thinking based simulation modeling and reengineering of SC processes. It helps to identify and define sustainability key performance indicators for the SC stakeholders that are fully compliant with regulations. It also prescribes KPI measurement process and their realization and thereby supports the 3rd level of the SSCM maturity model.

Once the initial transformation of various parts of the SC network is completed, the SMART system supports the application of the system dynamics approach. This enables the exploration, definition, description and measurement of the sustainability KPIs for the entire SC. This allows decision makers to pro-actively manage the comprehensive compliance levels and communicate within the SC members for achieving 4th level SSCM maturity and stepping towards 5th level maturity. The SMART system supports proactive measurement of the SC performance based on predefined leading target KPIs. It supports KPI comparison of current, industry standard, target, historical data and best performing SC unit (Fig. 4) for achieving 5th level of the SCM maturity model. The SMART system supports continuous iteration and refining and innovating of the SC process for taking up a sustainability leadership role in the SC industry for establishing a level 6 SC network. The level of certain KPIs in itself helps the user to understand their maturity in terms of sustainable SC. As one improves in terms of various sustainability initiatives and KPIs with respect to SC, the SMART system enables the organization and user to monitor in a systematic fashion more and more aspects and elements of the SC.

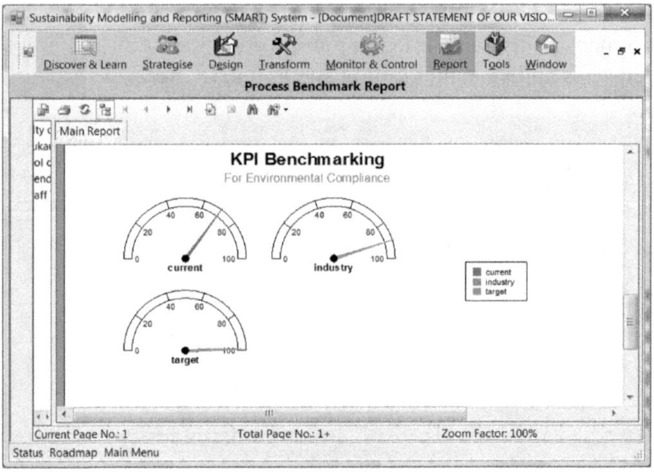

Fig. 4. KPI Reporting

6 Conclusion

Conceptual theory building has led to the development of a SSCM maturity framework. We recognized the need for more conceptual work in the field of SCM which justifies the chosen approach. We have accepted that organizations and their SC have to "link environmental, social, and economic goals within a broader strategic perspective" [5] and provides a strategic approach to link formerly disconnected and fragmented sustainability efforts to economic goals of SC [45]. Our framework explicitly links the members of a SC, with the goals of SCM, and the multi-dimensional nature of sustainability and also includes the main stakeholders of sustainability efforts. We furthermore include maturity considerations that guide SC towards higher order maturity levels. Thus, the developed frameworks and roadmap provide an over-arching view that combines multiple lenses in order to guide a more comprehensive SSCM strategy.

The described framework can be seen as a starting point for academics towards more research in this particular area. Due to the exploratory nature of this work and the early stage of the proposed framework, we cannot make claims about validity and precision. The SSCM strategy framework and roadmap therefore demand validation and refinement through more empirical research methods which can be of qualitative as well as quantitative nature. Thus, this research can potentially provide the background for further studies in the field. One possible avenue could be case study research in order to test the validity of our conceptual artifacts in a real-life SC environment. Additionally further refinement could be accomplished by taking an even wider theoretical background into account. A larger literature database coupled with interviews with experts in the field of SCM as well as sustainability could provide further insights. Survey research could also be applied in order to validate the framework as surveys allow the researcher to collect information "from the individual or unit directly responsible for managing a certain operation of interest" [46]. It could therefore be used to measure the aspects of the artifacts from an expert's perspective. A Delphi study with experts in the field of SSCM could also be used to refine and validate the framework and roadmap [47-48].

There are also implications for practitioners in the field of SCM. It can lead and direct sustainability efforts as it delivers a common ground and starting point for SC managers. The multi lens approach brought forward in the framework helps to direct inter-organizational communication. It furthermore helps to provide a common understanding of what constitutes a SSCM strategy and how the dimensions of sustainability link with the economic success of a company. Furthermore the multilayered roadmap can guide decision makers to elevate their SC from an 'ad hoc' approach to sustainability towards an 'extended' strategic sustainability concept ably supported by systems such as the Sustainability Modeling and Reporting System.

References

1. Hutchins, M.J., Sutherland, J.W.: An exploration of measures of social sustainability and their application to supply chain decisions. Journal of Cleaner Production 16(15), 1688–1698 (2008)
2. Kleindorfer, P.R., Singhal, K., Van Wassenhove, L.N.: Sustainable Operations Management. Production and Operations Management 14(4), 482–492 (2005)
3. McIntyre, K.: Delivering sustainability through supply chain management. In: Global Logistics - New directions in supply chain management, London (2007)
4. Senge, P., Carstedt, G., Porter, P.: Innovating our way to the next industrial revolution. MIT Sloan Management Review 44(2), 24–38 (2001)
5. Carter, C.R., Rogers, D.S.: A framework of sustainable supply chain management: moving toward new theory. International Journal of Physical Distribution & Logistics Management 38(5), 360–387 (2008)

6. Gunasekaran, A., Patel, C., Tirtiroglu, E.: Performance measures and metrics in a supply chain environment. International Journal of Operations & Production Management 21(1/2), 71–87 (2001)
7. Mabert, V.A., Venkataramanan, M.A.: Special Research Focus on Supply Chain Linkages: Challenges for Design and Management in the 21st Century. Decision Sciences 29(3), 537–552 (1998)
8. Mentzer, J.T., et al.: Defining Supply Chain Management. Journal of Business Logistics 22(2), 1–25 (2001)
9. Lambert, D.M., Cooper, M.C.: Issues in Supply Chain Management. Industrial Marketing Management 29(1), 65–83 (2000)
10. World Federation of Engineering Organisations: Declaration implementing Sustainable Development, Calcutta (2000)
11. Hall, J.: Environmental supply chain dynamics. Journal of Cleaner Production 8, 455–471 (2000)
12. Darnall, N., Jolley, G.J., Handfield, R.: Environmental Management Systems and Green Supply Chain Management: Complements for Sustainability? Business Strategy and the Environment 17(1), 30–45 (2008)
13. Svensson, G.: Aspects of sustainable supply chain management (SSCM): conceptual framework and empirical example. Supply Chain Management: An International Journal 12, 262–266 (2007)
14. Jayaraman, V., Klassen, R., Linton, J.D.: Supply chain management in a sustainable environment. Journal of Operations Management 25(6), 1071–1074 (2007)
15. Meredith, J.: Building operations management theory through case and field research. Journal of Operations Management 16(4), 441–454 (1998)
16. Merriam, S.B.: Case study research in education: a qualitative approach. The Jossey-Bass education series, p. 226 (1988)
17. Gill, J., Johnson, P.: Research methods for managers, London (1991)
18. Meredith, J.: Theory Building through Conceptual Methods. International Journal of Operations & Production Management 13(5) (1993)
19. Bullinger, H.J., Kuhner, M., Van Hoof, A.: Analysing supply chain performance using a balanced measurement method. International Journal of Production Research 40, 3533–3543 (2002)
20. Seuring, S., Müller, M.: From a literature review to a conceptual framework for sustainable supply chain management. Journal of Cleaner Production 16(15), 1699–1710 (2008)
21. Seuring, S.: Integrated chain management and supply chain management comparative analysis and illustrative cases. Journal of Cleaner Production 12(8-10), 1059–1071 (2004)
22. Linton, J.D., Klassen, R., Jayaraman, V.: Sustainable supply chains: An introduction. Journal of Operations Management 25(6), 1075–1082 (2007)
23. Preston, L.: Sustainability at Hewlett Packard: From Theory to Practice. California Management Review 43(3), 26–37 (2001)
24. Beamon, B.M.: Environmental and sustainability ethics in supply chain management. Science and Engineering Ethics 11(2), 221–234 (2005)
25. Woodhouse, E.J.: Curbing Overconsumption: Challenge for Ethically Responsible Engineering. IEEE Technology and Society Magazine 20, 23–30 (2001)
26. Brent, A.C.: Management of imported supply chain products: Incorporating country specific sustainability criteria in life cycle decision analysis. Department of Engineering and Technology Management (2002)
27. Lye, S.W., Lee, S.G., Khoo, M.K.: A design methodology for the strategic assessment of a product's eco-efficiency. International Journal of Production Research 39, 2453–2474 (2001)

28. Porter, M.E., van der Linde, C.: Green and competitive: ending the stalemate. Harvard Business Review 73(5), 120–134 (1995)
29. Zhu, Q., Sarkis, J.: An inter-sectoral comparison of green supply chain management in China: Drivers and practices. Journal of Cleaner Production 14(5), 472–486 (2006)
30. Mahler, D.: The Sustainable Supply Chain. Supply Chain Management Review (2007)
31. Clift, R.: Metrics for supply chain sustainability. Clean Technologies and Environmental Policy 5(3), 240–247 (2003)
32. Birchall, J.: Wal-Mart to seek greener supply chain. Financial Times (2008)
33. Baker, M.: What Wal-Mart's Sustainability Index is and isn't. Business Respect 157 (2009)
34. Karliner, J.: The Corporate Planet: Ecology and Politics in the Age of Globalization. Sierra Club Books (1997)
35. Carnegie Mellon University: Capability Maturity Model Integration (CMMI). In: CMMISM for Systems Engineering, Software Engineering, Integrated Product and Process Development, and Supplier Sourcing (CMMI-SE/SW/IPPD/SS), Pittsburgh, pp. 1–729 (2002)
36. Lockamy III, A., McCormack, K.: The development of a supply chain management process maturity model using the concepts of business process orientation. Supply Chain Management: An International Journal 9(4), 272–278 (2004)
37. Yusuf, Y.Y., et al.: Agile supply chain capabilities: Determinants of competitive objectives. European Journal of Operational Research 159(2), 379–392 (2004)
38. Venkatraman, N., Henderson, J.C.: Real strategies for virtual organizing. Sloan Management Review 40(1), 33–48 (1998)
39. Spectrum Innovation Group. Corporate Sustainability: Capability Maturity Model (2008)
40. Ahmed, M.D., Sundaram, D.: A Roadmap for Sustainable Business Transformation. The International Journal of Environmental, Cultural, Economic and Social Sustainability 5(2), 165–182 (2009)
41. Phaal, R., Farrukh, C.J.P., Probert, D.R.: Technology roadmapping-A planning framework for evolution and revolution. Technological Forecasting and Social Change 71(1-2), 5–26 (2004)
42. Scheer, A.-W., et al.: Business Process Change Management: ARIS in Practice. Springer, Heidelberg (2003)
43. The Natural Step, `http://www.naturalstep.org/en/our-story`
44. Ahmed, M.D.: Sustainable Business Transformation: roadmap, framework, architecture and implementation. Unpublished PhD Thesis, Department of Information Systems and Operations Management. University of Auckland (2009)
45. Porter, M.E., Kramer, M.R.: Strategy & society: The link between competitive advantage and corporate social responsibility. Harvard Business Review 84(12), 78–92 (2006)
46. Boyer, K.K., Swink, M.L.: Empirical Elephants-Why Multiple Methods are Essential to Quality Research in Operations and Supply Chain Management. Journal of Operations Management 26(3), 338–344 (2008)
47. Lummus, R.R., Vokurka, R.J., Duclos, L.K.: Delphi study on supply chain flexibility. International Journal of Production Research 43, 2687–2708 (2005)
48. Seuring, S., Müller, M.: Core issues in sustainable supply chain management - a Delphi study. Business Strategy and the Environment (2007)

Scenario Technique with Integer Programming for Sustainability in Manufacturing

Armin Fügenschuh[1], Pia Gausemeier[2],
Günther Seliger[2], and Semih Severengiz[2]

[1] Konrad-Zuse-Zentrum für Informationstechnik,
Takustr. 7, 14195 Berlin, Germany
fuegenschuh@zib.de
[2] Technische Universität Berlin,
Institut für Werkzeugmaschinen und Fabrikbetrieb (IWF)
Fachgebiet Montagetechnik und Fabrikbetrieb
Pascalstr. 8-9, 10587 Berlin
{gausemeier,seliger,severengiz}@mf.tu-berlin.de

Abstract. Scenario technique is a tool to reveal the knowledge of an expert group about possible and consistent scenarios of the future. The current use of this technique suffers from an inflexibility of the available software. To overcome this shortage, the core part of this solver, i.e. the scenario generation phase, has to be re-implemented. It is demonstrated that this problem can be formulated as an integer linear programming problem. To find all feasible solutions the solver SCIP was adapted. The solutions are then clustered and these clusters are verbally described. The approach is demonstrated by using data from the research project remanufacturing oriented production equipment development, in which scenarios for sustainable value creation were identified and described.

Keywords: Sustainable Manufacturing, Scenario Technique, Integer Programming, Solution Enumeration, Clustering.

1 Introduction

The acceleration over the past decades in demand for natural resources has reached a point where it is now considered to be a serious threat to the functioning of economies and societies. If the lifestyle of rapidly advancing nations becomes shaped by the predominant technologies of the industrialized countries, the global resource consumption will exceed every ecologically, economically and socially responsible level [1]. This is associated with environmental problems such as biodiversity loss, climate change, desertification and ecosystem degradation [2,3]. It also affects a fair wealth distribution economically, and the needs of future generations socially.

To tackle these issues, scenarios have to be set up for various fields of applications to show sustainable approaches for value creation. The term scenario, within the area of future research, was coined by Khan et al. in the 1960s. Their

W. Dangelmaier et al. (Eds.): IHNS 2010, LNBIP 46, pp. 320–331, 2010.

study "The Year 2000: A Framework for Speculations on the next Thirty-Three Years" is considered to be the driver for todays scenario technique [4]. As stated by Börjeson et al., scenarios can be classified into three different types: predictive, explorative, and normative scenarios [5].

According to Gausemeier et al., scenario creation builds on two main principles. Firstly, system thinking describes the environment of enterprises, stating that it must be perceived as a complex network of interrelated influencing factors. Secondly, influencing factors may have different projections for the future. Thus, taking different projections of these factors into account, different multiple futures can be considered [6,7]. Scenarios are defined as a generally intelligible description of a possible situation in the future, based on a complex network of influencing factors. This scenario technique approach can be classified into the category of exploratory scenarios. It consists of five phases: scenario preparation, scenario field analysis, scenario prognostic, scenario development and scenario transfer. This approach has been adapted to the usage of sustainability influencing factors and then been used for scenario creation.

1.1 Scenarios for Sustainability in Value Creation

Sustainable value creation nets attempt to achieve welfare through a balance between cooperation and competition. Their competitive strength derives from the productivity of their processes and the quality of their products. The participants in these nets also learn that they depend on each other in the global village. They try to achieve reciprocal benefits by working together in a cooperative manner. It becomes more and more obvious that material resources are the bottleneck, and that therefore both use and welfare in product life cycles have to be conceptualized with respect to a maximum in resource efficiency.

If four fifth of humanity possess less than one fifth of the global prosperity today, then there is clearly potential in demand and supply for developing these unsaturated global markets. Work is no longer the key factor, but natural resources. More use of fewer resources becomes a production paradigm.

The insights presented here are based on scenarios for sustainable value creation, which were identified in a previous project conducted in the Brazilian machine tool community. Different from the original outlines, the present approach includes criteria for sustainability and offers an alternative mathematical way for clustering scenario data.

1.2 Mathematical Background of Scenario Technique

The core element of the scenario technique is the computation of consistent scenarios. From an operations research point of view, this is a problem that can be formulated as a linear program with integrality constraints on the variables (ILPs, for short). The details of this model will be presented in Section 3 below. ILPs belong to a class of computationally difficult problems (that is, they are NP-hard), which means that no efficient algorithm (that is, an algorithm that finds a global optimal solution in polynomial running time) is known for their solution. One way to practically solve problems of this class is by enumeration schemes, where every possible assignment of values to the variables is tested – a divide-and-conquer approach. In this way the global optimal solution is guaranteed to be found in finite time, but due to the inherent complexity this approach is unpractical for more than 30 variables, even on fast computers (or clusters of computers). To speed up the solution process lower and upper bounds on the optimal solution are taken into consideration. In a maximization problem a lower bound is given by the objective function value of any feasible solution to the problem. Such feasible solutions can be obtained by heuristics, such as greedy algorithms and local search improvement strategies, or meta-heuristics such as genetic algorithms, tabu search, or simulated annealing, to name just a few. An upper bound value to a problem given as ILP is typically found by solving the linear programming relaxation, where the integrality constraints on the variables are dropped and a pure LP remains. The overall procedure of enumeration with lower and upper bounds is known as branch-and-bound or branch-and-cut method. It is today successfully applied to large scale optimization problems having several thousands variables and constraints. For further details we refer to [8] and the references therein.

In the case of exploring future scenarios, however, we are not interested in finding one particular solution, such as the global optimum. Instead it is desired to find *all* solutions to a given instance, and then to cluster the solutions into a few groups. The groups are then analyzed and this analysis is the foundation of the (non-mathematical) interpretation of the scenarios in the context of the application at hand. In order to become applicable we have thus to modify the branch-and-bound search, in a way that not only one, but all solutions are returned.

2 Scenarios for Remanufacturing Oriented Production Equipment Development

To meet the challenges of sustainability, Brazil and Germany established the Brazilian German Collaborative Research Initiative in Manufacturing Technology (BRAGECRIM) in 2008. The goal is to investigate how production technology can foster sustainable value creation. As a case study, this paper uses results of the joint research project entitled "Remanufacturing-oriented production equipment development".

Scenario technique was used to develop scenarios for the remanufacturing/reuse of machine tools and assembly equipment for Brazil in 2020. To consider sustainability related influence factors, the methodology was adapted through taking into account the Global Reporting Initiative (GRI) [9]. In a scenario workshop with governmental representatives, academic and industry experts a total of 30 projections were generated and then evaluated with regards to their consistency by means of a cluster analysis approach. Two consistent projection sets were identified, which provide and maintain information quality of manageable amounts. Finally, two remanufacturing scenarios were created. The data generated in this project provide the input for the current investigation. The following outlines, however, refer only to the insights. A detailed description of the reference project is available online [10].

3 A Mathematical Model for the Scenario Technique

In the sequel we demonstrate that the problem of defining a consistent scenario can be modeled as a binary integer linear programming problem.

A scenario is defined by a set of key factors $\mathcal{D}$. For each key factor several projections are possible that the key factor can assume, typically between 2 and 4. We denote by $\mathcal{P}$ the set of all projections of all key factors. Hence an element $p \in \mathcal{P}$ is in fact a tuple $p = (d, i)$ where $d \in \mathcal{D}$ is the corresponding key factor and i is the number of the actual projection of d.

For each pair of projection $(p, q) \in \mathcal{P} \times \mathcal{P}$ a consistency value $c_{p,q} \in \mathcal{I}$ is specified. We use $\mathcal{I} := \{1, 2, 3, 4, 5\}$. A value $c_{p,q} = 1$ means that p and q cannot occur together in a scenario, p and q exclude each other. A value of 2 means that it is unlikely but not impossible that both p and q occur. 3 means that p and q are not related to each other. Similarly in the other direction a value of 5 means that p and q will and will only occur together, and a value of 4 indicated that p and q are likely but not entirely sure to occur together. The matrix $C := (c_{p,q})_{p,q\in\mathcal{P}}$ is called *consistency matrix.*

3.1 A Consistency Model

We introduce binary decision variables

$$\forall p \in \mathcal{P} : x_p \in \{0, 1\}. \tag{1}$$

If $x_p = 1$ for some $p = (d, i) \in \mathcal{P}$ then key factor d will assume projection i in the scenario, and $x_p = 0$ otherwise. We introduce auxiliary decision variables

$$\forall p, q \in \mathcal{P} : y_{p,q} \in \{0, 1\}. \tag{2}$$

Here $y_{p,q} = 1$ if and only if p and q are both selected, that is, iff $x_p = x_q = 1$.

The constraints of the problem are formulated as follows. Exactly one projection per key factor must be chosen:

$$\forall d \in \mathcal{D} : \sum_{\substack{i\in\mathbb{N} \\ p=(d,i)\in\mathcal{P}}} x_p = 1. \tag{3}$$

If key factor d_1 assumes projection i_1 and key factor d_2 assumes projection i_2 then the corresponding auxiliary variable must also be switched on:

$$\forall p = (d_1, i_1), q = (d_2, i_2) \in \mathcal{P} : x_p + x_q \leq 1 + y_{p,q}. \tag{4}$$

Vice versa, if the auxiliary variable is active, then both projection selector variables must be switched on:

$$\forall p, q \in \mathcal{P} : y_{p,q} \leq x_p, \quad y_{p,q} \leq x_q. \tag{5}$$

Note that constraints (4) and (5) are a formulation in terms of linear constraints of the logical clause $x_p \wedge x_q \Leftrightarrow y_{p,q}$, which can also be seen as a pseudo-boolean constraint. Further details can be found in [11] and the references therein.

The objective rates the consistency of the scenarios as follows:

$$\sum_{p,q \in \mathcal{P}} c_{p,q} \cdot y_{p,q}. \tag{6}$$

Again we remark that the scenario technique is not using just one optimal solution but needs all feasible solutions. In the basic version of the scenario technique the objective function concept actually is of minor importance, since all feasible solutions are computed, independent of their objective function value. The value of the objective is only used as a level of confidence for a particular scenario.

3.2 Counting and Clustering Solutions

Computing the most consistent scenario, i.e., a global optimal solution to the integer programming problem

$$\xi = \max (6), \quad \text{s.t.} (1), (2), (3), (4), (5) \tag{7}$$

can be computationally carried out by applying an ILP solver. However, it is typically not enough for a successful application of the scenario technique to have just one solution, even if it is the global optimal one. In fact it is desired to obtain the entire solution space, or at least those solutions that have an objective function value that is within a certain range of the global optimal value ξ. A modification of the MIP solver SCIP to count all feasible solutions of an integer program was described by Achterberg, Heinz, and Koch [12]. Counting solutions is now a build-in feature of SCIP. We slightly modified this routine so that it does not only to count the solutions, but actually outputs them row-wise in a file.

Now standard scenario technique software applies some greedy-type clustering algorithm on this set of feasible solutions. The disadvantage of this clustering method is that it does not allow the incorporation of an evaluation of sustainability. Since this is exactly what we need for our analysis, we follow a different approach. To this end we specify how much each of the projections contributes to the welfare and sustainability in terms of social, economical, and ecological aspects. For each of the three categories we specify a value between 0 (no contribution) and 100 (high contribution). Denote by $d_{p,s} \in [0, 100]$ the contribution of projection p to category s (social, economical, or ecological). Then we can project each scenario x, i.e., each feasible solution of (7), into the three dimensional space by computing (s_1, s_2, s_3)-coordinates as follows:

$$\forall\, k \in \{1, 2, 3\} : s_k(x) := \sum_{p \in \mathcal{P}} d_{p,s_k} \cdot x_p. \tag{8}$$

In this way we can visualize all feasible solutions as a “cloud” of points, and identify a reasonable number of clusters. In general such visual approach is only possible if one deals with at most three categories. For more than three categories one has to resort to cluster algorithms in order to identify them (see Späth [13] for a survey). In particular, Lloyd's k-means clustering algorithm is suitable for this application [14].

Since each cluster consists of a huge number of individual scenarios (solutions of the consistency model), the output has to be aggregated to become meaningful for the human user. To this end we count the number of projections that are realized per cluster. That means we compute a value $agg_i(p)$ as the aggregation of projection p in cluster i as follows:

$$agg_i(p) := \frac{1}{|\mathcal{C}_i|} \sum_{s \in \mathcal{C}_i} x_p^s. \tag{9}$$

4 Results

For our study the data from the BRAGECRIM project was available for testing purposes. We have 10 key factors having 2-4 projections each, which gives a total number of 30 projections.

4.1 Scenario Generation

Each projection has a certain influence on the ecological, the economical, and social aspects. Our approach is to assign values from 0 (no influence) to 100 (strong influence) to each of the projections in each of the three categories. In order to clarify the extent of contribution of the different projections, and the choice of numerical value to reflect that extent, an example will be given through the key factor "sustainable development".

The possible projections associated with that factor are financial incentives, marketing and no promotion of sustainable development. Financial incentives implies a wide promotion campaign by the government and national as well as federal bodies, offering financial support and tax reductions to companies to reward remanufacturing and sustainability. The assigned value for that projection is $[100, 100, 100]$ ranking those measures very high in all three dimensions. The economical incentive makes it far more appealing for companies to favor processes with ecological benefits instead of just calculating production costs the traditional way. To accomplish that, it is up to the companies management to establish ways of training and encouraging employers to think and act in this way. Again, the numerical value just compares this particular projection to the following two, pointing out that incentives which support sustainability as a whole, will always affect all three given dimensions.

The second possible projection is marketing, assuming that the government is fostering public discussions and environmental education through advertisement campaigns. Those measures will also affect all three dimensions of sustainability, but compared to the first projection this one is valued $[50, 50, 75]$ to reflect the weaker impact on the different aspects. Changes motivated by marketing campaigns will primarily affect social facets, but for companies to strengthen their image they will certainly have to act upon those social changes. If consumers favor the use of products which have been labeled remanufactured, companies are motivated to obtain that certain label to increase their sales.

The third projection implies no promotion of sustainable development. Having no effect on sustainable development whatsoever, it is assigned a value of $[0, 0, 0]$.

Using this data we set up the model (7) using the modeling language Zimpl [15]. We introduce two additional families of constraints to this model in order to focus on those solutions having a high consistency. That is, we exclude all scenarios that have any total inconsistencies:

$$\forall p, q \in \mathcal{P}, c_{p,q} = 1 : y_{p,q} = 0. \tag{10}$$

And further we limit the number of weak inconsistencies by allowing at most W of them within the scenario:

$$\sum_{\substack{p,q \in \mathcal{P} \\ c_{p,q}=2}} y_{p,q} \leq W. \tag{11}$$

Here W is a parameter that can be controlled by the modeler. In our computational experiments we varied W in the range from 0 to 10 (there cannot be more than 10, since we have 10 key factors). Within the scenario technique framework it is desired not to describe one single projection of the future but at least two different consistent potential scenarios. In order to be consistent we have to limit the number of weak inconsistencies, hence a low value of W is preferred. A value of $W = 0$ or $W = 1$ turned out to be too restrictive in the sense that only very few scenarios fulfill this, and they are all in the same region of the solution space. For values of W larger than 3 on the other hand we obtain one single large cluster that fills the entire solution space. For $W = 2$ or $W = 3$ we obtained 3 independent clusters. These clusters will be described in the next section.

The model then has 434 all-binary variables, 1250 constraints and 2934 nonzeros. We applied the solver SCIP1.2 [16] with the modifications described in [12] to count all feasible solutions. As an LP solver we used Soplex1.4.2 [17]. As computer platform we used an Apple MacBookPro 2.33 GHz Intel Core 2 Duo processor and 3 GB 667 MHz DDR2 SDRAM running MacOSX10.6.2. The solver evaluated 2381 branch-and-bound nodes in 3.3 seconds and found 1148 feasible solutions.

All 1148 feasible solutions were evaluated using the impact weights. In this way we obtained a point in $\mathbb{R}^3$ for each scenario. The set of all points (the "cloud") is shown in Figure 1. The darker the point is drawn, the higher its objective function values, which means, the more consistent is the corresponding scenario. We remark that the most consistent scenarios all belong to cluster 3, which in addition is also the largest. Within the scenario technique framework, however, it is not valid to deduce that this will be "the" future, it still is only one possibility among others. Hence we do not focus solely on cluster 3 but describe also the other two clusters in the next section.

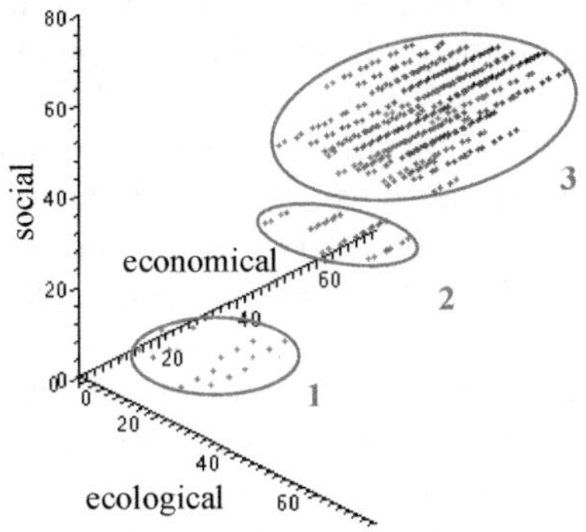

Fig. 1. The "cloud" of all scenarios, projected into $\mathbb{R}^3$ space. Three clusters are marked.

Having such visual representation we identified three clusters that are marked with (red) circles in Figure 1. For each cluster we carried out an aggregation, c.f. Section 3.2. In the next section we describe these three scenarios in detail.

4.2 Scenario Description

In the following, the three scenarios are described based on their projections. Overall, regarded broadly, scenario 1 displays a low level for all three sustainability dimensions (ecological, economic, and social). Especially the social dimension is insufficient here. Compared to scenario 1, scenario 2 shows a higher level of the sustainability dimensions, particularly regarding the social aspect, which is due to qualification perspectives. Scenario 3 shows the highest level on all three dimensions. Although it is quite similar to scenario 2, it includes financial incentives and higher reusability of the product itself.

In the following, scenario 1 and 2 are each described in detail and in their respective entirety. Since scenario 3 is in large parts identical to scenario 2, for this scenario only those parts are outlined in more detail, which are either different or additional to the aspects that occurred in scenario 2.

Scenario 1. Even though advantages of remanufacturing exist, Original Equipment Manufacturers (OEMs) stay out of the market. They neither remanufacture themselves nor subcontract companies to do the remanufacturing on their behalf. Therefore, the number of independent remanufacturers increases. However, due to the fact that the reusability of production equipment is mainly influenced during the product development process at the OEMs, a shift in the OEMs paradigms is hardly imaginable. The focus during product development will mainly be on the production process itself, rather than on the remanufacturing process. The increase in third party manufacturers leads to the outsourcing of manufacturing processes of OEMs to these parties. Therefore, users neither buy nor contract production equipment themselves, but contract third parties to produce certain products on their production equipment. Associations for remanufacturing/machine tools and assembly equipment have low influence on business and governmental decisions. Most initiatives, e.g. certifications, standards or regulations, promoted by the associations do not prove to be successful. No or only a low impact on the development of the market can be seen. The traditional profit strategy of selling products instead of services still dominates the market of machine tools and assembly equipment. The expected changes towards a more service-oriented value creation through Product Service Systems (PSS) based manufacturing equipment do not take place. Therefore, most of the machine tool producers are still focusing on selling products rather than on the sale of additional services. Manufacturing firms in general are not increasing the product variety anymore. Also the tendency of shortened product life cycles and increasing product changes is slowed or stopped. It is still necessary to change the manufacturing system designed for one generation of products to a system for the next generation of products, but due to increasing flexibility of manufacturing equipment this can be easily done within the existing system. There has neither been a major increase nor a decrease in the demand for remanufactured machine tools and assembly equipment. The demand stays the same, which means that it still focuses on specific areas, for example on some components of the control (retrofitting), and has not been widely spread within the industry. Even though sustainable development has been at the center of public discussion, government bodies are not actively engaged in any activities that would promote this development. With regards to end-of-life laws the Brazilian government has not released any laws that would affect remanufacturing. Thus, this forward projection of the status-quo does not provide any pressure from the law that would make manufacturers responsible for their products at the end-of-life. Concerning qualification that is needed to undertake remanufacturing activities, a forward projection of the status-quo is visible. There is no public promotion and offering of trainings related to remanufacturing. Thus, the knowledge keeps being restricted to the OEM.

Scenario 2. Due to expected advantages of remanufacturing, such as supply of spare parts, growing demand for remanufactured products or governmental financial incentives, OEMs enter the remanufacturing market through remanufacturing of their own products. All value-adding processes are executed by the OEMs, which leads to a share of up to 30% of the OEMs overall turnover. Some OEMs may also prefer to only manage the process and the control of these processes, for example by means of project management or price. However, the OEMs stay responsible for the product as the products wear their label. Despite differences in purchasing prices, manufactured as well as remanufactured machine tools and assembly equipment are both relatively expensive. Moreover, the overall lifetime of these products is decreasing. Therefore, OEMs will increase PSS, providing for example the possibility to contract production equipment. As a consequence, users will tend to mainly contract their production equipment. New design techniques have increased the interchangeability of components. In addition, the modularity of production equipment makes the remanufacturing operation easier. These operations can therefore be made by the OEM, the owner, or by third part companies at the shop floor of the owner or the supplier. This new practice reduces logistic costs and the time of the overall remanufacturing process, which in turn leads to an increase in productivity. Associations for remanufacturing/machine tools and assembly equipment support technical standards and legal regulations of remanufactured products. This results in an increased quality of products and services, and causes a positive influence on their demand.

Product accompanying information systems integrated into products are capable of monitoring relevant product and process data throughout the entire product life span. The available data enable a better overview of the product and its processes. This may foster higher product responsibility concerning advertising, customer health, safety, new products and services. Monitoring systems for predicting the life time of the parts are installed with electronic devices that use sensors. The demand for remanufactured machine tools and assembly equipment has increased. This is mainly due to the improved quality of remanufactured equipment and components. Certifications and new standards grant a certain quality level that is equal to that of equivalent new products. Moreover, new technologies and methods for disassembling have led to lower costs for remanufacturing of equipment and have with this increased its market share. Even though sustainable development has been at the center of public discussion, government bodies are not actively engaged in any activities that would promote this development. The creation of specific rules, which take into account the specificities of each individual sector, has become the choice with regards to end-of-life laws. This leads to an increase in responsibility of these companies for their products end-of-life treatment. However, these sectorial rules vary between the different sectors and can thus not direct companies towards the same goal. It may also be possible that government released general laws exist that relate to the end-of-life treatment of products. This causes companies to rethink their strategies regarding the management of their products end-of-life possibilities. These general laws, including goals and instruments, make companies from different sectors move towards the same goal. In addition, also sector specific regulations direct the companies. The public as well as business entities have realized the enormous potentials that come along with remanufacturing. Therefore, trainings in the areas of mechanics and electronics, which enable the fulfillment of certificate requirements, emerge. These trainings are mainly offered by associations, universities and companies. In addition to these special trainings, informal education has increased. This is due to the growth of public discussions with regards to sustainable development and in particular remanufacturing. Moreover, with a growth in remanufacturing business, an exchange with international experts has emerged.

Scenario 3. Associations for remanufacturing/machine tools and assembly equipment support certifications for remanufactured products, which assures the quality of remanufactured products. This asserts a positive influence on the demand for production equipment through the reduction of the overall technological gap and environmental impacts. Associations for remanufacturing/machine tools and assembly equipment may also support technical standards and legal regulations of remanufactured products. This results in an increase in the quality of products and services, which in turn has a positive influence on the demand for these products.

Manufacturing firms may also become increasingly interested in manufacturing systems that produce a variety of products. Product variety is continuously increasing: Customer demand is becoming more and more diverse, and the frequency of model change continues to rise. It is still necessary to change the manufacturing system designed for one generation of products to a system for the next generation of products. Thus, manufacturing companies have sought methods to rapidly change production equipment in economical ways. One method has been to increase reusability by employing adaptive components/parts of machine tools and assembly equipment. The government, national and federal bodies provide financial incentives in order to promote sustainable development. They offer financial support as well as tax reductions to the companies as a stimulus and reward to undertake activities with regards to remanufacturing. Alternatively, it is also likely that the government is fostering the public discussion on sustainable development. Environmental education through advertisement in different media and television is increasing. In addition, green labeling of products is introduced and thus further contributes to environmental education.

5 Summary and Conclusion

We gave an introduction to the scenario technique and demonstrated that the consistency problem inherent in this method can be formulated as an integer programming problem. For its solution we adapted the ILP and constraint programming solver SCIP, which generates all feasible solutions. Having defined suitable weights for the categories of ecological, economic and social impact, we mapped the set of solutions into a three dimensional space. We aggregated the solutions into three clusters and described them.

The evaluation of the scenarios to three categories is a common approach to estimate different sustainability levels. It is intended to create a predefined set of questions to receive more objective evaluation process of projections under sustainability criteria.

In conclusion, our approach is particularly valuable in situations where the black box projection methods of standard scenario technique solvers do not give satisfying results. While in these traditional approaches the user has no control of the way the individual scenarios are clustered, our approach allows using additional knowledge of the modeler. Using this knowledge can finally lead to new insights for the scenario to be analyzed, which were overlooked when applying only the classical method.

References

1. Seliger, G. (ed.): 12. Produktionstechnisches Kolloquium PTK 2007: Nachhaltige industrielle Wertschöpfungsnetze (2007)
2. Stern, N.: The Economics of Climate Change. In: The Stern Review, Cambridge University Press, Cambridge (2006)
3. Intergovernmental Panel on Climate Change (IPCC): Climate Change 2007: The Physical Science Basis. Summary for Policy Makers (2007)
4. Kahn, H., Wiener, A.J.: The Year 2000: A Framework for Speculations on the next Thirty-Three Years. Macmillan, New York (1967)
5. Börjeson, L., Höjer, M., Dreborg, K.H., Ekvall, T., Finnveden, G.: Scenario types and techniques: Towards a user's guide. Futures 38(7), 723–739 (2006)
6. Gausemeier, J., Fink, A., Schlake, O.: Scenario management: An approach to develop future potentials. Technology Forecasting and Social Change 59, 111–130 (1998)
7. Gausemeier, J., Plass, C., Wenzelmann, C.: Zukunftsorientierte Unternehmensgestaltung: Strategien. In: Geschäftsprozesse und IT-Systeme für die Produktion von morgen. Carl Hanser Verlag, München (2009)
8. Fügenschuh, A., Martin, A.: Computational Integer Programming and Cutting Planes. In: Handbooks in Operations Research and Management Science: Handbook on Discrete Optimization, vol. 12, pp. 69–122. Elsevier, Amsterdam (2005)
9. Severengiz, S., Widera, H., Seliger, G., Saavedra, Y.M.B., Ometto, A.R., Barquet, A.P., Forcellini, F.A.: Scenarios for the remanufacturing of machine tools and assembly equipment. In: The 7th CIRP Conference on Sustainable Manufacturing, IIT Madras, Chennai (December 2-4, 2009)
10. Seliger, G., Severengiz, S., Widera, H., Ometto, A.R., Barrera, Y., França, T., Puglieri, F., Forcellini, F.A., Barbosa, C., Barquet, A.P.: Scenarios for remanufacturing/reuse of machine tools and assembly equipment for brazil in, Technical report, Technische Universität Berlin, Universidade de São Paulo, and Universidade Federal de Santa Catarina, Online available at URL (2020), `http://www.mf.tu-berlin.de/uploads/media/Scenariosforremanufacturing.PDF` (retrieved November 20, 2009)
11. Berthold, T., Heinz, S., Pfetsch, M.E.: Nonlinear pseudo-boolean optimization: relaxation or propagation? In: Kullmann, O. (ed.) SAT 2009. LNCS, vol. 5584, pp. 441–446. Springer, Heidelberg (2009)
12. Achterberg, T., Heinz, S., Koch, T.: Counting solutions of integer programs using unrestricted subtree detection. In: Perron, L., Trick, M.A. (eds.) CPAIOR 2008. LNCS, vol. 5015, pp. 278–282. Springer, Heidelberg (2008)
13. Späth, H.: Cluster-Formation und -Analyse. Theorie, FORTRAN-Programme, Beispiele. R. Oldenbourg-Verlag, München (1983)
14. Lloyd, S.P.: Least Squares Quantization in PCM. IEEE Transaction on Information Theory 28(2), 129–137 (1982)
15. Koch, T.: Rapid Mathematical Programming. PhD thesis, Technische Universität Berlin, Technical Report ZIB-TR 04-58 (2004)
16. Achterberg, T.: Constraint Integer Programming. PhD thesis, Technische Universität Berlin (2007)
17. Wunderling, R.: Paralleler und Objektorientierter Simplex-Algorithmus. PhD thesis, Technische Universität Berlin, Technical Report ZIB-TR 96-09 (1996)

Modelling Post-carriage Transport Costs in Groupage Networks

Nicholas Boone and Tim Quisbrock

Chair of Logistics & Distribution,
University of Applied Sciences Ostwestfalen-Lippe, Liebigstraße 87, D-32657 Lemgo
nicholas.boone@hs-owl.de

Abstract. Groupage transports are frequently calculated based on freight rates assuming a declining rate with respect to volume and to transportation distance. This paper investigates the case of feeding many small shipments to customers via a freight forwarder network. It is argued that costs for post-carriage transport dominate other costs. Based on a theoretical estimate, simulation experiments for the case of a uniform distribution of customer locations around the subsidiary with a fixed shipment size per drop suggest that the aggregate costs grow progressively with increasing customer drop distances from the subsidiary. The resulting implications for transport modelling are discussed.

Keywords: cost-by-cause principle, freight forwarder, freight rates, network, overhead costs, pricing, simulation, transport.

1 Introduction

This paper considers the case of LTL (Less than Truck Loads) shipments in a groupage network. A typical physical groupage process includes the following steps:

- Pickup at the consignor and short distance *pre-carriage* to the forwarding agent's subsidiary or depot, that serves as a transshipment point
- Long distance (*on-carriage*) linehaul to the destination subsidiary, possibly passing further transshipment points and based on regular connections in the network
- Short distance *post-carriage* transport to the consignee.

In Western countries, wages and salaries typically account for the largest proportion of the transport cost, far more than fuel consumption or transshipment costs (cf. [4], [6], [10] and [16]). Large German groupage networks include 40 or more subsidiaries. Distribution areas of a single depot usually go up to a maximum of 80 to 100 km (cf. e.g. [10]).

Typically, the cost for the pre- and post-carriage transports comprises between 50 and 80 % of total transport cost, depending on network density, service levels and actual distance from consignee to consignor (cf. [16] and [25]). Therefore, *short distance transports dominate* the *network costs*.

Many theoretical logistics optimisation models either completely ignore transportation costs or assume them to be linear (cf. e.g. [9], [11], [13], [19], [24]). At best, some models are based on freight pricing in practice: Regular transports are

W. Dangelmaier et al. (Eds.): IHNS 2010, LNBIP 46, pp. 332–344, 2010.

priced based on standard freight rates, frequently assuming a declining rate with respect to volume and to transportation distance (cf. [5], [10], [15], [17], [20], [21], [22] and [23]).

However, it seems plausible, that drops that are farther away from the subsidiary than others would, on average, lead to tours with longer first and last 'legs' (distance from depot to first and from last drop back to depot), leaving less time for additional drops. This leads to the *assumption* that *costs* would actually *increase progressively* with the drop distance from the depot.

Grouping shipments from several shippers generates significant economies of scale compared to shipping individually. However, the result is a high proportion of overhead costs that cannot be directly allocated to the individual customer drop or pick-up, *making individual price calculations more complicated* (also cf. [12]).

The *objective of this paper* is to develop an approach to evaluate the relationship between distance and post-carriage transport costs according to the cost-by-cause principle. Following a theoretical estimate, the presented approach is examined by means of simulation experiments to quantify the aggregate cost effect of increasing customer drop distances.

2 Cost Model and Simulation Design

2.1 Objective and Framework

This paper assumes the case of a medium-sized manufacturer or retailer with a central warehouse, feeding many small shipments to customers via a freight forwarder network (*few-to-many*). The aggregate network volume is much higher than the sourcing manufacturer's.

In this case, costs for post-carriage transport dominate the other costs. Therefore, we try to determine costs related to the individual customer drop. Only few costs are directly related to the customer drop, some can at least be related directly to the respective pallets dropped:

- *Customer drop*: Some transport processes induce costs that are directly associated with the specific customer drop, e.g. waiting for contact person, ramp allocation or printing the respective delivery note etc.
- *Pallet*: Other processes are directly linked to an individual customer's pallet, like pre- and post-carriage loading/unloading, printing and applying pallet labels.

  Most costs, however, are overhead both to the drop and to the pallets at the drop:
- *Tour*: Some costs are directly associated with a single tour or route, but cannot be allocated to specific drops or pallets. They include scheduling, planning and printing the respective packing lists, ramp docking at the subsidiary, cleaning and refueling trucks after each tour.
- *Vehicle or Trailer*: Some costs are linked to vehicle or trailer mileage e.g. fuel consumption, depreciation due to wear and tear. Also, related fixed costs such as insurance, tax etc. are direct costs with regard to the vehicles or trailers.
- *Subsidiary*: Some costs are associated with the subsidiary, e.g. costs for maintaining all the subsidiary's assets and for managing and operating all overhead processes in the respective subsidiary.

- *Company/Network*: Finally, some costs cannot be allocated to specific drops or pallets, tours or even subsidiaries. They include central management, scheduling and operating the linehaul network and maintaining the overall IT systems etc.

In order to allocate these overhead costs to the individual customers, the following procedure is suggested, using *Fleischman*'s ring model as a loose analogy (cf. [7] and [27]):

- *Step 1*: Divide delivery area A into circles of increasing radius of r_k around the subsidiary for $k = 1$ to k_{max}, with r_{k_max} as the maximum distance of a customer drop from the subsidiary.
- *Step 2*: Set $k = 1$.
- *Step 3*: Plan tours for all customers within a radius of r_1 around the subsidiary and calculate all resulting direct and overhead tour, pallet and drop related costs.
- *Step 4*: Direct drop costs are allocated to the individual drop, direct pallet costs are allocated to drops according to their individual pallet demand.
- *Step 5*: All overhead costs are divided by the total number of pallets within the radius of r_1 and allocated to the drops according to their individual pallet demand.
- *Step 6*: Set $k = k + 1$.
- *Step 7*: Plan tours for all customers within a radius of r_k around the subsidiary and calculate all resulting direct and overhead tour, pallet and drop related costs.
- *Step 8*: Determine the marginal increase of customer drops, pallet demands and respective costs.
- *Step 9*: Divide marginal overhead costs by marginal increase of pallets from r_{k-1} to r_k and allocate overhead costs to drops according to their individual pallet demand. (This is in order to avoid that, in the case of fluctuating demand, a very large customer restricting the available pallet capacity for other customers would otherwise receive the same costs as very small customers on the exact same tour.)
- *Step 10*: Repeat from Step 6 until $k = k_{max}$.

2.2 Assumptions and Cost Model

In order to determine the costs, a simulation approach based on the following assumptions is developed:

- *Delivery area A*: Customer drop locations are uniformly distributed in A. The maximum distance of a drop i from the subsidiary is set at $\max\{d_{0i}\} = 100$ km.
- *Customer demand volume in pallets*: It is assumed that every customer receives $v_i = 1$ pallet. If the goods are handled on pallets, it is realistic to assume that there is hardly any difference in the process oriented costs incurred with regard to the actual pallet weight (also cf. [8], [21]).
- *Drop time*: Total time per drop consists of $t_{drop,flat} = 15$ minutes plus $t_{LU} = 5$ minutes for loading/unloading per pallet.
- *Cost per hour*: For the case of this simulation, all rates for drop, loading/unloading and driving related time are set at $c_{drop} = c_{LU} = c_{driv} = 80$ € per hour.
- *Mileage cost*: Fuel, wear and tear and other driving distance-related expenses lead to $c_{km} = 0.80$ € per km.

- *Tour related overhead cost*: All other overhead costs are charged via $c_{tour,flat} = 200$ € per tour.
- *Driving speed*: due to the uniform drop distribution with fairly large distances, an average driving speed of s = 50 km/h is assumed.

The following, process oriented cost model is suggested. *Direct costs* with regard to an *individual* customer drop i and to the corresponding volume in pallets are calculated as follows:

$$C_{drop,i} = t_{drop,flat} \cdot c_{drop} \tag{1}$$

$$C_{pall,i} = v_i \cdot t_{LU} \cdot c_{LU} \tag{2}$$

Total direct costs for *all drops* and the corresponding pallets up to a radius of r_k are then given by:

$$C_{drop}(r_k) = n(r_k) \cdot t_{drop,flat} \cdot c_{drop} \tag{3}$$

$$C_{pall}(r_k) = v(r_k) \cdot t_{LU} \cdot c_{LU} \tag{4}$$

Overhead costs related to the sum of all tour lengths (L_{km}) and to the total driving time for servicing all drops up to a radius of r_k are calculated as follows:

$$C_{driv}(r_k) = t_{driv}(r_k) \cdot c_{driv} \tag{5}$$

$$C_{km}(r_k) = L_{km}(r_k) \cdot c_{km} \tag{6}$$

The remaining overhead tour costs and the resulting total overhead costs up to r_k are given by:

$$C_{tour}(r_k) = m(r_k) \cdot c_{tour,flat} \tag{7}$$

$$C_{over}(r_k) = C_{driv}(r_k) + C_{km}(r_k) + C_{tour}(r_k) \tag{8}$$

Finally, from the second ring on, for k > 1 total overhead costs are allocated as 'marginal' costs per pallet to the additional pallets:

$$\Delta c_{over}(r_k) = \frac{C_{over}(r_k) - C_{over}(r_{k-1})}{v(r_k) - v(r_{k-1})} \tag{9}$$

For the sake of completeness, for customers in the first ring overhead costs per pallet are defined as:

$$\Delta c_{over}(r_1) := \frac{C_{over}(r_1) - 0}{v(r_1) - 0} = \frac{C_{over}(r_1)}{v(r_1)} \tag{10}$$

All in all, the individually assigned total cost for a specific drop i with $r_{k-1} < d_{0i} \leq r_k$ consists of:

$$C_i = C_{drop,i} + C_{pall,i} + v_i \cdot \Delta c_{over}(r_k) = t_{drop,flat} \cdot c_{drop} + v_i \cdot t_{LU} \cdot c_{LU} + v_i \cdot \Delta c_{over}(r_k) \tag{11}$$

This leads to total overall costs of:

$$C(r_k) = C_{drop}(r_k) + C_{pall}(r_k) + C_{over}(r_k) \tag{12}$$

Tours are planned based on *Clarke/Wright*'s Saving algorithm and a subsequently employed 2Opt-procedure (cf. [2]; for an overview of alternative algorithms cf. [14] and [18]).

2.3 Estimating the Length of Tours in Ring-Radial Transport Networks

Recently, an analytical approach to estimate the total length of tours in ring-radial transport networks was presented by *Boone/Quisbrock* (cf. [1]). For the case of a capacitated vehicle routing problem (CVRP) with volume restrictions R_{pall} and restricted driver working hours R_{time}, estimating the respective average first and last legs of all tours within r_k leads to:

$$\bar{d}_{0,first}(r_k) = \bar{d}_{last,0}(r_k) = \frac{2}{3} r_k \tag{13}$$

The average *distance between two drops* within a tour can be estimated by:

$$\delta(r_k) = \kappa \cdot \frac{\sqrt{A(r_k)}}{\sqrt{n(r_k)}} = \kappa \cdot \frac{\sqrt{A(r_{k\_max})}}{\sqrt{n(r_{k\_max})}} = \delta = \text{const.} \tag{14}$$

If only the volume restriction were active, the number of tours required would be:

$$\mu_{pall}(r_k) = \text{roundup}\left(\frac{n(r_k)}{R_{pall}}\right) \tag{15}$$

If, instead, only the drivers' working hours were restrictive, the amount of drops that an average single driver can service is given by:

$$v_{time}(r_k) = 1 + \frac{\left(R_{time} - \frac{\left(\bar{d}_{0,first}(r_k) + \bar{d}_{0,last}(r_k)\right)}{s} - \left(t_{drop,flat} + t_{LU} \cdot \frac{v(r_k)}{n(r_k)}\right)\right)}{\left(\frac{\delta}{s} + t_{drop,flat} + t_{LU} \cdot \frac{v(r_k)}{n(r_k)}\right)} \tag{16}$$

The total number of tours $m(r_k)$ required to service all drops up to r_k can then be estimated as follows:

$$\mu_{time}(r_k) = \text{roundup}\left(\frac{n_{drop}(r_k)}{v_{time}(r_k)}\right) \tag{17}$$

This leads to the expected total number of tours required to service the area $A(r_k)$:

$$m(r_k) = \max\{\mu_{pall}(r_k); \mu_{time}(r_k)\} \tag{18}$$

Finally, the expected total length of all tours in km for all customer drops up to r_k is given by:

$$\overline{L}_{km}(r_k) = m(r_k)\cdot\left(d_{0,first}(r_k) + d_{last,0}(r_k)\right) + \left(n(r_k) - m(r_k)\right)\cdot\left(\kappa\cdot\frac{\sqrt{A}}{\sqrt{n}}\right) \qquad (19)$$

In our case, the time required for driving can be calculated by dividing the tour length by the constant driving speed s. ([3] examines the correlation between realistic driving speeds and distances in a real world setting.). Customer drop times do not depend on the specific distance to the depot, but only on the pallet demand. They can easily be computed by determining the expected customers within $A(r_k)$.

2.4 Results of Various Parameter Pretests

Using equidistant radii, results of several pretests lead to dividing the delivery area A into k_{max} = 8 circles around the subsidiary with $\Delta r_{k+1} = r_{k+1} - r_k = r_1$ = 12.5 km. In order to avoid circles without drops, the total number of customer drops was set at $n(r_{k\_max})$ = 250. In this case, the expected amount of drops $n(r_1)$ = 3.91 and the probability of finding no customer in the first ring is only 1.95 %.

In order to calculate a lower bound, the theoretical case of a 'Mega'-Travelling Salesman tour was solved without either pallet restrictions per truck or a work time restriction for the drivers, $R_{pall} = R_{time}$ = infinite. For 250 customers with one pallet each, total time for all drops is 83.3 hours. Theoretically, the 'Mega-tour driver' would need a total of just under 128 working hours, of which he or she would spend 44.4 hours to drive approximately 2220 km.

For R_{time} = infinite, rather obviously, the less restrictive R_{time} becomes, the less net driving time is needed (only two 'legs'), see figure 1. As the actual driving distance L_{km}, the total time required for driving t_{driv}, total work hours t and the total number of tours n are practically 100 % correlated (see table 1), similar results apply to their respective relation to the driving time restrictions R_{time}.

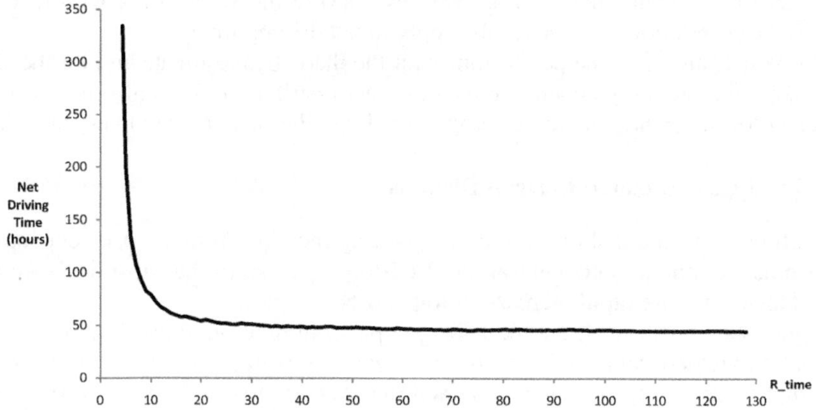

Fig. 1. Net driving time for 250 drops related to the work time restriction per driver without a pallet space restriction

Tests showed that similar results also apply to the case of R_{time} = infinite and increasing values of R_{pall}.

Based on several pretests, R_{pall} was set to *14 pallets per truck*, and R_{time} was set to *7 hours per driver*. This takes into account that, in practice, sometimes unpredicted, but necessary pick-ups at a drop point may occur. Also, unforeseen troubles (e.g. traffic jam, customer not available) spoil the originally planned tour, thus reducing the theoretical maximum working periods and rest periods in line with legal restrictions (cf. EU drivers' hours rules (EC) 561/2006).

Table 1. *Bravais/Pearson* correlation coefficients

	R_{time}	$L_{km}(r_{k\_max})$	$t_{driv}(r_{k\_max})$	$t_{total}(r_{k\_max})$	$m(r_{k\_max})$
R_{time}	1				
$L_{km}(r_{k\_max})$	-0,40276558	1			
$t_{driv}(r_{k\_max})$	-0,402775113	0,999999996	1		
$t_{total}(r_{k\_max})$	-0,402779418	0,999999995	0,999999989	1	
$m(r_{k\_max})$	-0,467880707	0,993848716	0,993848006	0,993849729	1

3 Simulation Results and Conclusions

3.1 Total Tour Length of the CVRP

Excluding test runs and sensitivity analyses, a total of 50 simulation experiments, corresponding to 50 work days, were run with the afore mentioned parameters. Table 2 shows the results of the simulation as well as the theoretical estimate of the total tour length. For a value of $\kappa = 0.79$, the estimate yields fairly good values, although it underestimates the actual number of tours and therefore the actual total tour length for very large radii. (This may be improved by varying the values of κ with r_k, cf. [1]). Due to the correlation, similar results apply to net driving time.

Drop times are the same per customer for the theoretical estimate and for the simulation and, therefore, only depend on the customer distribution. For uniformly distributed customer locations, $n(r_k)$ increases proportionally to the size of the delivery area $A(r_k)$.

3.2 Total Costs Related to Drop Distance

The following figure 2 shows the corresponding increase of total costs, depending on r_k. In practice, due to a concentration of customers close to the subsidiary, we would expect an even more rapid increase of total costs.

Figure 3 shows the average total cost per pallet in relation to the drop distance. The simulation yielded values of $1 \leq n(r_1) \leq 9$, with an average of 3.66 (compared to the expected value of approx. 4) customers drops. While $m(r_1) = 1$ for all of these values of $n(r_1)$, $m(r_2)$ remained one in 28 % of all simulation runs, with $m(r_2)$ at a maximum of two in all other cases. This is why total costs per pallet are so high for r_1, due to the allocated tour flat-rate.

Table 2. Simulated tour length L_{km} vs. theoretical estimate for $\kappa = 0.79$

k	r_k	Estimated tour length (km)	Simulated tour length (km) min	average	max
1	12.5	42	12	38	64
2	25.0	196	105	160	204
3	37.5	452	283	395	466
4	50.0	878	654	817	1,004
5	62.5	1,439	1,296	1,429	1,592
6	75.0	2,337	2,088	2,335	2,638
7	87.5	3,320	3,339	3,665	4,014
8	100.0	4,872	5,316	5,597	5,878

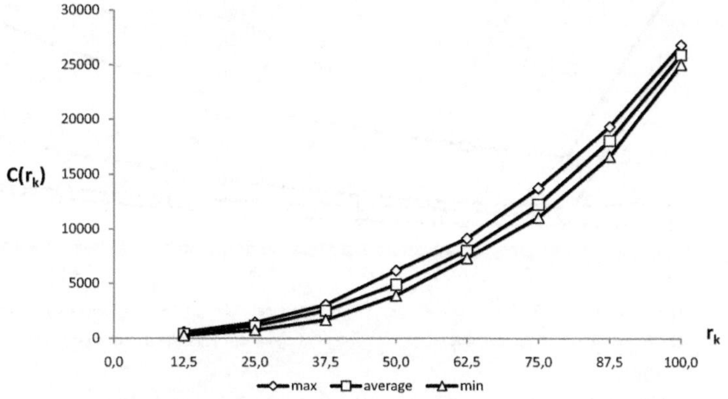

Fig. 2. Total costs related to drop distance

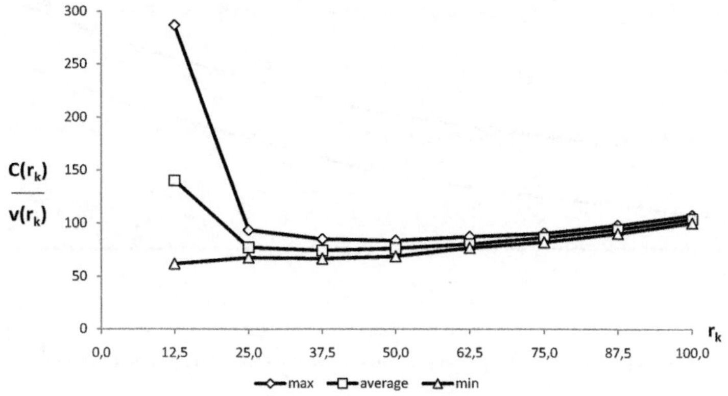

Fig. 3. Total cost/pallet related to drop distance

More interesting than the overall averages, however, are the characteristics of the 'marginal' costs.

3.1 Marginal Costs Related to Drop Distance

Total direct drop related costs are constant (cf. formula (1)). Due to the assumption that every customer receives one pallet each, total direct pallet related costs are also constant. (This is not a critical assumption: Test runs with uniformly distributed pallet demand $v_i \leq 7$ exhibited the same basic results.)

The following figure 4 shows the result of allocating the marginal indirect overhead costs according to the marginal pallets dropped. Test runs for the case of fluctuating pallet demand suggest that this procedure is more adequate than allocating the marginal overhead costs to the additional drops. (In this special case, due to $v_i = 1$ per drop, it makes no difference.)

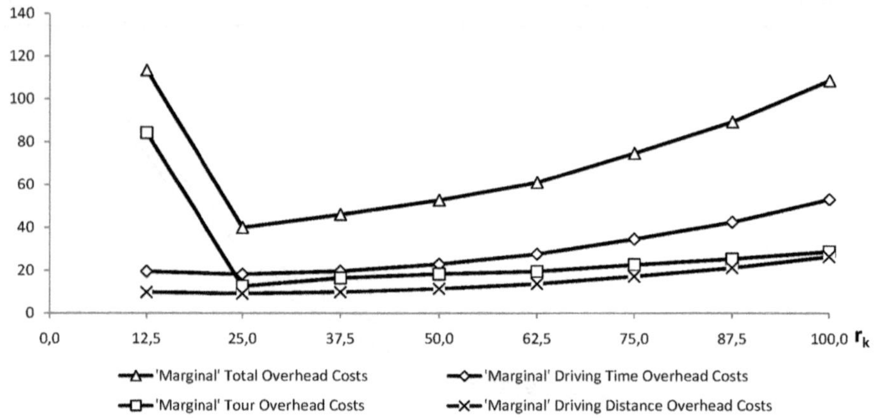

Fig. 4. Marginal overhead cost/pallet related to drop distance

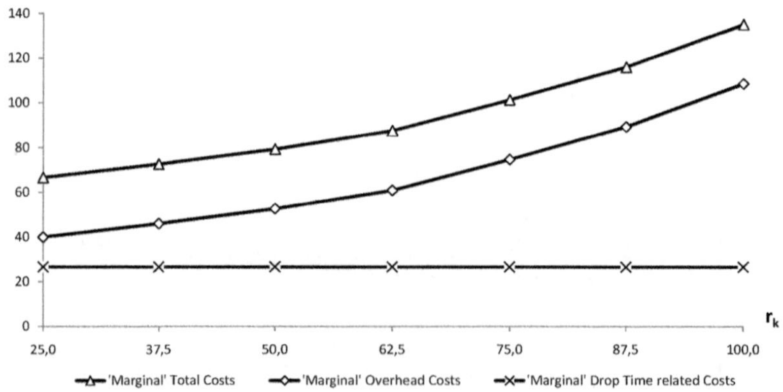

Fig. 5. Marginal total cost/pallet related to drop distance, starting from r_2

Again, due to the small number of drops $n(r_1)$, that still require at least one tour, inducing the allocated tour flat-rate, total overhead costs per pallet are so high for r_1 (and costs aren't 'marginal' for r_1, either, cf. (10)). Apart from this special case, one can easily see, that, even in the case of a uniform distribution for the drop locations, marginal costs per drop actually increase progressively, rather than decrease with growing distance.

As 'marginal' drop-time oriented costs are constant per pallet, marginal total costs exhibit the exact same structure as the marginal overhead costs (cf. fig. 5). If one took into account a more realistic customer concentration, it is very likely that the increase be steeper still.

4 Conclusion and Outlook

Our simulation results suggest that the classic transport modelling approaches can be misleading: Even for the 'benign' case of a uniform distribution of customer locations around the depot and for one pallet per drop, costs seem to increase much stronger than most models assume. It is argued that a realistic basis for pricing should take into account both a *direct drop-related* and a *pallet-related price component*, that is independent of the distance from subsidiary to drop. Additionally, *overhead costs* should be allocated according to the costs-by-cause principle, leading to *higher prices for more distant customers*.

If customer concentration in industrial zones were taken into account, delivering to outside customers would lead to an even stronger cost increase. (For the influence of population density on transport costs cf. [16].) Also, in the case of more restrictive driving times (e.g. due to time windows), costs are likely to grow even faster with regard to the distance.

One way to avoid post-carriage costs that are too high is to reduce delivery areas, e.g. by introducing additional subsidiaries or transshipment points.

Following *limitations* should be taken into account: The results presented are results for a specific set of cost parameters. Additional simulation runs suggest that, as long as driving distance and driving time related costs aren't 'too small' compared to the drop time related costs, the basic conclusions still hold. This, however, needs to be evaluated more systematically, in future.

As the approach presented here focused on post-carriage transports, the effects that the corresponding pickup and network linehaul have with regard to total cost per drop need yet to be examined. (For approaches to plan and how to coordinate linehaul networks cf. [25] and [26].)

Further research should include the cases of fluctuating customer pallet requirements, fluctuating customer densities or combined pick-up and drop situations. Finally, analysing practical aspects of subsidiary customer pricing strategy, intercompany transfer pricing and sub-contractor pricing in practice still require further empiric research activities. One of the reasons why prices frequently aren't based on a cost-by-cause allocation might be that many freight forwarders don't seem to really know their own costs (cf. [12] and [27]).

References

1. Boone, N., Quisbrock, T.: Why Size Does Matter - Estimating the Length of Tours in ringradial Transport Networks, Lemgo (2010)
2. Clarke, G., Wright, J.W.: Scheduling of vehicles from a central depot to a number of delivery points. Operations Research Quarterly 12, 568–581 (1964)
3. Ebner, G.: Controlling komplexer Logistiknetzwerke – Konzeption am Beispiel der Transportlogistik eines Multi-Standort-/Multi-Produkt-Unternehmens. Nürnberg (1997)
4. Erkens, E.: Kostenbasierte Tourenplanung, Bremen (1998)
5. Feige, D.: Optimierung von Frachtnetzstrukturen. In: Gabler Lexikon Logistik - Management logistischer Netzwerke und Flüsse, S. 186–193. Gabler/GWV Fachverlage, Wiesbaden (2004)
6. Fiedler, J.: Fahrzeugkostenrechnung und Kalkulation. In: Praxis des Controllings in Speditionen, pp. 45–69, Bildungswerk Spedition und Logistik e. V., Frankfurt a. M (2007)
7. Fleischmann, B.: Design of Freight Traffic Networks. In: Advances in Distribution Logistics, pp. 55–81 (1998)
8. Gfüllner, C., Raab, M.: Wider dem Tonnage-Denken im Nahverkehr – Prozessorientierung als Basis für eine leistungsgerechte (Sub-) Unternehmervergütung. In: DVZ Deutsche Verkehrszeitung, Ausgabe 87, p 3. Dt. Verkehrs-Verlag, Hamburg (2004)
9. Glagow, M., Riek, J., Zimmermann, J.: Tourenplanung mittelständischer Speditionsunternehmen in Stückgutkooperationen: Modellierung und heuristische Lösungsverfahren. In: Zeitschrift für Planung & Unternehmensführung, vol. 17, pp. 365–388. Springer, Heidelberg (2007)
10. Herry, M.: Transportpreise und Transportkosten der verschiedenen Verkehrsträger im Güterverkehr. In: Verkehr und Infrastruktur, No. 14. Kammer für Arbeiter und Angestellte für Wien, Wien (2001)
11. Holmberg, K., Ling, J.: A Lagrangean heuristic for the facility location problem with staircase costs. European Journal of Operations Research 97, 63–74 (1997)
12. Kapell, E.: Preise nicht verursachungsgerecht - Erstmals ZLU-Umfrage zu Frachtraten in Handel und Industrie. Lebensmittel Zeitung (2009)
13. Kopfer, H., Kopfer, H.W., Stache, U.: Strategien für die Auftragsdisposition in Speditionsunternehmen. In: Intelligent Decision Support – Current Challenges and Approaches, pp. 157–175. Gabler Verlag, Wiesbaden (2008)
14. Laporte, G.: The Vehicle Routing Problem: An overview of exact and approximate algorithms. Management Science 40, 1276–1290 (1994)
15. Mayer, G.: Strategische Logistikplanung von Hub&Spoke-Systemen. Gabler Verlag, Wiesbaden (2001)
16. Müller, S.: Pick-up and delivery als zentrales Element des Wertschöpfungsprozesses in der Stückgutspedition. Logistik Management 9(1), 34–49 (2007)
17. Oelfke, D., Oelfke, W.: Speditionsbetriebslehre und Logistik – Die wichtigen Prüfungsfragen mit Lösungen. Gabler Verlag, Wiesbaden (2003)
18. Richter, A.: Dynamische Tourenplanung - Modifikation von klassischen Heuristiken für das Dynamische Rundreiseproblem (DTSP) und das Dynamische Tourenplanungsproblem (DVRP) mit der Möglichkeit der Änderung des aktuellen Fahrzeugzuges. Univ. Dresden (2005)
19. Schmidthöfer, A.: Taktische Bestimmung von Tourgebieten für die speditionelle Tourenplanung, Univ. Augsburg (2003)

20. Stolletz, L., Stolletz, R.: Linearisierungsverfahren für Standortplanungsprobleme mit nicht-linearen Transportkosten. Zeitschrift für Planung & Unternehmenssteuerung (19), 261–280 (2008)
21. Stumpf, P.: Tourenplanung im speditionellen Güterfernverkehr, Nürnberg (1998)
22. Tempelmeier, H.: Die Bestimmung des optimalen Standorts eines Auslieferungslagers beientfernungsabhängiger Nachfrage und nichtlinearen Transportkosten. Zeitschrift für Operations-Research B 24, B167–B187 (1980)
23. Voth, M.: Leistungsprozesse Spedition und Logistik – Lernsituationen. Bildungsverlag EINS, Troisdorf (2008)
24. Webb, M.H.J.: Cost Functions in the Location of Depots for Multiple-Delivery Journeys. Journal of the Operations Research Society 19(3), 311–320 (1968)
25. Weddewer, M.: Verrechnungspreissysteme für horizontale Speditionsnetzwerke – Simulationsgestützte Gestaltung und Bewertung. Wiesbaden, Gabler (2007)
26. Wieberneit, N.: Linienverkehrsplanung in speditionellen Sammelgutnetzen. Univ.Augsburg (2008)
27. Wlček, H.: Gestaltung der Güterverkehrsnetze von Sammelgutspeditionen. Nürnberg (1998)

Appendix: Notation

$A(r_k)$	Circular delivery area around the subsidiary up to radius of r_k
C_i	Total costs for customer drop i
$C(r_k)$	Total costs up to a radius of r_k
c_{driv}	Cost per hour for driving
$C_{driv}(r_k)$	Total driving time related costs up to a radius of r_k
c_{drop}	Cost for drop per hour
$C_{drop,i}$	Direct costs for drop i
$C_{drop}(r_k)$	Total direct drop related costs up to a radius of r_k
c_{km}	Cost for driving per km
$C_{km}(r_k)$	Total driving distance related costs up to a radius of r_k
c_{LU}	Cost per hour for loading/unloading pallets
$\Delta c_{over}(r_k)$	Marginal overhead cost per pallet from radius r_{k-1} to r_k
$C_{over}(r_k)$	Total overhead cost up to a radius of r_k
$C_{pall,i}$	Total drop related pallet cost for customer i
$C_{pall}(r_k)$	Total direct pallet related costs up to a radius of r_k
$c_{tour,flat}$	Fixed transfer price for indirect overhead costs per tour
$C_{tour}(r_k)$	Total tour related overhead costs up to a radius of r_k
δ	Average distance between two drops in a shortest path tour
$\delta(r_k)$	Average distance between two drops in a shortest path tour within r_k
d_{0i}	Distance of customer drop i from subsidiary
$d_{0,first}(r_k)$	Averge first ‘legs’ of all tours within radius r_k
$d_{last,0}(r_k)$	Average last ‘legs’ of all tours within radius r_k
k_{max}	Maximum number of rings around subsidiary
$L_{km}(r_k)$	Total length of all tours up to r_k in km
m	Total number of tours

$m(r_k)$	Number of tours up to a radius of r_k
$\mu_{pall}(r_k)$	Estimated number of tours (up to r_k) due to R_{pal}
$\mu_{time}(r_k)$	Estimated number of tours (up to r_k) due to R_{time}
n	Number of customer drops
$n(r_k)$	Number of customer drops up to a radius of r_k
$v_{time}(r_k)$	Estimated number of drops per driver (up to r_k) due to R_{time}
r_k	Radius of the k_{th} circle around the subsidiary
R_{pall}	Pallet capacity restriction per truck
R_{time}	Working time restriction per driver
s	Driving speed
$t_{driv}(r_k)$	Total driving time up to a radius of r_k
$t_{drop,flat}$	Time per drop (constant)
t_{LU}	Time for loading/unloading per pallet
$t_{total}(r_k)$	Total working time of all drivers up to a radius of r_k
$v(r_k)$	Total drop pallets up to a radius of r_k
v_i	Customer demand volume in pallets for drop i

Discrete Lot-Sizing and Scheduling Including Deterioration and Perishability Constraints

Julia Pahl and Stefan Voß

University of Hamburg, Institute of Information Systems, D-20146 Hamburg
pahl@econ.uni-hamburg.de, stefan.voss@uni-hamburg.de
https://iwi.econ.uni-hamburg.de/IWIWeb/Default.aspx

Abstract. Constraints on the lifetime of items force organizations to carefully plan their production in cooperation with their supply chain partners up- and downstream along the chain. This is important because waiting times due to suboptimal planning give rise to increasing lead times and, consequently, to deterioration and thus decreasing quality of items so that, in the worst case, they cannot be used. Increased costs, delivery delays, quality decreases and unsatisfied customers are negative effects that can be avoided by accounting for product depreciation in the production process. We highlight the importance of including deterioration and perishability in planning decisions and present well-known discrete lot-sizing and scheduling models extended in this regard. The effects on plans derived by these models including depreciation is shown using a numerical example.

Keywords: Production Planning, Lot-Sizing, Scheduling, Deterioration, Perishability.

1 Introduction

Deterioration has a great influence not only on inventory management, but on every area of production where items are stocked or forced to wait due to uncertain demand, technical matters, variabilities or disruptions of the production process. Not surprisingly, scrap is one of the major causes of inventory loss regardless of the types of industries or products.

The aspect of deterioration and perishability of items came to prominence with the modeling of inventory in blood bank management and the distribution of blood from transfusion centers to hospitals. Later, the interest shifted also to other products, e.g., chemicals, food, various drugs, fashion clothes, technical components, newspapers etc. In accordance, the attention changed to a more general view on how to include deterioration and perishability constraints into mathematical models for production planning in different settings, e.g., regarding multiple facilities, warehouses, as well as supply chain perspectives.

In this paper we propose discrete lot-sizing and scheduling models accounting for depreciation effects of products in the production process. While depreciation effects are widely regarded in the area of stochastic and continuous

W. Dangelmaier et al. (Eds.): IHNS 2010, LNBIP 46, pp. 345–357, 2010.

inventory management, only very few models have been proposed for the deterministic, discrete case. Nevertheless, it may be useful to examine depreciation effects in non-complex, straightforward settings in order to develop a better understanding for more complex situations. Therefore, we analyze the *capacitated lot-sizing problem* (CLSP) extending it to deterioration and perishability effects. Furthermore, scheduling decisions are taken into consideration, too. The effects of depreciation are emphasized using a numerical example which is certainly not representable for practical cases due to its small size, but very useful regarding the effects on planning decisions.

The remainder of the paper is organized as follows. In Section 2 we give a review on the related literature on depreciation of products (Section 2.1) and lot-sizing and scheduling models (Section 2.2). In Section 3 we present the CLSP with perishability and deterioration further extending it to scheduling decisions. Section 4 provides a numerical example used to highlight computational results for two of the proposed scheduling models and Section 5 concludes with ideas for future research.

2 Literature Review

Depreciation effects are especially interesting in situations where items are forced to wait in the production process. The task of lot-sizing is to plan quantities of items produced together so as to minimize certain cost factors such as order or setup costs and inventory holding costs. For instance, the importance of accurately accounting for setup times and costs and their effects on capacity utilization has been highlighted by the discussion of just-in-time manufacturing [1] which requires small and frequent batch sizes in order to decrease WIP inventory. This influences capacity utilization, because setup times decrease machine capacity utilization and represent idle times.

Scheduling decisions play an important role regarding overall processing or lead times of items. If products are subject to enhanced deterioration or perishability it may be preferable to schedule them so that they do not spend much time waiting in inventory. This is even more true for multi-level situations regarding multiple (sequenced) machines. Consequently, when regarding depreciation effects scheduling decisions should be taken into account.

2.1 Depreciation of Parts and Products

All products deteriorate sooner or later depending on the considered time horizon. If the time horizon is chosen to be sufficiently short there may be no need to consider such product characteristics. Nevertheless, in practice and depending on the product, the time horizon for tactical and even operational planning is long enough, so that the influence of deterioration and perishability tends to play an important role.

Deterioration or *perishability* of goods is regarded as the process of decay, damage or spoilage of items such that they can no longer be used for their original purpose, i.e., they go through a change while stored and lose their utility partially or completely [2,3,4,5,6]. Deterioration is a continuous process so that items have a stochastic lifetime in contrast to perishable goods which are considered as items with a fixed, maximum lifetime [7]. This is true for products that become obsolete at some fixed point in time, due to various reasons or external factors, e.g., change in style, technological developments, outdates or external regulations (e.g. pharmaceuticals) that predetermine their shelf-lifes. Consequently, deterioration or perishability can be characterized by different aspects. [4] give three characteristics of decay of goods: *direct spoilage* (e.g. fresh food, vegetables), *physical depletion* (e.g. gasoline, alcohol), or the *loss of efficacy* in inventory (e.g. electronic components, medicine). A detailed discussion of a classification of deterioration and perishability is given in [8].

The lifetime of products can be regarded fixed or random: the first case implies that the products' lifetime is pre-specified and time-independent of deterioration factors. The utility of these products is given during a specific time period. When the products pass their lifetime, they perish completely. In the second case there is no specific lifetime for the products, instead there is depreciation over time. For instance, the lifetime of food items can be seasonal as, e.g., vegetables deteriorate faster in summer time. This kind of deterioration is called time-dependent. In the literature we find the integration of lifetime probabilities as, e.g., gamma distributions, Weibull, two-parameter Weibull, or exponential distributions. An alternative basis for classification is the *value loss* of the goods; see [9]. [10] emphasize the related utility loss and distinguishes products whose actual functionality deteriorates over time, e.g., fruits, vegetables, or milk, and those whose functionality does not degrade, but customers' perceived utility deteriorates over time which can be true for, e.g., fashion clothes, high technology products with a short life cycle, or items whose information content deteriorates, e.g., newspapers. For instance, [11] prove empirically that the willingness to pay for a product continuously decreases with its perceived actuality or utility; see also [12]. So this is somewhat fuzzy, as it clearly depends on customer preferences if a good still has some value, viz. if a customer still regards it as valuable or useful. A multitude of authors in the field of deterioration and perishability use the words interchangeably.

We make a clear distinction between perishable goods that cannot be used anymore and lose all their utility at once after a certain point in time whereas items that lose their utility gradually are regarded as being subject to continuous deterioration. For instance, Figure 1 depicts three functional relationships of the value of items in time where the first graph shows the course of perishability, the second the discrete and variational course of deterioration and the third three possible courses of continuous deterioration as defined above.

2.2 Simultaneous Lot-Sizing and Scheduling

Lot-sizing and scheduling decisions are generally considered in tactical and operational production planning. Often, these planning steps are executed sequentially which can transform formally feasible (tactical) lot-size plans to infeasible plans due to subsequent scheduling decisions that violate capacity constraints.

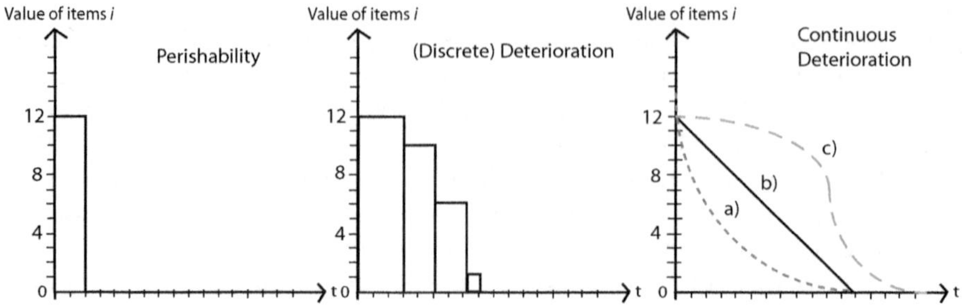

Fig. 1. Three examples for perishability and deterioration rates

Consequently, iterations of planning decisions and information are needed or, better, their simultaneous consideration. The classification of such models differentiates between single or multiple resources, single or multi-level structures of products, finite or infinite planning horizon, discrete or continuous time, and/or deterministic or stochastic input data. In this paper we concentrate on the deterministic, discrete single resource single structure case with finite planning horizon.

Discrete dynamic modeling assumes that the time dimension can be divided into "time buckets" that can be further segregated into "big bucket" and "small bucket" models. Big bucket models contain long time periods in which several different products can be set up and produced unlike small bucket models where time periods are rather short, so that startups, switch-offs and/or change-overs are modeled. Small bucket models can be further divided into those that allow at most one setup per period and those allowing more than one setup [13]. Simultaneous lot-sizing and scheduling models have been subject to extensive research for many decades. Consequently, we refer the reader to excellent surveys and reviews on (multi-level) lot-sizing and scheduling provided by [14], [15], [16], [17]. Nevertheless, discrete lot-sizing and scheduling decision models including deterioration and perishability constraints are rare. A short review on these models is given in the subsequent section.

2.3 Models Including Depreciation Constraints

Economic lot-sizing models (ELSP) including deterioration and perishability constraints are provided, e.g., by [18,19,20], where deterioration rates of inventory are dependent on the age of products. This is also true for the replenishment model of [21] where the shelf-life of items is constrained by two periods. [19] shows that the ELSP with perishability constraints is equivalent to a minimum cost network flow problem with flow loss. He analyzes special structural properties of the optimal solution in order to develop a dynamic programming algorithm able to solve the problem in polynomial time. Shortages and complete backorders are added to the model by [20]. [18] employ a Cobb-Douglas cost function in order to integrate effects of economies of scale.

Optimal utilization policies of time-varying deteriorating items are studied by [22] in the field of blood bank management. They define diverse age categories of items due to the fact that, in general, requested but unused blood preserves are returned to the blood bank after one or two days. Consequently, they do not begin to deteriorate in stock, but enter already aged. [23] studies the parallel machine scheduling problem with deteriorating jobs where the objective is to minimize the total completion time of jobs or total machine workload. Algorithms to solve the scheduling problem base on steepest descent search heuristics. [24] provides an *economic order quantity* (EOQ) model with deteriorating items and time proportional demand.

Our approach described in the next section integrates perishability that constrains the time items can be held in inventory. Thus, the inventory balance equation is subject to modifications. Furthermore, disposal costs are regarded in the objective function.

In order to consider deterioration and perishability constraints, we begin with the CLSP. We will extend it step by step in the course of our discussion considering the above mentioned aspects of lot-sizing and scheduling including sequence-dependent setup times and costs with depreciation.

3 Model Formulations

The CLSP presents a standard model for discrete, dynamic, capacitated multi-product lot-sizing where multiple products are planned on one machine or resource subject to finite capacity. Demand is deterministic and known in advance for each period t. Backlogging is not allowed and *first-in-first-out* (FIFO) is assumed regarding the stock outflow process. The objective is to find optimal production lot-sizes that minimize inventory and setup costs. Regarding perishability, we include the minimization of costs for disposal of perished items.

Given the variables, parameters, and indices in Table 1 that is valid for all presented models, the *CLSP with perishability constraints* (CLSP-P) is formulated as follows:

$$\min \sum_{i=1}^{M} \sum_{t=1}^{T} \left(K_i^s \delta_{it} + \pi_i I_{it} + \phi_i^D I_{it}^D \right) \tag{1}$$

subject to:

$$I_{it} = I_{i,t-1} + X_{it} - D_{it} - I_{it}^D \qquad \forall\, i,t \tag{2}$$

$$I_{it}^D \geq \sum_{\tau=0}^{t-\Theta_i} X_{i,\tau} - \sum_{\tau=0}^{t} D_{i\tau} - \sum_{\tau=0}^{t-1} I_{i\tau}^D \qquad \forall\, i, t \geq \Theta_i \tag{3}$$

Table 1. Variables, parameters and indices used in the models

Parameters	
K_i^s	setup costs of product i
ϕ_i^D	waste disposal cost factor for product i
π_i	inventory holding cost factor for product i
D_{it}	demand of product i in period t
ξ_i	resource consumption factor or production coefficient of product i
ϑ_i	setup time for product i
Cap_t	capacity that is available in period t
α_i^Θ	deterioration rate of product i
V	big number
Variables	
X_{it}	production of product i in period t
I_{it}	inventory holding of product i in period t
I_{it}^D	perished inventory of product i in period t
δ_{it}	setup indicator denoting if a product i is set up in period t
δ_{is}^s	setup state variable denoting if product i is set up at the end of period s ($\delta_{is}^s = 1$) or not ($\delta_{is}^s = 0$) used in scheduling models
Indices	
i	product with $i = 1, \ldots, M$
j	subset of products with $j \in M, i \neq j$
t	time period with $t = 1, \ldots, T$
τ	subset of periods
Θ_i	shelf-life of product i
s	micro-periods with $s = 1, \ldots, S$ used in scheduling models
u	subset of micro-periods used in scheduling models

$$\sum_{i=1}^{M} \xi_i X_{it} + \sum_{i=1}^{M} \vartheta_i \delta_{it} \leq \mathrm{Cap}_t \qquad \forall t \tag{4}$$

$$X_{it} \leq V \delta_{it} \qquad \forall\, i, t \tag{5}$$

$$I_{i0} = I_{iT} = 0 \qquad \forall\, i \tag{6}$$

$$X_{it}, I_{it}, I_{it}^D \geq 0 \qquad \forall\, i, t \tag{7}$$

$$\delta_{it} \in \{0, 1\} \qquad \forall\, i, t \tag{8}$$

The objective function (1) minimizes the sum of costs of setting up for product i in period t, inventory holding costs, and disposal costs of perished products. Equation (2) describes the inventory balance equation where inventory of product i in period t is composed by inventory of the previous period $t - 1$ and production less products that are used to satisfy demand in that period and less the products that are deteriorated in that period. Inequality (3) presents items that need to be disposed in period t. These are calculated by the sum of produced items in the interval $[0, t - \Theta_i]$, thus production from the beginning of the planning horizon until period t deduced by the shelf-life of items, less those items that are used to satisfy demand until period t and items that have been already disposed in the previous period $t-1$. In order to incorporate deterioration effects in the model, the parameter Θ_i is modified to Θ_{it}, so that changing deterioration rates over time are captured. Thus, the course of deterioration for items is defined via input data. This is valid also for the subsequent models. In our model formulation, we limit the mathematical presentation to the perishability case. Inequality (4) requires production and setup times to be smaller or equal available capacity in period t. Constraint (5) entails the machine to be set up for item i when production of that item takes place where V is a large number that needs to be chosen in a way that the production amount is not restricted. This constraint is also frequently formulated as follows:

$$X_{it} \leq \frac{\mathrm{Cap}_t}{\xi_i}\delta_{it} \quad \forall\, i,t \tag{9}$$

linking continuous variables X_{it} to the binary variables δ_{it} ensuring that whenever product i is produced in period t, thus $X_{it} > 0$, the related setup indicator δ_{it} is set, accordingly. Equality (6) denotes that initial and final inventory is zero. Non-negativity on the variables and the definition for the binary setup variable δ_{it} are given by (7) and (8), respectively.

Inequality (3) might be reformulated assuming that a certain amount of items in stock perish/deteriorate in a time period t independent of their age and how long they have been in stock. The inventory balance equation is reformulated, accordingly:

$$I_{it} = (1-\alpha_i^{\Theta})I_{i,t-1} + X_{it} - D_{it} \qquad \forall\, i,t \tag{10}$$

where α_i^{Θ} denotes the deterioration rate of product i. A certain percentage of inventory of the previous period $t-1$ perishes in stock and need to be subtracted from the inventory balance. For instance, this formulation is used by [19,20].

The CLSP is a "big bucket" model and, consequently, scheduling decisions are not included. Additionally, setup states are not carried over from one period to the next, thus, regardless if a product i is planned on the machine in period $t-1$ and period t, the model induces two setups. Moreover, a lot-size cannot be split and produced in two consecutive periods, but has to be produced in one period. Neither is it possible to setup the machine twice for the same product. In other words, a product i can be produced only once in a period t which constrains the maximal possible number of setups to the number of products (types) i. Additionally, lot-sizes X_{it} are constrained by Cap_t/ξ_i.

In order to include scheduling decisions, the CLSP-P is modified to a *discrete lot-sizing and scheduling problem* with perishability (DLSP-P). It inherits the assumptions from the CLSP despite that the macro-periods of the CLSP are subdivided into micro-periods denoted by $s = 1,\ldots,S$. The DLSP-P further assumes that at most one product type i is produced in a period s, so that full capacity is utilized for that product. This is denoted as the *all-or-nothing* assumption. Consequently, the capacity restriction of the CLSP, inequality (4), is reformulated considering micro-periods:

$$X_{is} = \frac{(\mathrm{Cap}_s - \vartheta_i)}{\xi_i}\delta^s_{is} \qquad \forall\, i, s \tag{11}$$

where δ^s_{is} is a setup state variable denoting that, e.g., no setup is required if the machine was left set up in the previous period for product i, thus $\delta^s_{is} = 0$. Consequently, the inventory balance equation is modified as (13) below. The setup state variables δ^s_{is} are linked to the setup costs $K^s_i \delta_{is}$ in the objective function by $\delta_{is} \geq \delta^s_{is} - \delta^s_{i,s-1} \quad \forall\, i, s$ denoting that if a setup in two consecutive periods takes place, a setup operation δ_{is} must have occurred. The complete DLSP-P is stated as follows:

$$\min \sum_{i=1}^{M} \sum_{s=1}^{S} \left(K^s_i \delta_{is} + \pi_i I_{is} + \phi^D_{is} I^D_{is} \right) \tag{12}$$

subject to

$$I_{is} = I_{i,s-1} + \frac{(\mathrm{Cap}_s - \vartheta_i)}{\xi_i}\delta^s_{is} - D_{is} - I^D_{is} \qquad \forall\, i, s \tag{13}$$

$$I^D_{is} \geq \sum_{u=0}^{s-\Theta_i} \frac{(\mathrm{Cap}_{s-\Theta_i} - \vartheta_i)}{\xi_i}\delta^s_{iu} - \sum_{u=0}^{s} D_{iu} - \sum_{u=0}^{s-1} I^D_{iu} \qquad \forall\, i, s \geq \Theta_i \tag{14}$$

$$\sum_{s=1}^{S} \delta^s_{is} \leq 1 \qquad \forall\, s \tag{15}$$

$$\delta_{is} \geq \delta^s_{is} - \delta^s_{i,s-1} \qquad \forall\, i, s \tag{16}$$

$$I_{i0} = I_{iS} = 0 \qquad \forall\, i \tag{17}$$

$$I_{is}, I^D_{is} \geq 0 \qquad \forall\, i, s \tag{18}$$

$$\delta^s_{is} \in \{0, 1\} \qquad \forall\, i, s \tag{19}$$

The preservation of setup states across periods gives a certain kind of continuity [25]. Nevertheless, still the all-or-nothing assumption underlies the model restricting lot-sizes to be produced within one period and not exceeding this period. This can be interpreted as a minimum lot size or batch. In practice, this might be required, e.g., due to technical issues in the production process. In fact, minimal lot sizes are the very hard constraints that lead to disposals, because enforcing production higher than demand due to capacity leading to increased inventory can cause disposals. Otherwise, disposals are always avoided independently how high setup costs might be due to the fact that these costs are due anyway within the production horizon in order to switch to production of another product type. Thus, producing at most the required demands is cost optimal. This is also highlighted in Section 4.

The problem of spoilage due to minimal lot sizes implied by the all-or-nothing assumption may be remedied by allowing production lot-sizes to take any value up to the capacity limit [26] further carrying over setup states without incurring additional setup costs [27]. Consequently, the inventory balance equation (13) is changed as follows:

$$I_{is} = I_{i,s-1} + X_{is} - D_{is} - I_{is}^{D} \qquad \forall\, i, s \tag{20}$$

with the production lot-size being constrained as in (11). Such model formulation is denoted *continuous setup lot-sizing problem* (CSLP) in the literature, although without perishability constraints. Still, in the CSLP with perishability (CSLP-P) production capacity may be lost due to the constraint that only one product type i can be produced per period which is given by inequality (11). This problem is readdressed by the *proportional lot-sizing and scheduling problem with perishability* (PLSP-P). Therefore, the production constraints (11) and capacity constraints are reformulated:

$$X_{is} \leq \frac{(\mathrm{Cap}_i - \vartheta_i)}{\xi_i}\left(\delta_{is}^{s} + \delta_{i,s-1}^{s}\right) \qquad \forall\, i, s \tag{21}$$

$$\sum_{i=1}^{M} \xi_i X_{is} + \sum_{i=1}^{M} \vartheta_i \delta_{is} \leq \mathrm{Cap}_s \qquad \forall\, s \tag{22}$$

where the remainder of the CSLP-P remains unaltered.

4 Numerical Results

In this section we highlight the effects of the all-or-nothing assumption calculating the DLSP-P, and the PLSP-P using exemplified data sets. We consider four types of products and a planning horizon of 10 periods. The capacity of one machine is 70 units per period. The data sets used are rather small, because the aim is to provide some insights regarding the mechanisms and hard constraints of minimum lot sizes that lead to disposals. Applying larger data sets to the models leads to increased computational calculation times as variables and constraints increase exponentially. This is straightforward and experienced by extending the data set to 7 products and 50 periods.

The models are implemented in AMPL/CPLEX and are calculated on a computer with an Intel Core(R) Pentium(R) 4 CPU processor with 2.42 GHz and 1 GB RAM.

Due to the relative small numerical example the calculating time for finding the optimal solution is negligible. When extending the example to seven products and 20 time periods and thus dealing with 550 variables and 423 constraints the optimal solution is still found within seconds, but time begins to be relevant. The optimal solution for the DLSP-P is found with the production/inventory plan as given in Table 3 resulting in total costs of 6.440. The calculation employs 289 MIP simplex iterations including one branch-and-bound node.

The PLSP-P gives a plan that does not generate disposals. Therefore, disposals that are zero for all products in all periods are not included in Table 4. For

Table 2. Demand data for the small numerical example

	Cost Factors						Demand in Periods									
Products	K_i^s	π_i	ϕ_i^D	ϑ_i	Θ_i	ξ_i	1	2	3	4	5	6	7	8	9	10
A	400	2	10	10	2	1	10	10	–	–	–	40	–	–	–	60
B	400	1.5	10	10	2	1	–	–	–	5	30	30	–	–	40	–
C	400	0.5	10	10	2	1	–	–	10	–	10	–	10	–	10	–
D	400	0.5	10	10	2	1	–	–	–	–	10	20	30	–	–	60

Table 3. Results of the DLSP-P model

		Periods									
		1	2	3	4	5	6	7	8	9	10
Production	A	60	-	-	-	-	60	-	-	-	60
	B	-	60	-	60	-	-	-	-	60	-
	C	-	-	60	-	-	-	60	-	-	-
	D	-	-	-	-	60	-	-	60	-	-
Inventory	A	10	-	-	-	-	-	-	-	-	-
	B	-	5	5	60	30	-	-	-	-	-
	C	-	-	10	10	-	-	10	10	-	-
	D	-	-	-	-	50	30	-	60	60	-
Disposal	A	40	-	-	-	-	20	-	-	-	-
	B	-	55	-	-	-	-	-	-	20	-
	C	-	-	40	-	-	-	40	-	-	-
	D	-	-	-	-	-	-	-	-	-	-

the PLSP-P, the optimal solution results in 3.740 total costs. CPLEX employed 16.879 MIP simplex iterations in order to find the optimal solution including 1.187 branch-and-bound nodes. This is due to the fact that the DLSP-P contains the all-or-nothing assumption which leads to full utilization of capacities even if demands are not equally high. Therefore, inventory does not decrease due to demand, so that items perish after their lifetime and need to be disposed. This is not true for the PLSP-P where production utilization is as high as the demand, so that items do not wait in inventory and the risk of perishability is rather low. When comparing the results for the PLSP-P to the PLSP without perishability we see that the number of setups in the PLSP are relatively small, inventory higher and, therefore, total costs smaller than in the PLSP-P. This is due to the fact that products not subject to perishability can be hold longer in inventory than perishable items which is straightforward. Furthermore, when less capacity is available for production, inventory increases. This behavior

Table 4. Results of the PLSP-P model

		Periods									
		1	2	3	4	5	6	7	8	9	10
Production	A	10	10	-	-	-	40	-	-	60	
	B	-	-	-	5	60	-	-	-	40	-
	C	-	-	20	-	-	-	-	20	-	-
	D	-	-	-	30	-	-	30	40	20	-
Inventory	A	-	-	-	-	-	-	-	-	-	-
	B	-	-	-	-	30	-	-	-	-	-
	C	-	10	10	10	-	-	-	10	-	-
	D	-	-	-	30	20	-	-	40	60	-

is true for all presented models except the DLSP-P where the number of products in inventory for all periods remains the same, but spoilage decreases. This is reasonable, because less capacity decreases minimum lot sizes implied by the all-or-nothing constraint.

5 Conclusion

Aspects of deterioration and perishability are important for tactical and operational production planning, because such effects might generate increased costs if not considered in production plans. We have integrated such effects in well known discrete lot-sizing models and compared resulting plans using a numerical example. The DLSP turns out to be not suited to problems where products are subject to deterioration and perishability due to the all-or-nothing-assumption which implies minimal lot sizes, because it enforces full utilization of capacities disregarding periodic demand rates and, thus, causing spoilage and related costs. Further research will be in the direction of sequence-dependent setup times and multiple levels including diverse product structures where items depend on each other further considering rework, thus the possibility to rework items that need perish during the production process.

References

1. Trigeiro, W., Thomas, L., McClain, J.O.: Capacitated lot-sizing with setup times. Management Science 35(3), 353–366 (1989)
2. Andijani, A., Al-Dajani, M.: Analysis of deteriorating inventory/production systems using a linear quadratic regulator. European Journal of Operational Research 106(1), 82–89 (1998)
3. Dave, U.: A probabilistic scheduling period inventory model for deteriorating items with lead times. Zeitschrift für Operations Research 30(5), A 229–A 237 (1986)

4. Ghare, P.M., Schrader, G.F.: A model for exponentially decaying inventories. Journal of Industrial Engineering 14, 238–243 (1963)
5. Shah, B., Shah, N., Shah, Y.: EOQ model for time-dependent deterioration rate with a temporary price discount. Asia Pacific Journal of Operational Research 22(4), 479–485 (2005)
6. Zhao, P.X.: An EOQ model for items with Weibull distribution deterioration. In: 2nd IEEE Conference on Industrial Electronics and Applications, pp. 451–454 (2007)
7. Ketzenberg, M., Ferguson, M.: Sharing information to manage perishables (ed. 2). Working Paper (2005), `http://smartech.gatech.edu/handle/1853/7098` (last call Janaury 2010)
8. Pahl, J., Voß, S., Woodruff, D.: Production planning with deterioration constraints: A survey. In: Ceroni, J. (ed.) The Development of Collaborative Production and Service Systems in Emergent Economies, Proceedings of the 19th International Conference on Production Research. IFPR, Valparaiso (2007)
9. Goyal, S.K., Giri, B.C.: Recent trends in modeling deteriorating inventory. European Journal of Operations Research 134(1), 1–16 (2001)
10. Ferguson, M.E., Koenigsberg, O.: Production and pricing decisions: How should a firm manage a product with deteriorating quality (ed. 2). Preliminary Draft, Internet-Source (2005), `http://smartech.gatech.edu/handle/1853/7078`
11. Tsiros, M., Heilman, C.M.: The effect of expiration dates and perceived risk on purchasing behavior in grocery store perishable categories. Journal of Marketing 69, 114–129 (2005)
12. Wee, H.M.: Economic production lot size model for deteriorating items with partial backordering. Computers & Industrial Engineering 24(3), 449–458 (1993)
13. Staggemeier, A.T., Clark, A.R.: A survey on lot-sizing and scheduling models. In: 23rd Annual Symposium of the Brazilian Operational Research Society (SOBRAPO) (2001)
14. Brahimi, N., Dauzere-Peres, S., Najid, N.M., Nordli, A.: Single item lot sizing problems. European Journal of Operational Research 168(1), 1–16 (2006)
15. Comelli, M., Gourgand, M., Lemoine, D.: A review of tactical planning models. Journal of Systems Science and Systems Engineering 17(2), 204–229 (2008)
16. Karimi, B., Ghomi, F.S., Wilson, J.: The capacitated lot sizing problem: a review of models and algorithms. Omega 31, 365–378 (2003)
17. Quadt, D., Kuhn, H.: Capacitated lot-sizing with extensions: A review. 4OR: A Quarterly Journal of Operations Research 6(1), 61–83 (2008)
18. Chu, L.Y., Hsu, V.N., Max Shen, Z.-J.: An economic lot-sizing problem with perishable inventory and economics of scale costs: Approximation solution and worst case analysis. Naval Research Logistics 52(6), 536–548 (2005)
19. Hsu, V.N.: Dynamic economic lot size model with perishable inventory. Management Science 46(8), 1159–1169 (2000)
20. Hsu, V.N.: An economic lot size model for perishable products with age-dependent inventory and backorder costs. IIE Transactions 35(8), 775–780 (2003)
21. Häselbarth, L., Scholl, A.: Dynamische Bestellmengenplanung für verderbliche Luxusgüter. Technical Report 13, Wirtschaftswissenschaftliche Fakultät, Friedrich-Schiller-Universität Jena (2003)
22. Pierskalla, W., Roach, C.: Optimal issuing policies for perishable inventory. Management Science 18(11), 603–614 (1972)

23. Gawiejnowicz, S.: Scheduling deteriorating jobs subject to job or machine availability constraints. European Journal of Operational Research 180(1), 472–478 (2007)
24. Dave, U.: On a discrete-in-time deterministic inventory model for deteriorating items with time proportional demand. Optimization 16(3), 449–461 (1985)
25. Suerie, C., Stadtler, H.: The capacitated lot-sizing problem with linked lot sizes. Management Science 49(8), 1039–1054 (2003)
26. Almada-Lobo, B., Klabjan, D., Carravilla, M.A., Oliviera, J.F.: Multiple machine continuous setup lot-sizing with sequence-dependent setups. Computational Optimization and Applications (2009)
27. Karmarkar, U.S., Kekre, S., Kekre, S.: Capacity analysis of a manufacturing cell. Journal of Manufacturing Systems 6(3), 165–175 (1987)

Developing and Maintaining Trust in Post-disaster Hastily Formed Networks

Peter Tatham[1] and Gyöngyi Kovács[2]

[1] Cranfield University, Centre for Human Systems, DACMT Shrivenham,
Swindon, SN6 8LA, United Kingdom
p.h.tatham@cranfield.ac.uk
[2] HUMLOG Institute,
P.O. Box 479, 00101 Helsinki, Finland
kovacs@hanken.fi

Abstract. Although there is a rich steam of literature championing the importance of trust in long-term business relationships, there has been relatively little interest in the area of the development and maintenance of trust in hastily formed networks that are formed at short notice and that operate for a limited period of time. Such networks can be found in the aftermath of a rapid onset disaster where individuals from a broad variety of organisations and geographic locations come together in order to bring relief to a disaster-stricken area. The aim of this paper is, therefore, to further the understanding of trust in hastily formed networks through the presentation of a theoretical model, a discussion of its practical application in a post-disaster scenario, and proposals for testing the approach in a live setting.

Keywords: Humanitarian Logistics, Trust in Networks, Supply Chain Management.

1 Introduction

The fields of inter-personal and inter- and intra-organisational trust have been the subject of significant academic research stretching back over at least three decades including, for example, special editions devoted to the subject in the journals *Organization Studies* [1] and *Organization Science* [2]. This has led to the development of frameworks that integrate psychological and behavioural concepts and, although further research is unquestionably required [3], such models offer useful insights into the antecedents of trust, as well as subsequent evolutional patterns.

Whilst the need for inter-organisational trust is a core concept in a broad spectrum of management literature [4] [5] [6] [7], much of this is devoted to understanding how appropriate <u>long term</u> relationships can be developed and maintained. This reflects issues such as the potential for reduced transaction costs [8], the relationship between trust and risk [9], trust and vulnerability [10], and as a substitute for contracts [11] or control [12] [13] [14]. By comparison, however, there are relatively few studies on trust in temporary networks even though such networks exist in a variety of settings

W. Dangelmaier et al. (Eds.): IHNS 2010, LNBIP 46, pp. 358–371, 2010.

such as the "minimal organizations" [15], "emergent multi-organizational networks" or "emergent response groups" [16] and hastily formed networks (HFNs) in a military context [17] and in disaster relief [18] [19].

Whilst the phrase "hastily formed network" provides a relatively clear mental model, the concept has been formally described by the HFN Research Group [20] as having five elements. It is a network of people (1) established rapidly, (2) from different communities, (3) working together in a shared conversation space, (4) in which they plan, commit to, and execute actions, (5) to fulfil a large, urgent mission. In following this definition of an HFN, it is important to note that such networks differ from the minimal organisations described by Weick [21] where the individual members (such as fire fighters) are likely to share a common aim, ethos, background, approaches and working practices. By contrast the members of an HFN in a humanitarian context, whilst sharing the same high level goals, may have not worked together previously nor have undergone the same training. For example, it is estimated that over 400 official international non-governmental organisations (NGOs) and over 5,000 associated staff were present in Indonesia in the immediate aftermath of the 2004 Southeast Asian tsunami [22], whilst the United Nations Joint Logistic Centre (UNJLC) formed in the aftermath of Cyclone NARGIS that struck Burma in 2008 was based in two separate countries (Burma and Thailand), and drew its 18 staff from 10 separate organisations [23].

The aim of this paper is, therefore, to further the understanding of the development and maintenance of inter-personal trust in hastily formed networks (HFNs) in the aftermath of a rapid onset disasters. To achieve this, the paper will begin by discussing the concept of HFNs in more in detail, before considering the ways in which, using a theoretical approach, trust can be formed in such networks. In doing so, it will discuss the problem of developing and maintaining trust from psychological, behavioural and physiological perspectives, drawing in particular on the concept of "swift trust" that has been developed by Meyerson *et al.* [24]. This will be followed by a discussion of the operationalisation of trust in an HFN, after which potential ways in which the model might be tested are considered.

2 The Development of Trust in Post-disaster Hastily Formed Networks

In the aftermath of such high profile events as the 2004 Indian Ocean tsunami, Pakistan earthquake and Hurricane Katrina in 2005, and Cyclone Nargis and the earthquake in Szechuan in 2008, there has been a significant movement towards improving the response to such rapid onset disasters. In this respect, the effectiveness of the relief operations is critically dependent on the speed of logisticians to be able to "procure, transport and receive supplies at the site of a humanitarian relief effort" [25]. Indeed, it can be argued that the humanitarian response is, in effect, a logistic operation given that an estimated 50-60% [26] to 80% [27] of the income of an NGO is spent in this broad area. Whilst the need for those responding to disasters to do so with speed is self-evident, perhaps less apparent are the sheer numbers of such organisations involved. For example the UK is estimated to have 3-4,000 internationally operating humanitarian organisations [28], whilst a recent analysis has suggested that over 30,000 such international NGOs exist world-wide [29]. It is unsurprising, therefore, that the arrival of a plethora of organisations within a disaster area can lead to huge challenges as evidenced by the 72 inter-agency co-ordination meetings reported to have been held weekly in Banda Aceh alone [30].

With this need for speed set against the influx of many organisations and individuals in mind, the need to achieve a high degree of inter-personal trust at an early stage in the response becomes more apparent. But, in considering how to conceptualise trust in the disaster relief context, it is appropriate to begin with a definition of the concept – although, as observed by Kramer [31] "a concise and universally accepted definition has remained elusive." On the other hand, there are many authors in this field who offer broadly similar definitions (e.g. [32] [33] [34] [35]) and these are reflected in the following which will be used within the paper:

Trust is present when the one party has a fundamental belief that the other can be relied upon to fulfil their obligations with integrity, and will act in the best interests of the other.

Importantly, this definition of trust focuses on inter-personal, rather than intra- or inter-organisational relationships. It is argued that decisions within organisations are made by individuals and, therefore, the level of intra/inter-organisational trust can be seen as a reflection of those individual relationships – although it is recognised that, in reality, the spectrum of possible and actual relationships is significantly more complex as demonstrated by Currall and Inkpen [36]. But how does such inter-personal trust develop? In decomposing the vast swathe of literature that discusses the concept of trust from a variety of perspectives, it is suggested that the key to understanding it in the context of HFNs is the recognition that, in this scenario, trusting inter-personal relationships need to be achieved in a very short time frame. For this reason, the work of Meyerson *et al.* [37] into the development of "swift trust" has significant applicability, and it has been expanded into a more general framework by Hung *et al.* [38] (see Fig. 1). Hung *et al.* [38] suggest that there are three different routes to trust: the peripheral, central, and habitual. If seen in sequence, the peripheral route refers to the early stages of a relationship in which the individuals forming the network cannot rely on their, as yet to be formed, perceptions of the trustee(s) ability, integrity and benevolence; rather trust at this stage is based on (peripheral) cues such as those provided by third parties. Hung *et al.* [38] suggest that adoption of this peripheral route to trust involves less cognitive effort than making one's own judgments and is, therefore, the preferred route in the initial stages of a relationship. However, once teams have formed, individuals are able to cognitively engage and this may lead to the development of trust through the central route. The final route is that of habitual trust which reflects the historical build up of successful trust transactions, and often leads to strong emotional bonds.

According to Hung *et al.* [34] the peripheral (i.e. swift) route, has five antecedent conditions that influence trust formation: (1) role, (2) third party information, (3) category, (4) rule and (5) dispositional. Importantly, it is argued that, in addition to the simple feedback loop generated by improved knowledge of the others in the relationship, the process is mediated by the perceived level of risk which, in turn, reflects the communications environment within the network. It will be appreciated that, in the context of HFNs in the aftermath of a disaster, there are likely to be many virtual elements to the network (including for example, head and regional offices, together with individuals who are on the ground in the disaster zones), and therefore this aspect of the framework is of particular importance.

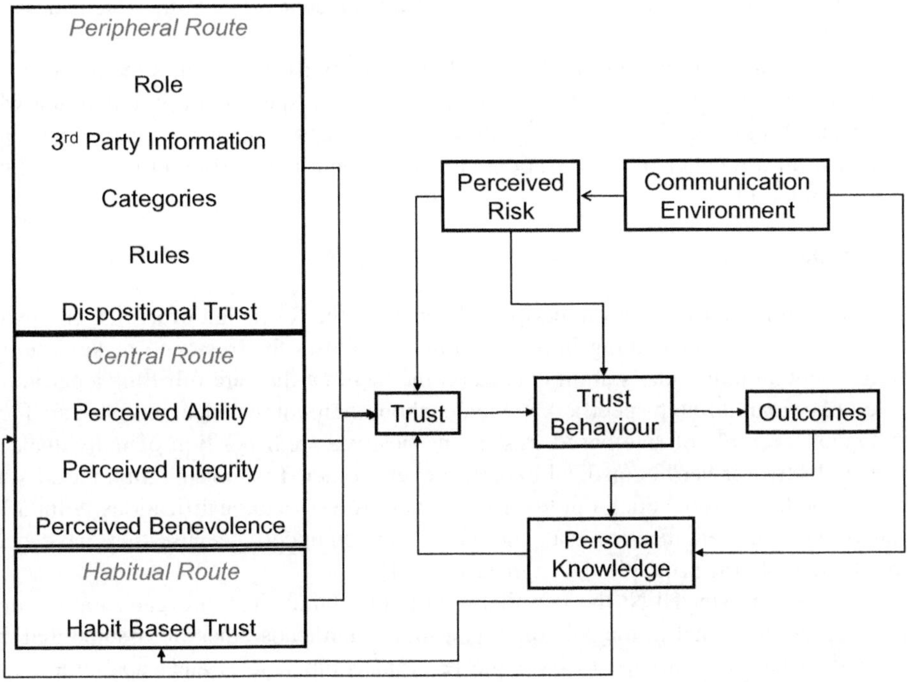

Fig. 1. Routes to trust formation (modified from Hung et al. [38, p.4])

3 Operationalising the Swift Trust Framework in a Humanitarian Logistic Context

In operationalising the framework shown in Figure 1, a number of overarching points must be considered. The first is the interdependence of the trusting relationships which means that, in essence, there is no one trustor or trustee; rather each member of the network engages in a relationship with the other members. Furthermore, the outcomes of any interchange between the parties will affect each party in, potentially, different ways. Secondly, the level of trust within a relationship is not static. Indeed, as Hung *et al.* [38] argue, over time, the relatively fragile swift trust can develop, via the central route, to the robust habitual form; on the other hand, trust can also decline. Furthermore, there is no absolute level of trust at any given time in a relationship, rather, parties may trust each other in relation to one issue, but not another. Finally, there is good evidence to suggest that, subject to any negative impacts of the perceived success of prior alliances [39], individual members of a network often act as if trust were in place and this leads to self-fulfilment. The very act of forming a network may of itself trigger an initial level of trust, where a positive assumption about the trusting behaviour of others becomes the baseline position [40]. However, Coppola *et al.* [41] and Ben-Shalom *et al.* [42] suggest that this baseline is also affected by the expectations of trust that members import from other settings with which they are familiar.

With the above points in mind, the following sections will discuss the antecedents of swift trust. This discussion is developed from the perspective of an individual who has been designated the leader of a post-disaster logistic team whose members are drawn from a number of organisations and who come from different countries (and, hence, cultures).

3.1 Role

In line with the concept of the peripheral route for the formation of swift trust, using roles as the basis for making initial assumptions allows the trustor to make assumptions about an individual's ability based on the fact that they are fulfilling a particular role rather than through specific knowledge about their competence, motives, etc [43]. A typical example of role-based trust is the positive predisposition of individuals to trust a doctor for her/his medical expertise even in their first consultation based simply on the fact that the doctor holds the relevant professional qualifications. Similarly, individuals trust engineering and, therefore, trust engineers because they have been taught to apply the principles of engineering [44].

In a post-disaster HFN, the fact that a particular humanitarian organisation is employing an individual in the role of, say, a logistician leads others to assume that the individual has been judged to have the relevant competencies and capabilities, and can therefore be trusted. For this means of developing trust to be effective, there is a very clear onus on humanitarian organisations to fulfil their side of this notional bargain – in other words only to employ staff who do, indeed, possess the relevant competencies etc. However, this quickly leads one to the prior question of what are the appropriate skills and attributes required of a humanitarian logistician and it should be noted that this remains an under-researched question [45]. Indeed, whilst there have been a number of models developed [46], the linkage between such proposed competencies and the logistics performance of an organisation has yet to be demonstrated [47].

That said, there is a general move towards the introduction of certification through such schemes as those provided by the UK Chartered Institute for Logistics and Transport (CILT) which are delivered in concert with the NGO RedR. Successful anchoring of the required skills and experience in an internationally recognised framework would provide valuable underpinning for the antecedent of "role" within the swift trust model and, hence, support the development of the desired interpersonal trust. However, in the absence of such internationally recognised standards, our exemplar team leader should still use this role-based route whenever possible and, where appropriate, promote the reliability and trustworthiness of individual team members to the group as a whole.

3.2 Third Party Information

Third party information enables the formation of trust based not on the, as yet, unidentified capabilities of an individual, but on their prior reputation. Thus, whilst the HFN itself does not have a shared history, some of the individuals within it may have carried out similar roles under different, but similar, circumstances. Thus, for example, whilst the leader may not know team member A, he or she does know team member B who, in turn, knows A. By drawing on team member B's (positive) view of A, it is possible to promote A's credentials – and hence their trustworthiness to the team at large.

This generic approach could be assisted by the development of a database of those members of humanitarian organisations who are on call for short term post-disaster duties which would include details of their qualifications and/or experience. However, in the absence of such a data base, it is suggested that humanitarian organisations have a responsibility to "advertise" the skills of their employees both within the organisation itself and between organisations. The aim here is not to develop an elitist mentality, rather to support the formation of trust by emphasising that individuals are likely to have the appropriate skills in advance of their demonstration of these.

3.3 Category

Further peripheral cues of trust are given by the membership of individuals in social groups or categories such as gender or race [48]. Within the context of disaster relief, this is potentially a highly divisive area – indeed, evidence of the negative effects of such categorisation has been noted by Zolin [49, p.7] who observed: "difficulties in establishing interpersonal working relationships between [the US Military] and [NGOs] due to perceived differences in organisational goals, strongly held negative organizational stereotypes and perceived ideological differences". Apart from trust judgements based on stereotypes based on gender, ethnicity, religion, race, or age (to name but a few of potential categories), organisational culture also forms a category in itself [50]. As an example, military organisations are often categorised as highly hierarchical organisations, whilst NGOs typically operate in a less structured organisation that prizes individualism. Unsurprisingly, therefore, such entities potentially present a clash of cultures – although the extent to which this is perceived, or can be seen as stereotypical rather than real, remains unclear.

Interestingly, the location of an organisation within such a categorisation must be seen in relative rather than absolute terms. Thus, within the broad humanitarian family, UN agencies are often seen as more hierarchical when compared to an NGO that is a network of individuals. This, in turn, helps to explain the potential tension between such organisations within the wider humanitarian aid supply network. Indeed, Dowty *et al.* [51] have used this to help explain not only the cultural dissonance between a number of organisations (e.g. FEMA and the US Coast Guard service) in their response to Hurricane Katrina, but these authors go further by suggesting that the propensity to take risk is link to culture. Thus, not only is there a potential clash between the approaches and *modus operandi* of organisations in different categories, but also their approach to the management of risk. In a disaster response scenario, it would seem likely that such differences would have a particularly negative effect on the building of trust between members of different organisations.

Given that such stereotypes undoubtedly exist, the implication of the swift trust model is that when the trustor and trustee belong (or perceive that they belong) to different categories, this will have a negative impact on the development of trust. This is an important area because separate experimental research has demonstrated a broad range of related psychological nuances including that images of smiling partners are trusted more than those who are not smiling [52], and individuals whose faces resemble the trustor are significantly more trusted [53]. In the context of an HFN, it is clearly incumbent on the leader to recognise the implicit potential for divisiveness, and to ensure that the team members appreciate that there is almost certain to be a difference in the ways in which individuals respond to challenges and interact. Prior appreciation of the near inevitability of culturally based misunderstandings will, it is argued, go a long way towards ensuring that these can be managed and do not unreasonably restrict the development of trust within the network.

3.4 Rule

The presence of rules (under which heading one can include processes and procedures) is deemed by Kramer [54] to be of significance in supporting the development of swift trust. Put simply, the suggestion here is that, by following "the rules", individuals are deemed by their peers to be trustworthy [55]. Indeed, Kramer [56] suggests that "explicit and tacit understandings regarding transactional norms, interactional routines and exchange practices provide an important basis for inferring that others in the organization are likely to behave in a trustworthy fashion." In short, the adherence to the organisation's rules is a guard against maverick behaviour which has the potential to reduce the level of inter-personal and inter-organisational trust. Indeed, this perspective has considerable resonance with the work of other researchers such as Grey and Garsten [57] and Maguire *et al.* [58] who conceptualise trust as enabling individuals to behave in a predictable way.

However, when it comes to initial the development of trust, rule-based behaviour refers to issues such as the normality of the situation and, potentially, the assurance of organisational structures [59]. But for humanitarian logisticians, "normality" is a highly fluid and uncertain situation and, hence, there is no single organisational structure they can be used as a template. That said, the general concept of "structuration", i.e. the development of common approaches, sets of rules, etc., has clear relevance to such HFNs as it would help to ensure that individuals who join the network from different organisations can make the transition with the minimum of effort. In this respect, co-ordinating initiatives such as the Logistics Cluster and the emerging Humanitarian Logistics Association (HLA) point towards the long term possibility of developing structures that can help the rule-based development of swift trust in HFNs.

On the other hand, it is acknowledged that, in the humanitarian context, the development of well documented processes and procedures (i.e. "rules") is counter-cultural. Those working within humanitarian organisations are, understandably, output and outcome focussed; their *raison d'être* is the relief of hardship and suffering of those affected by a disaster and adherence to "bureaucracy" is seen as a diversion from this real objective. On the other hand, when responding to a major disaster, NGOs are almost universally forced to use staff who are not part of their core teams, i.e. those from "on call" rosters and other augmentees. In all probability these additional resources will have had limited experience of working within the particular NGO and, therefore, will have even more limited exposure to that organisation's rules. This results in the potential for inadvertent maverick behaviour with its concomitant negative effect on the development of inter-personal trust. There is clearly a balance to be struck here, as it could be argued that such behaviour in the guise of strong leadership could be valuable in cementing relationships within a team. However, from the perspective of the swift trust model, there is clear benefit in the development and exposition of clear simple and easy to follow rules that will help ensure newcomers can fit into the organisation and become effective both speedily and with the minimum of effort.

3.5 Dispositional Trust

Dispositional trust is the final antecedent condition in the swift trust framework, and this relates to the general disposition of an individual to trust other people; in other words, some people are inherently more trusting than others. The existence of this phenomenon is well documented [60] [61] [62] but, as with the earlier discussion of "category", there is little that can be recommended to overcome the problem of differential dispositional distrust other than for the team leader to ensure that all members of the network are aware of this potential challenge to the development of trust.

4 Trusting Behaviour in Humanitarian HFNs

In considering the swift trust model, it is not only important to note the five antecedent conditions that give rise to trust, but also to distinguish between trust, and trusting behaviour. Hung *et al.* [63] depict trusting behaviour as being mediated by the perceived risk of potential gains (or losses) of acting on the basis of inter-personal trust. In the humanitarian context, perceived risks can encompass physical danger as well as the loss of reputation as a result of depending on the behaviour of other members of the HFN. In essence, the act of trusting is one in which the trustor is prepared to increase their vulnerability to the actions of others. It follows, therefore, that if level of perceived risk is greater than the level of trust, the individual is less likely to engage in trusting behaviour [63].

The second mediating issue is that of the communication environment, the importance of which has been emphasised by Weick [64] in his analysis of the Mann Gulch disaster in which 13 US fire fighters lost their lives. In this disaster one of the key failings was the near absence of communication between the team members, and the consequential reduction in the level of intra-team coordination. In short, the lack of communication in the early stages of the development of this temporary group heightened its vulnerability to disruption. When stressed by the advancing wild fire, the inter-team ties (which, in part, reflect the level of inter-personal trust) were insufficient to prevention fragmentation of the group and a reversion to self-interest (or perhaps more accurately, self-preservation). This point is equally emphasised by Drabek [65] whose review of emergency response organisations in the United States indicates that cross-agency communication was perceived to be the greatest weakness. In summary, there would appear to be broad support for the proposition that the clarity of the communications environment has an effect on the formation of trust. This perspective has been extended by Hung *et al.* [66] who suggest that "virtual" environments increase the perceived risk and, hence, reduce the propensity to convert trust into trusting behaviour.

Once again, in terms of mitigating these issues, the key would appear to lie in an understanding of the problem (i.e. the effect of the perception of risk) on the actions of individuals, and the role that the effectiveness of inter-personal communication has to play as an antecedent. Clearly a number of technological approaches (such as the use of video-conferencing) may help overcome the inherent defects of simple computer-based interaction (e-mails etc.) – subject of course, to its feasibility in a post-disaster scenario. As before, the role of the leader would seem to revolve around ensuring that team members are aware of this facet of the problem through appropriate guidance, training and education.

5 The Development and Maintenance of Trust from a Physiological Perspective

Up to this point, the discussion has approached the development and maintenance of inter-personal trust from the psychological and behavioural perspectives. However with the advent of improved neuroscience techniques it is becoming increasingly possible to consider the physiological aspects of these challenges. In particular, the work of Zack [67] and Kosfeld [68] has focussed on the role of a neuropeptide called oxytocin. As a result of a number of experiments using the trust game [69] it has been possible to demonstrate that the level of oxytocin rises in those who are trusted more, and these trustees reciprocate to a greater extent. It would appear that the underlying mechanism involves the action of oxytocin in triggering the release of dopamine. This leads to a pleasurable sensation that, in turn, rewards and reinforces the trusting behaviour.

This finding has been developed by Morhenn *et al.* [70] who considered the earlier finding (for what are believed to be evolutionary-based reasons) that the levels of oxytocin are higher in mammals that have been subject to repetitive stroking. Morhenn and her colleagues showed in a series of experiments that the level of oxytocin and the level of trusting behaviour rose when participants were given a massage prior to taking part in the trust game. Similar effects have been demonstrated by Keltner [71] who, whilst explaining the rules of the trust game, lightly touched some participants on the back as they were about to play and this led to increased levels of trust. Such results lead one to question whether such as simple action as a hand shake might have a similar effect. Clearly this research is at an early stage, but it is unquestionably an area in which further investigations may lead to a better understanding of the mechanisms underlying the development of swift trust.

6 The Danger of Excessive Trust

A final aspect of trust must also be discussed in this analysis of how it might be developed and maintained in an HFN. This has been labelled "the dark side of trust" by Gargiulo and Ertug [72], and these authors caution that the obverse of some of the benefits of increased trust should not be underestimated. They note that there are three types of detrimental effect, the first being that the trustor will, potentially, reduce their level of monitoring, vigilance and safeguarding in the particular relationship. Arguably, the whole point of a trusting relationship is that one can safely reduce such levels of monitoring in the knowledge that this will not be abused, but clearly there is a balance to be struck here and, therefore, a danger if an excessively trusting posture is adopted.

It would appear that there are two complementary mechanisms at work here: the reduced level of vigilance means that negative behaviour may not be detected in a timely fashion whilst, at the same time, the trustee may be in a position to take greater advantage of the trustor through the misuse of, for example, sensitive information in their possession. The second danger of excessive trust is that of complacency as a result of which the trustor will not notice a deterioration in the relationship or in the performance of the partner. Finally, Gargiulo and Ertug [72] suggest that there is a danger that one or both of the parties may take on unreasonable obligations to the other. Again, the authors argue that there is an optimal level for such obligations and, hence, a possibility that this may be inappropriately exceeded. It is recognised that revisiting an existing trusting relationship requires us to question people we have previously trusted and this is an unpleasant, but potentially necessary, experience [73].

7 Operationalising and Testing in Different Settings

A disaster relief operation typically includes a number of HFNs, and the ability of the members of these to work together has far-reaching consequences for the success or failure of the resultant response. With this in mind, this paper has focussed on the model of swift trust that originated in the work of Meyerson *et al.* [74] with the aim of suggesting how this might be operationalised in such a scenario. Clearly, it is necessary for the approach suggested in the paper to be tested in an empirical setting and it is initially considered that this should be achieved in two phases. The first could be undertaken in, for example, a university setting in which a class of students could be set a suitable exercise to be conducted in groups. In parallel with their task to resolve the particular leadership challenge, individual members of the group could be extracted and interviewed using the five swift trust antecedents as the broad agenda for the interviews. Indeed, if circumstances permit, additional realism could be injected by having some members of the group operating in a "virtual" way and the resulting levels of trust between the core team and its virtual members compared.

Based on what would, in effect, be a pilot of the main research effort, testing the swift trust framework in an operational setting would be the next stage. This, unfortunately, would require the advent of a significant rapid onset disaster in which multi-sourced teams such as the UNJLC are deployed. With the approval of their host agency/NGO, it is anticipated that researchers could undertake a number of interviews with key personnel both in the field and at the headquarters which would be, again, based around investigation of the impact of the five antecedent factors posited within the theoretical framework. Self-evidently, such research would require particularly careful management so as to be timely, and yet not interfere with the vital life-saving work of the organisation itself.

8 Summary

The swift trust model seems highly applicable to the humanitarian logistics context, and this leads to a number of important considerations that should be born in mind by the leader of a post-disaster team. However, there is only so much that the leadership of an HFN can do, and there would also appear to be an onus on the broader humanitarian community to play its part by developing pan-NGO mechanisms such as a standard course (and associated certification) for humanitarian logisticians, together with a register of those who are appropriately qualified. It is recognised that this is not a simpler task, not least because of the concomitant requirement for agreement on, for example, what qualities are needed to be a successful humanitarian logistician. Such an approach also implies the existence of some form of audit function and, indeed, agreement amongst NGOs that they should be subject to such a regime. Nevertheless, the fact that these are clearly tricky issues is not, of itself, a reason for not embarking on such a course of action.

In addition to testing the model in the aftermath of a disaster that has seen the formation of HFNs, a further strand of research would be to understand how one might identify a successful organisation (i.e. that in which a high level of inter-personal trust exists) from an unsuccessful one. In the case of the latter, not only would it be instructive to understand the nature of the perceived failings, but also the implications for the beneficiaries. It would be hoped that such an analysis, which reflects the reverse of the first line of enquiry, would help to triangulate the practical development of this model.

References

1. Bachmann, R., Knights, D., Sydow, J.: Trust and Control in Organizational Relations. Editorial for Special Edition of Organization Studies 22(2) (2001)
2. McEvily, W., Perrone, V., Zaheer, A.: Introduction to the special issue on trust in an organizational context. Organization Science 14, 1–4 (2003)
3. Currall, S.C., Inkpen, A.C.: On the complexity of organizational trust: a multi-level co-evolutionary perspective and guidelines for future research. In: Handbook of Trust Research. Edward Elgar, Cheltenham (2006)
4. Barratt, M.: Understanding the meaning of collaboration in the supply chain. Supply Chain Management: an International Journal 9(1), 30–42 (2004)
5. Christopher, M., Peck, H., Towill, D.: A taxonomy for selecting global supply chain strategies. International Journal of Logistics Management 17(2), 277–287 (2006)
6. Mentzer, J.T., DeWitt, W., Keebler, J.S., Min, S., Nix, N.W., Smith, C.D., Zacharia, Z.G.: Defining supply chain management. Journal of Business Logistics 22(2), 1–25 (2001)
7. Skjøtt-Larsen, T., Thernøe, C., Andresen, C.: Supply chain collaboration. Theoretical perspectives and empirical evidence. International Journal of Physical Distribution and Logistis Management 33(6), 531–549 (2003)
8. Ellickson, R.C.: Adverse possession and perpetuities law: two dents in the libertarian model of property rights. Washington University Law Quarterly 64, 723–738 (1986)
9. Das, T.K., Teng, B.-S.: Trust, Control, and Risk in Strategic Alliances: An Integrated Framework. Organization Studies 22(2), 251–283 (2001)

10. Peck, H.: Drivers of supply chain vulnerability: an integrated framework. International Journal of Physical Distribution and Logistics Management 35(4), 210–232 (2005)
11. Lu, S., Yan, H.: A model for evaluating the applicability of partnering in construction. International Journal of Project Management 25, 164–170 (2007)
12. Jarvenpaa, S.L., Knoll, K., Leidner, D.E.: Is anybody out there? Antecedents of trust in global virtual teams. Journal of Management Information Systems 14(4), 29–64 (1998)
13. Knights, D., Noble, F., Vurdubakis, T., Willmott, H.: Chasing Shadows: Control, Virtuality and the Production of Trust. Organization Studies 22(2), 311–336 (2001)
14. Maguire, A., Phillips, N., Hardy, C.: When 'Silence = Death', Keep Talking: Trust, Control and the Discursive Construction of Identity in the Canadian HIV/AIDS Treatment Domain. Organization Studies 22(2), 285–310 (2001)
15. Weick, K.E.: The collapse of sensemaking in organizations: The Mann Gulch disaster. Administrative Science Quarterly 38(4), 628–652 (1993)
16. Majcharzak, A., Jarvenpaa, S.L., Hollingshead, A.B.: Coordinating Expertise Among Emergent Groups Responding to Disasters. Organization Science 18(1), 147–161 (2007)
17. Ben-Shalom, U., Lehrer, Z., Ben-Ari, E.: Cohesion during military operations. Armed Forces and Society 32(1), 63–79 (2005)
18. Denning, P.J.: Hastily formed networks. Communications of the ACM 49(4), 15–20 (2006)
19. Uhr, C., Ekman, O.: Trust among decision makers and its consequences to emergency response operations. Journal of Emergency Management 6(3), 21–37 (2008)
20. Denning, P.J.: Hastily formed networks. Communications of the ACM 49(4), 15–20 (2006)
21. Weick, K.E.: The collapse of sensemaking in organizations: The Mann Gulch disaster. Administrative Science Quarterly 38(4), 628–652 (1993)
22. Völz, C.: Humanitarian coordination in Indonesia: an NGO viewpoint. Forced Migration Review (Special Issue July), 26–27 (2005)
23. WFP (World Food Programme): Logistics Cluster Myanmar, End of Mission Report, http://www.logcluster.org/23003/mm08a/LogsCluster_MMR_EndofMissionReportE%20_080810.pdf
24. Meyerson, D., Weick, K.E., Kramer, R.M.: Swift trust and temporary groups. In: Trust in Organizations: Frontiers of Theory and Research, pp. 166–195. Sage Publications Inc., Thousand Oaks (1996)
25. Thomas, A.: Why logistics? Forced Migration Review 18(4) (2003)
26. Blansjaar, M.: Private communication from the head of logistics in Oxfam. GB (2009)
27. van Wassenhove, L.N.: Humanitarian aid logistics: supply chain management in high gear. Journal of the Operational Research Society 57(5), 475–589 (2006)
28. Stoddard, A.: Humanitarian NGOs: challenges and trends, http://www.odi.org.uk/HPG/papers/hpgbrief12.pdf
29. Roberts, A.: NGOs: New Gods Overseas. The World in 2001. The Economist Publications 73 (2001)
30. Völz, C.: Humanitarian coordination in Indonesia: an NGO viewpoint. Forced Migration Review (Special Issue July), 26–27 (2005)
31. Kramer, R.M.: Trust and distrust in organizations: Emerging perspectives, enduring questions. Annual Review of Psychology 50, 571 (1999)
32. Kumar, N.: The power of trust in manufacturer-retailer relationships. Harvard Business Review 74, 92–106 (1996)
33. Lewicki, R.L., McAllister, D.J., Bies, R.J.: Trust and distrust: New relationships and realities. Academy of Management Review 23(3), 438–458 (1998)

34. Zaheer, A., McEvily, B., Perrone, V.: The strategic value of buyer-supplier relationships. International Journal of Purchasing and Materials Management 34(3), 20–26 (1998)
35. Spekman, R.E., Sweeney, P.J.: RFID: from concept to implementation. International Journal of Physical Distribution and Logistics Management 36(10), 736–754 (2006)
36. Currall, S.C., Inkpen, A.C.: On the complexity of organizational trust: a multi-level co-evolutionary perspective and guidelines for future research. In: Handbook of Trust Research. Edward Elgar, Cheltenham (2006)
37. Meyerson, D., Weick, K.E., Kramer, R.M.: Swift trust and temporary groups. In: Trust in Organizations: Frontiers of Theory and Research, pp. 166–195. Sage Publications Inc., Thousand Oaks (1996)
38. Hung, Y.-T.C., Dennis, A.R., Robert, L.: Trust in virtual teams: Towards an integrative model of trust formation. In: Proceedings of the 37th Hawaii International Conference on Systems Sciences, Track 1, vol. 1 (2004)
39. Gulati, R., Sytch, M.: Does familiarity breed trust? Revisiting the antecedents of trust. Managerial and Decision Economics 29, 165–190 (2008)
40. Meyerson, D., Weick, K.E., Kramer, R.M.: Swift trust and temporary groups. In: Kramer, R.M., Tyler, T.R. (eds.) Trust in Organizations: Frontiers of Theory and Research, pp. 166–195. Sage Publications Inc., Thousand Oaks (1996)
41. Coppola, N.W., Hiltz, S.R., Rotter, N.G.: Building trust in virtual teams. IEEE Transactions on Professional Communications 47(2), 95–104 (2004)
42. Ben-Shalom, U., Lehrer, Z., Ben-Ari, E.: Cohesion during military operations. Armed Forces and Society 32(1), 63–79 (2005)
43. Kramer, R.M.: Trust and distrust in organizations: Emerging perspectives, enduring questions. Annual Review of Psychology 50 (1999)
44. Kramer, R.M.: Rethinking trust. Harvard Business Review, 69–77 (June 2009)
45. Kovács, G., Tatham, P.H.: Gender, skills and logistics performance. In: Proceedings of 13th International Symposium on Logistics (ISL), Bansaen Thailand, July 6-8 (2008)
46. Mangan, J., Christopher, M.: Management development and the supply chain manager of the future. International Journal of Logistics Management, 16(2), 178-191 (2005)
47. Tatham, P.H., Kovács, G.: Logistics skills and performance. In: Proceedings of the Logistics Research Network (LRN), Liverpool, September 10-12 (2008)
48. Hung, Y.-T.C., Dennis, A.R., Robert, L.: Trust in virtual teams: Towards an integrative model of trust formation. In: Proceedings of the 37th Hawaii International Conference on Systems Sciences, Track 1, vol. 1 (2004)
49. Zolin, R.: Swift trust in hastily formed networks. The Hastily Formed Networks Research Group, `http://www.hfncenter.org/cms/files/swifttrustinHFN10-03-02.pdf`
50. Frosdick, S.: "Safety cultures" in British stadia and sporting venues: understanding cross-organizational collaboration for managing public safety in British sports grounds. Disaster Prevention and Management 4(4), 13–21 (1995)
51. Dowty, R.A., May, P.J., Beech, C.E., Wallace, W.A.: Responding to chaos: organizational cultures and the Katrina response. Madison (2006)
52. Scharlemann, J.P.W., Eckel, C.C., Kacelnik, A., Wilson, R.K.: The value of a smile: game theory with a human face. Journal of Economic Psychology 22, 617–640 (2001)
53. DeBruine, L.M.: Facial resemblance enhances trust. Proceeding of the Royal Society of London B, London 269, 1307–1312 (2002)
54. Kramer, R.M.: Trust and distrust in organizations: Emerging perspectives, enduring questions. Annual Review of Psychology 50 (1999)

55. Greenberg, P.S., Greenberg, R.H., Antonucci, Y.L.: Creating and sustaining trust in virtual teams. Business Horizons 50(4), 525–533 (2007)
56. Kramer, R.M.: Trust and distrust in organizations: Emerging perspectives, enduring questions. Annual Review of Psychology 50 (1999)
57. Grey, C., Garsten, C.: Trust, Control and Post-Bureaucracy. Organization Studies 22(2), 229–250 (2001)
58. Maguire, A., Phillips, N., Hardy, C.: When 'Silence = Death', Keep Talking: Trust, Control and the Discursive Construction of Identity in the Canadian HIV/AIDS Treatment Domain. Organization Studies 22(2), 285–310 (2001)
59. Hung, Y.-T.C., Dennis, A.R., Robert, L.: Trust in virtual teams: Towards an integrative model of trust formation. In: Proceedings of the 37th Hawaii International Conference on Systems Sciences, vol. 1(1) (2004)
60. Fukuyama, F.: Trust: The social virtues and the creation of prosperity. Free Press, New York (1995)
61. Webber, E.U., Hsee, C.K.: Culture and Individual Judgement and Decision Making. Applied Psychology: An International review 49(1), 32–61 (2000)
62. Zack, P.J., Knack, A.: Trust and Growth. The Economic Journal 111, 295–321 (2001)
63. Hung, Y.-T.C., Dennis, A.R., Robert, L.: Trust in virtual teams: Towards an integrative model of trust formation. In: Proceedings of the 37th Hawaii International Conference on Systems Sciences, vol. 1(1) (2004)
64. Weick, K.E.: The collapse of sensemaking in organizations: The Mann Gulch disaster. Administrative Science Quarterly 38(4), 628–652 (1993)
65. Drabek, T.T.: Managing the Emergency Response. Public Administration Review 45, 85–92 (1985)
66. Hung, Y.-T.C., Dennis, A.R., Robert, L.: Trust in virtual teams: Towards an integrative model of trust formation. In: Proceedings of the 37th Hawaii International Conference on Systems Sciences, vol. 1(1) (2004)
67. Zack, P.J.: The Neurobiology of Trust. Scientific American, 88–95 (June 2008)
68. Kosfeld, M.: Brain Trust. Greater Good Fall, 18–19 (2008)
69. Kramer, R.M.: Rethinking trust. Harvard Business Review, 69–77 (June 2009)
70. Morhenn, V.B., Park, J.W., Piper, E., Zack, P.J.: Monetary sacrifice among strangers is mediated by endogenous oxytocin release after physical contact. Evolution and Human Behavior 29, 375–383 (2008)
71. Zack, P.J.: The Neurobiology of Trust. Scientific American, 88–95 (June 2008)
72. Gargiulo, M., Ertug, G.: The dark side of trust. In: Handbook of Trust Research. Edward Elgar, Cheltenham (2006)
73. Kramer, R.M.: Rethinking trust. Harvard Business Review, 69–77 (June 2009)
74. Meyerson, D., Weick, K.E., Kramer, R.M.: Swift trust and temporary groups. In: Trust in Organizations: Frontiers of Theory and Research, pp. 166–195. Sage Publications Inc., Thousand Oaks (1996)

Humanitarian Cluster Leads as Fourth-Party Logistics Providers

Leif-Magnus Jensen

Department of Strategy and Logistics, BI Norwegian School of Management,
Nydalsveien 37, 0484 Oslo, Norway
leif-magnus.jensen@bi.no

Abstract. The concept of a fourth-party logistics provider is increasingly explored in the literature and has come to focus on firms that are able to organize and run significant parts of a supply chain. These firms are assigned increasing responsibility for design, recruitment of other firms and maintenance of supply chains due to very specific skills. Within the humanitarian sector there is considerable reorganization of the necessarily agile supply chains necessary to deliver relief during disasters. This reorganization has included the designation of clusters for different types of activity, and significantly leadership of the cluster to specific organizations. This article applies the concept of a fourth-party logistics provider on the cluster concept to help develop a concept for how the cluster leads themselves should operate.

Keywords: 4PL, cluster leads, humanitarian logistics, conceptual.

1 Introduction

Recently humanitarian aid has been reorganized according to the "cluster system" grouping organizations together in areas such as water and sanitation, camp management and logistics. The purpose of this reorganization has been to deal with a set of coordination issues including overlapping provision of relief and unclear responsibility for at risk populations. Each cluster has been assigned a global cluster lead, and although the cluster leads have been assigned a set of responsibilities, what is less clear is how they should handle them. That is, given the constraints inherent in the humanitarian system, how is the global cluster lead intended to fulfill its responsibilities in a setting with many independent organizations that often object to being directed. This paper suggests that the concept of a fourth-party logistics provider (4PL) can usefully be employed for a discussion of how the global cluster can balance the need for coordination with the need to achieve buy-in from the various organizations involved in humanitarian relief.

The purpose of the paper is to deal with the question: "How can the concept of a fourth-party logistics provider guide cluster leads in defining what role to take?" To answer this question it is natural to begin with a description of the cluster system itself, and then to tie this to a review of the literature on fourth-party logistics.

W. Dangelmaier et al. (Eds.): IHNS 2010, LNBIP 46, pp. 372–383, 2010.

2 Approach

This paper is largely conceptual in nature, but the description of the cluster system which is defined top-down as a coordination mechanism is supplanted by reviews and feedback from the cluster system. The cluster system itself is extensively documented. This review is not complete in the sense that some of the reviews written on the cluster system are internal to Non-governmental Organizations (NGOs) and other users of the system, but they do cover some of the most important review documents available. In addition, the basic description of the logistics cluster in particular is strengthened by referring to interviews carried out for a case study of the United Nations Joint Logistics Centre (UNJLC) and Logistics Cluster.

The core of the paper lies in taking the concept of a 4PL, expanding some of its components with supporting literature to give it a better anchoring within academic work, and then applying this vis-à-vis the cluster leads to give alternative scenarios for how the cluster lead can work.

3 The Cluster Concept

The cluster approach in humanitarian response is a result of the Humanitarian Review carried out as a consequence in part of coordination issues arising during the Asian Tsunami and Darfur crisis in 2004-2005 (see [1]). Among the key areas to be addressed were predictable leadership at the global and local level, strategic field level coordination and prioritization, and sufficient global capacity to meet current and future emergencies [2]. Clusters have only been defined for those areas where coordination issues were significant, so that for example food distribution which is largely carried out through the World Food Programme (WFP) is not part of the cluster system.

The basic cluster structure consists of 9 clusters, roughly corresponding to areas of specialty such as water and sanitation, camp management and security[1]. For each cluster a global cluster lead is defined with the responsibility for overall leadership of the cluster. The exact responsibilities of the cluster leads have been further defined:

- Facilitate the coordination between the cluster members
- Encourage joint working
- Ensure that responses are in line with existing guidelines and standards
- Collate and share information
- Identify gaps in the response
- Stand in as the "provider of last resort" when there are no other options[3]

The final of these six points, the provider of last resort has been somewhat controversial and has required further elaboration:

The 'provider of last resort' concept is critical to the cluster approach, and without it the element of predictability is lost. It represents a commitment of cluster leads to

[1] Agriculture, Camp coordination/management, Early recovery, Education, Emergency Shelter, Health, Nutrition, Protection, Water Sanitation Hygiene and the two supporting cluster Emergency Telecommunications and Logistics.

do their utmost to ensure an adequate and appropriate response. It is necessarily circumscribed by some basic preconditions that affect any framework for humanitarian action, namely unimpeded access, security, and availability of funding[4, p.1].

At the global level NGOs can enter into agreements with the cluster to become part of the working group of the cluster and gain some influence over ongoing developments. When the cluster system is mobilized for an intervention through the Inter-agency Standing Committee (IASC), a local cluster lead (often but not necessarily the same as the global) is mobilized. Attendance at cluster meetings is optional for agencies and NGOs. Although the cluster lead has some direct authority, for example when a common transport system is in operation, it is generally the case that it cannot in any way force local authorities or NGOs to take part in the cluster system. Even where and if this is the case, it seems clear that it is not desirable since it is likely to create a great deal of resentment.

Although the cluster concept is continually developing in part based on experiences from previous engagements, it is possible to see some general trends from the evaluations so far. A recurring issue in these deployments appears to be the question of how to carry out coordination, and to what degree the cluster through the cluster lead can and should be directive vis-à-vis other agencies and NGOs. This issue can be related to what we may call the role of the cluster, both globally and in a specific operation. The local cluster mobilized for a particular operation may be different than the global cluster for a sector, but will often be the same considering that large organizations present in many different areas are generally designated as global leads. The split between the global and local level does however demonstrate another level of complexity in the cluster concept with regards to leadership, since it is necessary to function at both levels, and decisions made at one level affects the other.

In several cluster interventions, there have been complaints that clusters have been too directive or authoritarian. In terms of efficiency the case can be made that organizations have previously had too much freedom in crafting their own response to a situation without considering the overall impact on the recipients. However, complaints from NGOs and others do point to two of the main challenges for the cluster concept, one more practical the other more. The more practical challenge is that both donors NGOs and other parties must be convinced of the benefits of the cluster system in order to participate fully. That is, using the clusters implies a certain level of added effort in terms of administration and coordination, and this effort must be compensated through better performance. If this is not the case then support for the cluster system is likely to wane. This is both an issue of how participants perceive the system is working and also about measurable performance which is a challenge in the sector as a whole.

The second main challenge is that the clusters, and certainly the cluster lead needs a concept for how to operate vis-à-vis the many interested parties. That is, the success of the cluster system in large part depends on whether NGOs and local governments participate. Whether they choose to do so will depend in part on the manner in which cluster leads conduct their business. A cluster lead which considers itself the "owner" of a cluster with NGOs and others simply as suppliers is unlikely to be successful. This is both because many actors in the humanitarian system have strong organizational identities and do not like to be "directed," but also because it does not reflect the reality on the ground. The cluster leads clearly do not have the authority to consider itself a traditional "channel captain" who is able to tell others what to do. This is why a concept such as a fourth-party logistics provider may be better suited.

4 Fourth-Party Logistics

The term "fourth-party logistics" was first coined by Accenture and defined accordingly:

A supply chain integrator that assembles and manages the resources, capabilities, and technology of its own organization with those of complementary service providers to deliver a comprehensive supply chain solution.[2]

Initially 4PL was a trademark but it has increasingly become a recognized term in business literature as well. The concept of a 4PL has come to mean a firm or organization carrying out certain specific tasks and taking a specific role in a supply system. Win [5] provides the following updated definition:

A 4PL is an independent, singularly accountable, non asset based integrator of a clients supply and demand chains. The 4PL's role is to implement and manage a value creating business solution through control of time and place utilities and influence on form and possession utilities within the client organization. Performance and success of the 4PL's intervention is measured as a function of value creation within the client organization [5, p.677].

This is clearly an important definition with regard to the cluster system because the value creation or at least the perception of usefulness will be critical for the success of the cluster system over time. The words "implement" and "manage" shows some very important characteristics of a 4PL – it is to a great degree responsible for making a supply chain operate. Win's definition is however very much in terms of a 4PL and its main client. Mukhopadhyay & Setaputra [6] has a further definition where the 4PL is very clearly placed as a form of intermediary:

A 4PL is treated as a strategic partner, rather than a tactical one and is a supply chain integrator that synthesizes and manages the resources, capabilities, and technology of its own organization with those of complementary service providers to deliver a comprehensive supply chain solution [6, p.718].

This definition is somewhat atypical for recent conceptions of 4PLs where it does not rely on having considerable assets or physical investments, but it is very much relevant to the cluster leads who have exactly this challenge of being able to match their own and complementary capabilities to deal with the needs of the cluster. These examples illustrate the relevance of the 4PL concept for the cluster leads, but a more detailed exploration of the elements of the original 4PL concept can give us a better insight into the match between 4PLs and cluster leads. To do this, I start with the original framework for a 4PL in Table 1:

---

[2] http://www.scmo.net/index.php?option=com_content&task=view&id=381&Itemid=25

Table 1. Four Key Elements of a 4PL (From Accenture definition, in [7]

Architect / Integrator	*Change leader:* - Supply chain visionary - Multiple customer relationship - Deal shaper and maker - Supply chain reengineer - Project management - Service, systems, and information integrator - Continuous innovation
Control Room (intelligence)	*Decision maker:* - Experienced logistician - Optimization of decision support - Management of multiple 3PL - Continuous improvement
Supply Chain Infomediary	*Information system:* - IT system integration - Real-time data capture - Convert data to information - Technical support
Resource Provider	*Assets:* - Transportation asset provider - Warehouse and property facility - Procurement service - Co-packing service

These four main themes are useful for describing a 4PL, but they can be improved by tying them more explicitly to existing literature which has more to say about the mechanisms and reasons behind 4PLs. To answer the question of how the global cluster leads fit with the 4PL concept then, it is necessary to discuss the four themes architect/integrator, control room, supply chain infomediary and resource provider in context. For each of the four themes it is also possible to relate them directly to experiences with the cluster leads, here using examples from the logistics cluster for consistency.

Architect/Integrator. The ability to be an architect or integrator is core to the description of a 4PL in the literature, and means having the competence to carry out supply chain design as well as the supporting skills within project management and customer management to enable the supply chain to work at an overall level. It can be based on either a strong competence within supply chain design as a result of extensive work as a 3PL or consultant in the industry, but also on being in an independent position making it easier to make overall arrangements that are beneficial for the supply chain in question. This makes the 4PL less vulnerable to allegations of favoring the use of its own resources – at least where it does not have heavy 3PL-style assets [5].

This fits quite well with the concept of a cluster lead. The cluster lead is meant to "make the system work" through focusing on coordination, primarily of other actors, although the cluster lead can be a full participant in the relief operation. For example, when WFP is the cluster lead it will naturally carry out a great deal of activities in the sector. These will typically be related to food logistics which is not the remit of the cluster system, but they are relevant to the cluster because it means WFP has considerable capacity and presence which can be useful for the logistics cluster. Existing activities in a crisis is however mainly tied to understanding in-field problems. That is, the most important role of the cluster lead in coordination is in getting other actors to cooperate in their activities. There are of course a number of ways of doing this. In a commercial setting the 4PL is able to negotiate contracts on behalf of a paying client, and often there are many potential 3PLs vying for these contracts, so that selection and maintenance of a "population" of suppliers is part of its activities. For the cluster leads, it is more a case that they have to deal with those organizations that are or choose to be present during a relief operation. However, effective organization, access to funding through the cluster system and the fact that the cluster system is more dependent on good and effective rather than complete participation means that there are in practice many parallels. Even so the degree of participation in itself is an important measure for the cluster leads.

Control Room. The control room places the 4PL as a decision maker with a focus on managing operations from day to day rather than necessarily carrying out transport and warehousing. The management of multiple 3PLs comes under this heading. In fact, the concept of a hub, at least when it is meant in operational terms is entirely relevant to this category [8]. Using the concept of a hub enables us to see more in detail some of the more structural effects of using a 4PL. Several aspects can be identified, including the reduction of the number of business ties or ongoing communication, the handling of risk and the more general concept of an intermediary.

The reduction of business ties or more broadly the amount of interaction needed in order to carry out business is based on a fairly simple mathematical model. If a set of producers A are selling a product to a set of customers B without an intermediary, then the number of interactions for all the customers to contact all the producers are AxB. This may be seen as an extreme example, but customers may generally want to be in contact with a number of producers to check price, product quality and so on. Inserting an intermediary can reduce the number of interactions to A+B [9]. This conclusion is based on the fairly strong assumptions that the same kind of business can be carried out through the intermediary as directly between customer and supplier, and that there is a lack of alternative mechanisms for matching supply and demand.

Risk management is a frequent task for intermediaries, for example financial intermediaries carrying financial risk or larger intermediaries absorbing fluctuations from many smaller customers thus evening out risk overall. Another category is the improvement of operational risk through the intermediary's specialist knowledge and continual focus on improving operations. In terms of the cluster concept risk is less prominent since all participants take considerable risk tied to for example the uncertainty of funding. In addition, it is often uncertain where the next disaster will strike and the operating environment often means that considerable risk must be taken already both for personnel and equipment.

Risk management can however be tied to a few tasks of the cluster, such as assisting with pre-positioning of supplies. This can help mitigate the risk of not having supplies available for a crisis because there are safety stocks. A further development here is developing virtual common stocks, i.e. stocks stored in a generic format, where the labeling for a particular NGO is carried out as the stocks are shifted. This also involves NGOs and agencies borrowing stock from each other, spreading risk across the organizations, thus increasing the overall ability to carry risk. Note however that many of these initiatives are taken either by NGOs themselves, or more generally through the United Nations Development Programme (UNDP), so they are not crucially dependent on the cluster lead itself. In this sense we should not over-emphasize the important of the cluster leads in risk management.

In terms of the clusters, it seems clear that in many respects the global lead can be considered an intermediary for certain tasks such as acting as a focal point for disseminating information. There are however features in humanitarian relief that complicate the picture. It was not the case that all organizations were communicating before the cluster concept. Indeed, the problem of coordination was partially insufficient contact between numerous participants. However, as a solution to the problem that everybody talking to everybody would be far too costly, the calculation above is consistent. A recurring issue will nevertheless be that operating the cluster concept has a cost and the benefits in terms of coordination will have to be substantial to justify this.

Resource provider. The ability to be a resource provider is one of the aspects of 4PLs that have changed somewhat since the original formulation, in that it is increasingly seen as an advantage if the 4PL itself does not have extensive physical resources within for example transportation and warehousing [5]. The reason is both that making use of such resources may require different competencies from the other tasks carried out by a 4PL, but equally important the incentive structure for the 4PL changes if it has physical assets that need to be utilized profitably. The argument goes that it is better to be "asset neutral" and simply to be good at buying the services of other firms such as 3PLs which can then focus on making the best possible use of physical resources [10]. For those tasks the 4PL carries out, the arguments are essentially tied to scale and specialization.

Advantages of scale are a common theme in business, production and distribution [11]. For an intermediary achieving scale is largely tied to accumulating volume in the type of task that the intermediary carries out, be it purchasing, logistics or pure transport. The scale can be achieved either in operations themselves, i.e. operating with more frequent deliveries due to large volume, or due to accumulating business, e.g. gaining concessions from suppliers due to large purchasing orders.

In the cluster concept and humanitarian response there are a number of tasks where achieving scale is beneficial, but it is not given that the cluster lead will be directly involved in these. For example, purchasing of relief supplies, managing of transport corridors or managing of camps are all areas where it is easy to see benefits of scale, but it is not given that a cluster lead must be directly involved in these. Certainly even if one cluster lead is involved in some of these tasks, others will not be. There are however some general tasks tied to the administration of a cluster and the cluster itself where scale does give some benefits. For example, establishing computer systems, identifying and disseminating some basic software and the development and use of standards for certain documents such as situation reports represent a recurring need.

However, for any situation many or most of the actual operations and activities will be carried out by individual NGOs or agencies at the scale they can achieve. This is a fundamental feature of the present organization of the sector, and it is not the purpose of the cluster concept to change this, but rather to help with coordination where it has been identified as a problematic issue. In general, tasks will be carried out by NGOs and agencies in accordance with particular grants of funding, limiting the flexibility to merge activities even where the organizations want this. Furthermore, many NGOs often have specific target groups necessarily limiting the scale of their activities or specific charters that make it difficult to cooperate with certain other organizations such as e.g. military forces. In this sense, it is possible for the cluster to identify and inform but not to force organizations to merge activities. This means that the scale argument only applies generally to a few overall tasks. However, since these tasks are recurring and crucially may impact on other tasks, e.g. setting up a system for situation reports quickly making it possible for NGOs and agencies to carry out their own tasks more efficiently, this can still be quite significant.

Supply Chain Infomediary. Acting as a supply chain infomediary means to "create electronic links between the supply chain members." [7, p.73]. That is, the 4PL not only takes on the intermediation role between different providers and users of their services, but also provides specialized infrastructure to make the day to day running of the system easier and more efficient. While this point is highly relevant in that certain cluster leads are already making specialized IT tools available for cluster members, it is perhaps better to make a more general point here. First is the value of the intermediary concept, i.e. the main task of a 4PL is to tie other organizations together in order to achieve an overall aim for the supply chain. Second, it shows the value of specialist knowledge possessed by the 4PL, even where it is not focused primarily on the typical 3PL tasks such as transport, handling and warehousing.

Specialist knowledge refers to the issue that some firms focus on certain skills and capabilities in order to carry out these more efficiently. This efficiency can be based on experience, particular investments in technology or people, or a combination of the two. These arguments can be found again in core competence literature or the resource-based view [12, 13], and is different from the scale argument in that the specialist carries out activities better even at the same scale as other firms. The effect of such specialist knowledge can be both carrying out the same activities at lower cost, and carrying them out with higher quality.

The concept of specialist knowledge requires some additional clarification in terms of the cluster concept. Within humanitarian response in general, it is reasonable to assume that it is the NGOs, agencies and local governments that are the experts and possess specialist knowledge within their own fields. This does not mean that there is no variation in competence between different organizations, but it is not the purpose of the cluster to provide basic knowledge on for example handling emergency shelter. The exception to this is that some general standards through for example the Sphere project[3] are better handled across a number of organizations. Furthermore and especially for smaller NGOs it is likely that the field of expertise is limited so there are a number of associated skills where the NGO is not specialized and could use outside help. For example, an NGO may focus on camp management, but have limited skill in handling inbound logistics to the camp. This means that there are gaps in each particular response where the global cluster lead can contribute, either directly through developing some skills that will be needed in each response, or by gap-filling.

[3] See http://www.sphereproject.org/

Table 2. Cluster leads and elements of 4PLs

	Positive fit with the concept	Lack of fit with the concept
Architect/Integrator	Cluster lead has overall responsibility for "making the system work" Constructing a supply chain in many cases Needs to match own and others' resources	Cannot ignore organizations that do not take part High participation in itself is a goal rather than picking best providers
Control Room (Focus on efficiency of interaction and risk management)	Global lead positioned as an intermediary cuts the amount of needed interaction Some global lead tasks are clearly about risk management	Operational control for the cluster lead most likely smaller than 4PL vision
Resource Provider	Can match "original" conception to a certain degree combining own and others' resources	Issue of neutrality, more recent 4PL conceptions stress a lack of heavy infrastructure as an advantage
Supply Chain Infomediary (Focus on specialist knowledge)	Developing standards and competence through repeated disasters Gathering and spreading information quickly during a disaster	Access to information may be limited and alternative sources are costly Assessment is an important task but may lead to the cluster leads evaluating themselves

To summarize, there are two obvious areas where the global cluster lead can contribute through the provision of specialist knowledge. One is through developing knowledge in specific areas where NGOs and other agencies typically do not have skills, but which are periodically required in response. Using the logistics cluster as an example, this would typically be logistics needs assessments, capturing and disseminating logistics information (through website and Geographical Information System (GIS) maps) and acting as a connection through common services set up through for example United Nations Humanitarian Air Service (UNHAS). The second will vary from response to response and deals with filling gaps. Here again there will be some degree of repetitiveness – i.e. some responses will require establishing temporary air traffic control administration. This is not the case in all disasters, but it is a recurring theme in certain types of disaster making it relevant for the cluster lead to maintain some experience in this. The task for the global lead is not to maintain unlimited capacity for these types of tasks but to either know who can be asked to take them on or at least maintain a certain minimal capacity in the organization. This will of course be easier for the larger global leads more easily able to second personnel during a crisis.

A final issue tied to specialist knowledge and the cluster lead role is that of evaluating and assessing the impact of a cluster response. This is a potential task for the cluster lead, and means developing ways of evaluating the effectiveness of the cluster response itself, though surveying recipients, reporting from participant organizations, and developing measurements of performance of a more general nature, i.e. % damage to goods, time to delivery etc. The advantage of having a cluster lead carry out this task is that it is repeatedly involved in similar situations and thus has the opportunity to develop measurements that work, and compare these over time to create benchmarks for typical performance. The downside of having the cluster lead carry out these types of tasks is that it is itself a party to the response, so it will to a certain extent be evaluating itself.

The main points made in the discussion of the four main aspects of a 4PL are summarized in Table 2 for convenience.

5 Discussion

This comparison of the Cluster concept with some of the most common 4PL conceptions has proved useful in several ways. First we see that the four elements of the 4PL concept – architect/integrator, control room, supply chain infomediary and resource provider are helpful to start with, but require some additional grounding in the literature. In this article we have emphasized some aspects of each of the four elements of the original 4PL conception, both to show how the concept has developed, but also in order to explore the cluster lead concept more. In this regard, it would be possible to carry out a more complete exploration of the concept since there are a number of aspects to each of the four 4PL elements listed in the original definition. However, exploring each such aspect in depth would be beyond the scope of this article. To deal with this only the most relevant parts have been discussed here. Naturally the relevance of the selection made can be debated.

The 4PL concept is useful in several ways in describing possibilities for the cluster leads. The main tasks of the cluster lead seem to fit relatively well within the categories for a 4PL. The integrator aspect matches well the cluster lead's responsibility for making a particular cluster "work" in a disaster response. The control room concept ties heavily into the operational control or daily running of the cluster, but comes up against significant limitations in practice since the cluster leads depend in large part on voluntary participation from NGOs and others. The supply chain infomediary fits well with the need to continually gather and disseminate important information for the cluster. Finally, the resource provider aspect shows the tension between combining the cluster lead's own resources and those of other organizations, and the ideal 4PL situation where it has a neutral position and can manage the use of other's resources.

In more general terms the 4PL concept helps to "unpack" the cluster concept in terms of specific actors. That is, ideally a 4PL acts in a certain way – it is independent with regard to the use of resources and finds the best combinations matching suppliers and customers. Furthermore, although a 4PL has some power with regard to suppliers since it is able to offer contracts on behalf of its customers, it is very much dependent on superior performance for its existence, since it only represents another layer of costs if it cannot provide cost savings. Although the parallel to the cluster leads is not complete, they too have the particular challenge that participation and overall success depends on offering a better way to organize the system which the users (e.g. NGOs and others) see as useful. The purpose of this reorganization is of course to benefit the recipients, but if NGOs are not convinced they will self-organize to provide aid to the recipients.

The concepts used here require further adaptation and application to the humanitarian setting. In particular, applying the framework to a specific intervention where the cluster concept was applied in practice will improve the framework itself. Using the framework will also make it possible to make more specific recommendations for the global cluster leads. This will also make it possible to formulate the four categories of a 4PL more in the context of the setting. For example, the cluster concept was applied during the Pakistan Earthquake in 2005. It was found that some clusters functioned much better than others (the Logistics Cluster performed quite well in this context). Thus a comparison of the response of different clusters to the same response can be the basis for developing the framework.

Finally, using the 4PL concept alludes to the importance of "intermediary" as a concept. That is, many of the advantages or benefits of using a 4PL is tied to the firm operating as a kind of intermediary between buyers and providers, with the addition that it is allowed to organize a part of the supply chain itself. There is an extensive literature on the concept of an intermediary, and it may be fruitful to pursue this as further research [14-16].

References

1. Adinolfi, C., et al.: Humanitarian Response Review. Office for the Coordination of Humanitarian Affairs, 112 (2005)
2. OCHA, Appeal for Building Global Humanitarian Response Capacity. Office for the Coordination of Humanitarian Affairs, 205 (2007)
3. IASC, Guidance note on using the cluster concept to strengthen humanitarian response. IASC (Inter-agency Standing Committee), 15 (2006)
4. IASC, Operational Guidance on the Concept of Provider of Last Resort, `http://www.humanitarianreform.org/humanitarianreform/Portals/1/Resources%20&%20tools/OPGUIDANCE-ProviderofLastResort-F.doc`
5. Win, A.: The value a 4PL provider can contribute to an organisation. International Journal of Physical Distribution and Logistics Management 38(9), 674–684 (2008)
6. Mukhopadhyay, S.K.: The role of 4PL as the reverse logistics integrator: Optimal pricing and return policies. International Journal of Physical Distribution and Logistics Management 36(9), 716–729 (2006)

7. Fulconis, F., Saglietto, L., Paché, G.: Exploring New Competences in the Logistics Industry: The Intermediation Role of 4PL. Supply Chain Forum 7(2), 68–77 (2006)
8. Jensen, L.-M.: The Role of Intermediaries in Changing Distribution Contexts: A Study of Car Distribution. In: Department of Strategy and Logistics. BI Norwegian School of Management, Oslo (2009)
9. Alderson, W.: Marketing Behavior and Executive Action: A Functionalist Approach to Marketing Theory. Richard D. Irwin, Homewood (1957)
10. Persson, G., Virum, H.: Growth Strategies for Logistics Service Providers: A Case Study. International Journal of Logistics Management 12(1), 53–64 (2001)
11. Chandler, A.D.: Scale and Scope: The Dynamics of Industrial Capitalism. The Belknap Press of Harvard University Press, London (1990)
12. Barney, J.B.: Firm Resources and Sustained Competitive Advantage. Academy of Management Review 17(1), 99–120 (1991)
13. Prahalad, C.K., Hamel, G.: The Core Competence of the Corporation. Harvard Business Review, 79–91 (1990)
14. Alderson, W.: Factors Governing the Development of Marketing Channels. In: Marketing Channels for Manufactured Products, pp. 5–34. Richard D. Irwin, Homewood (1954)
15. Howells, J.: Intermediation and the Role of Intermediaries in Innovation. Research Policy 35, 715–728 (2006)
16. Morris, L.J., Morris, J.S.: The changing role of middlemen in the distribution of personal computers. Journal of Retailing and Consumer Services 9, 97–105 (2002)

An Efficient Heuristic Algorithm for the Traveling Salesman Problem

Parham Azimi[1] and Peyman Daneshvar[2]

[1] Islamic Azad University (Qazvin Branch)
p.azimi@setarehiran.com
[2] Islamic Azad University (Naragh Branch)
p.daneshvar@gmail.com

Abstract. This paper presents a new heuristic algorithm based on combining branch and bound algorithm and a dynamic simulation model for the traveling salesman problem. The approach uses the simulation results for creating the best tours within the branch and bound tree. The most advantage of this approach lies in the learning profcedure both in simulation process and B&B algorithm. In order to test the efficiency of the proposed algorithm, several computational experiments were conducted over middle-scale and large-scale problems. As the computational results show the algorithm can be used easily in practice with reasonable accuracy and speed.

Keywords: Traveling salesman problem, simulation, mathematical programming, branch and bound algorithm.

1 Introduction

Travelling salesman problem (TSP) is one of the most challenging problems among combinatorial optimization methods. Its importance stems from the fact there is a plethora of fields in which it finds applications. Among all applications of this problem we can address some vast useful applications like optimizing the AGV (Automatic guided vehicles) routes in a production system and transportation problems like post office delivery problem or in a large warehouse transportation system. The other applications such as genome sequencing problem is very amazing where TSP rolls as a key factor in creating genome sequences using experimental data on proximity of individual pairs of markers. If we consider the markers as cities, a genome sequence can be viewed as a TSP path among the markers. The drilling problem is another famous application of TSP where the objective is to minimize the total distance travelled by a driller to drill some holes for CNC machines. The printed circuit boards which have a lot of holes for mounting components or electronic devices are another application of TSP. Even TSP has been used to design fiber-optic communication networks to coordinating military maneuvers to routing helicopters around oil rigs. Another applications were addressed by researchers are shop floor control (scheduling), product design (VLSI layout) and DNA fragment assembly. To review all details regarding the applications of TSP one can see [1].

W. Dangelmaier et al. (Eds.): IHNS 2010, LNBIP 46, pp. 384–395, 2010.

Since the TSP has proved to belong to the class of NP-hard problems [2], heuristics and metaheuristics occupy an important place in the methods so far developed to provide practical solutions for large instances and any problem belonging to the NP-hard class can be formulated with TSP. There is not any exact bounded polynomial algorithm to solve this problem. Therefore, one can see why there are several studies done in finding a good heuristic algorithm for solving this problem, a heuristic which guaranties the suitable speed and accuracy as well. Today, most researchers try to find better heuristics for solving a wide range of large-scale instances of the TSP. According to the latest one which was reported by Applegate, Bixby, Vasek and Cook [18] the largest TSP solved had 85,900 nodes while in 2005 the largest TSP problem ever solved had 13,059 nodes.

There are many publications on TSP, ranging back to at least the late 1940's. The TSP has held the interests of computer scientists and mathematicians because, even after more than half a decade of research, the problem has not been completely solved. The TSP falls in a distinguished category of hard problems while it has several applications. The objective of traveling salesman problem is to find a shortest closed tour which visits all vertices once and only once in an undirected graph G= (V, E), where V denotes the vertex set and E is the edge set.

In the case of the traveling salesman problem, the mathematical structure is a graph where each city is denoted by a point (or node) and lines are drawn connecting every two nodes (called arcs or edges). Associated with every line is a distance (or cost). When the salesman can get from every city to every other city directly, then the graph is said to be complete. A round-trip of the cities corresponds to some subset of the lines, and is called a tour or a Hamiltonian cycle in graph theory. The length of a tour is the sum of the lengths of the lines in the round-trip. Depending upon whether or not the direction in which an edge of the graph is traversed matters, one distinguishes the asymmetric from the symmetric traveling salesman problem.

The earliest exact algorithm returns to Dantzig et al. [10] works. They used a linear programming together with relaxation to solve TSP which is an Integer Problem in mathematical programming. B&B algorithm was used by researchers widely. Among them, we can address Eastman [11], Held and Karp [12], Smith et al. [13]. Some branch and cut (B&C) based on exact algorithms were developed by Crowder and Padberg [14]. Since solving the TSP optimally takes too long, instead one normally uses approximation algorithms or heuristics. Now the next literature review will be continued by grouping the heuristics and approximation algorithm which were developed by researchers so far:

1.1 Approximation Algorithms

The difference is that the approximation algorithms give us a guarantee as how bad solutions we can get, normally specified as c times the optimal value. One of the famous approximations algorithms is the one which was introduced by Arora [21], the algorithm guarantees a (1+1/c) approximation for every c>1. Generally speaking, the average speed of the heuristics is better than the approximation algorithms.

1.2 Tour Construction

Tour construction algorithms have one thing in common. They stop when a solution is found and never tries to improve it. The simplest and the most straightforward TSP heuristic in this class is Nearest Neighbor [15] and the complexity is $O(n^2)$. The next class is called Greedy algorithms in which they gradually construct a tour by repeatedly selecting the shortest edge and adding it to the tour as long as it doesn't create a cycle with less than N edges, or increases the degree of any node to more than 2 [15]. The complexity is$O(n^2 log_2^n)$. The third class is called Insertion Heuristics which starts with a tour of a subset of all cities and then inserting the rest by some heuristics [15]. The initial sub tour is often a triangle or a convex hull. The complexity is $O(n^2)$ or$O(n^2 log_2^n)$. The last class is called Christofides which was introduced by Professor Nicos Christofides [15]. He proved that his algorithm a worst-case ratio of 3/2 instead of 2 which is a normal ration in this class.

1.3 Tour Improvement

Once a tour has been generated by some tour construction heuristics, they try to improve the solution. There are several ways to do this, but the most common ones are the 2-opt and 3-opt local searches. However some metaheuristics like Tabu search [15], genetic algorithm [8] and Simulated annealing [7] also combined in this class to improve the current solution. The 2-opt algorithm basically removes two edges from the tour and reconnects the two paths created. The 3-opt algorithm works in similar fashion, but instead of removing two edges, they remove three ones. If a tour is 3-optimal, it is 2-optimal too. Some researchers used k-opt algorithms for the TSP. However it takes more time rather than 2-opt and 3-opt but better tours reported [15].

1.4 Branch and Bound

Branch and bound algorithms are always used in combinatorial optimization problems, it can be easily applied to TSPs no matter it is the Asymmetric TSP (ATSP) or Symmetric one (STSP). A method for solving the ATSP using a depth-first B&B algorithm is studied in [22] and improved later by the other researchers [15]. The algorithm starts with the original ATSP and solves the Assignment Problem (AP). The AP is to connect each city with its nearest city such that the local cost of all connections is minimized. So the AP is the relaxation of the ATSP, thus acting as a lower bound to the optimal solution of the ATSP.

1.5 Ant Colony Optimization

Researchers are often trying to mimic nature when solving complex problems [4], [5], [6]. The idea was quite successful when applied to the TSP, giving optimal solutions to small problems quickly [6]. Ants leave a trail of pheromones when they explore new areas, which is a guide for other ants to find it. In this method, we start with a group of ants, typically 20 or so. They are placed in random cities and are then asked to move to another city. The ant that picked the shortest tour will be leaving a trail of pheromones inversely proportional to the length of the tour. This process is repeated until a tour being short enough is found.

1.6 The Held-Karp Lower Bound

A common way of measuring the performance of TSP heuristics is to compare its results to the Held-Karp (HK) lower bound. This lower bound is actually the solution of the linear programming relaxation of the integer programming formulation of the TSP [15].

One of the advantages of our algorithm is that it relaxes the 0-1 programming model of the TSP and used the Held-Karp lower bound for empowering the B&B algorithm.

It should be mentioned that several miscellaneous techniques have been used and one can find a comprehensive review on the approaches used on TSP in [15].

2 The Integer Programming Model for Solving TSP

To formulate the asymmetric TSP on m nodes, one introduces zero-one variables

$$x_{ij} = \begin{cases} 1 & if\ i \to j \quad is\ in\ the\ tour \\ 0 & Otherwise \end{cases} \qquad i \neq j, i, j = 1,2,\dots,m\,. \tag{1}$$

And given the fact that every node of the graph must have exactly one edge pointing towards it and one pointing away from it, one obtains the classic assignment problem. These constraints alone are not enough since this formulation would allow "sub tours", that is, it would allow disjoint loops to occur. For this reason, a proper formulation of the asymmetric traveling salesman problem must remove these sub tours from consideration by the addition of "sub tour elimination" constraints.

Problem 1

$$Min.Z = \sum_{i=1}^{m}\sum_{j=1}^{m} c_{ij}x_{ij}$$

S.t.

$$\begin{aligned}
&(1) \quad \sum_{j=1}^{m} x_{ij} = 1 && \forall i = 1,2,\dots,m \\
&(2) \quad \sum_{i=1}^{m} x_{ij} = 1 && \forall j = 1,2,\dots,m \\
&(3) \quad \sum_{i\in A}\sum_{j\in A} x_{ij} \leq |A| - 1 && \forall A \subset \{1,2,\dots,m\} \\
&(4) \quad x_{ij} = \{0,1\} && \forall i,j = 1,2,\dots,m\,.
\end{aligned} \tag{2}$$

Where A is any nonempty proper subset of the cities 1,...,m. The cost c_{ij} is allowed to be different from the cost c_{ji}. Note that there are m*(m-1) zero-one variables in this formulation. To formulate the symmetric traveling salesman problem, one notes that the direction traversed is immaterial, so that $c_{ij} = c_{ji}$. Since direction does not now matter, one can consider the graph where there is only one arc (undirected) between every two nodes. Thus, we let $x_j = \{0,1\}$ be the decision variable where j runs through all edges E of the undirected graph and cj is the cost of traveling that edge. To find a tour in this graph, one must select a subset of edges such that every node is contained in exactly two of the edges selected. Thus, the problem can be formulated as a 2-matching problem in the graph G=(V,E) having m(m-1)/2 of zero-one variables, i.e. half of the number of the previous formulation. As in the asymmetric case, sub tours must be eliminated through sub tour elimination constraints. The problem can therefore be formulated as:

Problem 2

$$Min.Z = \frac{1}{2}\sum_{j=1}^{m}\sum_{k\in J(j)} c_k x_k$$

S.t.

(1) $$\sum_{k\in J(j)} x_k = 2 \qquad \forall j = 1,2,\dots,m$$

(2) $$\sum_{j\in E(A)} x_j \le |A| - 1 \qquad \forall A \subset \{1,2,\dots,m\}$$

(3) $$x_j = \{0,1\} \qquad \forall j \in E\ . \tag{3}$$

Where J(j) is the set of all undirected edges connected to node j and E(A) is the subset of all undirected edges connecting the nodes in any proper, nonempty subset A of all cities. Of course, the symmetric problem is a special case of the asymmetric one, but practical experience has shown that algorithms for the asymmetric problem perform, in general, badly on symmetric problems. Thus, the latter need a special formulation and solution treatment. No consider the asymmetric problem i.e. Problem 1. Obviously this is an integer programming model which needs a heuristic approach to be solved in a reasonable time especially in large-scale problems.

3 The Heuristic Algorithm for Solving Problem 1

Since Problem 1 is a pure 0-1 programming model, a branch and bound algorithm together with a dynamic simulation model have been developed to solve the problem. At first, the algorithm starts with relaxation of constraints (4) and then replaces them by the following ones:

$$0 \le x_{ij} \le 1 \qquad \forall i,j = 1,2,\dots,m\ . \tag{4}$$

The resulting sub model which we call it Problem 2 is a linear programming one and it is very easy to be solved by mathematical programming soft wares. Now the heuristic model:

3.1 The Termination Rule

Consider the B&B tree and assume each node at the tree as a sub problem of the Problem 2. If the maximum acceptable error according to the optimal solution of Problem 2 is defined as Δ for estimating the optimal solution of Problem 1 by the user, i.e.:

$$\Delta = \frac{|Z^*_{Problem\ 2} - Z^*_{Current\ Node}|}{Z^*_{Problem\ 2}} . \qquad (5)$$

Then the algorithm terminates when we receive a suitable solution at a node in the B&B tree where the error is less or equal than Δ. The current solution which has been found by simulation model is the near optimum solution for Problem 1.

3.2 The Upper Bound and the Lower Bound of the Algorithm

It should be noticed that in a branch and bound algorithm, defining an upper bound and a lower bound can increase the efficiency of the algorithm which results in shorter length of the tree. The lower bound for Problem 1 is the optimum solution of Problem 2, since this problem had a relaxation and therefore it has better optimum solution than Problem 1.

For defining the upper bound we use the simulation model of Problem 1, i.e. a simulation model will be developed according to TSP by any software. All constraints in problem 1 should be programmed in the simulation model. The simulation model is a discrete one so when using this model, a dummy entity (as the traveler) is being sent to the TSP starting a node. For the purpose of making the simulation as smartest as possible, the entity selects the next node according to a probability distribution which arises from optimum solution of Problem 2. Since in optimum solution of Problem 2, all variables are real numbers between 0 and 1 and according to constraints 1 in Problem 1, the summation equals to one then the value of a variable can be seen since the probability of the variable to get the value of 1 or in a TSP, this is the probability distribution of going from node to the other one. This is a key on efficiency of the algorithm as a learning system. Each time we going deeper in the B&B tree and solve a sub problem, the value of the each variable in optimum solution is considered as a probability in the corresponding simulation model. For example when $x_{ij} = 0.9524$ then we assume that the probability of selecting the rout between the node i and node j in the optimal solution is 0.9524. So it uses an empirical distribution function for selecting any rout by using a table in the simulation model and this distribution function is being updated whenever a node in the B%B tree is being solved. This is another advantage of this algorithm in comparison to other works where the simulation model just runs once.

The next important factor in the simulation model is the number of entities that should be generated and sent to the simulation model. However since there is no defined distribution function in this regard then a heuristic sample size is suggested according to the experiences achieved through solving the sample problems. However the higher sample size can be lead to a better efficiency of the algorithm. The suggestion sample size is based on the total variables exist in Problem 1 as follows:

$$n \geq 0.3 * \frac{m(m-1)}{2} . \qquad (6)$$

For calculation the total distance of the total cost in the simulation model, whenever the entity selects the next node according to the empirical distribution function defined, the cost of such a travel between two nodes will be added to the attribute of the entity, so when each entity ends a tour, the value of the attribute is the total cost travelled. It should be mentioned that at the start point, the initial value of the attribute is set to 0. Now we run the simulation model with the mentioned sample size according to the above mentioned formula. Then we choose a sample entity which has the lowest cost in its attribute. The value of this attribute is the upper bound of problem 1, since we are not sure that this is the optimal solution of Problem 1.

Another advantage of using a simulation model is that the best suggested solution (tour) by simulation models is integer. So it really helps us in B&B tree to find a feasible solution for Problem 1. In previous researches which used B&B algorithm, one cannot find such a good capability to create integer variables.

3.3 Fathoming Rule

Whenever the optimum solution of the current node is equal or greater than the last updated upper bound, that node will be fathomed. This is another learning system which used in the model for shortening the B&B tree.

3.4 Branching Strategy

This strategy defines which node should be selected among the list of open nodes. This strategy has great influence on branch and bound efficiency. According to our algorithm, among the open nodes, a node with the lowest Z^* is being selected for branching. This strategy has been selected to reach to better solutions as maximum as possible.

3.5 Separation Rule

Separation rule means the selection of a variable in the selected node to separate on. Assume to the "Jump Tracking Strategy" node j is selected. After selecting a node according to branching strategy, we select a variable which has the maximum value in the optimum solution. If there is any tie, we put these variables to a memory assigned for candidate variables.

Now we run the correspondent simulation model and find the best tour among the samples. If the candidate variable has been used in the minimal tour in the simulation model, we select this variable for separation on otherwise we update the candidate memory be selecting the second maximum and keep the mentioned step until a variable will be selected. In case of a tie up, we select a variable which has been used in the minimal tour got from simulated model, again if there is any tie up, a variable is being selected randomly among the variables inside the candidate memory.

Suppose variable x_{ij} has been chosen from the current node, now one can separate the current node into two new nodes. Each new node is a sub problem of Problem 2 with two new constraints:

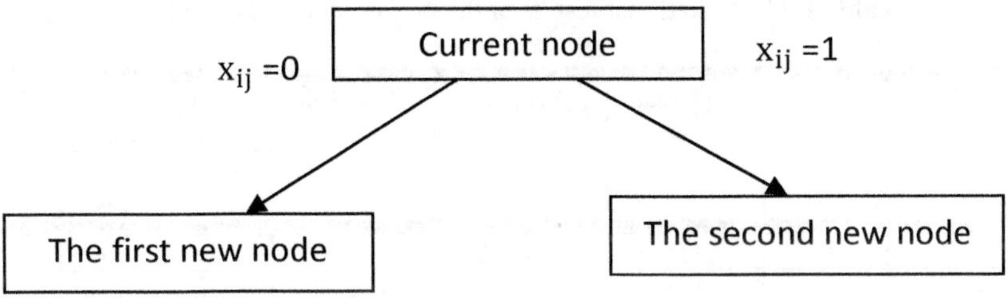

Fig. 1. Node Relation

Finally, the branch and bound algorithm for Problem 2 is as follows.

Step 0: Initializing. Suppose that a TSP with m nodes is given and we want to find the minimal tour. Now consider the Problem 2. A node in a branching tree is active if its corresponding problem has not been either solved or subdivided yet. Let A denoted the current list of active notes. Compute the upper bound and the lower bound according to 3-2.

Step 1: Branching. If A is empty or the termination rule meets according to 3-1, then stop. The optimal solution has been found. Otherwise, select a node j from the active list A according to "Jump Tracking Strategy" which was described in 3-4. If the objective function of the current solution is equal or greater than the last updated upper bound then fathom the current node and select the next node.

Step 2: Selecting. Separate the current node into two new sub problems according to "Separation Rule" described before and then it's necessary to update the upper bound if the minimal suggested tour by simulation model is less than the current upper bound. In each new node solve its corresponding problem. Add new sub problems to A and go to the step 1.

4 Computational Results

In this section the experimental results are summarized. The branch and bound algorithm coded with Microsoft Visual Basic, the corresponding relaxation problems were solved by OSL V.3.0 and simulation models ran by Show Flow 2.0. All computations were run on a PC with 2.6 GHz CPU and 1GB RAM. All TSP graphs have been generated randomly using the simulation software. It should be mentioned that the graphs relates to the real data of an AGV (automatic guided vehicles) routing problem in a spare part warehouse. In the following table, the computational results of the algorithm have been shown among different TSPs. In the computations, a sample of 100 instances was taken for each TSP and the average results have been shown, i.e. the algorithm was repeated 100 times of each problem name. In the table, The first column is the problem name, the second column is the total number of the nodes, the 3rd column is the exact optimal solution of the given problem gained by Lingo 9.0 , the 4th column is the average solution found by the algorithm in different runs, The 5th column is the best solution found by the algorithm in different runs, The 6th column

Table 1. Computational results of the proposed algorithm where Δ=0.02

Problem	Nodes	Optimal Sol.	Average Sol.	Best Sol.	Error Percentage		Time in Sec.
					On Ave.	On Best	
TSP1	70	375	378.33	375.00	0.89	0.00	0.29
TSP2	95	421	424.87	422.00	0.92	0.24	3.43
TSP3	97	388	393.45	388.00	1.40	0.00	4.11
TSP4	101	3,569	3,581.61	3,571.14	0.35	0.06	4.39
TSP5	120	1,040	1,043.90	1,042.00	0.38	0.19	5.11
TSP6	150	789	793.23	789.00	0.54	0.00	6.10
TSP7	170	990	1,009.56	998.00	1.98	0.81	10.76
TSP8	200	56,789	57,012.00	56,854.21	0.39	0.11	12.37
TSP9	300	2,136	2,175.64	2,164.00	1.86	1.31	14.20
TSP91	318	5,788	5,810.20	5,795.49	0.38	0.13	22.67
TSP10	400	4,057	4,120.66	4,104.00	1.57	1.16	54.87
TSP11	500	10,367	10,411.00	10,367.00	0.42	0.00	321.23
TSP12	600	7,321	7,442.00	7,412.52	1.25	1.05	459.65
TSP13	700	8,909	9,085.33	9,033.00	1.98	1.39	775.28
TSP14	800	7,698	7,811.41	7,786.00	1.47	1.14	812.34
TSP15	900	12,605	12,778.48	12,605.00	1.38	0.00	922.03
TSP16	1000	18,933	19,255.80	19,057.00	1.70	0.65	1,003.11
TSP17	1100	14,321	14,508.54	14,486.00	1.31	1.15	1,221.55
TSP18	1200	25,876	26,156.98	26,011.00	1.09	0.52	1,227.05
TSP19	1300	22,043	22,480.74	22,368.00	1.99	1.47	1,311.32
TSP20	1400	31,422	31,998.70	31,864.00	1.84	1.41	1,411.64
Average			11,633.75	11,593.90	1.22	0.68	479.09

is the error percentage based on both the average and the best solution and the last column is the average computation time for each problem when was solved by the algorithm:

$$\textit{The optimal solution} \leq \textit{The best solution} \leq \textit{The average solution}.$$

For the first comparison of the computation results, Liu et al. [16] computational results for TSP have been chosen. They used an overlapped neighborhood based local search algorithm to solve TSPs. The idea of the algorithm is to cut a TSP tour into overlapped blocks and then each block is improved separately. According to their proposed algorithm, the average error is estimated at 0.97% while the average error of the proposed algorithm is 0.68%. It shows the strength of the algorithm for creating suitable tours. It should be mentioned that the only large-scale problem which was checked by them had 1400 nodes and most TSPs had less than 200 nodes. As one can see on Table 1, a wide range of graphs specially large-scale graphs has been tested by the proposed algorithm and the results are acceptable even in large-scale graphs for example when they have more than 1000 nodes. However since the sample graphs are bigger than the ones used by [16] so the total average time in Table 1 is higher than their works, but If we consider similar graphs based on node size, their calculation time for a 200 nodes graph is 12.68 seconds but the time in proposed algorithm is 12.37 or for a graph with 1400 nodes their calculation time is 1,549.93 but the time used here is 1,411.64 seconds on average. Table 2 shows a comparison between the proposed algorithm (PA) and the one which was developed by Lie et al. (LIE) [16]:

Table 2. The comparison of computational results between PA and LIE algorithms

Number of Nodes	Time		Error Percentage on Average		Error Percentage on Best case	
	PA	LIE	PA	LIE	PA	LIE
101	4.39	4.28	0.35	3.64	0.06	2.54
150	6.10	7.80	0.54	3.12	0.00	1.13
200	12.37	12.68	0.39	3.21	0.11	1.44
318	22.67	33.41	0.38	4.14	0.13	2.43
1400	1411.64	1549.93	1.84	3.39	1.41	2.32

For the second evidence to prove the efficiency of the proposed algorithm, we refer to the research done by Dorigo and Gambardella [17]. They used an ant colonies method for solving TSPs. According to their computations, the average error in their algorithm is 1.89% but the time cannot be compared since they just mentioned the time needed for creating a tour. In comparison of problem size, the maximum size problem which was solved had 1577 nodes. The next comparison set based on the works done by Brest and Zervonik [19] where they used an Insertion heuristic for the asymmetric TSP while their proposed algorithm had complexity of $O(n^4)$ in the worst case time. For a 400 nodes graph their algorithm consumed 76.39 seconds while in our algorithm for a 400 nodes graph, it uses 54.87 seconds and the largest graph tested had 443 nodes.

And for the final comparison we refer to the works done by Liu and Zeng [20] where they used some reinforcements to the previously genetic algorithms and generated a new version of this algorithm for the TSP. For best comparison, for a graph with 1002 nodes, their algorithm needed 1,442.38 seconds while in the proposed algorithm, it needs 1003.11 seconds and the optimal solution is being found and for a graph with 1432 nodes they reported 2,157.34 while in the algorithm, the computational results show 1,411.64 seconds for a 1400 nodes graph.

Therefore in terms of speed and accuracy the proposed algorithm is strong enough to be compared with other works even in large-scale problems.

5 Conclusions

This paper presents a new heuristic algorithm to solve a given TSP. The algorithm mixes branch and bound algorithm together with a dynamic suitable simulation model to find the best possible solution for a given TSP. If we compare the computational results of applying the algorithm with other works which used ant colonies algorithm and neural network methods, one can see that it's an efficient one in terms of speed and accuracy for solving even large-scale problems. However it should be mentioned that one may use other metaheuristic methods like genetic algorithm together with our approach to reach better results. In using the proposed algorithm, we cannot mention any special limitation since all symmetric and asymmetric TSPs can be solved by it, however using a fast efficient simulation software like ED 8.0 can help the user to find the solution as quickly as possible. The future directions based on the proposed algorithm could be emphasizing on finding better lower and upper bounds which were used here to shorten the B&B tree and reaching to the acceptable solutions faster. The other direction could be working on simulation model to develop more efficient one. The simulation model rules as a key in our model to get the best solutions, so the more effective simulators will result in the more accurate solutions with lower speeds. Finally, this approach can be easily used in wide range of combinatorial optimization problems which are NP-hard, a suggested field to use this approach is Graph Labeling Problems like graceful labeling problems and magic total labeling problems.

References

1. Applegate, D.L., Bixby, R.E., Chvatal, V., Cook, W.J.: The traveling salesman problem: A computational study. Princeton University Press, Princeton (2006)
2. Garey, M.R., Johnson, D.S.: Computers and intractability: A guide to the theory of NP-completeness. W.H. Freeman, San Francisco (1979)
3. `http://www.keck.caam.rice.edu/tsp/usa13059`
4. Mascato, P.: On Evolution, Search, Optimization, Genetic algorithm and Martial arts: Towards Mimetic algorithms. Caltech Concurrent Computation program, 79–158 (1989)
5. Mühlenbein, H.: Parallel Genetic algorithms in combinatorial optimization. Computer Science and Operations research, 441–456 (1992)
6. Dorigo, M., Maniezzo, V., Colorni, A.: The Ant system: An Autocatalytic optimizing process. Technical Report 91-016. Politecnico di Milano, Italy (1991)
7. Lin, F., Kao, C., Hsu, C.: Applying the genetic approach to simulated annealing in solving some NP-hard problems. IEEE Transactions on Systems, Man, and Cyberspace 23, 1752–1767 (1993)
8. Goldberg, D.: Genetic algorithms in Search, Optimization & Machin Learning. Addison Wesley, Reading (1989)
9. Matsuda, S.: Set theoretic comparison of mappings of combinatorial optimizations problems to Hopfield neural networks. Systems and Computers in Japan 27, 45–49 (1996)
10. Dantzig, G.B., Fulkerson, D.R., Johnson, S.M.: Solution of a large-scale traveling salesman problem. Operations Research 2, 393–410 (1954)
11. Eastman, W.L.: Linear programming with pattern constraints, Ph.D. Thesis. Harvard University, Cambridge (1958)
12. Held, M., Karp, R.M.: The traveling salesman problem and minimum spanning trees: Part II. Mathematical programming 1, 6–25 (1971)
13. Smith, T.H.C., Srinivasan, V., Thompson, G.L.: Computational performance of three sub tour elimination algorithms for solving asymmetric traveling salesman problems. Annals of Discrete Mathematics 1, 495–506 (1977)
14. Crowder, H., Padberg, M.W.: Solving large-scale symmetric traveling salesman problems to optimality. Management Science 26, 495–509 (1980)
15. Bodin, L.D., Golden, B.L., Assad, A.A., Ball, M.: Routing and scheduling of vehicles and crews: the state of art. Computers and Operations Research 10, 63–212 (1983)

16. Liu, S.B., Ng, K.M., Ong, H.L.: A new heuristic algorithm for classical symmetric traveling salesman problem. World Academy of science, Engineering and Technology 27, 267–271 (2007)
17. Dorigo, M., Gambardella, L.M.: Ant colonies for the traveling salesman problem. Bio Systems 21, 36–45 (1997)
18. Applegate, D.D., Bixby, R.E., Chvátal, V., Cook, W.J.: The Traveling Salesman Problem, a computational study. Princeton series in Applied Mathematics (2007)
19. Brest, J., Zerovnik, J.: A heuristic for the asymmetric Travelling Salesman problem. In: The 6th Metaheuristics International Conference (MIC 2005), Vienna (2005)
20. Liu, F., Zeng, G.: Study of genetic algorithm with reinforcement learning to solve the TSP. Expert Systems with Applications 36, 6995–7001 (2009)
21. Arora, S.: Polynomial time approximation schemes for Euclidian TSP and other geometric problems. Journal of the ACM 45, 753–782 (1998)
22. Zhang, W.: Depth-First Branch and Bound versus Local Search: A case study. Information Sciences Institute and Computer Science Department, University of Southern California (2000)

Control of Disassembly Systems Based on the Division of Labour by Means of Dynamically Adapting Routing Plans

Gert Zülch and Jan Hrdina

Ifab-Institute of Human and Industrial Engineering,
Karlsruhe Institute of Technology, Kaiserstrasse 12
D-76131 Karlsruhe, Germany
{gert.zuelch,jan.hrdina}@kit.edu

Abstract. Disassembly and recycling activities are becoming ever more important as a result of legislative changes. The disassembly of end of life products will come to have an increasingly industrial character focusing on the planning and control of economical disassembly processes. As a consequence, the predominant form of work organisation will be based on the division of labour as this system allows for the efficient usage of various disassembly tools and methods in contrast to the complete-disassembly workstations mainly in use today. But the control strategies derived from assembly are only partially suitable for these disassembly systems as they do not sufficiently consider the specific characteristics of disassembly. Any control method suitable for this field of application needs to incorporate new approaches to the flexibilisation of workflows in disassembly systems.

Keywords: Disassembly, Disassembly Control, Control Strategy, Simulation, Optimisation.

1 Disassembly Systems Based on the Division of Labour

Changes in the legislative framework originating in particular from the German recycling management and waste act [1] (also see [2], [3], [4], [5]) are increasingly requiring the goods manufacturing industry to provide for the recycling of end of life products. Disassembly is at the heart of many recycling concepts created for this purpose. It is an integral part of efficient recycling management [6] and is not only required to recycle a product at the end of its useful life, but also for maintenance purposes during the entire product life cycle [7].

These shifts will presumably lead to an increasing amount of products which will need to be disassembled. At the same time, compared to goods manufacturing, disassembly is characterised as low value adding – a characteristic which is reinforced by the uneconomical sequences of operation often found within disassembly processes as opposed to industrial assembly processes [8]. It is, therefore, essential to reduce disassembly costs through the systematic planning and control of the operations sequences

W. Dangelmaier et al. (Eds.): IHNS 2010, LNBIP 46, pp. 396–407, 2010.

[9]. In addition, the profitability of disassembly processes needs to be improved by giving it an increasingly industrial character.

For technical and economical reasons, automation however, can only be an option for specific end of life products to be disassembled in large volumes [10]. Specifically for the recovery of secondary products (spare parts recovery), an automated disassembly of components is mostly too complex, which means that manual disassembly will continue to play an important role. Given the rising quantities of products to be disassembled imposed by legislation, this will lead to disassembly systems being increasingly based on the division of labour. Specialised workstations utilising tools and methods that are suitable for specific disassembly operations will be required, and the potential for rationalisation will have to be realised [11].

All of the above indicate a move away from complete-disassembly workstations which, in addition to the afore-mentioned disadvantages, are often characterised by a poor work situation (e.g. dirty workstations, physical stress, high noise levels) [12]. Modern disassembly systems, in comparison, feature ergonomic workstations equipped with adequate disassembly technology. Such systems include disassembly lines with interlinked manual and automated workstations, some of which may have links to separate back-up workstations [13].

Automated material flow systems can be used to integrate the disassembly systems into the global logistical concept of a manufacturing company. Old vehicle disassembly and the preparation of specific components in particular, is an important field of application for these disassembly systems. The different workstations for specific disassembly operations are arranged along a disassembly line, following the flow shop principle. In comparison to complete-disassembly workstations, this type of system in general does not require cost-intensive disassembly equipment and tools to be provided more than once, thus increasing the profitability of the disassembly process.

Disassembly systems based on the division of labour do not only require more complex system planning, but also a more sophisticated control of the disassembly orders in comparison to systems consisting of complete-disassembly workstations. Priority is given to high utilisation of the workstation within the line. Given that the large share of manual disassembly operations is mainly responsible for the high costs of product disassembly, it is desirable for all workstations to be equally utilised, where possible, in order to keep the personnel costs low and to ensure that the disassembly processes are economical.

Therefore, control of the disassembly processes needs to be adapted to the requirements of a disassembly system based on the division of labour. The main objective of disassembly control is to distribute the operations in such a way that ensures an efficient and economical disassembly sequence resulting from high system utilisation. It is also important to prioritise those disassembly operations which potentially account for the highest possible contribution margins. The control process also needs to be flexible enough to adapt to disassembly-specific events (cf. Chapter 2).

2 Dynamically Adapting Routing Plans

Deterministic operation times exist for assembly processes but not for disassembly operations. As the successful implementation of disassembly operations carries specific risks that do not occur in assembly, disassembly planning is made more complicated because the condition of the end of life products can rarely be predicted [14]. Joints, for example, are not always easily separable and it is impossible to know the exact level of component wear and tear prior to disassembly as the strain bearing on components differs with the intensity of usage. It is, therefore, of no use to establish a fixed disassembly routing plan for the sequence of operations. The variability of end of life product conditions, for example with regards to wear, corrosion or damage, which have an impact on the separability of joints and the recyclability of parts and component assemblies, must be dynamically accounted for in the routing plan. Existing operation documents however, are strictly deterministic and are therefore only partially suitable for disassembly if they can be used at all. In contrary to disassembly processes, workflows in assembly systems do not change and always follow a previously established assembly sequence.

The work content of disassembly operations is not standardised due to the above-mentioned uncertainty regarding the characteristics of the end of life product to be disassembled and cannot be determined a priori. The disassembly method needs to be chosen in line with the individual condition of an end of life product, and only then can the required operation times be determined. The scope of disassembly operations as well as the level of disassembly are not necessarily previously defined but need to be individually determined depending both on the condition of the end of life product and the consideration of economical aspects.

It was within this context that Schiller [14] presented an approach as early as 1998 which took this uncertainty into consideration. The disassembly information system covered by his approach is able to react both dynamically and adaptively to changing conditions (e.g. disassembly operations which cannot be performed), changing objectives (overhaul, removal of harmful substances) and the level of knowledge which increases with the amount of time spent disassembling (e.g. recurring contamination of parts or damaged fasteners). A so-called stochastic disassembly network (SDN) describes the dismantling of an end of life product including possible operation sequences (cf. [14], [15]). The SDN is a digraph with nodes representing disassembly states with possibly various conditions and arcs representing disassembly operations (activity-on-arc network; cf. Fig. 1).

The SDN is developed by means of the analysis and evaluation of end of life products based on sample disassemblies [16]. On the basis of the results, the SDN of a product type serves to determine the optimum disassembly sequence and depth. The disassembly sequence for a specific end of life product is then determined with the SDN on the basis of a break-even analysis including the revenues achievable with component assemblies and parts yet to be disassembled and the required disassembly costs. This optimum sequence is then communicated to the worker via a multimedia routing plan set up at the disassembly workstation. The system is informed by the worker of any disassembly operations which cannot be performed as indicated by the SDN, and a new optimum disassembly sequence is identified in accordance with the respective disassembly condition of the end of life product [14]. The probabilities of end of life product conditions and their different levels are linked to feasibility probabilities for the required disassembly operations which describe the probability of a disassembly operation being carried out successfully [17], [18].

This procedure developed by Schiller [14] allows for the best possible disassembly sequence to be determined for individual end of life product conditions for single disassembly workplaces. His principle of determining a dynamically adapting routing plan on the basis of stochastic disassembly networks is also used in the following to control disassembly systems based on the division of labour [18] (cf. Chapter 3.2).

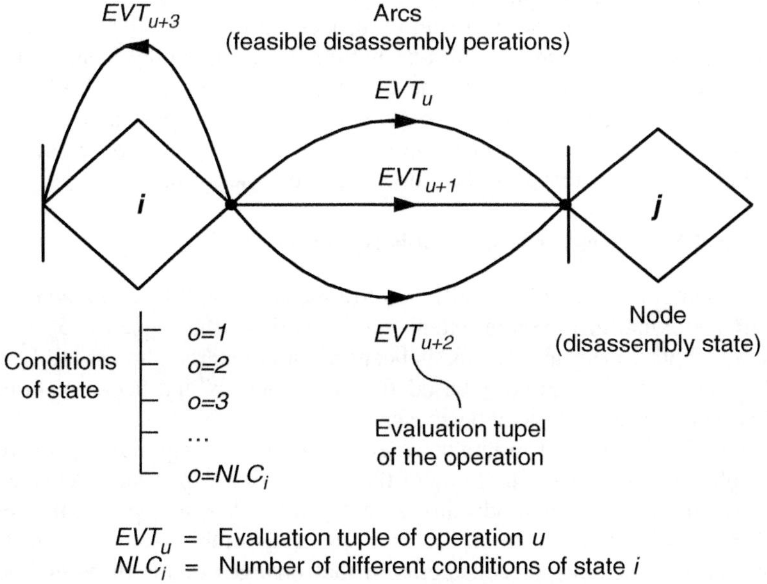

Fig. 1. The basic elements of the stochastic disassembly network (SDN; following [14], [15])

3 Method to Control Disassembly Systems Based on the Division of Labour

Control strategies developed for parts manufacturing and assembly (e.g. kanban control, load-dependent order release, performance agreements, etc.) can only be used for disassembly control to limited extents as they do not make allowances for the specific characteristics of disassembly processes (cf. Chapter 2). There is an extended range of control parameters available for disassembly processes unlike those for conventional manufacturing in order to achieve an efficient disassembly process [19].

3.1 Extending Manufacturing Control by Disassembly-Specific Parameters

Controlling a work system entails "the initiation, monitoring and ensuring of task implementation with regards to quantity, deadline, quality, cost and working conditions" (transl. from [20]). Firstly, the control of disassembly systems based on the division of labour involves the initiation of existing orders. Secondly, the implementation of these orders needs to be monitored, and an efficient sequence of operations has to be ensured by means of intervening control measures [20] if unplanned conditions within the disassembly system require them. These unplanned conditions may occur if operation times for end of life products deviate from the initial routing plan resulting in unevenly distributed work content at different workstations and thus, jeopardising the profitability of the disassembly system.

The disassembly control method developed here aims at ensuring high and equal utilisation at all workstations in the disassembly line. It is also geared towards a maximisation of the value added by disassembly processes [19]. For a control method to be able to intervene in the disassembly process in order to adapt it, it needs to be able to recognise unplanned conditions within the disassembly system in order to communicate necessary changes to the workers (and, in case of hybrid disassembly systems, to the automation) in the form of a new routing plan and amended work instructions. This requires a suitable information system which processes the information and makes it available to the worker in a suitable form.

3.2 Control Measures for Disassembly Systems

In the following, disassembly-specific control measures will be explained. Disassembly is often flexible to a certain extent with regards to the sequence of disassembly operations (control measure 1). This is because, in comparison to assembly, there are a smaller number of technology-based forced sequences and because different disassembly methods and tools can be chosen.

The selection of the appropriate disassembly method is important in this regard. If, for example, utilisation is high at one of the workstations, the time-intensive but non-destructive disassembly method can be replaced by a faster yet partly destructive method in order to improve the utilisation of the disassembly system (control measure 2). It needs to be taken into consideration that changes of the disassembly method may negatively impact the possibilities for re-use/alternate use or the reprocessing/further processing of end of life product components.

Depending on the purpose of the disassembly (e.g. complete disassembly vs. selective disassembly of specific components) it may not be necessary to follow a specific sequence of disassembly orders whereas orders can contain different components ordered by one or several customers, meaning that their sequence can be varied (control measure 3). One end of life product can contain different customer orders. This flexibility can be used to reduce the loss of time caused by difficulties of load balancing between workstations.

The level of disassembly can also be varied (control measure 4) in order to reduce fluctuations between the utilisation of workstations. This means that the extent of disassembly can be determined on a case by case basis depending. If utilisation is high at one workstation, the level of disassembly and thus, the workload of the station can be reduced. Disassembly will take less time, and idle times will be avoided at the downstream stations. The focus will then be on disassembly operations that add the highest value and lower value-adding operations will be skipped. If utilisation at one workstation is low, the level of disassembly can be increased so that disassembly operations will be performed which would be uneconomical in the case of normal system utilisation, but which still generate some added value albeit a little value.

Another option to avoid idle times resulting from a poor balancing of the disassembly stations is to remove the fixed allocations of work operations to specific workstations. Depending on utilisation, disassembly operations can be moved from one workstation to an adjoining one as long as it is technically feasible (control measure 5). For that purpose, the allocation of disassembly operations to workstations determined during the planning process of the disassembly work system must be monitored during disassembly. If there are coordination-induced losses, this allocation can be modified and adapted to the situation prevailing in the disassembly system, or to the end of life products in the system. However, this is only possible if allowances have been made for disassembly operations to be moved between stations during planning of the disassembly system. For example, the workstation where the disassembly operation is to be relocated must have the technical equipment required to do the job. This control measure is, therefore, mostly suitable for comparatively simple, manual disassembly operations and for workstations which allow for the subsequent collection and sorting of disassembled component assemblies and parts destined for re-use/alternate use or reprocessing/further processing.

In addition to the above-mentioned options, the workers can help each other in case of difficulties (control measure 6). In such a case, the disassembly operations will not be allocated to other workstations as with control measure 5, but will stay at their initial station. The workers either split up the disassembly operations (e.g. remaining disassembly operations from a specific point within the tact until the end of the tact), or specific disassembly operations (e.g. in case of disassembly-specific unfavourable events, cf. Chapter 2) are carried out by two workers in parallel. In the first case, the additional worker needs to have the skills required to perform the respective disassembly operations. In the second case, the second worker supports the first worker so that the feasibility probability is increased (cf. Chapter 2) and/or less time is required to perform the task.

3.3 Integrating the Developed Control Measures into a Global Concept

The main task of the new control method consists of integrating the above-described disassembly-specific control measures into a global concept (cf. Fig. 2) which requires an IT connection to the disassembly system that is to be controlled. This is the only way to ensure that all potentials for the improvement of the disassembly process profitability are realised. For the effectiveness of the developed approach to be verified, the control method was integrated into the *OSim* simulation procedure (*O*bject *Sim*ulator [21], [22]) developed at ifab-Institute of Human and Industrial Engineering of the Karlsruhe Institute of Technology. The result is the extended *OSim-DPS* procedure (*OSim* for *D*isassembly *P*lanning and *S*cheduling) taking additional, disassembly-specific modelling aspects into consideration [19].

In *OSim-DPS*, an end of life product is modelled in the form of an activity-on-node network. Activity networks include all activities in a logical order that need to be performed to process an order. Activity networks are, therefore, similar to precedence graphs used in assembly planning [23], which are defined as "directed graphs with a logical order of activities used to describe sequences" (transl. from [24]). Activity networks, which were previously modelled in *OSim* as AND graphs only, have been extended to include the AND/OR constraint for other feasible disassembly operation

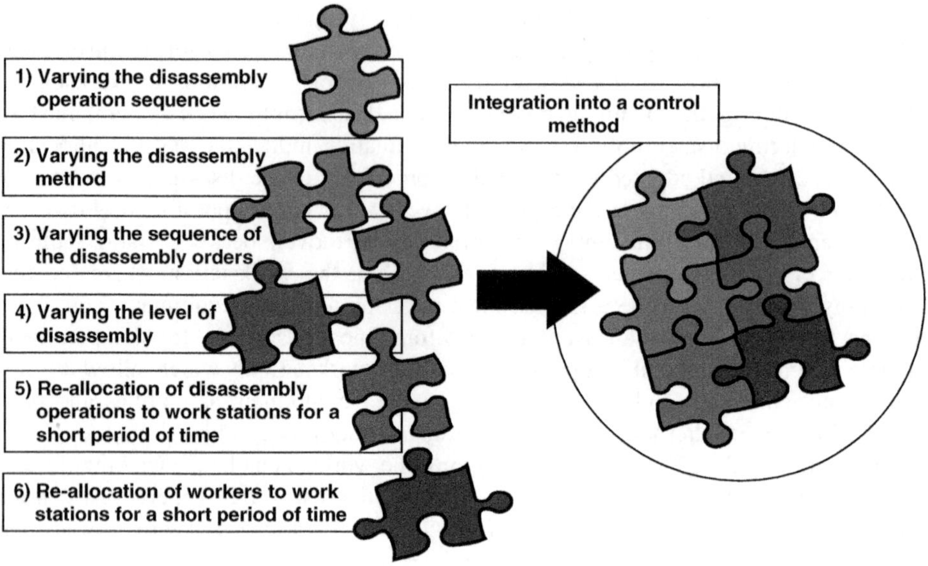

Fig. 2. Compiling the control measures into a global control concept, and integrating it into the simulation procedure *OSim-DPS*

sequences. In order to ensure the correct modelling of disassembly operations, the probability with which they can be used (cf. Chapter 2) was additionally taken into consideration. To that end, stochastic disassembly nodes were developed.

The level of detail included in this specialised activity network, i.e. the way the disassembly operations are structured, can be determined as required (to certain extents) by means of the above-described concept. However, the activity network modelled for the planning and control of a disassembly system should not be excessively complex. It is, therefore, advisable to combine several consecutive small-scale partial operations into one large disassembly operation ensuring that the required partial operation routing is maintained [25]. Furthermore with regards to disassembly, it is usually sufficient to subdivide materials into fractions of compatible secondary material groups referred to as recycling groups [26].

Figure 3 shows how disassembly operations can be modelled in the form of a specialised activity network which indicates the end of life product structure consisting of end of life parts and end of life component assemblies in addition to the various disassembly paths. The end of life product to be disassembled represents the starting point of the disassembly process. End of life component assemblies and end of life parts are recovered after the respective disassembly operations have been carried out successfully. The term "residual product" refers to that part of the end of life product which is passed on within a disassembly system that is based on the division of labour from one workstation to the next. "Should components" include end of life component assemblies, or end of life parts which are meant to be disassembled according to the routing plan. They are labelled "should" instead of "must" components because uncertainties

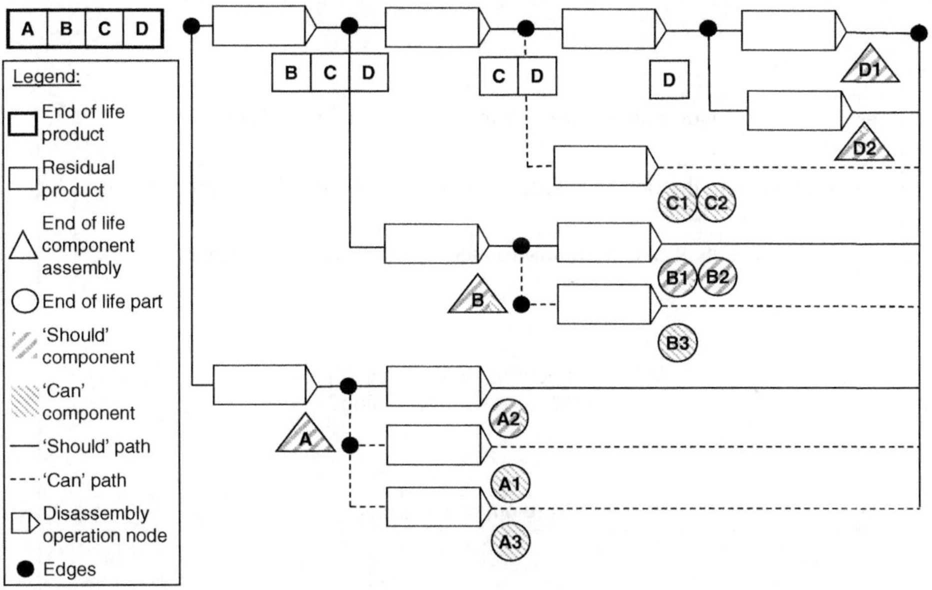

Fig. 3. Modelling disassembly operations in the form of an activity network

(cf. Chapter 2) need to be factored in meaning that there can be no a priori assumption that disassembly of these parts would be feasible. "Can components" are not specified as parts to be disassembled within the routing plan, for example when disassembly would be uneconomical. They may still be disassembled during downtime of the disassembly line for example, but only after considering all the relevant costs for this decision (variable costing). The "should" and "can" paths are determined in accordance with whether an end of life component assembly or an end of life part, is characterised as a "should" or "can" component as defined during the modelling of the respective disassembly node.

The information contained in the activity network (possible disassembly tools, feasibility probabilities, expected revenues, etc.) is used to control the division of labour disassembly system. Based on this information, an SDN is generated for each workstation, and the optimal disassembly strategy is determined for all workstations separately. The strategy contains the disassembly operations to be carried out (successively) and takes the form of a dynamically adapting routing plan (cf. Fig. 4).

The optimal path is calculated considering all possible control measures (cf. Chapter 3.2) indicated in the SDN as feasible disassembly operations (single arcs or partial paths) [14], [15]. In the case of disassembly systems based on the division of labour, there is also a prioritisation of all disassembly operations that are required to transfer the residual product to the subsequent workstation in order to ensure, where possible, that there are no idle times due to the work piece being overdue. In addition, there is a lower-ranking prioritisation of "should components". "Can components" are ranked according to the break-even analysis.

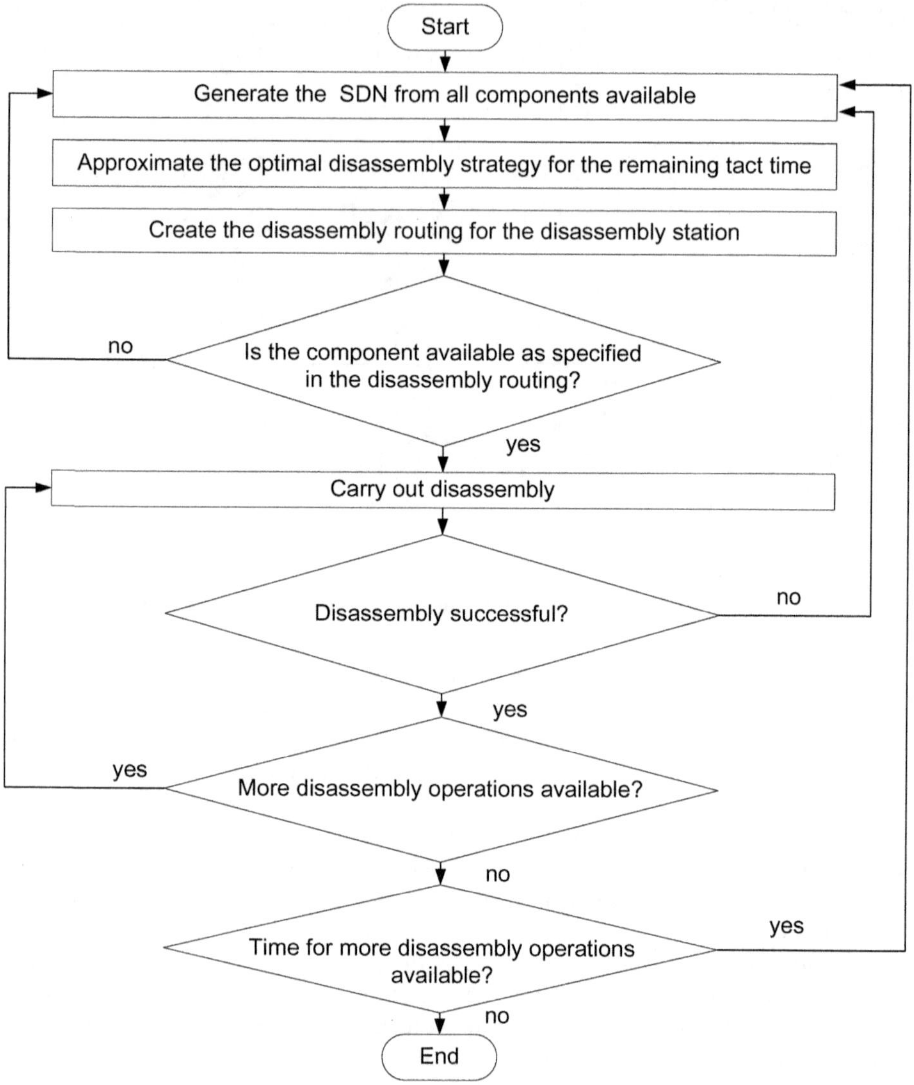

Fig. 4. Flow chart of reactive disassembly control for systems based on the division of labour

The routing plan, once created, will be maintained until the occurrence of a condition which makes it impossible to follow the order of disassembly operations as laid out in the routing plan (e.g. the component that is to be disassembled is missing or cannot be disassembled), or until all disassembly operations have been performed. In either of these cases, the disassembly system is adapted to the current situation, and a new routing plan will be created (dynamically adapting routing plan).

4 Verification by Means of a Pilot Example

Currently, simulation studies are being carried out with industrial partners with the purpose of proving the effectiveness of the method using real-life data. The design of the disassembly systems under examination is based on the flow shop principle. All industrial partners use balanced lines, but have to carry out disassembly programs where problems with specific characteristics of disassembly processes (cf. Chapter 2) occur frequently.

Beyond the existing disassembly lines, there is also the option of disassembly systems with separate workstations that are part of a non-balanced overall system structure. Simulation is used to plan different alternatives and evaluate them in quantitative terms. It is expected, that the described method to control disassembly systems will increase the efficiency of these disassembly systems.

5 Potentialities for Further Increasing the Efficiency of Disassembly Systems

This paper highlights the necessity to provide a specialised method available for the control of disassembly systems based on the division of labour. Disassembly-specific conditions, due to which traditional control methods used in assembly can only be applied to disassembly processes to limited extents, are identified. The presented approach consists in the development of disassembly-specific control measures and their integration into a control tool which is able to overcome unplanned events which frequently occur during the disassembly of an end of life product. This is to be achieved through the implementation of dynamically adaptive routing plans.

With regards to disassembly systems based on the division of labour it is not only important to control the different disassembly orders but also to plan their sequence for a specified planning horizon in order to achieve good system efficiency. The fact that disassembly operations differ substantially from assembly, in that the level of disassembly may change, is a planning parameter that should be taken into consideration for order sequencing in the form of changes to the order release logic [27], [28], [29] capitalising on additional potential for success.

References

1. KrW-/AbfG: Gesetz zur Förderung der Kreislaufwirtschaft und Sicherung der umweltverträglichen Beseitigung von Abfällen (Kreislaufwirtschafts- und Abfallgesetz). Act of September 27, 1994. In: Bundesgesetzblatt I, p. 2705. In the version of August 11 (2009)
2. The European Parliament and the Council of the European Union: Directive 2002/96/EC of the European Parliament and of the Council of 27 January 2003 on waste electrical and electronic equipment, WEEE (2008)
3. ElektroG: Gesetz über das Inverkehrbringen, die Rücknahme und die umweltver¬trägliche Entsorgung von Elektro- und Elektronikgeräten (Elektro- und Elektronikgerätegesetz). Act of March 16, 2005. In: Bundesgesetzblatt I, p. 762 (2007)

4. AltfahrzeugG: Gesetz über die Entsorgung von Altfahrzeugen (Altfahrzeug-Gesetz). Act of June 21, 2002. In: Bundesgesetzblatt I, p. 2199
5. AltfahrzeugV: Verordnung über die Überlassung, Rücknahme und umweltverträgliche Entsorgung von Altfahrzeugen (Altfahrzeug-Verordnung). Decree of June 21, 2002. In: Bundesgesetzblatt I, p. 2214. In the version of July 31, 2009
6. Hentschel, C.: Beitrag zur Organisation von Demontagesystemen. Fraunhofer IPK, Berlin (1996)
7. Seliger, G.: Nachhaltige Technologien. In: Erfolg in Netzwerken, pp. 235–243. Springer, Berlin (2002)
8. Spur, G.: Demontagesysteme im Recyclingprozeß. wt - Produktion und Management 86, 428 (1996)
9. Rautenstrauch, C.: Fachkonzept für ein integriertes Produktions-, Recyclingplanungs- und Steuerungssystem (PRPS-System). deGruyter, Berlin (1997)
10. Boks, C., Tempelman, E.: Future disassembly and recycling technology - Results of a Delphi study. Futures - The Journal of Forecasting and Planning 30, 425–442 (1998)
11. Ruprecht, R.: Disassembly and recycling. Industrial Robot 23, 4 (1996)
12. Kiesgen, G., Emenako, M.E., Slawik, F.: Development of partially-automated disassembly plants for regaining reusable parts. Industrial Robot 23, 11–15 (1996)
13. Schnauber, H., Kiesgen, G., Slawik, F., et al. (eds.): Das Element Produktwiederverwendung im Qualitätskreislauf. Ergebnisse des EUREKA Forschungsprojektes CLEAN-TECH (EU 1104). Bochum (1997)
14. Schiller, E.F.: Ein Beitrag zur adaptiv-dynamischen Arbeitsplanung in der Demontage. Shaker Verlag, Aachen (1998)
15. Zülch, G., Schiller, E.F., Schneck, M.: Adaptive Dynamic Process Plans - A Basis for a Disassembly Information System. In: ZüstECO-Performance 1996, pp. 245–253. Verlag Industrielle Organisation, Zürich (1996)
16. Müller, R., Zülch, G.: Planung von Demontageabläufen auf der Basis von Probedemontagen. In: Kolloquium zur Entwicklung umweltgerechter Produkte, pp. 12–15. Technische Universität Darmstadt (1998)
17. Zülch, G., Schiller, E.F., Müller, R.: A Disassembly Information System. In: Life Cycle Networks, pp. 400–412. Chapman & Hall, London (1997)
18. Zülch, G., Schiller, E.F., Müller, R.: Adaptiv-dynamische Arbeitspläne als Grundlage für ein Demontage-Informationssystem. In: Kolloquium zur Kreislaufwirtschaft und Demontage, pp. 386–390. Technische Universität, Hochschule der Künste, Berlin (1997)
19. Zülch, G., Hrdina, J., Schwarz, R.: Development of a Computer Aided Procedure to Control Division of Labour Based Disassembly Systems. In: Manufacturing Systems and Technologies for the New Frontier, pp. 217–220. Springer, London (2008)
20. REFA-Verband für Arbeitsstudien und Betriebsorganisation (edt.): Planung und Steuerung - Teil 1 (Methodenlehre der Betriebsorganisation). Hanser, München (1991)
21. Jonsson, U.: Ein integriertes Objektmodell zur durchlaufplanorientierten Simulation von Produktionssystemen. Shaker Verlag, Aachen (2000)
22. Zülch, G., Fischer, J., Jonsson, U.: An integrated object model for activity network based simulation. In: Proceedings of the 2000 Winter Simulation Conference, vol. 1, pp. 371–380. The Institute of Electrical and Electronics Engineers, Piscataway (2000)
23. Grobel, T.: Simulation der Organisation rechnerintegrierter Produktionssysteme. In: ifab-Forschungsberichte aus dem Institut für Arbeitswissen¬schaft und Betriebsorganisation der Universität Karlsruhe, Band 3. Karlsruhe (1992)
24. Brinkmeier, B.: Prozeßorientiertes Prototyping von Organisationsstrukturen im Produktionsbereich. Shaker Verlag, Aachen (1998)

25. Zülch, G., Schwarz, R.: Industrialisierung des Recyclings durch Rücknahmeverpflichtungen – Neue methodische Herausforderungen an die Demontageplanung und steuerung. In: Technologiegetriebene Veränderungen der Arbeitswelt, pp. 273–253. GITO-Verlag, Berlin (2008)
26. Weber, J., Meedt, O.: Methodik der recyclinggerechten Produktgestaltung auf Basis des Recyclinggraphen. In: Proceedings of the 10th International Conference on Engineering Design, pp. 987–1007. Industrielle Organisation, Zürich (1995)
27. Zülch, G.: Der strategische Steuerungsraum - Ein Ansatz zur Systematisierung von Strategien der Fertigungssteuerung. In: VDI-Zeitschrift, vol. 131, pp. 58–65. Düsseldorf (1989)
28. Zülch, G.: Systematisierung von Strategien der Fertigungssteuerung. In: Organisationsstrategie und Produktion, pp. 151–178. gfmt-Verlags KG, München (1990)
29. Zülch, G.: Strategien zur Steuerung von Fertigungssystemen. In: Fortschrittliche Betriebsführung und Industrial Engineering, vol. 40, pp. 52–58. Darmstadt (1991)

Integrated Production Program and Human Resource Allocation Planning of Sequenced Production Lines with Simulated Assessment

Lothar März[1], Hansjörg Tutsch[2], Stefan Auer[1], and Wilfried Sihn[1]

[1] Institute for Management Science, Technical University of Vienna, Theresianumgasse 27, 1040 Vienna, Austria
{lothar.maerz,stefan.auer,wilfried.sihn}@aproperplan.eu
[2] Flexis AG, Schockenriedstrasse 46, 70565 Stuttgart, Germany
hansjoerg.tutsch@aproperplan.eu

Abstract. The personnel flexibility of car manufacturers is used mostly as reactive means, in order to be able to realize the planned production program. The approach of trying to combine production planning with human resource planning is pursued within the scope of the Eurostars research project "Advanced Production Program and Personnel Assignment Planning" (A ProPer Plan). This approach combines the advantages of car sequencing with the advantages of the mixed modeling approach by coupling optimization with simulation. The solution constitutes an iterative planning process, where an optimized sequence is examined by simulation for feasibility regarding personnel flexibilities. The contribution points out, how such an integrated solution was realized and which results could be obtained with real time data of manufacturers of vehicles.

Keywords: Car Sequencing, Personnel Assignment Planning, Simulation.

1 Introduction

The European automotive industry is characterized by complex products, which enable wide customer individuality of vehicles. Production planning therefore has to distribute the different vehicle variants on the assembly line throughout the day so that bottleneck resources are evenly utilized and capacity peaks can be avoided. In this stage there already are complex planning logics, which take into account block constraints, space constraints and transition restrictions. By integrating human resource allocation planning the bottleneck and capacity can be influenced deliberately, which can lead to relocation of the bottleneck if human resources can be relocated in order to relieve an acute bottleneck. The integrated planning of human resource allocation and production therefore expands the freedom of planning.

The content and results of this article were compiled within the framework of the research project “Advanced Production Program and Personnel Assignment Planning” (A ProPer Plan; funded within the Eurostars Programme which is powered by EUREKA and the European Community). The goal of A ProPer Plan is to create a

W. Dangelmaier et al. (Eds.): IHNS 2010, LNBIP 46, pp. 408–419, 2010.

tool, which allows the integration of the human resource allocation plan with the short-term production plan and to achieve optimal production results with high adherence to delivery dates.

2 Tasks of Production Scheduling

2.1 Synchronized Production Lines

This subject encompasses synchronized production lines where orders are planned in a sequence. Synchronized flow production is a flow-oriented production system where parts are being moved by means of a transportation system (usually an assembly line) through the production stations arranged in sequence, in which the machining time is restricted by a cycle time (Boysen, 2005a, p. 8). Synchronized flow production is mostly encountered in the automotive and electrical industries. The project focuses on the planning of the final assembly in vehicle and vehicle component plants where variant flow production with low automation and high labor intensity exists.

Sales prognosis, installation rates, and monthly production and sales quantities serve as input for production planning. Restrictions which have to be taken into account are minimum line load resulting from the model mix problem (provision for the production factor resources), the capacity of plants with regard to annual working hours (staff) and potential bottlenecks on the supplier side (material). Planning results are detailed, for example weekly production plans for the production plants. The task of operational production planning is to decide the types and quantities of model variants to be assembled from the existing model portfolio (Meyr, 2004). That way the order backlog of a month is being broken down into single day and shift plans using production planning (Monden, 1993, p. 67). In doing so the predetermined capacity has to be complied with using the available production cycles and the availability of incoming components must be considered (Scholl, 1999, p. 111).

Production planning is usually continuous. The allocation of orders to week or day periods or to shifts is also called slotting. Continuous planning can be added to slotting until the sequence is fixed, which balances the orders based on capacity and material criteria. Individual orders can thereby be moved to a period different from the initial planned production period by taking into account other detailed restrictions. This shifting, which results from an adjustment, is also called balancing. Fixing the order sequence assigns a decided production cycle to each order from the order pool. The methodology for implementing and the objective of sequencing can vary and lead to three different classes of optimization models, which are outlined in the following section.

2.2 Sequencing

Comprehensive scientific literature exists for the planning task of sequencing in the automotive industry. In place of them we want to point to the works of Decker (1993), Lochmann (1999), Scholl (1999), and Boysen et al. (2009). In principle one can differentiate between three methods.

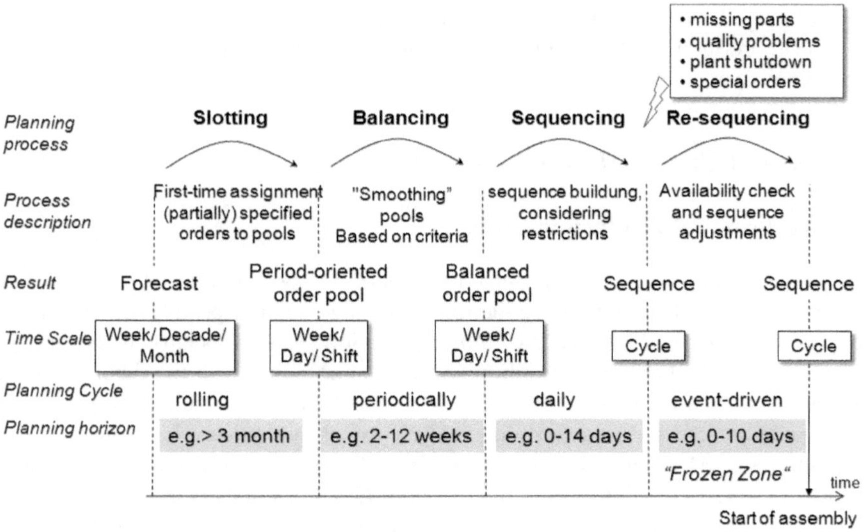

Fig. 1. Planning processes of sequenced assembly lines

Level Scheduling. The so-called level scheduling, which descended from the well known Toyota Production System (Monden, 1993), targets the distribution of demand for each material as evenly as possible. Different models and algorithms were proposed. An overview of the research papers for level scheduling can be found in Boysen (2005a, p. 217, 2005b).

Mixed-Model Sequencing. The so-called mixed-model sequencing aims to prevent capacity overloads in the flow system. Sequence-related capacity overload of the workstations or workers are to be minimized by exact scheduling of the different variables at the stations and by taking into account the variant dependent machining time, length of workstations, and cycle times (Wester and Kilbridge, 1964; Thomopoulos, 1967; Bard et al., 1992, 1994; Tsai 1994; Scholl et al., 1998; Boysen, 2005b).

Car Sequencing. Also capacity oriented, car sequencing tries to avoid these detailed entries by prohibiting partial sequences with capacity overload via H_0:N_0 sequence rules. A sequence rule of 1:3 relating to the sunroof option states that of three back-to-back components only one can have the sunroof option. Otherwise capacity overload occurs. Solution techniques for car sequencing can be found in Drexl and Kimms (2001), Gagné et al. (2005), Drexl et al. (2006), Gravel et al. (2006), as well as Fliedner and Boysen (2008).

While level scheduling is mostly used at Asian automotive manufacturers like Toyota (Monden, 1993) and Hyundai (Duplaga et al., 1996), capacity overload oriented methods are used more at Western manufacturers like Renault (Gagné et al., 2005). Nonetheless, Western European automotive manufacturers focus on offering customized design options for vehicles (Sihn et al., 2009). BMW offers its vehicles in up to 10^{32} variants as a result of a multitude of individually selectable options (Meyr, 2004). In order to avoid the high data collection effort, almost all of the Western European companies use the car sequencing method.

2.3 Production Cycle

In sequenced production lines the cycle determines the productivity. The cycle is a parameter of sequenced production lines and implies the periodic constant feed of orders in production. The period length is the cycle. This requires, however, a constant cycle at least over the particular planning horizon of a shift. This can be assumed for the majority of assembly lines in the passenger car and power train production, whereas in the truck assembly cycle times dependent on vehicle length are prevalent.

Equally spaced surface sections, which are travelled per cycle, can be calculated for uniform cycle lengths. These are called stations. The length of a station depends on cycle and cycle speed. The depth of a station depends on the product to be produced (engine, axle, passenger car body, etc.). The arrangement into stations has organizational reasons in that it facilitates the assignment of human resources, materials, resources, etc. on the lines into sections. As will be shown, the assignment of tasks on the station level is not always helpful, because station overlapping processes and organizational forms are difficult to display.

The workload of a cycle results from the ratio of the sum of the process times (work content) to the cycle time and is stated in percentages. This workload is distributed among the required number of workers. Depending on the product variants and configurations, the workloads sometimes vary considerably per station. The goal of human resource allocation planning is to utilize the workers as equally as possible across the sequence of variant rich products.

3 Capacity and Human Resource Planning Functions

Human resource demand planning and human resource allocation planning are part of capacity planning, which has to meet market demand with adequate production capacity. While human resource demand planning clarifies how many human resources with which qualifications are necessary, the task of human resource allocation planning is to utilize the existing staff optimally. When balancing capacity, the capacity flexibility plays a key role, because the actual market behavior cannot be forecasted accurately and the demand fluctuations have to be absorbed within bandwidth. Because the assembly lines in the automotive industry are human resource-intensive, this is where the biggest leverage for flexibility in the operational hours can be assumed.

The working hour flexibility varies with lead times, which is necessary for implementing capacity adjustment measures: Capacity increases through new line installations can only happen long-term, because 2 to 3 years of lead time is necessary. Annual hours of work and weekly hours of work flexibility can be varied mid-term through agreements with the workers council. Short-term model mix fluctuations can be balanced with human resources that can be assigned to different workstations (so-called “floaters“). Floaters can be utilized for different reasons and under different demand scenarios.

– Use of a floater to balance the processing time at workstations when capacity overload threatens.
– Assignment of floaters to big orders with high expenditure, who will attend to the whole assembly line or just parts of it.
– Assignment of floaters for production stations that require special skills which are only rarely needed (technologically contingent floater assignment).
– Floater as backup capacity for work absences, vacation, illness, etc.

The first two assignment scenarios use the free capacity of a floater at the time of occurrence. The two latter floaters functions don't serve to support the sequencing via a (operational short-term) capacity increase. Therefore they are not pictured in the following model, but are only considered as input parameters (number of floaters). Sometimes a short-term capacity increase is also possible through production during breaks.

4 Correlation between Production Program and Human Resource Allocation Planning

In car sequencing, balancing based on key features (called heavy items) already takes place when assigning orders to order inventory. These product features are gradually being broken down into more detail in the subsequent planning steps. The bridging of sales demand (market demand) and production supply (production elements) occurs via product features. The sequencing rules, which are described on product features level, represent implied capacity restrictions within production. For this reason sales demands are described in terms of product features and the restrictions resulting from capacity and material related limitations of the production system are translated into the language of the product features.

Balancing is now carried out subject to the planning horizon and the available product feature data. The product features from the production plan are extracted from the available order data. Only the variant information of products, which passes a possible capacity limit, is important. Looking at the past is very important here: frequency distribution of variants is known from experience. If a variant occurs more often and as a result load peaks occur, a sequencing rule is added which can be taken into account during the next sequencing run. A forward-looking prognosis of the load distribution of assembly lines through the currently planned sequence, like in mixed model sequencing, is not possible.

The following links arise between production planning and human resource and human resource allocation planning:

– Mid- and long-term human resource planning provides the capacity framework for mid- and long-term production planning.
– In the short- to mid-term horizon, human resource planning follows production planning and only influences it indirectly.
– In the short- and mid-term horizon, the consideration of human resource restrictions happens through indirect rules and strategies in production planning, e.g.:
  – Equal distribution strategies for labor intensive tasks

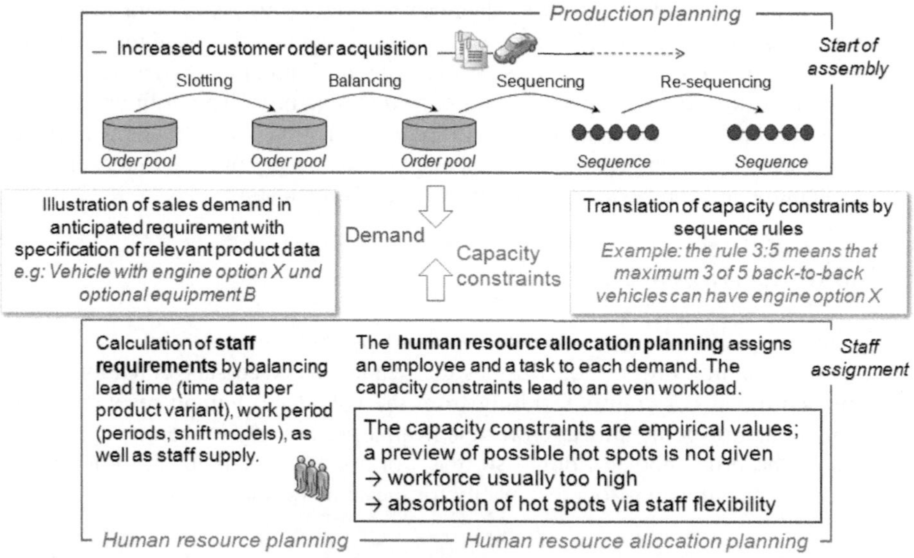

Fig. 2. Relationship between areas of planning

  - Spacing rules for labor intensive tasks
  - Transition rules for labor intensive tasks
- No direct human resource restrictions are taken into account in production planning.
- Little or no direct coordination between human resource allocation planning and production planning happens in short-term planning and order processing.
- During temporary production bottlenecks, the response is usually human resource measures and not an adjustment of the production plan.

These observations suggest that an integrated planning approach, which allows direct influence of human resource aspects on the short- and mid-term production planning process, is able to open up additional optimization potentials. Provision for human resource flexibility in production planning requires that

- The impact of production planning on the capacity demands per order are known
- The skills and availability of human resources as well as flexibility of their deployment are known, and
- The interaction of process demands and the availability of human resources in the dynamic sequence of the order sequence are analyzed.

The process data per order is necessary for this kind of planning. Because the orders are already specified at the time of sequencing, the capacity demand per workstation and order and therefore the impact of the order sequence on capacity demand can be determined. The discrete event simulation lends itself to the prognosis of load distribution insofar as further analysis of floater assignment and situational decisions are to be conducted. Mapping in a simulation model also has the advantage of simplifying scenario generation, and the resulting data can be used as an assessment function for higher-ranking optimization. This approach was implemented by combining a revised flexis AG sequencing optimization solution with a simulation model validated by means of real-time data from an automotive manufacturer. The following section describes the basic model foundations and the interaction with the sequencer. The benefits of car sequencing are hereby expanded by the aspect of human resource allocation planning and checked for feasibility with dynamic testing.

5 Solutions Approach

The sequencing solution gets a list of the valid production constraints as well as a continuously running order pool. Besides the technically necessary constraints, there are now additional constraints resulting from the calculated human resource allocation. The human resource allocation results from a preliminary human resource based analysis of the order pool. In doing so, the workload included in the order pool is balanced with the human resource capacity for the same period.

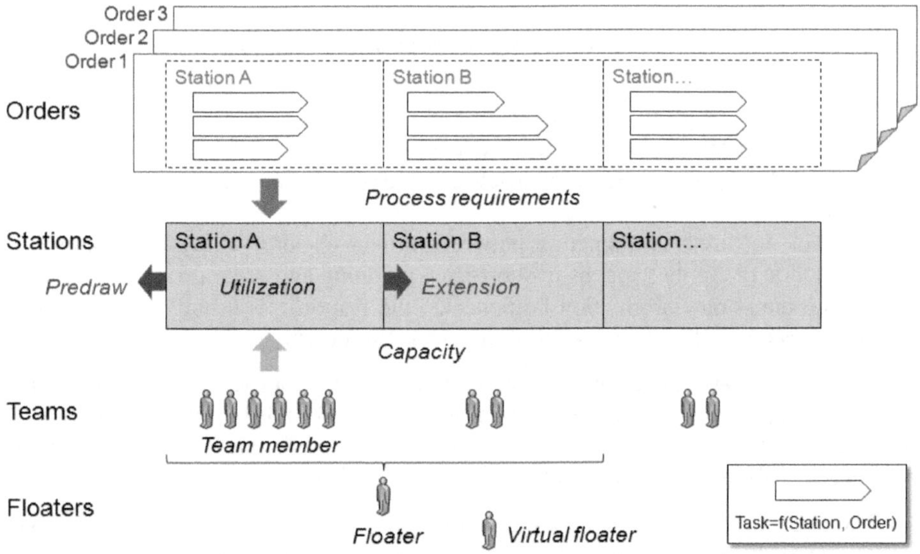

Fig. 3. Model approach

The calculation of an optimized sequence is then carried out, taking into consideration all technical and human resource related production constraints. Subsequently, the calculated sequence regarding the human resource scenarios at the individual station groups is reviewed in a simulation. The simulation model maps the variant dependent assembly processes per order and takes into consideration the existing human resource capacities. Each order consists of a multitude of executions which result in different process times depending on the product variant and (optional) equipment. One or more employees are necessary per execution. This process and organizational data is available per order. The order sequence is the result of sequencing and serves as input for the simulation.

Human resource allocation takes place during simulation runtime, namely at the moment when the execution within the simulation is one cycle before to start. The employees which have been scheduled for the execution and have been combined as a team are allocated. The company's goal is to engage its employees in the same activities, because experience has shown that in doing so a high level of quality and therefore less rework is necessary. Each execution is assigned to a team. If the sum of the process requirements is higher the sum of the team capacity, the team has the possibility to predraw or to extend the work in predecessor or successor cycles within given time constraints. If there are not enough unused employees available the team has to use floaters. Floaters are assigned to groups, which are responsible for a number of teams. If a floater is not available the simulation model generates one (virtual floater). The simulation calculates the feasibility of the sequence at hand and calculates the additionally required employees in case the assigned human resources are not enough. This information can now be relayed back to the previous sequencing task. This starts an iterative process where the sequence solution together with the defined human resource allocation is passed to the simulation and which now reveals the expected workload with the help of the fully specified order data.

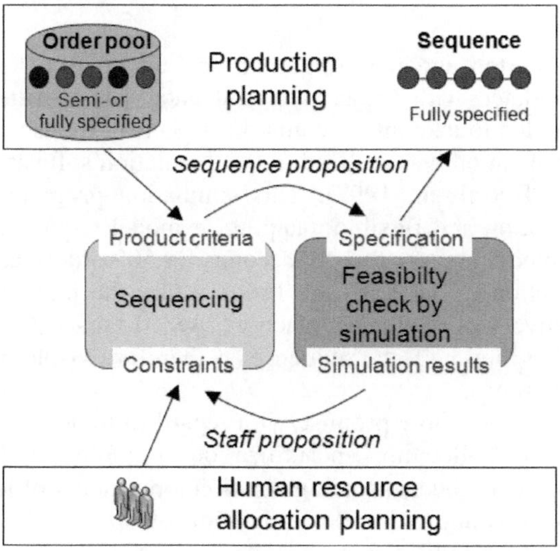

Fig. 4. Iterative planning processes

The implementation of the planning system follows these planning principles:

- The integrated planning system expands the proven workflow of production planning by a human resource optimization component which provides dynamic human resource optimization rules and parameter adjustments for sequencing.

- By analyzing the work content for the order pool and the order sequence and by comparing them with the existing human resource capacities and deployment scenarios, the system calculates the best response possibilities to the detected bottlenecks.
- To create an optimal order sequence also with respect to human resource allocation, relevant rules and parameter calibrations are generated for the chosen response options and taken into consideration in the next optimization run.
- The simulation of the assembly processes, based on of the work schedule data, each variant and its lead time per execution are taken into consideration. In doing so a highly exact forecasting of employee workload for a given sequence is possible.
- This course is repeated until a solution is found which is optimal under production as well as human resource planning aspects.

This approach combines the advantages of car sequencing with the advantages of the mixed modeling approach. The advantage in car sequencing lies in the restriction on few product features in order to limit the data to the relevant criteria. With this restriction, optimization algorithms are capable of solving the sequencing task in finite time and acceptable quality. A further analysis of labor times per variant and station allows the prognosis of load profiles. Additional potential is tapped through the repetitive interplay between optimization of features and the feasibility prognosis based on labor time data. An efficiency increase of the human resource allocation is thus possible.

6 Results

The sequencing solution and the simulation application were first developed independently and validated with the help of test data. The existing flexis sequencing solution was expanded to take into account the aspects of human resource restrictions. The simulation application was realized in the simulation software development environment of SLX (Henriksen, 1997). This simulation program distinguishes itself through fast calculation and flexible mapping of model requirements. The run times of the simulation model are less than 10 seconds for 500 orders and 100 executions. A post-process animation in the Software Proof allows the transactions to be visually inspected. The delivery of data takes place via ASCII tables. An iterative integration into an optimization loop with the sequencer is therefore viable, while also regarding solution generation times.

The simulation results show promise with regard to transparency and analysis options of transactions in allocation and utilization of employees. It was demonstrated that in combination with the sequencing solution a balancing of load fluctuations and a reduction of overloading and under utilization of up to 30% is possible. With it comes an increase of staff utilization.

The results, however, are based on (realistic) test data which have not been tested in a real life scenario. The results were therefore presented to automotive manufacturers to get feedback from users on the solution approach and to ask for real time data. So far one manufacturer of sport utility vehicles and one of commercial vehicles have agreed to provide real time data of the final assembly line.

In the first step, the data for a simulation of the assembly were defined. The data acquisition is still in progress. Meanwhile, the first simulations of two or three stations show the possibilities of simulation to analyse and evaluate the human resource allocation planning. There are several parameters which have fundamental influence on process requirements and capacity demands in every station. One can distinguish following primary key drivers.

Production programme: The sequence of orders determines the process requirements over time. To build an order sequence is in the responsibility of order management.

Staff assignment: The prognosis of the process requirements in every station offers new possibilities to plan the work schedule of floater. A floater has to observe his working area to react in case of overwhelming process requirements at a station. The visualisation of the process requirements and the expected utilization of the team member of every station given in advance, the floater know when he will be needed (time, station). It is the task of the production management to plan the floater.

Process planning: A work plan consists of a number of process steps; the sum of all process time steps represents a task time, depending on vehicle type and station. Every task is assigned to a station. The tasks can be divided by splitting process steps in two tasks or they can be merged by aggregating two tasks to one task, depending on what the line planner has to take into account like technology restrictions, etc. Nevertheless, in most cases, the aggregation of process steps are not done in a similar way, more often it depends on how a planner is used to build tasks. But there is a great influence on the staff utilization in choosing different methods to build a task. If the task times are too long, it can happen that a staff member cannot start a task because of lack of sufficient rest cycle capacity. The division of small tasks leads to better fits, but can be critical, if the number of tasks is too high, because of the negative influence on the working quality.

Cycle time: The cycle time determines the productivity. The assignments of tasks to stations can vary if the cycle time will change.

The production programme, the staff assignment, the process planning and the cycle time are correlated mutually. If you change e.g. the cycle time, it will be necessary to review the utilization of your staff and perhaps to rearrange the processes of your work plans. To do this, you must have the knowledge about the technologies, the restrictions and the no-goes of an assembly line and its organisation. That's why the improvements in planning of the production programme and human resources allocation planning forces all planning experts to work together. The planning process is iterative whereas the simulation serves as a decision support system by visualizing the results of utilization and productivity.

The developed solution for integrated human resources and production planning will be analyzed extensively by evaluating the results with these industry partners and testing their deployment potential and suitability.

7 Summary and Outlook

European automotive manufacturers continuously invest large sums into initial and continuing training of employees which lead to above average flexibility of production workers in international competition. At present, however, it is only used as a reactive tool in implementing the planned production program in the production halls. This way a high-cost human resources flexibility is retained instead of integrating the planning of production and human resource allocation and reaching an overall optimum.

The approach of trying to combine production planning with human resource planning is pursued within the scope of the Eurostars research project "Advanced Production Program and Personnel Assignment Planning" (A ProPer Plan). In doing so, the causal correlations between production program and production factors are detected. Existing sequencing solutions are expanded by human resource related restrictions and an iterative solution loop with simulative evaluation is integrated. The simulation allows the feasibility testing of the proposed sequence and clearly shows where under utilization and overload will occur.

Altogether, the following benefits of integrated human resources and production planning ensue:

- Transparency increase of human resource allocation and assembly process
- Balancing of load fluctuations and reduction of cold spots by up to 30%
- Increase of staff utilization
- Proof of buildability of a production program

By combining car sequencing and the mixed model approach wisely, further potential in and applications for integrating production program and human resource allocation planning can be found. The result is the logistical and economical optimization of the production process supported by simulation. The payoff is a significant increase in the hours per vehicle efficiency.

Within this project of integrating production programme and human resource allocation planning in the short term planning horizon several topics for further research where uncovered. The main problem of several European OEM's that took part in the project is the missing integration of the different planning tasks combining the long-, mid- and short-term planning horizon in one integrated planning environment. To solve these problems together with the automotive industry several research activities are going to be initiated.

References

1. Bard, J.F., Dar-El, E., Shtub, A.: An analytical framework for sequencing mixed model assembly lines. International Journal of Production Research 30, 35–48 (1992)
2. Bard, J.F., Shtub, A., Joshi, S.B.: Sequencing mixed-model assembly lines to level parts usage and minimize line length. International Journal of Production Research 32, 2431–2454 (1994)
3. Boysen, N.: Variantenfließfertigung. Deutscher Universitätsverlag, GWV, Wiesbaden (2005a)
4. Boysen, N.: Reihenfolgeplanung bei Variantenfließfertigung: Ein integrativer Ansatz. Zeitschrift für Betriebswirtschaft 75, 135–156 (2005b)
5. Boysen, N., Fliedner, M., Scholl, A.: Production planning of mixed-model assembly lines: Overview and extensions. Production Planning & Control 20, 455–471 (2009)
6. Drexl, A., Kimms, A.: Sequencing JIT mixed-model assembly lines under station-load and part-usage constraints. Management Science 47, 480–491 (2001b)

7. Drexl, A., Kimms, A., Matthießen, L.: Algorithms for the car sequencing and the level scheduling problem. Journal of Scheduling, 153–176 (2006)
8. Duplaga, E.A., Hahn, C.K., Hur, D.: Mixed-model assembly line sequencing at Hyundai Motor Company. Production and Inventory Management Journal 37, 20–26 (1996)
9. Fliedner, M., Boysen, N.: Solving the Car Sequencing Problem via Branch & Bound. European Journal of Operational Research 191, 1023–1042 (2008)
10. Gagné, C., Gravel, M., Price, W.L.: Solving real car sequencing problems with ant colony optimization. European Journal of Operational Research (2005)
11. Gravel, M., Gagné, C., Price, W.L.: Review and comparison of three methods for the solution of the car sequencing problem. Journal of the Operational Research Society 56, 1287–1295 (2006)
12. Henriksen, J.O.: An Introduction to SLX. In: Andradóttir, S., Healy, K.J., Withers, D.H., Nelson, B.L. (eds.) Proceedings of the 1997 Winter Simulation Conference, Piscataway, pp. 559–566 (1997)
13. Lochmann, M.: Operative Planung und Steuerung von Mehrprodukt-Fließlinien. Deutscher-Universitäts-Verlag, Wiesbaden (1999)
14. Meyr, H.: Supply chain planning in the German automotive industry. OR Spectrum 26, 447–470 (2004)
15. Monden, Y.: Toyota Production System: An integrated approach to just-in-time. Norcross (1993)
16. Scholl, A.: Balancing and sequencing of assembly lines, Heidelberg (1999)
17. Scholl, A., Klein, R., Domschke, W.: Pattern based vocabulary building for effectively sequencing mixed-model assembly lines. Journal of Heuristics 4, 359–381 (1998)
18. Sihn, W., Palm, D., Schmitz, K., Leitner, R.: Automotive Region Central and Eastern Europe: Auswirkungen der Krise. Studie der Fraunhofer Austria Research GmbH (2009)
19. Thomopoulos, N.: Line balancing – Sequencing for mixed-model assembly. Management Science 14, B59–B75 (1967)
20. Tsai, L.-H.: Mixed-model sequencing to minimize utility work and the risk of conveyor Stoppage. Management Science, 485–495 (1995)
21. Wester, L., Kilbridge, M.: The assembly line model-mix sequencing problem. In: Proceedings of the third international conference on Operations Research, pp. 247–260 (1964)

Simulation of Container Traffic Flows at a Metropolitan Seaport

Hendrik Reefke

Department of Information Systems and Operations Management,
University of Auckland Business School,
Private Bag: 92019, Auckland 1142, New Zealand
h.reefke@auckland.ac.nz

Abstract. Lack of integration, holism, and customizability are perennial problems that have beset research and practice in port simulations. This study investigates the container movements at a metropolitan sea terminal. The operations are mapped out and a logical model of the transportation flows is constructed. A simulation is developed in Arena that describes the transportation flows in an accurate but parsimonious manner. The study addresses several research gaps by providing a highly integrated, customisable and upgradeable simulation model which describes port operations in a more holistic fashion. Several experiments focussing on different modal splits for inter-terminal transport are carried out. The model provides a practical background to study the influence of operational changes, e.g. the influence of larger sea vessels on handling operations; effects of a larger share of rail transport; effects of increasing container throughput on throughput time and storage capacity; and the evaluation of sustainability measures.

Keywords: Port; Logistics; Simulation; Intermodal; Container; Transportation.

1 Introduction

The container system used today developed about five decades ago and spread rapidly over all areas of ocean borne trade. It drastically improved the efficiency of berth operations and decreased personnel requirements. International transportation became substantially cheaper, and it significantly affected the development of today's world trade [1]. Specialized container terminals developed, transforming port layout and operations to support and facilitate container shipping. Along with the increased usage of containers, specialized container handling equipment emerged. In particular over the past ten years maritime traffic has increased significantly and sea-goods transport and hinterland operations changed rapidly with regard to technological, organizational and economical aspects [1].

1.1 Research Motivation and Objectives

Research on intermodal container transport is generally limited especially when it comes to the quantification of potential benefits of intermodal freight transport [2].

W. Dangelmaier et al. (Eds.): IHNS 2010, LNBIP 46, pp. 420–431, 2010.

Intermodal terminals are often managed without information technologies and therefore solely depend on the experience of their staff [3]. Systems with different modes of transportation require precise planning and harmonization to avoid hold-ups in the transportation flow [4] which might prove especially relevant in light of the growth of the shipping industry. Thus, it is one study objective to analyze the operations at a seaport, the handling equipment, the operational interconnections and the intermodal linkages to gain a full understanding of the transportation flow.

Building a model of current port operations allows researchers to simulate how an actual system would be affected by changes without interfering with the real system. New policies and strategies can be evaluated where the infrastructure and operations in place prove to be insufficient. There is an apparent research gap when it comes to holistic simulations for container terminals [5-6]. Thus, one objective is to develop a model which integrates all operations in a comprehensive fashion. Additionally terminals are often seen as standalone systems. Another objective is therefore to design a highly customizable model which considers the port as part of a bigger system and functions as a starting point for further research studies.

This study provides a comprehensive overview of the existing systems and results in an expandable simulation model. It can be used to support decisions at the port regarding modal choices and to prevent bottlenecks in the transportation chain. As a proof of concept several scenarios aimed at evaluating the modal split for inter-terminal transport are presented.

1.2 Paper Outline

Section 2 gives an overview of container terminal operations and simulations. *Section 3* outlines the modeled system and presents the complete simulation modeled in Arena. *Section 4* shows the applicability of the model for three simulation experiments. *Section 5* discusses the usefulness of the developed simulation model for decision support and shows its unique advantages. Concluding statements about the limitations as well as future research opportunities are provided in *Section 6*.

2 Containerization

Containers represent the standard unit-load-concept for international freight. Transportation of containers involves e.g. manufacturers, freight haulers, shipping lines, transfer facilities and customers. Container terminals connect various modes of transportation, e.g. rail, truck and sea vessels [1, 7-8]. Liner shipping companies connect seaports around the world and control a growing share of the transportation market. This industry is characterized by an ongoing transformation process, e.g. mergers and bankruptcies [1]. Container shipping achieved expansion rates of 10 % per annum between 1990 and 2005 due to e.g. trade liberalization, increasing suitability of goods for containers, and more efficient port operations [9]. Around 95% of all cargo by weight is transported by ship and the incurred transport cost accounts for only 1% of the cost of the final product [3]. Demand is expected to grow but will be met by an expansion of shipping capacities which will keep freight rates at nearly the same level [9].

2.1 Container Terminals and Handling Equipment

Container terminals can be operated on an exclusive, preferential or common user basis [10]. Fixed costs are high due to complex container handling equipment and a constant need for infrastructure improvements. Potential new entrants are therefore confronted with high initial investments [11].

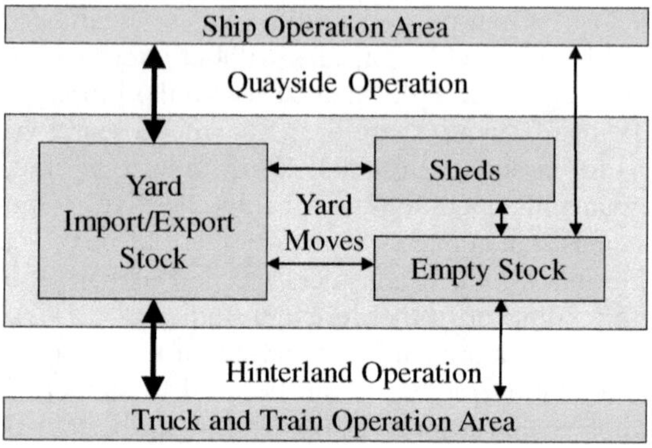

Fig. 1. Operation areas of a seaport container terminal and flow of transports [8]

The basic container terminal consists of quayside and hinterland operations which are in constant interaction as the containers move through them [8]. The hinterland operations facilitate the connection to the inland. Containers usually arrive via truck or train, are unloaded by internal vehicles, and then moved to a stock within the terminal yard. Quayside operations facilitate the loading and unloading operations of sea vessels. Arriving ships are assigned to a particular berth, where cranes load export containers and unload import containers. Container movements within the terminal depend on the design of the terminal and can include transports between empty stock, packing centre, and import and export container stocks [8].

The use of handling equipment at container terminals is determined by factors like container throughput, operating strategies, operating space constraints, layout of roads and tracks, or the degree of container standardization [4]. Gantry cranes on rails located at quayside are used to un/load cargo and containers. Performance is theoretically between 50-60 moves per hour, whereas the practical operational capacity is around 22-30 containers per hour. In the near future quay cranes with a theoretical capacity of 100 moves per hour will be possible [12]. Quay cranes efficiency greatly determines overall productivity of a port [13]. Stacking gantries on rubber tires or rail-mounted gantries are used to move containers in a container yard or to un/load rail cars. They can span up to 12 rows of containers and can stack them 4-10 high with a performance of about 20 moves per hour [14]. Straddle carriers can straddle a row of containers three to four high and are used to lift and move containers within the container yard or to un/load trucks. They are not locally bound and therefore guarantee access to all containers in the yard. Straddle carrier operations typically have densities of 500-600 TEU per ha of terminal area [14]. Side loaders are designed to transport containers in the yard to and from storage stacks, trucks or railcars. Automated guided vehicles and internal trucks are used for horizontal transportation within the container terminal [4, 8].

2.2 Port Operations

Container terminals are complex operational systems. The managerial decisions regarding long-, medium-, and short term issues can be categorized into terminal design, operative planning, and real-time control [7]. Uncertain future demand, incomplete knowledge and large capital investment make these decisions challenging.

Construction and equipment are substantial investments for terminal operators and require long term planning and forecasting. Typical design issues here are multimodal interfaces, terminal layout, determination of berth capacities, and selection of equipment, IT systems and control software [7]. Many container terminals facilitate multi modal transport with linkages to rail or trucks. The hinterland transportation is distinctive to every port and the modal split largely determines the type of equipment and the terminal layout. Terminal performance depends on capacity and arrangement of storage areas, quays, and transportation methods. Container storage areas are divided into stacks for e.g. export, import, reefers, and empty containers. The size of the stacking area has to meet demand and should allow for future expansions. In case of space constraints port operators can expand the terminal or develop designated inland depots [15] which must be coordinated with the corresponding inland transport systems. Total system costs are made up of construction, transportation, and usage of infrastructure [16]. Selection of terminal equipment is crucial for performance and is often based on factors like performance figures, degree of automation, space restrictions, costs and existing equipment [8].

Operative planning refers to the short term preparation of operational tasks and logistic processes and can be further sub-categorized into storage and stacking policies, crane assignment and split, berth allocation, and stowage planning [7]. Such planning activities are especially challenging in transshipment hubs [17]. The average dwell time partially determines the maximum capacity of a terminal [15]. In large container terminals it is about three to five days but can be as low as a few hours. Due to limited storage space, stacking of containers is common practice [18] which increases the need for efficient logistic processes. The main aim is to minimize reshuffling of containers, whilst maximizing storage utilization. To facilitate storage decisions detailed information regarding vessel assignment, discharge port, and container weight has to be available. Crane assignment and split refers to the allocation of quay cranes to a ship and its sections. Large vessels are operated with up to five cranes whereas feeder ships may require only one or two cranes. The number of cranes determines the possible container throughput of a terminal. Hence, effective crane split becomes imperative to reach objectives like minimization of ship delays, maximization of ship performance, or economic utilization of cranes [7-8, 19]. Berth allocation starts about two or three weeks before ship arrival. Quay cranes have to be compatible with ship length and width and travelling distances for all containers in the yard should be minimal. The dwell time of the ship has to be considered with regard to other arriving vessels and their specific requirements. Main objectives of stowage planning are maximization of ship utilization and minimization of time at berth. Shipping lines plan the position of containers in the ship, i.e. they have to meet the requirements of all ports and the constraints of the vessel. Stowage planning systems at the terminal have to take into account positioning and movement of containers in the ship and the yard. Stowage plans are prepared in advance, but there is a development towards planning in real time with the goal to eliminate crane waiting times, queuing, and reshuffling. One attempt to overcome congestion is a consignment strategy, i.e. containers with the same destination, content, and loading time are stored in dedicated storage areas [7-8].

Real time control at a seaport container terminal encompasses the coordination of transportation activities for landside and quayside vehicles, the assignment of storage slots to containers, and the planning of schedules and operation sequences for quay and stacking cranes [7]. Usually import and export containers are temporarily stored close to the quay area and may be moved between successive stacks during their dwell time in the yard [20]. Processes at a container terminal can be divided into e.g. arrival of the ship, unloading and loading of the ship, transport of containers, the stack, inter terminal transport, and connection to other modalities [18]. Information about the contents of a ship is typically sent to the terminal just a few hours before the ship calls at berth. Based on this information terminal dispatchers generate a detailed sequence for each quay crane stating how the containers have to be unloaded and discharged into the yard. Typically two to four possible storage locations are provided for each container and additionally the crane operator requires specific information about the containers which still have to be unloaded [21]. After the ship calls at the designated berth all containers assigned for transshipment are un/loaded and moved to a storage buffer below the gantry which is one of the bottlenecks of the waterside operations [22]. The random access system is commonly used in European and Australasian ports and lets customers deliver and pick up containers directly to/from train or ground storage [4].

2.3 Simulation at Container Terminals

A main objective at terminals is the minimization of handling time through maximization of quay crane throughput and efficient transportation strategies. For simulations it is common practice to decompose a decision problem into simpler problems [5-6], such as berthing, scheduling, routing and inventory problems. Such models cannot be a full representation of reality and might suggest solutions for certain operational areas, which are not beneficial to overall terminal operations [6].

Simulations can be applied to optimization areas at container terminals which can be classified into ship planning, terminal logistics, and transport optimization [23]. Ship planning is the task of organizing the un/loading operations. Common tasks here are the allocation of quay crane work shifts, computation of the bay plan, assignment of yard destinations, development of load plans and the supervision of un/loading operations [24]. As part of terminal logistics, operators must cope with increasing container throughput. Stack capacity of a terminal is basically the product of length, width, and stacking height, but is also influenced by container sizes or requirements for reefers and hazardous goods. By increasing stack height capacity can be extended. However, stack height can negatively influence logistic performance as it increases the possibility of containers blocking each other. Stack response time is also influenced by factors such as crane capacity, the position of the desired container, and the need for reshuffling [12, 14]. With regard to transport optimization total process time depends on crane un/loading time, handling equipment travelling time, and waiting times [25]. Terminal efficiency is especially influenced by quay crane productivity [26]. Interruptions can be avoided by maintaining a stock at the cranes [27]. Quay crane scheduling determines the un/loading sequences with the objective to minimize ship turnaround times and crane idle times which can be caused by interferences between cranes [13]. Scheduling of handling equipment is a challenge for efficient discharging and loading operations. The quay cranes usually determine transportation and handling schedules for trucks and yard cranes. The complexity of scheduling decisions involved should not be underestimated as traditional scheduling techniques cannot directly be applied to container terminal operations [25].

3 Simulation Model

This research aims to study and simulate the transportation flows at a metropolitan seaport with connection to an inland port. Simulation models should display the modeled systems as accurately but also as simply as possible. Too much detail can lead to overcomplicated models but on the other hand models can also be over-simplified. This might remove some uncertain elements but might not represent the system realistically [28]. Thus, a sufficient balance has to be found.

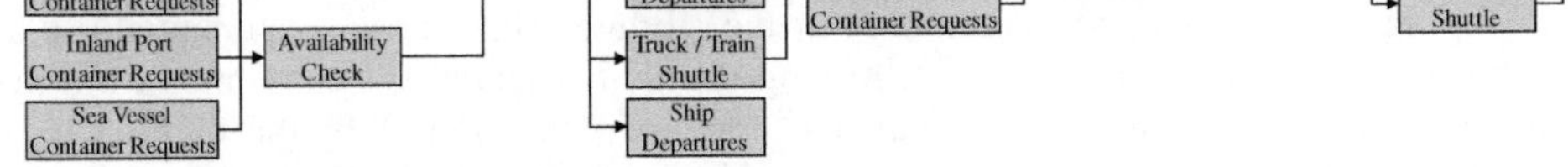

Fig. 2. Simplified transportation flows at sea terminal and inland port

Figure 2 shows the logical connection between the elements of the transportation chain, i.e. the left half displays the container flow at the downtown terminal, whereas the right depicts the inland port connections. The two locations are linked with a shuttle connection. There are four possible inflows of containers into the downtown terminal, i.e. customer container arrivals via truck, shuttle arrivals via truck or train from the inland port, train arrivals and incoming sea vessels. A number of carriers pick up the containers from their incoming vehicles and transport them to their destination within the terminal. There are also four outflows of containers in the downtown terminals, i.e. customer pick-up, truck or train shuttle to the inland port, train departures, and departing sea vessels. All these outflows are triggered by one of three request modules, i.e. customer container request, inland port container request, and sea vessel container request. It is assumed that all containers are directly available. The container request module is then linked to the carrier assignment module, i.e. carriers are assigned to retrieve the requested containers from the downtown storage yard and take them to the appropriate pick-up area. In the case of the truck and train shuttle module, the containers are transported to the inland port, which is shown by the corresponding linkage.

The overall structure of the inland port terminal is similar to the downtown terminal area, but there are no ship arrivals and container throughput and capacity are significantly lower. One possible inflow of containers into the inland port terminal is containers arriving via shuttle from the downtown terminal. The other possible inflow is customer container arrivals via truck. From there the incoming containers are assigned to a carrier and transported to their destination within the inland storage area. There are also two outflows, i.e. customer pick-up and truck/train shuttle to the downtown terminal. These are triggered by two request modules, namely customer container request and downtown terminal container request. Similar to the downtown terminal the requested containers are assumed to be available so that the requests are directly linked to the inland carrier assignment module. Here one of the inland carriers is assigned to every container request in order to retrieve the corresponding container from the storage yard and move it to the pick-up areas. The truck and train shuttles move containers from the inland port to the downtown terminal.

3.1 Container Movements

Customers bring in export containers, which will be stored in the storage area at the downtown terminals until they can be loaded onto a departing sea vessel. There is a significant fluctuation in container arrivals. The input distribution used takes into account the variations of the container flow throughout the day, i.e. separate values for every hour. It furthermore distinguishes between weekdays and weekends. Containers leaving on trucks are import containers that were previously unloaded from an incoming ship and then moved to the storage areas. After arrival at the terminals the incoming import containers are stored at the storage areas for an exponentially distributed dwell time of three days. Incoming shuttle trucks deliver containers from the inland port to the downtown terminal. The containers transported by the shuttle trucks to the downtown terminal were dropped off by customers at the inland port destined for export. When containers are requested for pick-up at the inland port, they first have to be moved from the downtown terminal to the inland port which can be done by shuttle trucks or shuttle train.

Trains can arrive from different destinations and bring higher quantities of containers to the terminal at a time. This characteristic makes the throughput lumpier and can therefore draw a significant amount of resources. About 14% of all containers travel via rail with up to 150 in and out moves per day of which 40% are 20 ft containers and 60% are 40 ft containers. The amount of transhipments accounts for approx. 10%. Trains arrive generally five times per day. The turnaround time is generally about 40 minutes to unload a train and about 2 hours to unload and reload a train. Shuttle trains bring in export containers from the inland port. They are able to complete the journey in 30 minutes, are up to 25 wagons long with two TEU per wagon and can make four round trips per day.

Ships arrive at the downtown terminal to drop off import containers and load export containers. The can arrive in prescheduled timeslots or in irregular intervals. The number of containers a vessel loads and unloads can vary greatly which makes the container-flow rather lumpy. Container must be handled quickly to decrease ship turnaround times. On average 1.58 ships arrive per day, i.e. one ship every 15.19 hours. Transhipment containers enter the container terminal from an incoming ship and have to be stored in the storage yard until they can be loaded onto a leaving vessel. There are great differences between arrivals or departures, which results in varying storage levels for transhipment containers. They compete with other containers for crucial resources, i.e. container carriers, storage space, and quay cranes.

3.2 Port Infrastructure

The cranes at the berths have to load and unload incoming vessels. The two downtown berths have 8 cranes overall and average roughly 24 container moves per hour gross. The port employs container carriers within the storage yard to store incoming containers and to retrieve outgoing containers, thus for almost any transportation job within the terminal. A specific job allocation strategy is used, i.e. if a carrier is idle, the next transportation job is immediately allocated to that carrier with preference to the carrier which is closest to the container that needs to be transported next. A maximum of 45 carriers can be operated at any one time but they are only used within their specific operational area. All carriers can be employed if necessary operating with an average transportation time of five minutes per job. The container storage area is used to store incoming containers until they are routed to their next destination. Container slots on average are stacked 1.8 boxes high and have a dwell time of around 3 days. The container storage yard at the downtown terminals has a limited capacity but the model assumes an unlimited capacity to ensure an uninterrupted container flow. The storage levels are logged throughout a simulation run which allows for decisions about necessary storage area extensions.

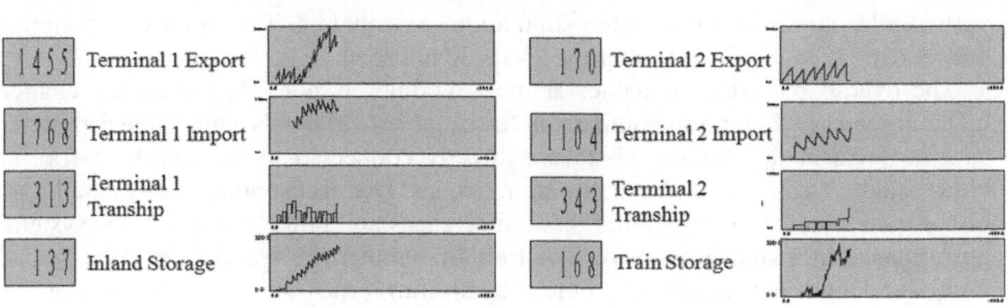

Fig. 3. Storage Areas at 300hrs into Simulation

3.3 Model Validation

All input data is based on empirical observations and is transformed into appropriate distributions which were verified with statistical measures that test the goodness of fit. For testing, a total number of 100 replications of 395 days are carried out. This includes a warm-up time of 30 days for every run which proved sufficient to initialize the system. After the warm-up phase the statistic collector functions are cleared, so that a total simulation time of 365 days is used. The output values of the model were compared with available data from the modeled port to check how closely the model output represents the real system. Several counter-modules were incorporated into the model, in order to capture the number of entities leaving and entering the system. In order to evaluate how closely the model matches the real system, the difference of each value pair is calculated in number of entities and percentage. Comparing the container throughput the discrepancy between the model behavior and the real system range from -5.81% to 1.67%, while most figures can be found in the range between +/- 1 to 3 %. The characteristics of the model were also validated by system experts through a qualitative evaluation of the model behavior.

4 Design and Results of Scenario Experiments

The transportation of containers between two locations can be a crucial concern for terminal operators which can be driven by public as well as economical interests. The model was used to run three scenarios with focus on the container transport between the downtown terminal and inland port. Different ratios between truck and rail shuttle transport were employed, i.e. 100% truck based (scenario 1), an even split between the two transportation modes (scenario 2) and 100% train based (scenario 3).

The simulation experiments are designed to analyze the effects that potential changes to the inter-terminal transport methods would have on the current system. The container throughput records show the intended behavior, i.e. the split between transportation modes in each scenario is as intended and overall throughput volumes are similar throughout the three scenarios. The results can therefore be employed for the comparison of other operational aspects and model behavior.

When comparing the storage areas it becomes evident that the results differ significantly for the minimum and maximum values, due to the usage of random numbers in the model design. However, in the averages these variations are evened out and show very similar numbers for all three simulation experiments. The figures vary slightly between the scenarios but the storage levels do not seem to be affected.

The overall behavior of queues at most modules is not affected by the changes made in scenario 2 and 3 in comparison to scenario 1. The only noticeable differences can be found at the queues which are directly connected to the shuttle traffic, i.e. either shuttle truck queues or shuttle train queues. Due to the comparably small number of transported TEU by shuttle these differences are unlikely to affect the system's performance in a significant way. The time that containers spend in queues depends on queue length and processing time. The waiting times are therefore correlated to queue lengths and only slight variations in terms of differences between waiting times can be noticed. The three simulation experiments have no major impact on the time entities spend at the selected modules.

The utilization of container carriers at the downtown terminal varies only marginally between the scenarios whereas there is an increase in the utilization of the train carriers from scenario 1 to 3. An increase in train usage also increased the utilization of carriers at the inland port slightly, i.e. it went up by 0.7% from scenario 1 to scenario 2 and by another 1.25% in scenario 3. Thus, the usage of shuttle trains seems to slightly increase the utilization of the inland carriers.

It can be concluded that under the modeled conditions a solely train based shuttle service is a viable option. No obstacles such as increasing queue lengths or waiting times were identified. The utilization of train and inland port carriers increases only slightly, which would be unlikely to negatively influence the transport operations. Overall the effects are rather small due to the low number of transported containers.

5 Research Contributions

As pointed out in section 1.1, the main research objectives were to develop an open, upgradeable simulation model to support decision processes at the port under investigation. The aim was to follow a holistic, comprehensive approach, taking into account the special nature of operations for intermodal transport.

Decision support systems can combine planning and management techniques with simulation models and statistical data analysis tools [24]. An applicability of the simulation model can be seen here. The developed simulation model can be employed for optimization areas at container terminals as pointed out in section 2.3. Thus, quay crane allocation plans as well as un/load plans can be evaluated. The effect of different stack heights for response times can be simulated and future requirements for capacity extensions can be evaluated by simulating higher container throughput figures. Simulation of scenarios can provide insights with regard to congestion, i.e. inadequacy in container handling, the ratios of empty vs. full containers and of import vs. export containers; overall throughput; or the height of stacks.

This simulation is highly customizable due to its modular design which allows for easy adjustment of parameters. Updates can be incorporated depending on simulation requirements that may arise. Despite being operational in nature, this model can be employed to support strategic decisions due to the holistic approach. In contrast to most other simulations, the model is a more complete representation of operational reality and thus overcomes the shortcomings of problem decomposition. In simulation experiments the full impact of operational changes on all modeled port functions can be evaluated which prevents overlooking of negative impacts.

6 Conclusions

An in-depth analysis of the transportation flows and their characteristics led to the construction of a simulation model which evolved over a number of iterations. This model can be used for a number of simulation experiments and can furthermore be employed as decision support tool for port managers. Due to the high flexibility of the utilized simulation program, Arena, the model is open for further extensions and improvements, thus it is upgradeable to suit other research needs.

Cost considerations could not be incorporated into the model due to time constraints and data sensitivity issues. This makes it currently difficult to quantify operational changes to the system in terms of monetary value. In order to enable further research concerned with evaluating the economic performance of the port, it would therefore be necessary to include e.g. storage costs, in yard transportation costs, or costs associated with truck and train transport. Another limitation is the detail and therefore accuracy of the simulation model, e.g. data availability was an issue. The model is based on some general assumptions and may therefore be seen as constrained in the level of detail and correctness it provides.

One research opportunity is to evaluate economical factors associated with the transportation flow either on a purely operational or more tactical level. The simulation model allows for the addition of cost functions to the various modules. The possible linkage of an operational evaluation and an economic evaluation would allow for a direct assessment of the feasibility of operational decisions, e.g. assessing the usage of equipment or space availability. Sustainability measures could also be added to the modules to support decisions regarding e.g. emission levels or fuel consumption. The output of this operational model can also be used to feed into a separate model for strategic decisions. The model is upgradeable e.g. it could be expanded to incorporate more detailed container flow at the terminal areas, incorporate all satellite inland locations or connections to main customers. Another model extension could be seen with regard to the uneven distribution of import and export containers which is a logistic challenge for not only shipping lines, but also terminal operators. Storage areas for empty containers outside of the terminal and the influence of empty containers on the container handling operation could be incorporated. The trend towards larger ship sizes and the effects on the modeled seaport can also be investigated. Larger ships demand more handling equipment, more storage space and put more pressure on the terminal operations.

References

1. Levinson, M.: The Box: How the shipping container made the world smaller and the world economy bigger. University Press, Princeton (2006)
2. Intermodal Freight Transport Committee: Proposed Intermodal Freight and General Freight Research Ideas, in Working Paper (2005)
3. Rizzoli, A.E., Fornara, N., Gambardella, L.M.: A simulation tool for combined rail/road transport in intermodal terminals. Mathematics and Computers in Simulation 59(1-3), 57–71 (2002)
4. Kozan, E.: Optimising container transfers at multimodal terminals. Mathematics and Computer Modelling 31, 235–243 (2000)
5. Gambardella, L.M., Rizzoli, A.E.: The Role of Simulation and Optimisation in Intermodal Container Terminals. In: European Simulation Symposium, Istituto Dalle Molle di Studi sull' Intelligenza Artificiale (2004)
6. Dell'Olmo, P., Lulli, G.: Planning Activities in a Network of Logistic Platforms with Shared Resources. Annals of Operations Research 129, 155–169 (2004)
7. Günther, H.-O., Kim, K.-H.: Container terminals and terminal operations. OR Spectrum 28, 437–445 (2006)
8. Steenken, D., Voß, S., Stahlbock, R.: Container terminal operations research - a classification and literature review. OR Spectrum 26, 3–49 (2004)
9. Heymann, E.: Container shipping - Overcapacity inevitable despite increasing demand, in International topics - Current Issues. Frankfurt am Main (2006)

10. Martin, J., Thomas, B.J.: The container terminal community. Maritime Policy & Management 28(3), 279–292 (2001)
11. Notteboom, T.E.: Consolidation and contestability in the European container handling industry. Maritime Policy & Management 29(3), 257–269 (2002)
12. Duinkerken, M.B., Evers, J.J.M., Ottjes, J.A.: A Simulation Model for Integrating Quay Transport and Stacking Policies on Automated Container Terminals. In: European Simulation Multiconference, Delft University of Technology (2001)
13. Moccia, L., et al.: A Branch-and-Cut Algorithm for the Quay Crane Scheduling Problem in a Container Terminal. Naval Research Logistics 53, 45–59 (2005)
14. Saanen, Y.A., Valkengoed, M.V.: Comparison of Three Automated Stacking Alternatives by Means of Simulation. In: 37th Winter Simulation Conference, Orlando (2005)
15. Kia, M., Shayan, E., Ghotb, F.: The importance of information technology in port terminal operations. International Journal of Physical Distribution & Logistics Management 30, 331–344 (2000)
16. Koh, Y.-K.: Optimal investment priority in container port development. Maritime Policy & Management 28(2), 109–123 (2001)
17. Lee, L., et al.: An optimization model for storage yard management in transshipment hubs. OR Spectrum 28(4), 539–561 (2006)
18. Vis, I.F.A., de Koster, R.: Transshipment of containers at a container terminal: An overview. European Journal of Operational Research 147(1), 1–16 (2003)
19. Bish, E.K.: A multiple-crane-constrained scheduling problem in a container terminal. European Journal of Operational Research 144, 83–107 (2003)
20. Ottjes, J.A., et al.: Simulation of multiterminal system for container handling. OR Spectrum 28, 447–468 (2006)
21. Bish, E.K., et al.: Analysis of a new vehicle scheduling and location problem. Naval Research Logistics 48, 363–385 (2001)
22. Böse, J., et al.: Vehicle Dispatching at Seaport Container Terminals Using Evolutionary Algorithms. In: 33rd International Conference on System Sciences, Hawaii (2001)
23. Dorndorf, U., et al.: Ports o'Call For O.R. Problems. OR/MS Today 34(2), 36–40 (2007)
24. Gambardella, L.M., Rizzoli, A.E., Zaffalon, M.: Simulation and Planning of an Intermodal Container Terminal. Simulation on Harbour and Maritime Simulation (Special Issue), 1–10 (1998)
25. Ng, W.C., Mak, K.L., Zhang, Y.X.: Scheduling trucks in container terminals using a genetic algorithm. Engineering Optimization 39(1), 33–47 (2007)
26. Tongzon, J.L.: Determinants of Port Performance and Efficiency. Transportation Research 29A(3), 245–252 (1995)
27. Evers, J.J.M., et al.: Agile Logistic Systems: the case of a container terminal. In: GLORI Conference on Dynamic Logistics, Delft University of Technology (1998)
28. Kelton, W.D., Sadowski, R.P., Sturrock, D.T.: Simulation with Arena. McGraw-Hill series in industrial engineering and management science, vol. xxiii, p. 630 (2007)

Simulation of ITSM Processes as Training Tool Set

Achim Schmidtmann

Fachhochschule Dortmund, Emil-Figge-Str. 42, 42277 Dortmund, Germany
achim.schmidtmann@fh-dortmund.de

Abstract. As IT Service Management becomes more and more common in actual IT organizations, the need of trained personnel raises sharply. To cover theses needs, companies can either train their employees or they hire new employees, who possess these skills already. But what competence is required and how should it be trained? Employees will need some knowledge of the IT Service Management systems, they are going to work with, but even more they have to study and understand the fundamental processes of IT Service Management. Today several simulations respectively simulation games are used to give participants a general comprehension of IT Services and a slight insight into some processes. Nevertheless, this experience is quite far from the real processes and without further training a lot of time and effort is needed to completely transform the IT organization. The closing of this gap is addressed in this paper. It proposes training with realistic and more detailed simulations based on an IT Service Management process model as a valuable solution.

Keywords: IT Service management, processes, simulation, training.

1 Introduction

The last years have shown a constant increase of IT Service Management (ITSM) in the perception of IT Managers. More and more ITSM-introduction and –conversion (reengineering) projects according to standards like IT Infrastructure Library (ITIL) [1] are run in IT organizations. Quality and quantity of IT Services are the main objectives of these undertakings. To optimize those, IT activities and processes have to be standardized and adjusted to Service requirements. Through these changes the Service should become much more target-oriented, customer friendly, and cost optimized.

Surveys have shown that in these projects there is often an internal resistance to change. During implementations of formal process models of IT Service Management employees respectively customers are not prepared to be involved in the new processes. What they are especially missing is an understanding of the value of the end state of the initiative [2].

So training of employees has to take up an important part in these projects. Next to integrating the personnel in each step of the project, so that they are always informed about activities, changes, and the actual status, there has to be a thorough approach to educate them in the fundamental IT Service Management processes. And these processes should not be taught on a general level but on a detailed and very realistic scale adjusted to the particular organization. This can be done with simulations based on an

W. Dangelmaier et al. (Eds.): IHNS 2010, LNBIP 46, pp. 432–442, 2010.

IT Service Management process model of the organization. The ITIL model can be used as a reference to create this organizationally specific process model which simultaneously demonstrates its value.

The remainder of this paper is organized as follows. First, IT Service Management and in particular the service processes in general as well as defined by the ITIL framework are discussed. Then, in Section 3, the requirements and characteristics of process simulation are described. In Section 4 it is explained, how simulations can be used for training for all staff working directly with the process as a participant or user and in Section 5 the conclusions and an outlook are presented.

2 IT Service Management

Until a few years ago technology was the determining factor in the IT, this has now changed almost everywhere into services. Services have become an integral part of business processes in most of the organizations. While providing these services disturbances occur and IT support resolves the issues with a goal to minimize the downtime. So ITSM aims to improve systems' responsiveness to these unforeseen disturbances [3].

IT services are based on IT infrastructure components like hardware, software, procedures, documentation and skills which in particular represent the human part of this foundation. To provide a better IT Service, these components, people certainly included, must be managed efficiently and effectively.

2.1 IT Service Management Processes

The processes which are defined to manage the IT services as well as the infrastructure required for providing the IT services are referred to as Service Management processes. Typically, these processes are broken down into two core groups, one which includes the operational and another one which includes the tactical and strategic part.

Basically all Service Management processes are dependent upon each other, as there are various interconnections. This raises the requirement, that the employees involved not only know their process, but also the neighboring very closely, and are able to estimate impacts. Experience shows that the effectiveness of Service Management implementation is adversely impacted in absence of a correct implementation approach.

Process modeling and analysis is already used to guarantee a high quality of management of efficient IT services. This approach is based on an identification and definition of the core processes and then combines execution and automation of those processes with an analysis to find improvement opportunities [4].

2.2 ITIL Processes

The Information Technology Infrastructure Library (ITIL) was established in 1989 by the United Kingdom's former Central Computer and Telecommunications Agency (CCTA) to improve its IT organization. ITIL is now managed by the UK's Office of Government Commerce (OGC) and has become the de facto standard in IT Service

Management (ITSM). The ITIL framework is based on a set of good practices for lowering the cost, while improving the quality of IT services delivered to users [1].

Although, the newest version, ITIL v3, is fundamentally different from previous versions and shows an evolution from process to service model [5] by focusing on the service life cycle, the interdependent IT Service Management processes still represent the core of it.

The ITIL v3 publications offer for each of those processes a process model as a reference for implementing or introducing this process. Regardless whether the focus is on process development or process reengineering, these reference models will be very useful. They also form the basis for eventual simulations carried out.

3 Process Simulation

Since the nineties process oriented analysis of the performance of companies gets more and more attention. Nowadays, business processes are regularly evaluated and redesigned to keep pace with the changes in market and economic. The process landscape today is more complex than ever, more dynamic, and difficult to master. But business process reengineering, introducing new processes, or process changes is always a difficult and lengthy plan. With the interest particularly in relation to IT service management in these the value of modeling and simulating business processes has increased attention.

Process simulation is used in many contexts with the general aim of obtaining deeper insight into their functioning. This includes process simulation which is used for the design, development, analysis, and optimization of technical and business processes. With a simulation valid source information about the relevant selection of key characteristics and behaviors can be acquired. In addition, simplifying approximations and assumptions within the simulation can be tested. Simulation runs can be used to show the eventual real effects of alternative conditions and courses of action. Key issue of the simulation outcomes is their fidelity and validity. So process modeling combined with simulation helps people in understanding the behavior of complex and cross functional business processes by revealing bottlenecks, cost and quality drivers and activity based costs.

In manufacturing simulation has been used successfully for many years. The global competitive pressures have forced manufacturers to develop increasingly efficient and effective process designs [6]. To achieve this goal, simulation is used in the above mentioned fields. Furthermore, simulation in manufacturing is, however, also used in the areas of testing, training and education.

The following list is intended to summarize the varied use of process simulation in short again:

- Visualization
- (What-if) Analysis
- Optimization
- Validation
- Reporting
- Testing

- Training
- Education

To further clarify, what is being studied using these methods, another list should give some insight:

- Utilization of resources: space, equipment and human
- Queues
- Cycle times
- Delivery reliability
- Cost
- Scenarios
- Vulnerabilities
- Identification of bottlenecks

As studies show, a wide range of simulation tools exist, but only a few of them are applicable for process simulation [7]. If so they support understanding, analyzing, and designing processes but have no emphasis on training and education of those responsible for process execution. One of these tools is SYMIAN, a decision support tool for the improvement of incident management performance. It is a discrete event simulator that permits to test possible corrective measures for the IT support organization before the actual implementation. SYMIAN has proved to be effective in the performance analysis and optimization of the incident resolution time [8]. Another approach based on Petri-Nets models and simulates service processes in terms of availability levels to gain a priori estimations during design time on the potential impact and interaction of availabilities of services involved in provisioning processes [9]. However, both simulation tools were developed for process owners and process managers, and not for those carrying out the process.

4 Training through Simulation

In recent years various business simulations respectively business simulation games have found their way into the business education. Business simulations that include a dynamic model which enables to experiment with business strategies in a risk free environment provide a useful complement to the normal and static case study discussions. Besides, they train the participants in various skills like strategic thinking, financial analysis, market analysis, teamwork, and leadership.

In establishing these training simulations, the key question is what and who has to be trained? In many cases they are designed for managers who rather need a general understanding than a deep insight into a function or process. In other cases simulations are used to give a short introduction into a new topic which is also done on an elevated level. But especially when it comes to implementation, redesign or modification of IT service management processes, the target group should be all personnel working with these processes and they need to be trained in depth. Gaining the confidence of these process participants or users is the key to success of simulation tools. Typically process participants understand the process and its problems but not the solution proposed. Users with their understanding of business process can check for scope for improvements by trying simulations with different combinations of dependencies and resources. Process simulations help users design innovative processes using their internal knowledge of the organization and processes [10]. On the other hand they support users in better understanding changes of processes and their dependencies.

4.1 ITSM Simulators

As it has been realized a few years ago, that ITSM and especially ITIL is gaining ground inside IT, but it's still a “tough sell” to management and other departments, vendors developed simulators to demonstrate the value of these approaches. Through these simulations "cross-functional" groups of senior IT and business management are introduced to the value of formal process models like ITIL [11].

The simulation tools show, by increasing external forces on the 'system,' the advantages to be gained through the use of predefined and described processes. The simulations also educate on the elements and relationships between the components of an ITSM respective ITIL model and the importance of communications. This is all done in a relaxed atmosphere of fun and games.

Examples of these simulations are:

- BMC Software's virtual airport,
- Gaming Works Apollo 13 simulation,
- HP's Formula 1 racing team game,
- Simagine’s ITSM game Control-IT
- Maternas FortFantastic,
- Getronics Silverstar Business Simulation, and
- Microsoft McKinley simulation.

The true benefit of these simulations is that they can provide a great level of awareness for all levels of management. In doing this they raise the internal IT and Business buy-in to ITSM and are a perfect vehicle to "bring to life" ITSM and ITIL. Besides, they are an ideal communications and team building tool to be used as part of a larger ITSM implementation program, but it has to be kept in mind that such a program is far from being a game and these simulations somehow simplify and embellish the reality. What they cannot accomplish is the concrete and detailed preparation for the work with new or changed processes. They provide a general understanding but not a deep insight, which is necessary for the efficiency of those processes.

4.2 ITSM Process Simulation

Such a very specific training needs a simulation which is based on the real IT Service Management processes in the organization. Therefore, the best solution for that kind of training is a simulation based on a thorough business process model of the organization which contains each activity, decision and alternative. It has to be a complete process map tailored to the organization. A simulation of such a character provides a comprehensive and very detailed preparation of employees to their changing world of work with the new or changed business processes.

As mentioned above there is a number of tools which enable both modeling and simulation. In this approach the object-oriented tool for graphical modeling, documentation and analysis of business processes, organizations and information systems BONAPART® has been chosen to clarify how process participants and users can be trained on a detailed IT Service Management process. This decision is based on the fact, that BONAPART® offers a wide variety of business process modeling methods to use and map the organization's processes as well as complex simulation functionality.

4.3 Training through Process Simulation

Below, an example of a training session on the basis of a BONAPART® process simulation is illustrated. The background of this training is an ITIL-introduction project in the IT division of a large company. Because of the detailed approach it is important to accurately select a training group which consists of employees who work closely together on one or different but connected processes. It is then the task of training to familiarize these employees with the amended and new process components in detail.

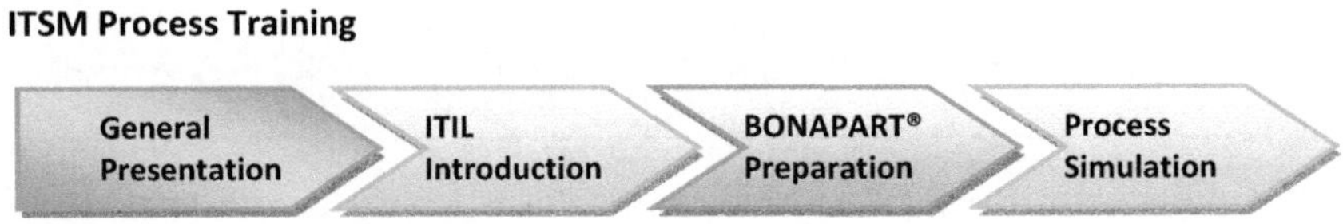

Fig. 1. Structure of the training

The structure of the training as shown in fig. 1 should contain the following parts: It should start with a general presentation of the ITSM project and its objectives to get everybody on board. In a second step the participants should be introduced to the ITIL framework and its processes as well as its functions. Next, the training group should get some instructions on how to work with the tool BONAPART® which is a preparation for the last phase. The final session of the training deals with the core subject of this paper, an intense analysis, discussion and adaption of the relevant ITSM processes exposed in BONAPART® based on the ITSM process and organization model, its settings and various simulation runs.

To clarify this final and main step of the training the Incident Management Process, probably the most important process of ITIL v3 phase Service Operation [12], has been chosen. Fig. 2 shows a narrow excerpt of this process based on the ITIL reference model. In this example the Communication Structure Analysis is the used business process modeling method. The training for this process starts with an explanation and deep insight into this process. Participants analyze the model and thoroughly walk through it. In particular, they learn the individual process steps, decisions and responsibilities respectively responsible agents that are stored in the model. They also have the opportunity to bring ideas, corrections and improvements in the process. However, this "static experience" of the Incident Management Process is only the first part.

In the next step of the training the participants can easily run simulations based on this process model. Herewith, they can configure various scenarios e.g. different capacities, costs or cycle times and validate these concerning the quality of change and

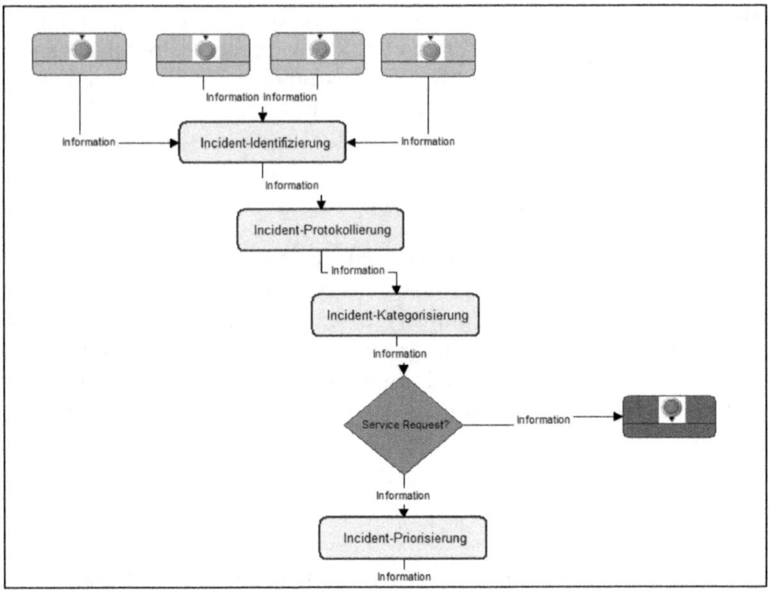

Fig. 2. An excerpt of the Incident Management Process according to ITIL modeled in BONAPART®

their set objectives. In these scenarios the participants can define the number and distribution of the incoming data which are in the case of this process the different kind of incidents. Similar to this, they can configure the decision steps mostly based on experience value.

The two figures below show examples of possible settings of the simulation process step Incident-Protokollierung (Incident Logging). In fig. 3 a possible agent setting and in fig. 4 a plausible time setting with an equipartition is shown. Accordingly, the participants of the training can configure all the process steps and identify in this way, how the process can be optimally adapted to their requirements for process times, number of agents, etc.

In carrying out these simulations, there are three alternative procedures:

- all training participants work together on one simulation model and in all process steps
- all training participants are working on one simulation model, but everyone is responsible for another process step and configures this
- all training participants work on their own simulation models and in a second phase, the results of the individual simulations are analyzed, evaluated, and incorporated in the optimal settings for the joint model

The validation of these simulation runs is done by checking the utilization of the concerned activities as pictured in fig. 5 as well as the utilization of the concerned personnel resources as described in fig. 6.

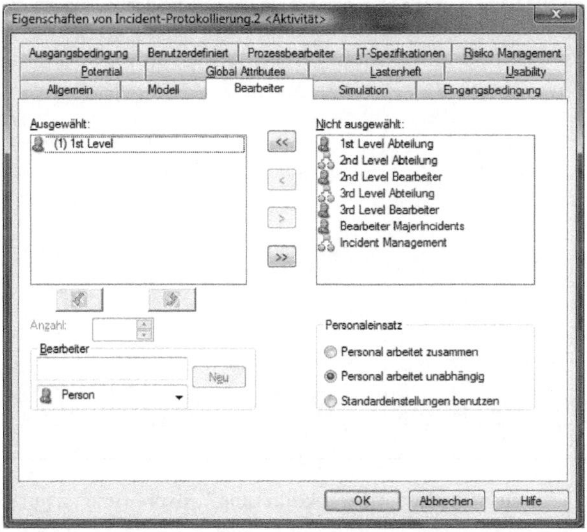

Fig. 3. Agent setting of the process step Incident-Protokollierung in BONAPART®

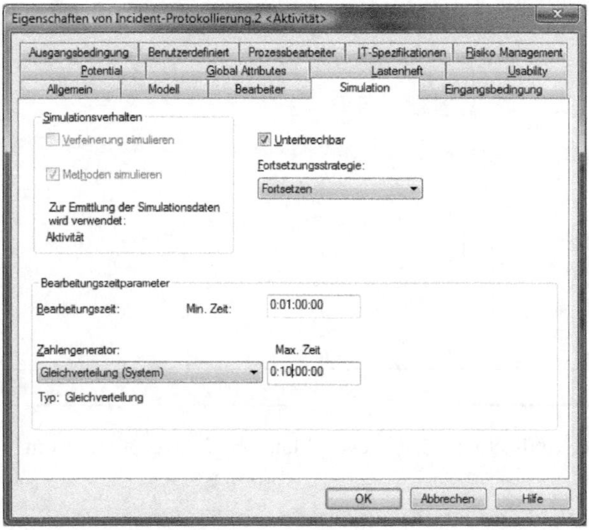

Fig. 4. Time setting of the process step Incident-Protokollierung in BONAPART®

Through these illustrations training participants easily understand vulnerabilities of the process and its settings and identify possible bottlenecks. Moreover, they get the opportunity to test various settings and discover better solutions. Not least, the participants are in this way developers and optimizers to their own processes, which leads to a close and the motivation strongly supported bond.

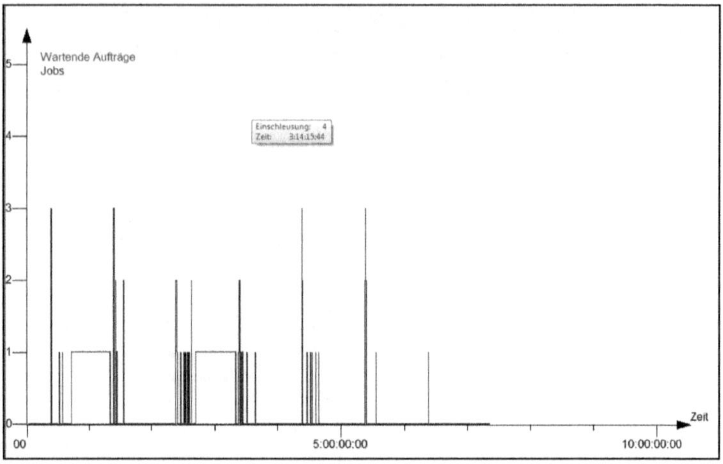

Fig. 5. Jobs at activity "Incident-Protokollierung" over time from simulation run in BONAPART®

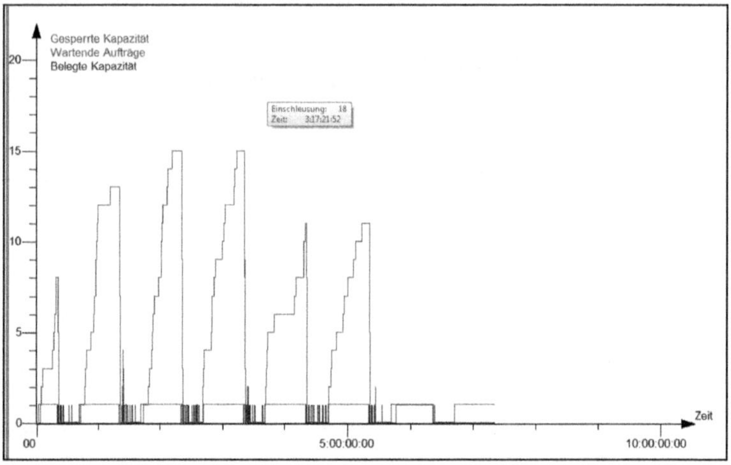

Fig. 6. Resource utilization First Level Manager 2 over time from simulation runs in BONAPART®

Next to this more independent training which is accompanied by the coaches there is also another part which has more the character of an exercise. Therein coaches build faults into the model or simulation parameters that must be solved. Or they simply leave out settings, as in a gap in the text, which must be added. In this way, the training participants will be trained in more depth the processes and moreover get to directly know possible weaknesses. This method might be preferable for employees who have no process or process model experience.

In addition to individual processes the BONAPART® model also represents the interfaces and linkages between the various IT Service management processes, so that these can be simulated under the training and simultaneously can be analyzed and optimized. Precisely these transitions between the processes are not easy to imagine and by this means can be understood much easier and in detail.

5 Conclusion and Outlook

Overall, this modeling and simulation approach has shown the special opportunity to deal in depth with each IT service management processes and their many interactions. This is in contrast to the usual training with ITSM simulators which primarily aim at a general understanding of IT service management. By working with the process models and various simulations based on it, and the independent setting of parameters and execution of simulation runs, a particularly high learning effect is guaranteed.

Participants of this training get much more involved in the processes and acquire a quicker and better understanding of them. Besides, they get already used to those processes and even are able to improve them through the outcomes of the simulation. Finally, this will strongly expedite the ITSM projects and lead to a much higher success rate through well-trained and very motivated employees.

Up to now this simulation approach has been developed and tested as part of an information system Master Course. It has not been checked during an ITSM project in an organizational environment. The students in this course did only have limited previous knowledge of IT Service Management and its processes. But by using the simulation and participating in the above explained training it has been shown that the IT Service Management model and its processes are understood fast and easy. Therefore, this approach is also very useful in university education and will be continued and enhanced.

Acknowledgments. The author wishes to thank the students in his Master Course "Sicherheits- und Servicemanagement" at the University of Applied Science and Arts Dortmund for helpful discussions and valuable comments on process simulation issues. In addition he thanks BTC Business Technology Consulting AG to put a number of demo versions of BONAPART® at his disposal.

References

1. Köhler, P.T.: ITIL. Das IT-Servicemanagement Framework. Springer, Heidelberg (2007)
2. Gliedman, C.: ITIL Simulators Demonstrate: The Value of Process Models. Forrester Research, Inc. (2006)
3. Lee, J.H., Han, Y.S., Kim, C.H.: IT Service Management Case Based Simulation Analysis & Design: Systems Dynamics Approach. In: Proceedings of the 2007 International Conference on Convergence Information Technology, pp. 1559–1566 (2007)
4. Shen, B.: Support IT Service Management with Process Modeling and Analysis. LNCS. Springer, Heidelberg (2008)
5. Hubbert, E.: ITIL V3: The Evolution From Process to Service Model. Forrester Research, Inc. (2007)
6. Greasley, A.: Simulation modelling for business. Innovative business textbooks. Ashgate Publishing, Ltd. Hants (2004)

7. Jansen-Vullers, M., Netjes, M.: Business process simulation – a tool survey, http://www.daimi.au.dk/CPnets/workshop06/
8. Bartolini, C., Stefanelli, C., Tortonesi, M.: SYMIAN: A Simulation Tool for the Optimization of the IT Incident Management Process. LNCS. Springer, Heidelberg (2008)
9. Bartsch, C., Mevius, M., Oberweis, A.: Simulation of IT Service Processes with Petri-Nets. In: Feuerlicht, G., Lamersdorf, W. (eds.) ICSOC 2008. LNCS, vol. 5472, pp. 53–65. Springer, Heidelberg (2009)
10. ALAGSE: Institutionalize innovation through business process modeling and simulation, http://www.alagse.com/pm/p9.php
11. DeJean, D.: Making ITIL fun and games. Microsoft Logfile (2006), http://blogs.computerworld.com/node/3413
12. Office of Government Commerce (OGC): ITIL V3 – Service Operations. TSO (2007)

Author Index

GPSR Compliance

The European Union's (EU) General Product Safety Regulation (GPSR) is a set of rules that requires consumer products to be safe and our obligations to ensure this.

If you have any concerns about our products, you can contact us on ProductSafety@springernature.com

In case Publisher is established outside the EU, the EU authorized representative is:

Springer Nature Customer Service Center GmbH
Europaplatz 3
69115 Heidelberg, Germany

Batch number: 10844619

Printed by Printforce, the Netherlands